T0273803

UNITEXT

La Matematica per il 3+2

Volume 141

The **UNITEXT - La Matematica per il 3+2** series is designed for undergraduate and graduate academic courses, and also includes advanced textbooks at a research level.

Originally released in Italian, the series now publishes textbooks in English addressed to students in mathematics worldwide.

Some of the most successful books in the series have evolved through several editions, adapting to the evolution of teaching curricula.

Submissions must include at least 3 sample chapters, a table of contents, and a preface outlining the aims and scope of the book, how the book fits in with the current literature, and which courses the book is suitable for.

For any further information, please contact the Editor at Springer: francesca.bonadei@springer.com

THE SERIES IS INDEXED IN SCOPUS

***

UNITEXT is glad to announce a new series of free webinars and interviews handled by the Board members, who rotate in order to interview top experts in their field.

Click here to subscribe to the events!

https://cassyni.com/events/TPQ2UgkCbJvvz5QbkcWXo3

Simone Secchi

A Circle-Line Study of Mathematical Analysis

 Springer

Simone Secchi
Department of Mathematics and
Applications
University of Milano Bicocca
Milan, Italy

ISSN 2038-5714   ISSN 2532-3318   (electronic)
UNITEXT
ISSN 2038-5722   ISSN 2038-5757   (electronic)
La Matematica per il 3+2
ISBN 978-3-031-19737-6   ISBN 978-3-031-19738-3   (eBook)
https://doi.org/10.1007/978-3-031-19738-3

Cover illustration: The circular structure of the book. Painted by Francesca Vettori

This Springer imprint is published by the registered company Springer Nature Switzerland AG
The registered company address is: Gewerbestrasse 11, 6330 Cham, Switzerland

This book is dedicated to Francesca, for her constant and doubtless support

Preface

My first encounter with Mathematical Analysis was on the celebrated *Principles of Mathematical Analysis* by Walter Rudin. I immediately fell in love with this discipline, so elegant and polished in Rudin's pages. Seventy years ago, when that book was written, the common approach to Mathematical Analysis was deeply influenced by the *topological revolution* introduced by the Bourbaki group. Metric spaces were the basement of the building, and a general effort towards abstraction and generality was evident.

Nowadays many things have changed. I have been teaching Calculus and Mathematical Analysis for more than 15 years, and I have read many books on this subject. A current trend is to isolate Calculus from Analysis, and a rather disappointing corollary of this approach is that the topological revolution has been forgotten. The field of real numbers is so rich that many students do not understand that limits and continuity are not necessarily related to algebraic structures. Sequences are strange objects that appear and disappear without any good reasons, since they are not used to describe the topology of the real line.

In this book I have tried to place the puzzle pieces where they have to stay, starting from an exposition of *aware Basic Analysis* in $\mathbb{R}$ and reaching the rigorous development of more advanced branches of Analysis. In my own experience, students learn limits, continuity and the main properties of the real line, but they hardly realize that a fair amount of analysis ideas are just general topological facts applied to the topological space $\mathbb{R}$. So to speak, our students would need a return ticket to modern mathematical analysis, but they just get a one-way ticket to Cauchy's calculus. Figure 1 gives an idea of our view of modern analysis (and mathematics as well!).

Our journey starts with some *naïve* set theory and an axiomatic introduction of the real numbers. Natural numbers are introduced as a subset of $\mathbb{R}$, and mathematical induction is fully justified. I believe that there is no concrete alternative to this approach, since sets and numbers are those mathematical objects which can be *used* before learning their coherent definitions. Let me be clear: when you are an experienced working mathematician, you may very well believe that real numbers should be constructed from the rational numbers, and natural numbers should be

Fig. 1 The tree of this book

defined in terms of sets. But this viewpoint would prevent any student from *using* numbers on a daily basis. Students are not supposed to think of $\sqrt{2}$ as a Dedekind cut or as an equivalence class of sequences of rational numbers, when they first meet limits, derivatives and integrals. And I would bet that no experienced analyst ever learned analysis (or fell in love with analysis) this way. It is a matter of fact that many students will never feel the need of studying Dedekind cuts, and they will be happy to know that $\mathbb{R}$ can be constructed in one way or another.

A first taste of topology on the real line then comes into play, and this is enough to have a rigorous definition of all the main tools of analysis in one dimension: limits, sequences and series, derivatives and Riemann integrals. Our first definition of the derivative is formally the same definition that one would give in normed vector spaces. We prove that the derivative is the usual limit of the incremental ratio, but we stress that a Calculus definition must be changed as early as we want to differentiate functions of two variables. Linearization is the keyword of differential calculus, and we use this in full strength.

Metric spaces are not overestimated in this book. It is a common opinion that metric spaces are the natural environment for doing analysis. This is surely supported by some recent theories, but we prefer to present them for what they are: a special case of topological spaces. Most textbooks avoid the generality of topological spaces, and students often feel that mathematical analysis can exist only if distances can be measured. This is largely false, as we will see. Furthermore, I totally dislike the habit of keeping analysis and topology disjoint: some branches of Functional Analysis do need abstract topological structures.

The chapter on differentiation may look short. Since this is not a Calculus textbook, we do not spend time to compute dozens of elementary derivatives. The purpose of the chapter is to present the *theory* of differentiation in one real variable, so I omit any discussion about the physical meaning of derivatives. I assume that the average reader of my book knows that "velocity is the derivative of motion, and acceleration is the derivative of velocity." Although Mathematics is continuously stimulated by other sciences, it is indeed a science itself. My students are not supposed to define derivatives as slopes of a line and integrals as energies.

The chapter on Riemann integration is probably longer than usual: I present a few results that most textbooks only propose in the special case of continuous functions. Indeed, I think that the class of Riemann integrable functions should be kept distinct from the subclass of continuous functions as long as possible. The most important feature of modern mathematics is the capability to arrange objects according to their properties: should we deal with differentiable functions only, since they form an interesting subclass of continuous functions? I do not think so.

A chapter on the so-called elementary functions is necessary in any rigorous exposition of mathematical analysis. On one hand, it is true that differentiating the cosine function before providing a definition of the cosine function is nonsense. On the other hand, however, it would be unrealistic to remove any example which involves elementary functions from a textbook. In my humble opinion, this is a problem without solution: we must teach differential and integral calculus before teaching the construction of the elementary functions. But a formal approach to such functions is not enough for a good analyst, who needs to learn why elementary functions actually exist.

Our journey can now proceed backwards, so that we look again at the basic ideas from a more advanced viewpoint. In particular, I propose a quick-and-dirty review of axiomatic set theory. Set theory is, somehow, *what every mathematician pretends to know*. Students are not exposed to the difficulties of the axiomatization of this discipline, and they usually ignore what entities are really primitive. A function is more often than not proposed as a black box that converts a number into another number. Once sets have been introduced or assumed to exist, there is no need to suppose that functions are a primitive notion. This may be convenient, but it is definitely unnecessary.

Axiomatic Set Theory looks close to philosophy, since different and non-equivalent approaches have been proposed over the decades. I present a review of John Kelley's set theory, which I find particularly suited to the analyst's mindshape. For the sake of completeness, I also list the axioms of Zermelo and Fraenkel, which are probably dominant among experts.

I then introduce some general topology: open and closed sets, neighborhoods, and of course limits. I am proud to define nets, a generalization of sequences that completely characterize the topology of any space. Convergence is usually the most appreciated idea of mathematical analysis, and we see that the whole Calculus can be described in terms of converging nets. Compact sets and connected sets are defined, so that the reader can appreciate several classical theorems about continuous functions of a single variable as particular cases of abstract properties. Compactness is ubiquitous in analysis, and I hope that our survey may help readers understand its use. The reader will notice that the chapter on topology is not written in a systematical way, and ideas appear when they are needed. The realm of analysis is not so polished as the realm of geometry, because analysis is more concerned with borderline cases.

I have wondered for a long time whether a chapter on abstract differentiation in normed vector spaces was a good idea. It needs some Functional Analysis to be understood, and this is not a book on Functional Analysis. Anyway I came up with the conclusion that linearization is seldom taught in a unitary way: students differentiate functions of one variable with one definition, then functions of two variables with another definition, the functions on manifolds with again a different definition. In this respect I wanted to show that differentiation is the same in $\mathbb{R}$, $\mathbb{R}^2$, $\mathbb{R}^3$, and in any vector space that allows distances to be measured. A Global Inversion Theorem is proved in Hausdorff spaces, and this is a nice result that is seldom proposed in Analysis books. My gratitude goes to Antonio Ambrosetti and Giovanni Prodi, who popularized abstract differential calculus in the Italian mathematical community.

The two chapters on integration and measure theory are a journey inside the journey. I present a flavor of Integration Theory based on the Daniell approach: the integral becomes a suitable extension of a linear functional with some weak continuity condition. This extension leans on an elementary integral which we may imagine as the Cauchy integral of continuous functions with compact support. I believe that this functional-analytic construction may be of great utility to young mathematicians who do not need all the pathologies of Geometric Measure Theory.

The subsequent chapter returns to the integral via a different path, based on suitable families of sets called σ-algebras. I would like the reader to understand that there is a path which connects measurable sets and integrable functions, and that we can decide in what direction we prefer to go along this path. But even in this second chapter, I completely avoided the Carathéodory machinery of outer measures, and the concrete Lebesgue measure is constructed via a Riesz Representation Theorem *à la* Rudin.

This book contains a few figures, often realized in a sketchy way. The use of personal computers would surely allow us to produce perfect figures, but I wanted to draw the same pictures I would draw on the blackboard. Most drawings were made by Dr. Francesca Vettori, to whom I am indebted.

Who will read this book? I hope that it can be useful to students of Mathematics and Physics who wish to go further than standard Calculus. If compared to other similar textbooks, the main difference here is that we offer a second glance—and indeed more than a glance—on every traditional topic. Of course, I also hope that some colleagues may find the book useful for preparing their lectures.

Instructors will surely need to complement this text with some additional examples. In my opinion, the same example can be enlightening for a student but obscure for another student. A good instructor knows his/her audience, and can suitably illustrate abstract ideas with concrete examples. Exercises appear after some important results, so that the reader is invited to solve them before proceeding further. Several chapters end with a short collection of problems, which may be solved by collecting the ideas of the whole chapter to which they refer.

I would like to express my gratitude Dr. Francesca Bonadei and Dr. Francesca Ferrari of Springer Nature for their support during the preparation of the manuscript.

I am obviously responsible for any misprint appearing in this book. If you are reading it and you have just found a (serious) error in the text, feel free to contact me at `simone.secchi@unimib.it`.

Milano, Italy   Simone Secchi
July 2022

Acknowledgements

Every scientific book is the result of several months of hard work, but also of several years of study. First of all I need to acknowledge the support of my family during the last year. I also want to express my gratitude to my university for providing me with all the necessary tools that were used during the preparation of the manuscript.

Contents

Part I
First Half of the Journey

Chapter 1
An Appetizer of Propositional Logic

Abstract Mathematics is based on the language of proposition logic: every state-ment is a combination of logical propositions, and theorems are simply true statements that can be deduced according to the rules of logic. We follow the first chapter of Mendelson (Introduction to mathematical logic. CRC Press, Boca Raton, 2015).

1.1 The Propositional Calculus

Sentences are just statements to which it is possible to attach a binary value: true (T) or false (F). For example, "Roses are flowers" is a sentence, "dogs have five legs" is another sentence. But "Any cat is" is not. Sometimes sentences depend on free variables, as in "The integer n is a prime", or "The real number x is irrational". The variable n and x are free in the sense that they may take any (admissible) value: compare with "For every integer n, $n + 1 > n$". In this sentence, the variable n is quantified by "For each", and is not a free variable. Another example is "There exists a positive real number r such that $r^2 = 2$".

Definition 1.1 (Negation) If A is a sentence, its negation $\neg A$ is the sentence governed by the following table:

A	$\neg A$
T	F
F	T

© The Author(s), under exclusive license to Springer Nature Switzerland AG 2022
S. Secchi, *A Circle-Line Study of Mathematical Analysis*,
La Matematica per il 3+2 141, https://doi.org/10.1007/978-3-031-19738-3_1

Definition 1.2 (Conjunction) If A and B are sentences, then their conjunction $A \wedge B$ is the sentence governed by the following table:

A	B	$A \wedge B$
T	T	T
T	F	F
F	T	F
F	F	F

Definition 1.3 (Disjunction) If A and B are sentences, then their disjunction $A \vee B$ is the sentence governed by the following table:

A	B	$A \vee B$
T	T	T
T	F	T
F	T	T
F	F	F

Remark 1.1 Although the mathematical conjunction agrees with the use of "and" in everyday language, the mathematical disjunction reflects a use of "or" which may differ from the use in common language. To be explicit, we may formulate a golden rule: $A \vee B$ corresponds to "either A, or B, or both". In common language we tend to understand "either A or B, but not both."

Definition 1.4 (Implication) If A and B are sentences, the sentence $A \implies B$ is defined by the following table:

A	B	$A \implies B$
T	T	T
F	T	T
T	F	F
F	F	T

Remark 1.2 Logical implication may be written in different ways: $A \supset B$ was common in Logic textbooks a few years ago, but also $A \rightarrow B$ is often found. The symbol $A \implies B$ is pronounced "If A, then B", and we also call it a conditional.

Definition 1.5 (Logical Equivalence) If A and B are sentences, the sentence $A \iff B$ is the sentence governed by the following table:

A	B	$A \iff B$
T	T	T
F	T	F
T	F	F
F	F	T

Logical equivalence is also denoted by $A \equiv B$.

1.2 Quantifiers

As we said before, sentences may contain one or more free variables.

Example 1.1 The sentence $A(x, y)$ defined by "the real number x is strictly smaller than the real number y" is sentence with two free variables x and y. From a logical viewpoint, $A(x, y)$ is indistinguishable from $A(\alpha, \beta)$ or $A(\clubsuit, \spadesuit)$. Of course we cannot replace free variables with symbols that are already taken: $A(x, y)$ is not the same as $A(1, 4)$ or $A(\cos, \log)$. However $A(\pi, e)$ may be acceptable, provided that we do not understand π as the number 3.14159... and e as the Napier number 2.718281... As a stronger example, think of $A(i)$: is $i^2 = -1$ as in Complex Analysis, or is i a free variable?

The truth value of a sentence depending on free variables may depend on the choice of these variables. If $A(x, y)$ is defined by "the real number x is strictly smaller than the real number y", then $A(1, 2)$ is certainly true, while $A(4, 0)$ is false.

Definition 1.6 (Universal Quantifier) The universal quantifier $\forall$ means "for all", or "for each".

Definition 1.7 (Existential Quantifier) The existential quantifier $\exists$ means "there exists".

Important: $\exists$ vs. $\exists!$

In mathematics, "there exists" always means "there exists at least one". The sentence "there exists a solution $x \in \mathbb{R}$ to the equation $x^2 = 1$" is true, although we know that there exist *exactly* two real solutions to the equation $x^2 = 1$. Since existence and uniqueness is often important, the symbol $\exists!$ is reserved for the sentence "there exists a unique".

The syntax of sentences with quantifiers is not completely universal. The sentence "For each x the sentence $A(x)$ holds" can be written in different ways:

$$\forall x \; A(x)$$

$$(\forall x) A(x)$$

$$\forall x, \; A(x)$$

$$(x) \; A(x).$$

The last one is clearly the most economic, and the first one is affordable. Logicians tend to avoid brackets as far as they may, and also commas are seen as inessential objects. It is a matter of facts that most mathematicians use brackets freely on their blackboards, and commas are also ubiquitous.

Remark 1.3 The existential quantifier is not a primitive symbol, since the sentence $\exists x \; A(x)$ is logically equivalent to (and actually defined as) $\neg(\forall x \; \neg A(x))$. As a consequence, the negation of $\exists x \; A(x)$ is precisely $\forall x \; \neg A(x)$, and the negation of $\forall x \; A(x)$ is precisely $\exists x \; \neg A(x)$.

If we are given a sentence $A(x_1, \ldots, x_n)$ defined by n free variables and one or more of them is quantified by either $\forall$ or $\exists$, the quantified variables become bound variables. Bound variables essentially disappear from the arguments of the sentence.

Example 1.2 Suppose that $A(x_1, x_2)$ means "$x_1 - x_2 = 0$". The sentence A contains two free variables, but the sentence $\exists x_2 \; A(x_1, x_2)$—which means "there exists x_2 such that $x_1 - x_2 = 0$"—contains one free variable. The sentence $\forall x_1 \exists x_2 \; A(x_1, x_2)$ does not contain any free variable, and means "For every x_1 there exists (at least one) x_2 such that $x_1 - x_2 = 0$".

We follow the first chapter of Mendelson (Introduction to mathematical logic. CRC Press, Boca Raton, 2015).

Chapter 2
Sets, Relations, Functions in a Naïve Way

Abstract We start our journey with naïve set theory. In the second half of the book we will provide a rigorous foundation of these ideas.

We begin this book in the worst possible manner: we introduce a meaningless definition.

Definition 2.1 (Sets) A *set* is a collection of *elements*.

Important: Sets Remain Undefined

It should be clear in the reader's mind that the previous sentence is far from being a mathematical definition. A set is defined through the word "collection", but we do not provide any primitive definition of collections. In other words, we are assuming that the concept of set is already present in our minds. More formally, we can say that our set theory is based on two primitive objects: sets and elements.

We write $x \in X$ to mean that x is an element of the set X, and we say that x is an element of X, or that x belongs to X. We will avoid the reversed symbol $X \ni x$, since $\ni$ is sometimes used in mathematics with a different meaning.

The typical way of constructing a set is as follows:

$$X = \{x \mid \text{some proposition about } x\}.$$

The variable x is a dummy variable, in the sense that it can be replaced by any other symbol without affecting the validity of the definition of the set X.

Example 2.1 To clarify the use of dummy variables, consider

$$\{x \mid x \text{ is a cat}\} = \{C \mid C \text{ is a cat}\}.$$

S. Secchi, *A Circle-Line Study of Mathematical Analysis*,
La Matematica per il 3+2 141, https://doi.org/10.1007/978-3-031-19738-3_2

On both sides we are introducing the set of all cats, no matter how we name the generic cat.

By definition, $X = \{x \mid x \in X\}$. Two sets X and Y are equal when they share the same elements: $x \in X$ if and only if $x \in Y$.

Definition 2.2 (Empty Set) The *empty set* is

$$\emptyset = \{x \mid x \neq x\},$$

Exercise 2.1 Prove that and $\emptyset$ contains no element at all. *Hint:* for every x, the statement $x \neq x$ is false.

It should be remarked that the definition of the empty set is meaningful, in the sense that it does not rely on some intuitive knowledge. The empty set could be equally defined by means of any statement which is false, for instance

$$\emptyset = \left\{x \in \mathbb{R} \;\middle|\; x^2 = -1\right\}$$

$$= \{n \in \mathbb{N} \mid n \text{ is neither odd nor even}\}$$

$$= \{f \mid f \text{ is a function which is both bounded and unbounded}\}$$

Example 2.2 Why don't we define the *opposite* of the empty set, namely

$$\mathcal{U} = \{x \mid x = x\}?$$

This object would contain anything, since anything is equal to itself by definition of equality. It would be desirable to have such a "set". wouldn't it? Unfortunately $\mathcal{U}$ cannot be a set, as Russel showed in his celebrated *paradox*. Let us consider $R = \{x \mid x \notin x\}$, the set of all sets which do not belong to themselves. What can we say about the relation $R \in R$?

Well, if $R \in R$, then R is a set which does not belong to itself, so that $R \notin R$. Viceversa, if $R \notin R$, then R is not a set which does not belong to itself, hence $R \in R$. Formally, $R \in R$ if and only if $R \notin R$. The consequence of this logical equivalence is that sets cannot be described unrestrictedly, and the universe $\mathcal{U}$ cannot be a set in the naïve sense. We will see in the second part of this book that Axiomatic Set Theory can be used to speak of sets without facing Russel's paradox. But most mathematicians think of sets naïvely, and so will we do for the moment. The only recommendation is to avoid any use of the universe.

Definition 2.3 (Subsets) If A and B are sets, then A is a subset of B if and only if each element of A is an element of B: in symbols,

$$\forall x (x \in A \Rightarrow x \in B).$$

In this situation we write $A \subset B$ or $B \supset A$. A set A is a *proper* subset of B if $A \subset B$ and $A \neq B$. We remark that $A = B$ if and only if $(A \subset B) \wedge (B \subset A)$.

Important: Proper Inclusion

It must be observed that $A \subset B$ is compatible with $A = B$. Since many mathematicians do not like this occurrence, the notation $A \subseteq B$ is often found in the literature, so that $A \subset B$ means $A \subseteq B$ and $A \neq B$. In this book we will never understand $\subset$ in this restrictive sense.

Definition 2.4 (Union and Intersection) The *union* of two sets A and B is the set $A \cup B$ of all points that are element of either A or B (or both):

$$A \cup B = \{x \mid (x \in A) \vee (x \in B)\}.$$

The *intersection* of two sets A and B is the set $A \cap B$ of all points that are elements of both A and B:

$$A \cap B = \{x \mid (x \in A) \wedge (x \in B)\}.$$

Two sets A and B are *disjoint* if $A \cap B = \emptyset$.

Definition 2.5 (Complement) The *absolute complement* of a set A is the set $\complement A = \{x \mid x \notin A\}$. We remark that $\complement\complement A = A$. The *relative complement* of a set A with respect to a set X is $X \setminus A = X \cap \complement A$.

Figures 2.1, 2.2, and 2.3 describe visually the basic operations on sets.

Definition 2.6 (Singleton) The set that contains only the element x is denoted by $\{x\}$ and called *singleton x*.

Fig. 2.1 Intersection of two sets

Fig. 2.2 Union of two sets

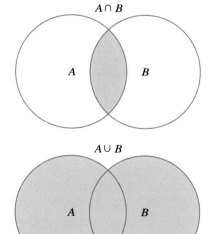

Fig. 2.3 Difference of two
sets

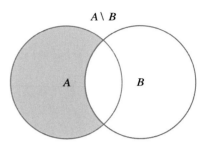

Let us suppose that for each element α of a set A, which is called the *index set*, we are given a set X_α. We can extend our definition of union and intersection as follows:

$$\bigcup \{X_\alpha \mid \alpha \in A\} = \bigcup_{\alpha \in A} X_\alpha = \{x \mid \exists \alpha (\alpha \in A \wedge x \in X_\alpha)\} \qquad (2.1)$$

$$\bigcap \{X_\alpha \mid \alpha \in A\} = \bigcap_{\alpha \in A} X_\alpha = \{x \mid \forall \alpha (\alpha \in A \wedge x \in X_\alpha)\}. \qquad (2.2)$$

A particular case arises when the index set is a collection $\mathcal{A}$ of sets, and in this case we can write

$$\bigcup \{A \mid A \in \mathcal{A}\} = \{x \mid x \in A \text{ for some } A \in \mathcal{A}\}$$

and similarly

$$\bigcap \{A \mid A \in \mathcal{A}\} = \{x \mid x \in A \text{ for each } A \in \mathcal{A}\}.$$

Exercise 2.2 For each positive real numbers α and β, let $Q_{\alpha,\beta}$ be the rectangle $[0, \alpha] \times [0, \beta]$ in the plane. Describe the sets

$$\bigcap \{Q_{\alpha,\beta} \mid \alpha > 0, \ \beta > 0\}, \quad \bigcup \{Q_{\alpha,\beta} \mid \alpha > 0, \ \beta > 0\}.$$

Theorem 2.1 *Let A be an index set, and for each $\alpha \in A$ let X_α be a subset of a fixed set Y. Then*

(a) If B is a subset of A, then

$$\bigcup \{X_\beta \mid \beta \in B\} \subset \bigcup \{X_\alpha \mid \alpha \in A\},$$

and

$$\bigcap\{X_\beta \mid \beta \in B\} \supset \bigcap\{X_\alpha \mid \alpha \in A\}.$$

(b) $Y \setminus \bigcup\{X_\alpha \mid \alpha \in A\} = \bigcap\{Y \setminus X_\alpha \mid \alpha \in A\}$, *and* $Y \setminus \bigcap\{X_\alpha \mid \alpha \in A\} = \bigcup\{Y \setminus X_\alpha \mid \alpha \in A\}$.

Proof

(a) If $x \in \bigcup\{X_\beta \mid \beta \in B\}$ then there exists $\beta \in B$ such that $x \in X_\beta$. By assumption $\beta \in A$, and thus $x \in \bigcup\{X_\alpha \mid \alpha \in A\}$. If $x \in \bigcap\{X_\alpha \mid \alpha \in A\}$ then $x \in X_\alpha$ for each $\alpha \in A$, so that in particular $x \in X_\beta$ for each $\beta \in B$. Thus $x \in \bigcap\{X_\beta \mid \beta \in B\}$.

(b) If $x \in Y \setminus \bigcup\{X_\alpha \mid \alpha \in A\}$ then $x \in Y$ and x is not an element of any X_α, $\alpha \in A$. Hence x belongs to Y and for each $\alpha \in A$ there holds $x \notin X_\alpha$. This means that $x \in \bigcap\{Y \setminus X_\alpha \mid \alpha \in A\}$. Reversing this argument we prove the first identity. Now, if $x \in Y \setminus \bigcap\{X_\alpha \mid \alpha \in A\}$ then $x \in Y$ and there exists $\alpha \in A$ such that $x \notin X_\alpha$. Hence $x \in \bigcup\{Y \setminus X_\alpha \mid \alpha \in A\}$. Reversing this argument we prove the second identity.

$\square$

An *ordered pair* is a new object (x, y) characterized by the following property: two ordered pairs (x, y) and (u, v) are equal if and only if $x = u$ and $y = v$. Actually an ordered pair may be defined in terms of sets as follows.

Definition 2.7 (Ordered Pair)

$$(x, y) = \{\{x\}, \{x, y\}\}.$$

Exercise 2.3 Prove that indeed $(x, y) = (u, v)$ if and only if $x = u$ and $y = v$. *Hint:* by assumption $\{\{x\}, \{x, y\}\} = \{\{u\}, \{u, v\}\}$. Consider first the case $x = y$, then deal with the general case.

Definition 2.8 (Relations) A *relation* is a set of ordered pairs: a relation is therefore a set whose elements are ordered pairs.

If R is a relation, we usually write $x R y$ instead of the more formal $(x, y) \in R$, and we say that x is related to y via R.

Definition 2.9 The *domain* of a relation R is the set $\{x \mid \exists y((x, y) \in R)\}$. The *range* of a relation R is the set $\{x \mid \exists x((x, y) \in R)\}$. The *field* of a relation R is the union of the domain and of the range of R.

One of the simplest relations is the set of ordered pairs (x, y) such that x is a member of a fixed set A, and y is a member of a fixed set B. This relation reduces

Fig. 2.4 A cartesian product

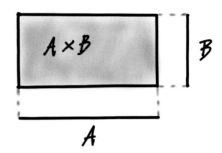

therefore to

$$A \times B = \{(x, y) \mid (x \in A) \wedge (y \in B)\},$$

and is called the *cartesian product* of A and B: see Fig. 2.4. It is clear that any relation is a subset of the cartesian product of its domain and its range.

Remark 2.1 The identification of sets and relations usually sounds strange to students. In this book we will never think of relations or functions like *black boxes* which transform elements of some set into elements of some other set.

The *inverse* of a relation R, denoted by R^{-1}, is the relation obtained by swapping each of the ordered pairs belonging to R. Formally,

$$R^{-1} = \{(y, x) \mid (x, y) \in R\},$$

or equivalently $y R^{-1} x$ if and only if $x R y$.

The *composition* of two relations R and S is

$$R \circ S = \{(x, z) \mid \exists y((x, y) \in S \wedge (y, z) \in R)\}.$$

We remark that, roughly speaking, first comes S, then comes R, and not viceversa. The domain of $R \circ S$ is the domain of S, while the range of $R \circ S$ is the range of R. This will be of crucial importance when we introduce functions.

Definition 2.10 Suppose that R is a relation and X is the set of all points that are elements of either the domain or the range of R. We say that R is

- reflexive, if each element of X is in relation R with itself;
- symmetric, if $x R y$ whenever $y R x$;
- antisymmetric, if $x R y$ and $y R x$ imply $x = y$;
- transitive, if $x R y$ and $y R z$ imply $x R z$.

Definition 2.11 An equivalence relation is a reflexive, symmetric and transitive relation. An order relation is a reflexive, antisymmetric and transitive relation.

It is customary to use the symbol $\sim$ for equivalence relations, and $\leq$ for order relations.

A *function* is a relation such that no two distinct members have the same first coordinate. More explicitly, a relation f is a function if for each element x of its domain there exists a unique element y of its range such that $(x, y) \in f$, see Fig. 2.5. Uniqueness means that if $(x, y) \in f$ and $(x, z) \in f$, then $y = z$. For a function it is customary to abandon the general notation $(x, y) \in f$ (or xfy) in favor of $y = f(x)$. Then $f(x)$ is the *image* of the element x of the domain of f. In mathematical analysis a function $f \subset X \times Y$ is denoted by the (more complicated) symbol

$$f : X \rightarrow Y, \quad x \mapsto f(x).$$

A function $f : X \rightarrow Y$ is *injective* if distinct points of X have distinct images in Y. Equivalently, $f(x_1) = f(x_2)$ implies $x_1 = x_2$. A function $f : X \rightarrow Y$ is *surjective* if the range of f coincides with Y. Equivalently, for each $y \in Y$ there exists $x \in X$ such that $f(x) = y$. Finally, a function $f : X \rightarrow Y$ is *bijective* if it is both injective and surjective.

Exercise 2.4 Let X and Y be sets. Prove that the map $f : X \times Y \rightarrow Y \times X$ defined by $f(x, y) = (y, x)$ for each $(x, y) \in X \times Y$ is a bijection. In this sense, $X \times Y$ and $Y \times X$ are *essentially* the same object.

If A is a set and f is a function, the set

$$f(A) = \{y \mid \exists x (x \in A \wedge f(x) = y)\} = \{f(x) \mid x \in A\}$$

Fig. 2.5 Intuition of a function

is called the *image* of the set A under f. Similarly, if B is a set and f is a function,

$$f^{-1}(B) = \{x \mid \exists y(x \in B \wedge f(x) = y)\}$$

is called the *pre-image* of B under f. We notice that $f^{-1}(B)$ is just the image of the set B under the inverse relation f^{-1}. Clearly $f(A)$ is a subset of the range of f, while $f^{-1}(B)$ is a subset of the domain of f.

Theorem 2.2 *If f is a function and A and B are sets, then*

(a) $f^{-1}(A \setminus B) = f^{-1}(A) \setminus f^{-1}(B)$;
(b) $f^{-1}(A \cup B) = f^{-1}(A) \cup f^{-1}(B)$;
(c) $f^{-1}(A \cap B) = f^{-1}(A) \cap f^{-1}(B)$.

More generally, if we are given a set X_α for each member α of a non-empty index set C, then

(d) $f^{-1}(\bigcup\{X_\alpha \mid \alpha \in C\}) = \bigcup\{f^{-1}(X_\alpha) \mid \alpha \in C\}$;
(e) $f^{-1}(\bigcap\{X_\alpha \mid \alpha \in C\}) = \bigcap\{f^{-1}(X_\alpha) \mid \alpha \in C\}$.

Proof We prove part (e), leaving the rest of the proof as a simple exercise. A point x is an element of $f^{-1}(\bigcap\{X_\alpha \mid \alpha \in C\})$ if and only if $f(x)$ is an element of this intersection, in which case $f(x) \in X_\alpha$ for each $\alpha \in C$. But the latter condition is equivalent to $x \in f^{-1}(X_\alpha)$ for each $\alpha \in C$, i.e. $x \in \bigcup\{f^{-1}(X_\alpha) \mid \alpha \in C\}$. □

Remark 2.2 Any function f is invertible *as a relation*. However the inverse relation f^{-1} need not be again a function: this happens if and only if for each y there exists a unique x such that $yf^{-1}x$, i.e. $f(x) = y$. We have proved that the relation f^{-1} is a function if and only if f is a bijective function. It is customary to say that a function $f : X \to Y$ is *invertible* if it is bijective.

Remark 2.3 Any injective function $f : X \to Y$ can be somehow inverted, in the sense that we can define a function $g : f(X) \to X$ such that $g(y) = x$ if and only if $f(x) = y$. In general the domain of g is a proper subset of Y, but the rule which defines g is exactly the same rule which defines f^{-1}. Many mathematicians do not require surjectivity in order to define invertible functions. This is fairly reasonable, since $f(X)$ is the largest subset of Y on which we can define the inverse function of the injective function f.

Exercise 2.5 Let $f : X \to U$ and $g : Y \to V$ denote two functions. Prove that $(x, y) \mapsto (f(x), g(y))$ defines a function $f \times g : X \times Y \to U \times V$, which we call the Cartesian product of f and g. Prove the following statements:

(i) if f and g are injective, then so is $f \times g$;
(ii) if f and g are surjective, then so is $f \times g$.

> **Important: Sets or Subsets?**
>
> Most surveys of naïve set theory for mathematical analysis only deal with subsets of a given *universe*. We followed another route, and this may have been surprising. The use of a given universe is motivated by Russel's paradox, but for the moment this remains irrelevant to us. As we will see, denying the set of all sets is not the only escape from Russel's paradox.

2.1   Comments

We have presented a quick survey of Set Theory from a non-axiomatic viewpoint. Most textbook in Mathematical Analysis contain similar information, with only minor differences in the language. As an example, functions are typically defined as *rules* of assignment instead of special relations between two sets. A standard reference is [1], a book which goes however much beyond the level suggested by the title.

Before proceeding further, we should stop and think about notation. It is a matter of facts that most instructors discourage the abstract use of

$$\{x \mid P(x)\} \tag{2.3}$$

for the definition of a set. In this book we may seem to be lazy, since such a notation is allowed and even typical. Let us try to elaborate on this issue.

From a very abstract viewpoint, (2.3) contains the troublesome formula

$$\{x \mid x = x\},$$

which leads to the paradox of the *universe*. On the contrary, the more precise formula

$$\{x \mid x \in \mathcal{U} \land P(x)\},$$

often written as $\{x \in \mathcal{U} \mid P(x)\}$, is admissible, since it defines a subset of a (given) set $\mathcal{U}$. Nowadays, most introductory discussions about (naïve) Set Theory are based on axiomatic theories which discard arbitrarily large sets, like ZF (Zermelo-Fraenkel), and this accounts for the recommendation against the use of $\{x \mid P(x)\}$.

On the contrary, we will discuss a different Axiomatic Theory of Sets which allows large objects (called *classes*). In some sense, we should say that $\{x \mid x = x\}$ exists as a class, but not as a set. Since the algebra of classes is quite similar to the algebra of sets, at a first stage we forget the distinction and we allow a more relaxed notation.

Reference

1. P.R. Halmos, *Naive Set Theory* (Dover Publications, 2017)

Chapter 3
Numbers

Abstract Classical mathematical analysis is actually analysis over the field of real numbers. In a later chapter we will construct the set $\mathbb{R}$ of real numbers from the axiom of set theory, as a *completion* of the set $\mathbb{Q}$ of rational numbers. These are in turn constructed from the set $\mathbb{Z}$ of (signed) integers, which are constructed from the set $\mathbb{N}$ of positive integers, or *natural* numbers. However, this approach is time-consuming, and we prefer to quickly introduce real numbers axiomatically.

Important: Warning

The existence of the natural numbers 1, 2, 3, 4, ... will be taken as granted. This is a reasonable compromise in a first approach to Analysis. In the second half of the book we will show that numbers can be defined in terms of sets.

3.1   The Axioms of $\mathbb{R}$

The *system* of *real numbers* is a set $\mathbb{R}$, together with a distinguished subset P and two functions from $\mathbb{R} \times \mathbb{R}$ into $\mathbb{R}$, called operations. The first operation is the *sum*, and is denoted by $(x, y) \mapsto x + y$. The second operation is the *product*, and is denoted by $(x, y) \mapsto xy = x \cdot y$.

Remark 3.1 The subset P should be thought of as the subset of positive numbers. The choice of isolating a subset P instead of introducing an order relation is clearly idiosyncratic.

S. Secchi, *A Circle-Line Study of Mathematical Analysis*,
La Matematica per il 3+2 141, https://doi.org/10.1007/978-3-031-19738-3_3

Axiom 1 For each $x \in \mathbb{R}$ and $y \in \mathbb{R}$ there results $x+y = y+x$ and $xy = yx$.

Axiom 2 For each x, y and z in $\mathbb{R}$, there results $x + (y + z) = (x + y) + z$ and $x(yz) = (xy)z$.

Axiom 3 For each x, y and z in $\mathbb{R}$, there results $x(y + z) = xy + xz$.

Axiom 4 There exist two distinct elements 0 and 1 in $\mathbb{R}$ such that $0 + x = x$ and $1x = x$ for each $x \in \mathbb{R}$.

Axiom 5 For each $x \in \mathbb{R}$ there exists a unique $-x \in \mathbb{R}$ such that $x + (-x) = 0$. If $x \in \mathbb{R}$ and $x \neq 0$, there exists a unique $x^{-1} \in \mathbb{R}$ such that $xx^{-1} = 1$. We will write $x - y$ instead of $x + (-y)$, and x/y instead of xy^{-1} for $y \neq 0$.

One summarizes the first five axioms by saying that $\mathbb{R}$ is a *field*. Abstract fields are algebraic structures, and we will not discuss them in this book. It is noteworthy that these axioms allow us to recover the basic algebraic rules of manipulation of numbers.

Theorem 3.1 *If x, y, z and w are real numbers, and $w \neq 0$, then $x + z = y + z$ implies $x = y$. Furthermore, $xw = yw$ implies $x = y$.*

Proof Indeed

$$x = 1x = x1 = x(ww^{-1}) = (xw)w^{-1}$$
$$= (yw)w^{-1} = y(ww^{-1}) = y1 = 1y = y.$$

Similarly,

$$x = x + (z + (-z)) = (x + z) + (-z)$$
$$= (y + z) + (-z) = y + (z + (-z)) = y.$$

$\square$

Theorem 3.2 *If x, y, z and w are real numbers such that $z \neq 0$ and $w \neq 0$, then*

1. $x0 = 0$;
2. $-(-x)) = x$;
3. $(w^{-1})^{-1} = w$;
4. $(-1)x = -x$;
5. $x(-y) = -xy = (-x)y$;
6. $(-x) + (-y) = -(x + y)$;
7. $(-x)(-y) = xy$;
8. $(x/z)(y/w) = (xy)/(zw)$;
9. $(x/z) + (y/w) = (wx + zy)/(zw)$.

The proof is left as an exercise. For instance, $x0 + x0 = x(0+0) = x0 = 0 + x0$, thus $x0 = 0$. Or $x + (-x) = 0 = (-x) + (-(-x)) = -(-x) + (-x)$, thus $x = -(-x)$.

Axiom 6 The three sets $\{0\}$, P and $-P = \{x \in \mathbb{R} \mid -x \in P\}$ are pairwise disjoint, and their union is $\mathbb{R}$:

$$\mathbb{R} = P \cup \{0\} \cup (-P).$$

Remark 3.2 Equivalently, we could require that $\mathbb{R} = P \cup (-P)$ and $P \cap (-P) = \{0\}$.

Axiom 7 If x and y are element of P, then $x + y \in P$ and $xy \in P$.

We call P the *subset of positive real numbers*. The set $-P$ is the subset of *negative real numbers*. It is customary to write $x < y$ instead of $y - x \in P$, or equivalently $x - y \in -P$. Moreover $x \leq y$ will mean that either $x < y$ or $x = y$. Finally, $y \geq x$ is the same as $x \leq y$, and $y > x$ is the same as $x < y$.

Theorem 3.3 *For each x, y and z in $\mathbb{R}$, we have*

(i) $x < y$ and $y < z$ imply $x < z$;
(ii) exactly one of the relations $x < y$, $x = y$, $x > y$ holds;
(iii) $x < y$ implies $x + z < y + z$;
(iv) $x < y$, $z > 0$ imply $xz < yz$;
(v) $x < y$, $z < 0$ imply $xz > yz$;

(vi) $1 > 0$ *and* $-1 < 0$;
(vii) $z > 0$ *implies* $1/z > 0$;
(viii) $0 < x < y$ *implies* $0 < 1/y < 1/x$.

Proof (i) Since $y - x \in P$ and $z - y \in P$, their sum $z - x \in P: x < z$. (ii) according to Axiom 6, the number $y - x$ must lie exactly in one of the three sets P, $\{0\}$, $-P$. The proofs of (iii), (iv) and (v) are similar. (vi) We want to show that $1 \in P$. If not, since $1 \neq 0$, $1 \in -P$ and $-1 \in P$. Thus $1 = (-1)(-1) \in P$. (vii) Suppose $z > 0$ and $1/z < 0$. Then $1 = z \cdot (1/z) < 0z = 0$, a contradiction. Finally, $(1/x) - (1/y) = (y - x)/(xy) \in P$ as a product of two positive numbers, and (viii) follows.   □

It is now clear that $\leq$ is an order relation on $\mathbb{R}$. This order is called *total* since it satisfies (ii) of Theorem 3.3. In view of the first seven axioms, we may say that $\mathbb{R}$ is a totally ordered field. Unfortunately this is not enough to distinguish $\mathbb{R}$ from other different sets, like $\mathbb{Q}$. We need the last axiom, that we call *Dedekind completeness*.

Axiom 8 Let A and B be subsets of $\mathbb{R}$ such that $A \neq \emptyset$, $B \neq \emptyset$, $A \cap B = \emptyset$, $A \cup B = \mathbb{R}$ and for each $a \in A$, $b \in B$ there results $a < b$. Then there exists a unique element $x \in \mathbb{R}$ such that

(i) $a \in \mathbb{R}$ and $a < x$ imply $a \in A$;
(ii) $b \in \mathbb{R}$ and $x < b$ imply $b \in B$.

The element x is often called the *separator* of A and B. Since either $x \in A$ or $x \in B$, but of course not both, it follows that either $A = \{t \in \mathbb{R} \mid t \leq x\}$ and $B = \mathbb{R} \setminus A$, or $B = \{t \in \mathbb{R} \mid x \leq t\}$ and $A = \mathbb{R} \setminus B$.

3.2   Order Properties of $\mathbb{R}$

Definition 3.1 Let E be a non-empty subset of $\mathbb{R}$. A number $b \in \mathbb{R}$ is an upper bound for E [resp. a lower bound], if $x \leq b$ [resp. $b \leq x$] for each $x \in E$. The set E is bounded from above [resp. from below] if an upper [resp. lower] bound for E exists. The set E is bounded if it is bounded both from above and from below. A number $s \in \mathbb{R}$ is called the *supremum* (or least upper bound) of E, and we write $s = \sup E$, if s is the smallest upper bounds of E. Similarly a number $s \in \mathbb{R}$ is called the *infimum* (or greatest lower bound) of E, and we write $s = \inf E$, if s is the largest lower bound of E.[1]

---

[1] The symbols glb E for inf E and lub E for sup E are old-fashioned.

Concretely, $s = \sup E$ if and only if s is an upper bound for E, and for each upper bound b of E there results $s \le b$. A similar statement holds for the infimum of E, and is left as an exercise.

Example 3.1 Consider $E = \{x \mid (\exists n \in \mathbb{N})\, (x = 1/(n+1))\}$. With the convention that $\mathbb{N} = \{0, 1, \ldots\}$, we claim that $\inf E = 0$ and $\sup E = 1$. Indeed, $1 \in E$ with the choice $n = 0$, and clearly $1/(n+1) < 1$ for each $n \ge 1$. On the other hand, 0 is a lower bound for E, since $x \ge 0$ for each $x \in E$. We fix any $\varepsilon > 0$ and prove that ε is not a lower bound for E. Indeed, there exists $n_0 \in \mathbb{N}$ so large that $1/(n_0+1) < \varepsilon$: any positive integer larger that $1/\varepsilon - 1$ will suffice.[2] Hence $0 = \inf E$.

Definition 3.2 Let E be a non-empty subset of $\mathbb{R}$. If $b \in E$ is an upper bound of E, then we call it the *maximum* of E, denoted by $b = \max E$. Similarly, if $b \in E$ is a lower bound of E, then b is called the *minimum* of E, denoted by $b = \min E$.

Definition 3.3 For each $x \in \mathbb{R}$ we define the *absolute value* of x to be

$$|x| = \begin{cases} x & \text{if } x \ge 0 \\ -x & \text{if } x < 0. \end{cases}$$

It follows that $|x| = \max\{x, -x\}$.

Example 3.2 The absolute value has a few algebraic properties that follow directly from the definition. Clearly $|x| \ge 0$ for each x. Furthermore, $|xy| = |x||y|$ for each x and y. The proof is trivial if both x and y are positive. If $x < 0$ and $y < 0$, then $xy > 0$, and so $|x||y| = (-x)(-y) = xy = |xy|$. If $x < 0$ and $y > 0$, then $xy < 0$ so that $|x||y| = (-x)y = -xy = |xy|$. Every other case can be reconducted to these ones.

Theorem 3.4 (Triangle Inequality) *For each real numbers x and y, there results*

$$|x + y| \le |x| + |y|.$$

Proof We consider two cases. The first one is $x + y \le 0$, so that $|x + y| = x + y \le |x| + |y|$. The second case is $x + y < 0$, so that $|x+y| = -(x+y) = (-x) + (-y) \le |x| + |y|$. $\qquad\square$

Exercise 3.1 Prove the following *reversed triangle inequality*: for each real numbers x and y, there results $||x| - |y|| \le |x - y|$. *Hint:* write $x = (x - y) + y$ and deduce that $|x| \le |x - y| + |y|$. Now swap x and y.

Theorem 3.5 (Supremum Principle) *Every non-empty set $E \subset \mathbb{R}$ that is bounded from above has a supremum in $\mathbb{R}$.*

---

[2] This proof actually leans on Theorem 3.11.

Proof Let B the set of all upper bounds of E, and set $A = \mathbb{R} \setminus B$. Then $B \neq \emptyset$, and if $x \in E$, then $x - 1 \in A$, so that $A \neq \emptyset$. Let $a \in A$, $b \in B$; then a is not an upper bound of E, so there exists $x \in E$ such that $a < x \leq b$ and thus $a < b$. It follows now from Axiom 8 that there exists a unique real number s such that $a \leq s$ for each $a \in A$ and $s \leq b$ for each $b \in B$. If $s \in A$, then there would exist $x \in E$ with $s < x$. Setting $a = (s + x)/2$ we would derive $a \in A$ because $a < x \in E$, and $s < a$: contradiction. Therefore $s \in B$, and is an upper bound of E. Since s is smaller than any upper bound b of E, it follows that $s = \sup E$. $\square$

Corollary 3.1 *Every non-empty set $F \subset \mathbb{R}$ that is bounded from below has an infimum in $\mathbb{R}$.*

Proof Let E be the set of all lower bounds of F. It follows that $E \neq \emptyset$ and that each element of F is an upper bound of E. Hence there exists $s = \sup E$. Let us now prove that $s = \inf F$. Indeed, it is clear that s is larger than any each lower bound of F. To conclude we need to check that s is a lower bound of F. Suppose not. Then there exists $y \in F$ such that $y < s$. Then y is not an upper bound of E, so there exists $x \in E$ with $y < x$. Since $y \in F$, this is impossible, and s is a lower bound of F. $\square$

The following is a useful characterization of the supremum. The reader is invited to prove this result, and to provide a similar statement for the infimum.

Theorem 3.6 *Let E be a non-empty subset of $\mathbb{R}$. The real number s is the supremum of E if and only if*

1. *for each $x \in E$, $x \leq s$;*
2. *for each $\varepsilon > 0$ there exists $x \in E$ with $s - \varepsilon < x$.*

Exercise 3.2 Consider subsets A and B of $\mathbb{R}$. If $A \subset B$, prove that $\inf B \leq \inf A \leq \sup A \leq \sup B$. *Hint:* Trivially, $\inf A \leq \sup A$. Pick $a \in A$ and observe that $a \in B$. Hence $a \leq \sup B$. Since $a \in A$ is arbitrary, deduce that $\sup A \leq \sup B$. Now complete the proof.

3.3 Natural Numbers

Although any mature mathematician should be aware that natural numbers are set-theoretic objects, many analysts prefer to consider them as real numbers. This is how we introduce them now, postponing a set-theoretic definition to a later chapter.

Definition 3.4 (Inductive Sets) We say that a set $I \subset \mathbb{R}$ is *inductive* if and only if

(i) $1 \in I$ and
(ii) $x \in I$ implies $x + 1 \in I$.

Definition 3.5 The set $\mathbb{N}$ is the smallest (in the sense of set inclusion) inductive subset of $\mathbb{R}$, i.e.

$$\mathbb{N} = \bigcap \{I \mid I \text{ is an inductive subset of } \mathbb{R}\}.$$

Remark 3.3 Plainly 0 is not a natural number according to our definition. This is just a matter of choice, since one could replace 1 by 0 in the definition of inductive sets. To be honest, names are just names.

Important: 0 or 1?

Whether $\mathbb{N}$ must contain 0 is an exhausting discussion. Many—if not most—mathematicians tend nowadays to include 0, but there is a (funny) issue. We have defined the natural numbers by picking 1 as the *first* element, and then by adding it recursively. If we choose 0 as the first element, we cannot add it recursively, since 0 is the neutral element of the addition in $\mathbb{R}$. To sum up, starting with 0 breaks the construction of $\mathbb{N}$ as those numbers obtained by adding up the first element as many times as we wish.

Example 3.3 The set $\mathbb{R}$ is inductive (trivially). The set

$$\mathbb{R}_+ = \{x \in \mathbb{R} \mid x \geq 0\}$$

is also inductive.

Exercise 3.3 Prove that each half-line $[a, +\infty)$, where $a \in \mathbb{R}$, is an inductive set.

Proposition 3.1 $\mathbb{N}$ *is an inductive set.*

Proof Indeed $1 \in \mathbb{N}$, since 1 is an element of each inductive subset of $\mathbb{R}$. Now let $x \in \mathbb{N}$, so that x is an element of each inductive subset I of $\mathbb{R}$. Then $x + 1 \in I$, and since I is arbitrary we get $x + 1 \in \mathbb{N}$. $\square$

The following is a mere restatement of the inductive property of $\mathbb{N}$. Anyway, it deserves a special name for historical reasons.

Theorem 3.7 (Induction Principle) *Suppose $S \subset \mathbb{N}$ is such that $1 \in S$ and $x \in S$ implies $x + 1 \in S$. Then $S = \mathbb{N}$.*

Proof Indeed S is an inductive set of $\mathbb{R}$, then $\mathbb{N} \subset S$. But $S \subset \mathbb{N}$ by assumption, and equality follows. $\square$

Example 3.4 We prove that $1 + 2 + \cdots + n = \frac{1}{2}n(n + 1)$ for each natural number n. We formally put

$$A = \left\{ n \in \mathbb{N} \;\middle|\; 1 + 2 + \cdots + n = \frac{1}{2}n(n + 1) \right\},$$

and show that $A = \mathbb{N}$. Since $A \subset \mathbb{N}$, it is enough to prove that A is inductive. For $n = 1$ we have $1 = \frac{1}{2} \cdot 1 \cdot 2$, and $1 \in A$. We suppose that $n \in A$, and we check that $n + 1 \in A$. Indeed,

$$1 + 2 + \cdots + n + n + 1 = \frac{1}{2}n(n + 1) + (n + 1)$$

$$= (n + 1)\left(\frac{1}{2}k + 1\right)$$

$$= \frac{1}{2}(n + 1)(n + 2),$$

and we conclude that $n + 1 \in A$. Hence A is an inductive set, thus $A \supset \mathbb{N}$. Since $A \subset \mathbb{N}$, we necessarily have $A = \mathbb{N}$.

Example 3.5 We prove that $2^n > n^2$ for each integer $n \geq 5$. Indeed, our statement is equivalent to $2^{m+4} > (m + 4)^2$ for each integer $m \geq 1$. Let

$$A = \left\{m \in \mathbb{N} \;\middle|\; 2^{m+4} > (m + 4)^2\right\}.$$

By direct computation, $1 \in A$. Suppose that $k \in A$; then

$$((k + 1) + 4)^2 = ((k + 4) + 1)^2$$

$$= (k + 4)^2 + 2(k + 4) + 1$$

$$< (k + 4)^2 + 2(k + 4) + (k + 4)$$

$$= (k + 4)^2 + 3(k + 4)$$

$$< (k + 4)^2 + (k + 4)(k + 4)$$

$$= 2(k + 4)^2 < 2 \cdot 2^{k+4} = 2^{(k+1)+4}.$$

Thus $k + 1 \in A$, and $A = \mathbb{N}$ by induction.

Theorem 3.8 (Binomial Theorem) *For every real numbers a and b and for every $n \in \mathbb{N}$ there results*

$$(a + b)^n = \sum_{k=0}^{n}\binom{n}{k}a^k b^{n-k},$$

where $\binom{n}{k} = \frac{n!}{k!(n-k)!}.$

Exercise 3.4

(i) Prove that for every $n \in \mathbb{N}$ and every integer $k \in \{1, \ldots, n\}$ there results

$$\binom{n}{k - 1} + \binom{n}{k} = \binom{n + 1}{k}.$$

(ii) Using mathematical induction, prove Theorem 3.8.

Theorem 3.9 *For each $n \in \mathbb{N}$ there results*

  (i) $1 \leq n$;
  (ii) $n > 1$ *implies* $n - 1 \in \mathbb{N}$;
 (iii) $x \in \mathbb{R}$, $x > 0$, $x + n \in \mathbb{N}$ *imply* $x \in \mathbb{N}$;
 (iv) $m \in \mathbb{N}$, $m > n$ *imply* $m - n \in \mathbb{N}$;
  (v) $a \in \mathbb{R}$, $n - 1 < a < n$ *imply* $a \notin \mathbb{N}$.

Proof (i) The set $\{x \in \mathbb{R} \mid x \geq 1\}$ is an inductive set, hence it contains $\mathbb{N}$. (ii) Let $S = \{1\} \cup \{n \in \mathbb{N} \mid n - 1 \in \mathbb{N}\}$. Plainly $1 \in S$, and if $n \in S$ then $(n + 1) - 1 = n \in S \subset \mathbb{N}$. Hence $n + 1 \in S$, and S is inductive. It follows that $S = \mathbb{N}$. Let now T be the set of points $n \in \mathbb{N}$ such that (iii) holds. It follows from (ii) with $x + 1$ instead of n that $1 \in T$. Let $n \in T$ and $x > 0$ be such that $x + (n + 1) \in \mathbb{N}$. Now $x + 1 > 0$ and $(x + 1) + n \in \mathbb{N}$. Hence $x + 1 \in \mathbb{N}$. Again (ii) implies $x \in \mathbb{N}$. This proves that $n + 1 \in T$, and thus $T = \mathbb{N}$. (iv) follows from (iii) upon setting $x = m - n$. Finally, if (v) were false, then we would have $a \in \mathbb{N}$, $n < a + 1$ and $(a + 1) - n < 1$. From (iv) it would follow that $a + 1 - n \in \mathbb{N}$, and this contradicts (i).   □

The following is a fundamental property of the natural numbers. It looks almost trivial, but we encourage the reader to keep in mind the well-order property, since it will come back in a more complicated way.

Theorem 3.10 ($\mathbb{N}$ Is Well-Ordered) *Any non-empty subset of $\mathbb{N}$ has a minimum.*

Proof Let A be a non-empty subset of $\mathbb{N}$, and suppose it has no smallest element. Set

$$S = \{n \in \mathbb{N} \mid \forall a (a \in A \Rightarrow n < a)\}.$$

It is clear that $1 \in S$, otherwise 1 would be the smallest element of A. Assume that $n \in S$. If $n + 1 \notin S$, there exists $a \in A$ such that $a \leq n + 1$. Since $n \in S$, $n < a$, and thus a is a natural number lying between n and $n + 1$. The previous theorem yields $a = n + 1$, and therefore a is the smallest element of A. This contradiction shows that $n + 1 \in S$. By induction, $S = \mathbb{N}$. Let a be any element of A: then $a < a$, a contradiction. We conclude that A has a smallest element.   □

The next result shows a deep interplay between $\mathbb{R}$ and $\mathbb{N}$.

Theorem 3.11 (Archimedean Property of $\mathbb{R}$) *If a and b are real numbers and $a > 0$, then there exists $n \in \mathbb{N}$ such that $na > b$. As a particular case, $\mathbb{N}$ is not bounded from above.*

Proof If the conclusion is false, then the set $E = \{na \mid n \in \mathbb{N}\}$ is bounded from above by b. Let $s = \sup E$. It follows from the definition of supremum that there exists $n \in \mathbb{N}$ with $na > s - a$. Then $(n + 1)a \in E$ and $(n + 1)a > s$, in contradiction with the choice of s. Choosing $a = 1$ yields the last statement.   □

We may now add a sign to the natural numbers.

A real number x is an *integer number* if either $x = 0$, or $x \in \mathbb{N}$, or $-x \in \mathbb{N}$. The set of integer numbers is denoted by $\mathbb{Z}$.

Theorem 3.12 *Let $x \in \mathbb{R}$. Then there exists a unique integer n such that $n \leq x < n + 1$, $x - 1 < n \leq x$.*

Proof Suppose the conclusion holds true for different integers $n < m$. Then $n < m \leq x$ and $x < n + 1$, thus $n < m < n + 1$. Hence there would exist an integer m between the consecutive integers n and $n + 1$, a contradiction to (v) of Theorem 3.9.

Let a be the smallest element of $\mathbb{N}$ that is greater than $|x|$. If $x \geq 0$, we take $n = a - 1$. If $x < 0$, we take $n = -a + 1$ or $-a$ according as x is an integer or not.   □

Definition 3.6 The unique integer n of the previous theorem is called the integral part of the real number x, and is denoted by $[x]$.

A *rational number* is any real number of the form a/b, for some integers a and b, $b \neq 0$. The set of rational numbers is denoted by the symbol $\mathbb{Q}$. The elements of $\mathbb{R} \setminus \mathbb{Q}$ are called *irrational* numbers.

The reader will easily check that $\mathbb{Q}$ satisfies Axioms 1–5, so that it is a field itself. A classical result shows that irrational numbers exist.

Theorem 3.13 $\sqrt{2} \notin \mathbb{Q}$.

Proof Assume that $\sqrt{2} = m/n$ for some integers $m, n, n \neq 0$. We may assume that m and n are coprime, in the sense that the fraction m/n cannot be further reduced. Then $2 = m^2/n^2$, i.e. $m^2 = 2n^2$. Hence m^2 is an even number, and so is m. Hence $m = 2k$ for some integer k, and $m^2 = 4k^2$. Thus $4k^2 = 2n^2$, or $2k^2 = n^2$. This yields that n^2 is even, and so is n. But m and n are coprime, a contradiction.   □

Thus rational numbers do not exhaust $\mathbb{R}$. However, there are no "holes" between rational numbers.

Theorem 3.14 (Density of $\mathbb{Q}$ in $\mathbb{R}$) *If x, y are real numbers with $x < y$, there exists $z \in \mathbb{Q}$ such that $x < z < y$.*

Proof The Archimedean property yields $b \in \mathbb{N}$ such that $b > (y - x)^{-1}$. Then $b^{-1} < y - x$. Let $a = [bx] + 1 \in \mathbb{Z}$. Hence $a - 1 \leq bx < a$, i.e. $a/b < x + b^{-1}$ and $x < a/b$. We conclude that $x < a/b < x + b^{-1} < x + (y - x) = y$. The conclusion follows with $z = a/b$.   □

Exercise 3.5 Prove that $\mathbb{Q}$ intersects any open interval of $\mathbb{R}$. This property will be called *topological density* of $\mathbb{Q}$ in $\mathbb{R}$.

3.4 Isomorphic Copies

Theorem 3.15 *Every ordered field contains sets isomorphic to the natural numbers, the integers, and the rational numbers.*

Proof Let us consider any ordered field $(X, +, \cdot, \leq)$. By induction we may define a function $\varphi \colon \mathbb{N} \to X$ recursively as follows:

$$\varphi(1) = 1$$

$$\varphi(n + 1) = \varphi(n) + 1.$$

Let p and q be different natural numbers, say $p < q$. There exists $n \in \mathbb{N}$ such that $q = p + n$: we claim that $\varphi(p) < \varphi(q)$.

Indeed, for $n = 1$ we just have $q = p + 1$ and $\varphi(q) = \varphi(p) + 1 > \varphi(p)$. For a general $n \in \mathbb{N}$ we have $\varphi(p + n + 1) = \varphi(p + n) + 1 > \varphi(p + n)$, and so $\varphi(p + n) > \varphi(p)$ implies $\varphi(p + n + 1) > \varphi(p)$. By induction $\varphi(p + n) > \varphi(p)$ and we see that φ is injective.

Again by induction we can show that $\varphi(p + q) = \varphi(p) + \varphi(q)$ and $\varphi(pq) = \varphi(p)\varphi(q)$. Thus φ is a bijective function from $\mathbb{N}$ onto a subset of X which preserves sums, products and the order relation. By taking differences of natural numbers we obtain $\mathbb{Z}$ as a subset of X, and by taking quotients of integers we obtain $\mathbb{Q}$ as a subset of X. The proof is complete. □

3.5 Complex Numbers

As we have seen, up to isomorphisms we may always assume that $\mathbb{N} \subset \mathbb{Z} \subset \mathbb{Q} \subset \mathbb{R}$. We close this chapter on number systems with a quick introduction to the filed of complex numbers. For the first time we define a set *larger* than $\mathbb{R}$.

Definition 3.7 The field of complex numbers is the set $\mathbb{C}$ of ordered pairs of real numbers, together with two operations. The sum of two complex numbers $z = (a, b)$ and $w = (c, d)$ is defined to be

$$z + w = (a + c, b + d).$$

The product of z and w is defined to be

$$zw = (ac - bd, ad + bc).$$

The set $\mathbb{C}$ is a field under these operations. Indeed the complex number $(0, 0)$ satisfies $(a, b) + (0, 0) = (a, b)$ for any $(a, b) \in \mathbb{C}$. The complex number $(1, 0)$ satisfies $(a, b)(1, 0) = (a, b)$ for any $(a, b) \in \mathbb{C}$. If $z = (a, b)$, then $-z = (-a, -b)$

is such that $z + (-z) = (0, 0)$. If $z = (a, b) \neq (0, 0)$, then the number

$$z^{-1} = \left(\frac{a}{a^2 + b^2}, -\frac{b}{a^2 + b^2} \right)$$

is such that $zz^{-1} = (1, 0)$, namely the multiplicative inverse of z.

If $z = (a, b)$ is a complex number, we call a the *real part* of z, and b the *imaginary part* of z. The symbols are $a = \Re z$, $b = \Im z$. The *modulus* of z is

$$|z| = \sqrt{a^2 + b^2},$$

and the *conjugate* of z is

$$\bar{z} = (a, -b).$$

Definition 3.8 The imaginary unit is the complex number $i = (0, 1)$.

The main reason for introducing i is that $(a, b) = (a, 0) + (b, 0)(0, 1)$. If we identify $(a, 0)$ with a, $(b, 0)$ with b, we can formally write $(a, b) = a + bi$.

Exercise 3.6 Let $z = (a, b)$ and w be complex numbers. Prove that

1. $\bar{z} = a - bi$
2. $\overline{z + w} = \bar{z} + \bar{w}$
3. $\overline{zw} = \bar{z}\bar{w}$
4. $z + \bar{z} = 2\Re z$ and $z - \bar{z} = 2i\Im z$
5. $z\bar{z} = |z|^2$
6. $z^{-1} = \bar{z}/|z|^2$ provided that $z \neq 0$.

Proposition 3.2 *Let z, w be complex numbers. Then*

(i) $|z| > 0$ *unless* $z = 0$, *and* $|0| = 0$;
(ii) $|zw| = |z| \, |w|$;
(iii) $|\bar{z}| = |z|$;
(iv) $|\Re z| \leq |z|$, $|\Im z| \leq |z|$;
(v) $|z + w| \leq |z| + |w|$.

Proof (i) is obvious from the properties of real numbers. Let $z = a + bi$, $w = c + di$. Then $|zw|^2 = (ac - bd)^2 + (ad + bc)^2 = (a^2 + b^2)(c^2 + d^2) = |z|^2|w|^2$. Since the modulus cannot be negative, (ii) follows. (iii) is trivial. To prove (iv) we just remark that $a^2 \leq a^2 + b^2$, so that $|a| \leq \sqrt{a^2 + b^2}$. The same holds the imaginary part b. Finally, $z\bar{w} + \bar{z}w = 2\Re(z\bar{w})$. Hence

$$|z + w|^2 \leq (z + w)(\bar{z} + \bar{w})$$

$$= z\bar{z} + z\bar{w} + \bar{z}w + w\bar{w}$$

$$= |z|^2 + 2\Re(z\bar{w}) + |w|^2$$

$$\leq |z|^2 + 2|z\overline{w}| + |w|^2$$
$$= |z|^2 + 2|z||w| + |w^2|$$
$$= (|z| + |w|)^2.$$

We conclude again by the positivity of the modulus. □

Theorem 3.16 (Cauchy-Schwartz Inequality) *Let* $a_1,\ldots,\ a_n$ *and* $b_1,\ldots,b_n$ *be complex numbers. Then*

$$\left|\sum_{k=1}^{n} a_k \overline{b_k}\right|^2 \leq \sum_{k=1}^{n} |a_k|^2 \sum_{k=1}^{n} |b_k|^2$$

Proof Define $A = \sum_{k=1}^{n} |a_k|^2$, $B = \sum_{k=1}^{n} |b_k|^2$, $C = \sum_{k=1}^{n} a_k \overline{b_k}$. Then

$$0 \leq \sum_{k=1}^{n} |Ba_k - Cb_k|^2 = \sum_{k=1}^{n} (Ba_k - Cb_k)(B\overline{a}_k - \overline{C B_k})$$

$$= B^2 \sum_{k=1}^{n} |a_k|^2 - B\overline{C} \sum_{k=1}^{n} a_k \overline{b_k}$$

$$= BC \sum_{k=1}^{n} \overline{a}_k b_k + |C|^2 \sum_{k=1}^{n} |b_k|^2$$

$$= B^2 A - B|C|^2$$

$$= B(AB - |C|^2).$$

We conclude that $AB \geq |C|^2$, since $B \geq 0$. □

Important: Order Properties of the Complex Numbers

We might suspect that the order properties of $\mathbb{R}$ could "pass over" to the set $\mathbb{C}$, since real numbers can be identified with complex numbers whose imaginary part is zero. However this is impossible: indeed we know that the product of two positive real numbers is always positive, and in particular $x^2 \geq 0$ for each $x \in \mathbb{R}$. Hence i^2 should be a positive number, but $i^2 = -1$. Mathematical analysis in the field of complex numbers is deeply influenced by this fact.

3.6 Polar Representation of Complex Numbers

Every complex number $z = (x, y) = x + iy \neq (0, 0)$ can be written in the form

$$z = r \left(\cos \vartheta + i \sin \vartheta \right),$$

for some numbers $r \geq 0$ and $\vartheta \in [0, 2\pi)$. Indeed, $r = |z| = \sqrt{x^2 + y^2}$ (since $\cos^2 \vartheta + \sin^2 \vartheta = 1$), and ϑ is defined as follows:

(a) if $x \neq 0$, then

$$\vartheta = \arctan \frac{y}{x} + k\pi,$$

where

$$k = \begin{cases} 0 & \text{if } x > 0 \text{ and } y \geq 0 \\ 1 & \text{if } x < 0 \text{ and } y \in \mathbb{R} \\ 2 & \text{if } x > 0 \text{ and } y < 0. \end{cases}$$

(b) If $x = 0$ and $y \neq 0$, then

$$\vartheta = \begin{cases} \frac{\pi}{2} & \text{if } y > 0 \\ \frac{3}{2}\pi & \text{if } y < 0. \end{cases}$$

It is impossible to associate an angle ϑ to the complex number $z = 0$ in a coherent way, as Fig. 3.1 shows.

The representation

$$z = |z| \left(\cos \vartheta + i \sin \vartheta \right)$$

is called the *trigonometric form* of the complex number $z \neq 0$. The number $\vartheta \in [0, 2\pi)$ is called the *principal argument* of z, and it is often denoted by $\operatorname{Arg} z$.

Exercise 3.7 Prove that two complex numbers

$$z_1 = |z_1| \left(\cos \vartheta_1 + i \sin \vartheta_1 \right)$$
$$z_2 = |z_2| \left(\cos \vartheta_2 + i \sin \vartheta_2 \right)$$

such that $z_1 \neq 0$ and $z_2 \neq 0$ are equal if and only if $|z_1| = |z_2|$ and $\vartheta_1 - \vartheta_2 = 2k\pi$ for some integer k.

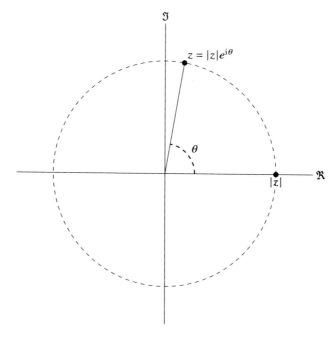

Fig. 3.1 Trigonometric intuition of a complex number

Proposition 3.3 *Suppose that*

$$z_1 = |z_1| \, (\cos \vartheta_1 + i \sin \vartheta_1)$$
$$z_2 = |z_2| \, (\cos \vartheta_2 + i \sin \vartheta_2) \, .$$

Then

$$z_1 z_2 = |z_1| |z_2| \, (\cos(\vartheta_1 + \vartheta_2) + i \sin(\vartheta_1 + \vartheta_2)) \, .$$

Proof Indeed,

$$
\begin{aligned}
z_1 z_2 &= |z_1| |z_2| \, (\cos \vartheta_1 + i \sin \vartheta_1)(\cos \vartheta_2 + i \sin \vartheta_2) \\
&= |z_1| |z_2| \, (\cos \vartheta_1 \cos \vartheta_2 - \sin \vartheta_1 \sin \vartheta_2 + i(\sin \vartheta_1 \cos \vartheta_2 + \sin \vartheta_2 \cos \vartheta_1)) \\
&= |z_1| |z_2| \, (\cos(\vartheta_1 + \vartheta_2) + i \sin(\vartheta_1 + \vartheta_2)) \, .
\end{aligned}
$$

$\square$

Theorem 3.17 (De Moivre's Formula) *If $z = |z|(\cos \vartheta + i \sin \vartheta)$ and $n \in \mathbb{N}$, then*

$$z^n = |z|^n \left(\cos(n\vartheta) + i \sin(n\vartheta) \right).$$

Proof It is enough to recall that $z^n = z \cdots z$ (n times) and to apply Proposition 3.3.

$\square$

3.7   A Construction of the Real Numbers

When I began to write this book, I did not consider any discussion about the construction of $\mathbb{R}$ in the first part. As I said before, it seems that a purely axiomatic definition of the real numbers is more than enough as a first approach to Mathematical Analysis. But then some colleagues convinced me that an authoritative approach ("Believe me, real numbers do exist!") may not be the best choice for an instructor: the pace of teaching is not the pace of logic.

In this section we (the reader and I) will meet the basic ideas of a popular construction of $\mathbb{R}$ from $\mathbb{Q}$. Let me try to introduce the topic.

Imagine you know and use rational numbers (as fractions), but you have no idea about irrational numbers. What is, for example, $\sqrt{2}$? Well, Dedekind proposed to identify real numbers with *subsets* of $\mathbb{Q}$, in an appropriate way. Roughly speaking,

$$\sqrt{2} = \left\{ r \in \mathbb{Q} \,\middle|\, r < \sqrt{2} \right\}.$$

This definition should be prosecuted by the law, of course: nothing should be defined recursively! But look at the following proposal:

$$\sqrt{2} = \left\{ r \in \mathbb{Q} \,\middle|\, r \leq 0 \text{ or } r^2 < 2 \right\}.$$

This looks much better, doesn't it? A (real) number is a set of rational numbers; but what sets?

Definition 3.9 (Dedekind Cuts) A subset L of $\mathbb{Q}$ is a Dedekind cut, if

(a) $L \neq \emptyset$ and $L \neq \mathbb{Q}$;
(b) for every $x \in L$ there exists $y \in L$ such that $x < y$;
(c) if $x \in L$ and $y < x$, then $y \in L$.

In our minds, a cut is a proper, non-empty subset of $\mathbb{Q}$ which has no largest element (condition (b)) and looks like a half-line starting from $-\infty$ (condition (c)). The term *cut* can be explained as follows: if L is a Dedekind cut, then $\mathbb{Q}$ is cut by L in two parts, L and $\mathbb{Q} \setminus L$, such that any element of L is smaller than any element of $\mathbb{Q} \setminus L$.

Example 3.6 For every $r \in \mathbb{Q}$, the set

$$L_r = \{x \in \mathbb{Q} \mid x < r\}$$

is a Dedekind cut. However, our initial discussion suggests that this type of subsets does not exhaust the class of cuts.

We are tempted to define $\mathbb{R}$ as the collection of all Dedekind cuts. The main issue is that $\mathbb{R}$ must be an ordered field.

We can now abandon the letter L (as in Left) to denote Dedekind cuts, and use Greek letters instead. Hence α, β, γ, ... will be real numbers, i.e. cuts.

Definition 3.10 (Order) Let $\alpha \in \mathbb{R}$ and $\beta \in \mathbb{R}$. We say that $\alpha \leq \beta$ if and only if $\alpha \subset \beta$.

It is not difficult to convince oneself that $\leq$ is indeed a reflexive, antisymmetric and transitive relation on $\mathbb{R}$. In other words, it is an order relation. As a first step, we can be satisfied.

Now algebra comes into play, since $\mathbb{R}$ must be endowed with two operations.

Definition 3.11 (Sum) Let $\alpha \in \mathbb{R}$ and $\beta \in \mathbb{R}$. The sum of α and β is defined as

$$\alpha + \beta = \{p + q \mid p \in \alpha, \ q \in \beta\}.$$

Of course we should check that the set $\{p + q \mid p \in \alpha, \ q \in \beta\}$ is a Dedekind cut, but this is a straightforward exercise.

Exercise 3.8 Let $\alpha \in \mathbb{R}$. Show that

$$-\alpha = \{r \in \mathbb{Q} \mid \exists s (s > r \wedge -s \notin \alpha)\}$$

is a Dedekind cut, and that $\alpha + (-\alpha) = 0$, where $0 = \{r \in \mathbb{Q} \mid r < 0\}$.

The product should be defined carefully, since we the product of negative numbers is expected to be a positive number.

Definition 3.12 (Product) Let $\alpha \in \mathbb{R}$ and $\beta \in \mathbb{R}$. The product of α and β is firstly defined in the case $\alpha \geq 0$, $\beta \geq 0$ as

$$\alpha\beta = \{pq \mid p \in \alpha, \ q \in \beta\}.$$

Then

$$\alpha\beta = \begin{cases} -((-\alpha)\beta) & \text{if } \alpha < 0 \text{ and } \beta \geq 0 \\ -(\alpha(-\beta)) & \text{if } \alpha \geq 0 \text{ and } \beta < 0 \\ (-\alpha)(-\beta) & \text{if } \alpha < 0 \text{ and } \beta < 0. \end{cases}$$

Clearly enough, we have adjusted the signs so that the elementary properties of the algebraic product with respect to the order relation are satisfied.

Exercise 3.9 The real number 1 is defined as

$$1 = \{r \in \mathbb{Q} \mid r < 1\},$$

where the number 1 in the right-hand side is the *rational* number $1 = 1/1$. Show that for any $\alpha \in \mathbb{R}$, $\alpha \neq 0$, there exists $\alpha^{-1} \in \mathbb{R}$ such that $\alpha\alpha^{-1} = 1$.

We will not prove in detail that $\mathbb{R}$ with these two algebraic operations is a field: the proofs are straightforward but boring, and can be a good exercise for the reader. Instead, a proof of (some version of) completeness is more interesting. Let us state it as follows.

Theorem 3.18 *The ordered field $\mathbb{R}$ (obtained from Dedekind cuts) has the upper bound property: any non-empty subset, bounded from above, has a least upper bound in $\mathbb{R}$.*

Proof Let $A \subset \mathbb{R}$ be a non-empty set, and let $\beta \in \mathbb{R}$ an upper bound for A. Let us set

$$\gamma = \bigcup \{\alpha \mid \alpha \in A\}.$$

This definition is meaningful, since the elements of A are sets. We are going to show that $\gamma = \sup A$.

Pick any $\alpha_0 \in A$, so that $\alpha_0 \neq \emptyset$. Since $\alpha_0 \subset \gamma$, $\gamma \neq \emptyset$. Then $\gamma \subset \beta$ by construction, hence $\gamma \neq \mathbb{Q}$. Let now $p \in \gamma$, hence there exists $\alpha_1 \in A$ such that $p \in \alpha_1$. If $q < p$, then $q \in \alpha_1$ and thus $q \in \gamma$. If we finally choose $f \in \alpha_1$ such that $r > p$, we see that $r \in \gamma$: we have proved that γ is a Dedekind cut, or equivalently $\gamma \in \mathbb{R}$.

The very definition of γ implies that $\alpha \subset \gamma$ for every $\alpha \in A$. To prove that γ is the least upper bound of A, we fix $\delta < \gamma$. Hence there exists $s \in \gamma$ such that $s \notin \delta$. Moreover, there exists $\alpha \in A$ such that $s \in \alpha$. Hence $\delta < \alpha$, and δ is not an upper bound for A. The proof is now complete. $\qquad\square$

Dedekind cuts are probably the most elementary model of the real numbers as a complete ordered field. It is important to keep in mind that any construction of $\mathbb{R}$ must start from a consistent definition of the rational numbers. From a logical viewpoint, the axiomatization of $\mathbb{R}$ is equivalent to the axiomatization of $\mathbb{N}$: we have defined $\mathbb{N}$ as a subset of $\mathbb{R}$, but we have also proved that $\mathbb{R}$ can be constructed from $\mathbb{Q}$, and hence from $\mathbb{N}$.

3.8 Problems

3.1 If r is a rational number different than zero, and if x is an irrational number, prove that both $r + x$ and rx are irrational numbers.

3.2 Fix a real number $b > 1$.

1. If m, n, p and q are integers, $n > 0$, $q > 0$, and $r = m/n = p/q$, prove that

$$\left(b^m\right)^{1/n} = \left(b^p\right)^{1/q}.$$

   In particular we may define $b^r = (b^m)^{1/n}$.
2. If r and s are rational numbers, prove that $b^{r+s} = b^r b^s$.
3. If $x \in \mathbb{R}$, define $B(x) = \left\{b^t \mid t \in \mathbb{Q},\ t \le x\right\}$. Prove that

$$b^r = \sup B(r)$$

   whenever r is rational. In particular we may define $b^x = \sup B(x)$ for each $x \in \mathbb{R}$.
4. Prove that $b^{x+y} = b^x b^y$ for each real numbers x and y.

This problem provides a rigorous definition of the power of a real number.

3.3 Let $z \in \mathbb{C}$ be a complex number with the property that $|z| = 1$. Compute

$$|1 + z|^2 + |1 - z|^2.$$

3.4 Compute the infimum and the supremum of the set

$$E = \left\{ \frac{1}{2^k} + \frac{1}{3^m} + \frac{1}{5^n} \ \middle|\ \{k, m, n\} \subset \mathbb{N} \right\}.$$

3.9 Comments

The chapter on real numbers is always the most important one in textbooks about (Real) Mathematical Analysis, and often the less self-contained one. The reason is that an rigorous *definition* of $\mathbb{R}$ requires a strong background in Set Theory and in Abstract Algebra. We will see that natural numbers stem directly from Set Theory, while rational numbers can be defined in terms of natural numbers with an algebraic construction. These steps are usually omitted, since most students are satisfied with intuitive definitions like

Natural numbers are 0, 1, 2, 3, . . .

or

Rational numbers are just quotient of integers.

Needless to say, the first definition is based on the hope that every student can decide whether an object is a natural number, and the second definition is meaningless until quotients are defined. In other words, rational numbers are quotient only when rational numbers already exist, or when *real* numbers already exist.

However, it turns out that such an intuition about numbers *does* suffice to develop Calculus of one or more real variables. As a result, I believe that only few graduate students can construct $\mathbb{R}$ from $\mathbb{Q}$, and only a small minority of them can define $\mathbb{N}$ in terms of sets. I was one of those students, and this is why this book contains a chapter on Axiomatic Set Theory.

The books [1] and [2] are good sources about numbers.

References

1. W. Rudin, *Principles of Mathematical Analysis*. International Series in Pure and Applied Mathematics, 3rd edn. (McGraw-Hill Book Co., New York, 1976)
2. K.R. Stromberg, *An Introduction to Classical Real Analysis* (AMS Chelsea Publishing, 2015)

Chapter 4
Elementary Cardinality

Abstract What does it mean that two sets have the same number of elements? This may appear clear if we can write down all the members in a finite list. The answer becomes complicated if the sets contain infinitely many elements. In this chapter we propose a definition of cardinality in an elementary fashion.

4.1 Countable and Uncountable Sets

Definition 4.1 (Sequences) A sequence is any function whose domain is of the form $\mathbb{N} \setminus F$, for some finite subset F of $\mathbb{N}$. If X is a set, a sequence in X is any function which takes values in X and whose domain is of the form $\mathbb{N} \setminus F$, for some finite subset F of $\mathbb{N}$.

If s is a sequence, it is customary to abridge the notation $s(n)$ to s_n. Hence we will also write $\{s_n\}_n$ for a sequence, but we remark that n is a dummy variable: $\{s_n\}_n = \{s_j\}_j = \{s_k\}_k = \dots$

Important: Notation for Sequences

Since a sequence is a function, one might wonder why we make so many efforts to avoid the natural use of functional notation. This sounds as a reasonable question, because historical habit remains the only answer. Sequence are often denoted by $(s_n)_n$ or $\langle s_n \rangle_n$, to distinguish the sequence from the *set* of its values.

We try to illustrate our definition of sequences.

Theorem 4.1 *Let N be a subset of $\mathbb{N}$. The following statements are equivalent:*

(a) $N = \mathbb{N} \setminus F$ for some finite subset F of $\mathbb{N}$;
(b) N contains an interval of the form $\mathbb{N} \cap [n_0, +\infty)$ for some $n_0 \in \mathbb{N}$.

Proof If (a) holds, we call $n_0 - 1$ the largest positive integer which does not belong to N. Then (b) holds. Conversely, we suppose that (b) holds and we consider the finite set $\{1, 2, \ldots, n_0 - 1\}$. Thus at most finitely many positive integers do not belong to N, and (a) holds. $\square$

In other words, our sequences may be considered as functions from an unbounded interval $\mathbb{N} \cap [n_0, +\infty)$ for some $n_0 \in \mathbb{N}$. In the Comments at the end of the chapter we will discuss again our definition.

Definition 4.2 (Subsequences) Let s be a sequence, and let $k : \mathbb{N} \to \mathbb{N}$ a sequence of positive integers with the property that $k_n < k_{n+1}$ for each $n \in \mathbb{N}$. Then the composition $s \circ k$ is called a subsequence of s. Explicitly, $s \circ k = \{s_{k_n}\}_n$.

Remark 4.1 In a subsequent chapter we will see that a weaker condition on the sequence k could be assumed in order to define subsequences. The strong monotonicity $k_n < k_{n+1}$ is however more popular in the literature.

Definition 4.3 (Equal Cardinality) Two sets A and B are equinumerous (or have the same cardinality), if there exists a bijective function $F : A \to B$. In this case we will write $A \sim B$, or even $\#A = \#B$.

It is an easy exercise in set theory to check that $\sim$ is actually an equivalence relation between sets. We will use this fact in the rest of the chapter.

Definition 4.4 We say that a set A has cardinality n, if $A \sim \{1, 2, \ldots, n\}$. By extension, the cardinality of the empty set is zero. A set A is finite, if there exists a positive integer n such that A has cardinality n. Otherwise it is called infinite. A set A is countably infinite if $A \sim \mathbb{N}$, and it is countable if it is either finite or countably infinite. If A is not countable, we say that A is uncountable.

Important: Finite or Countable?

The use of the adjective "countable" is not completely universal. Several mathematicians actually think of countable sets as countably infinite sets. Hence they would not say that $\{5, 7, 11, 23\}$ is a countable set. In my opinion, such an agreement is popular among analysts, who seldom work with finite structures. For this reason, it may happen that in this book the word countable can be used instead of countably infinite. The reader should not have any trouble in recognizing such an abuse of language.

Exercise 4.1 Prove that the Cartesian product of two finite sets is a finite set. *Hint:* this is essentially a "matrix" proof. If X has n members and Y has m members, you can write down $X \times Y$ as a matrix of n rows and m columns. Then just... count the entries of this matrix.

A countably infinite set S can always be described as $S = \{s_1, s_2, \ldots\}$, where s is the bijective function that describes the fact that $A \sim \mathbb{N}$. In this sense, a countably infinite set can be seen as a *labeled* list of points.

Theorem 4.2 *Every subset of a countable set is countable.*

Proof Let S be a countable set, and let $A \subset S$. If A is finite, there is nothing to prove. We may therefore assume that A is infinite, and S is infinite as well. We select a sequence $s = \{s_n\}_n$ of distinct points such that $S = \{s_1, s_2, \ldots\}$. We define a function as follows: let k_1 be the smallest positive integer such that $s_{k_1} \in A$. If $k_2, k_3, \ldots, k_{n-1}$ have been selected, we choose k_n as the smallest positive integer $> k_{n-1}$ such that $s_{k_n} \in A$. It is evident that $k_n < k_{n+1}$ for each n. The composition $s \circ k$ is defined on $\mathbb{N}$ and its range is A. Since $s_{k_n} = s_{k_m}$ implies $k_n = k_m$ (because the points $s_1, s_2, \ldots$ are distinct) and this implies $n = m$, we see that $s \circ k$ is injective. The proof is complete. $\qquad\qquad\square$

Theorem 4.3 *The cartesian product $\mathbb{N} \times \mathbb{N}$ is countably infinite.*

Proof For each $(m, n) \in \mathbb{N} \times \mathbb{N}$ we set $f(m, n) = 2^m 3^n$. This is an injective function whose range is contained in $\mathbb{N}$. Since this range is countable by the previous theorem and $\mathbb{N} \times \mathbb{N}$ is clearly infinite, the proof is complete. $\qquad\square$

What about the cardinality of $\mathbb{Q}$? To answer this question we need some preliminary result about unions of countable sets.

We say that a family F of sets is a collection of disjoint sets, if any two elements of F are disjoint.

Theorem 4.4 *If F is a countable collection of disjoint sets, say $F = \{A_1, A_2, \ldots\}$, such that each A_n is countable, then $\bigcup F = \bigcup_{n=1}^{\infty} A_n$ is also countable.*

Proof For each n, let $A_n = \{a_{1,n}, a_{2,n}, a_{3,n}, \ldots\}$. Call $S = \bigcup_{n=1}^{\infty} A_n$. Every element x of S must lie in some A_n, thus $x = a_{m,n}$ for some pair of integers (m, n). This pair is uniquely determined, since F is a collection of disjoint sets. This defines a function $f : S \to \mathbb{N} \times \mathbb{N}$ via $f(x) = a_{m,n}$. We have just seen that f is injective, so its range is countable. We conclude that S is also countable. $\qquad\square$

We want to remove the assumption that F should be a collection of disjoint sets. This is possible, but it requires some attention.

Theorem 4.5 *If F is a countable collection of countable sets, then the union of all the members of F is also countable.*

Proof We need to reduce to the case of a collection of disjoint sets. A standard way to achieve this result is as follows: put $B_1 = A_1$, and, for $n > 1$,

$$B_n = A_n \setminus \bigcup_{k=1}^{n-1} A_k.$$

Clearly $G = \{B_1, B_2, B_3, \ldots\}$ is a disjoint collection. Setting $A = \bigcup_{n=1}^{\infty} A_n$, $B = \bigcup_{n=1}^{\infty} B_n$, we show that $A = B$. If $x \in A$, then $x \in A_k$ for some k. Let n be the smallest k with this property, so that $x \notin A_k$ for $k < n$. This implies $x \in B_n$, and in turn $x \in B$. Viceversa, if $x \in B$, then $x \in B_n$ for some n, and in particular $x \in A_n$ for the same n. The proof is complete.   □

Corollary 4.1 *The set $\mathbb{Q}$ of rational numbers is countably infinite.*

Proof We call A_n the set of all positive rational numbers whose denominator is n. The set $\mathbb{Q}$ is therefore equal to $\bigcup_{n=1}^{\infty} A_n$, a union of countable sets. The result follows from the previous theorem and the trivial remark that $\mathbb{Q}$ is an infinite set.   □

We already know that $\mathbb{R} \neq \mathbb{Q}$ as sets. We can now show that $\mathbb{R}$ has actually more elements than $\mathbb{Q}$.

Theorem 4.6 *The set $\mathbb{R}$ is uncountable.*

Proof Since the interval $(0, 1) = \{x \in \mathbb{R} \mid 0 < x < 1\}$ is a subset of $\mathbb{R}$, it suffices to show that $(0, 1)$ is uncountable. Suppose not, so that there exists a sequence $s = \{s_n\}_n$ whose range is $(0, 1)$. We show that this is impossible by constructing a real number in $(0, 1)$ which is not a term of the sequence s. As a starting point, we assume that each real number can be uniquely written as an infinite decimal, and in particular $s_n = 0.u_{n,1}u_{n,2}u_{n,3} \ldots$ Each $u_{n,i}$ is a digit, i.e. an element of $\{0, 1, 2, 3, 4, 5, 6, 7, 8, 9\}$. Consider the number $y = 0.v_1 v_2 v_3 \ldots$ where

$$
v_n = \begin{cases} 1 & \text{if } u_{n,n} \neq 1 \\ 2 & \text{if } u_{n,n} = 1. \end{cases}
$$

We claim that no term of the sequence $\{s_n\}_n$ can equal y. Indeed y differs from s_1 in the first digit, differs from s_2 in the second digit, and in general differs from s_n in the n-th digit. But $0 < y < 1$ by construction, and this contradicts the assumption that $(0, 1)$ is countable.   □

Example 4.1 Every open subset (a, b) of $\mathbb{R}$ has the same cardinality as $\mathbb{R}$. Indeed, we choose a number $c \in (a, b)$ and we define $f : (a, b) \to \mathbb{R}$ as

$$
f(x) = \begin{cases} \frac{x-c}{b-x} & \text{if } c \leq x < b \\ \frac{x-c}{x-a} & \text{if } a < x \leq c. \end{cases}
$$

It is easy to check that f is a bijective map.

Exercise 4.2 Let P be the set of all positive real numbers. Prove that $(0, 1)$ and P have the same cardinality by using the function $f : (0, 1) \to P$ defined by

$$
f(x) = \begin{cases} x & \text{if } 0 < x \leq 1/2 \\ \frac{1}{4(1-x)} & \text{if } 1/2 < x < 1. \end{cases}
$$

Exercise 4.3 Prove that any infinite set contains a countably infinite subset. *Hint:* let X be an infinite set. Pick any $x_1 \in X$. Since X is infinite, there exists $x_2 \in$

$X \setminus \{x_1\}$. For the same reason, there exists $x_3 \in X \setminus \{x_1, x_2\}$, and so on. In this way we construct a subset $\{x_j \mid j \in \mathbb{N}\}$ of X which is clearly countably infinite.

Let us call $\mathfrak{c}$ the cardinality of $\mathbb{R}$ and $\aleph_0$ for the cardinality of $\mathbb{N}$. From our discussion it is clear that

$$\aleph_0 < \mathfrak{c},$$

in the sense that there exists an injective function from $\mathbb{N}$ into $\mathbb{R}$, but there cannot exist a bijection between these two sets.

Important: Question

Is there any set whose cardinality is strictly larger than $\aleph_0$ and strictly smaller than $\mathfrak{c}$?

The answer is more than difficult: it is actually impossible! To be more precise, let us state the following

Continuum Hypothesis There exists no set whose cardinality κ satisfies $\aleph_0 < \kappa < \mathfrak{c}$.

Although David Hilbert proposed a proof that the continuum hypothesis was actually true, it soon turned out that his proof was incorrect. Some years later, Gödel showed that the continuum hypothesis cannot be disproved in the framework of any consistent theory of sets. The debate was closed in 1963 by Paul Cohen, who showed that the continuum hypothesis cannot be proved in the framework of any consistent theory of sets, either. Roughly speaking, and since we always assume to have a consistent Set Theory at our disposal, the continuum hypothesis remains independent: it is a matter of taste whether we want to include it among our axioms. Luckily enough, it is rather hard to single out a milestone of Mathematical Analysis which depends on the continuum hypothesis. For this reason, we will not pursue further this topic in the book.

4.2   The Schröder-Bernstein Theorem

We have decided that two sets have the same cardinality if a bijective map exists which takes one set onto the other. A celebrated result by Schröder and Bernstein simplifies our task.

Theorem 4.7 (Schröder-Bernstein) *If there is a one-to-one function on a set A to a subset of a set B and there is also a one-to-one function on B to a subset of A, then A and B have the same cardinality.*

Proof Suppose that $f : A \rightarrow B$ and $g : B \rightarrow A$ are two injective maps. We may assume without loss of generality that $A \cap B = \emptyset$. We say that a point x of either A or B is an ancestor of a point y if and only if y can be obtained from x by successive application of f and g, or of g and f. Now we split A into three subsets: A_E consisting of all points of A which have an even number of ancestors, A_O consisting of all points of A which have an odd number of ancestors, and A_I consisting of all points of A which have infinitely many ancestors. The set B can be split in the same way. We finally define $F : A \rightarrow B$ as follows:

$$F = \begin{cases} f & \text{on } A_E \cup A_I \\ g^{-1} & \text{on } A_O \end{cases}$$

is a bijective map.   □

Remark 4.2 How do we interpret the previous proof? We have actually constructed the map F by an inductive process:

$$E_0 = A \setminus g(B)$$
$$E_1 = g(f(E_0))$$
$$E_2 = g(f(E_1))$$
$$\dots$$
$$E_{n+1} = g(f(E_n)),$$

and so on. Then we set $E = \bigcup_n E_n$. The function F is constructed in such a way that $F = f$ on A, and $F = g^{-1}$ on $A \setminus E$.

We present a second proof of this important result in Set Theory. We need a preliminary tool.

Lemma 4.1 *Let $\mathfrak{X}$ be an ordered set such that every non-empty subset has a greatest lower bound. If $\mathfrak{f} : \mathfrak{X} \rightarrow \mathfrak{X}$ is such that*

1. there exists $x \in \mathfrak{X}$ such that $\mathfrak{f}(x) \leq x$;
2. for every $x \in \mathfrak{X}$, $y \in \mathfrak{X}$, $x \leq y$ implies $\mathfrak{f}(x) \leq \mathfrak{f}(y)$,

then $\mathfrak{f}$ has a fixed point, i.e. there exists $a \in \mathfrak{X}$ such that $\mathfrak{f}(a) = a$.

Proof The set

$$A = \{x \in \mathfrak{X} \mid \mathfrak{f}(x) \leq x\}$$

is non-empty, hence there exists a greatest lower bound $a \in X$ for A. If $x \in A$, then $a \leq x$, hence assumption 2 implies $\mathfrak{f}(a) \leq \mathfrak{f}(x) \leq x$. Thus $\mathfrak{f}(a) \leq a$, since $a = \inf A$. Using again 2, we see that $\mathfrak{f}(\mathfrak{f}(a)) \leq \mathfrak{f}(a)$, hence $\mathfrak{f}(a) \in A$ and so $a \leq \mathfrak{f}(a)$. The proof is complete.   □

Proof (of Theorem 4.7) Let $f: X \to Y$ and $g: Y \to X$ be injective functions. We claim that there exists a subset A of X such that $g(Y \setminus f(A)) = X \setminus A$. Once this claim is proved, the construction of a bijective application of X onto Y is easy.

Let us define $F: 2^X \to 2^X$ such that

$$A \mapsto X \setminus g(Y \setminus f(A)).$$

Lemma 4.1 can be applied with $X = 2^X$, ordered by inclusion $\subset$, and $\mathfrak{f} = F$, since F satisfies condition 2. Condition 1 is also satisfied, since 2^X contains a largest element. Thus $F(A) = A$ for some $A \subset X$, and the proof follows. $\square$

A remarkable fact is that given a set A, one can always construct another set whose cardinality is different than the cardinality of A. We call $\mathcal{P}(A)$ the set of all subsets of A.

Theorem 4.8 (Cantor) *If $A \neq \emptyset$, then there exists no surjective map $f: A \to \mathcal{P}(A)$. In particular, A and $\mathcal{P}(A)$ do not have the same cardinality.*

Proof Let $f: A \to \mathcal{P}(A)$; we will prove that the set

$$S = \{x \in A \mid x \notin f(x)\}$$

does not belong to the image of f. Suppose that $S \in f(A)$, so that $S = f(s)$ for some member $s \in A$. If $s \in S$, then $s \notin f(s) = S$; if $s \notin S$, then $s \in f(s) = S$. In any case we reach a contradiction. $\square$

Exercise 4.4 Suppose that $A = \{x\}$. What is the cardinality of $\mathcal{P}(A)$? Think carefully!

4.3 Problems

4.1 A complex number z is an algebraic number if there exist integers $a_0, \ldots, a_n$, not all zero, such that

$$a_0 z^n + a_1 z^{n-1} + \cdots + a_{n-1} z + a_n = 0.$$

Prove that the set of all algebraic numbers is countable. *Hint:* given $N \in \mathbb{N}$, there exist only finitely many equations with $n + |a_0| + \cdots + |a_n| = N$.

4.2 Is the set $\mathbb{R} \setminus \mathbb{Q}$ countable?

4.3 Prove that a set E is infinite if and only if E has the same cardinality of a proper subset of E. *Hint:* one direction is Exercise 4.3. Conversely, if $f: E \to E$ is an injective function and $a \in E \setminus f(E)$, define recursively $a_1 = f(a), a_{n+1} = f(a_n)$.

4.4 Comments

The rigorous definition of sequences is more problematic than we might suspect. Most textbooks propose to call sequence in a set X any function from $\mathbb{N}$ to X. But a problem immediately arises: with this definition the function $n \mapsto \sqrt{n^2 - 9}$ should not be termed sequence. Our definition clearly absorbs the previous one.

A more refined definition appears in [1]: a sequence in a set X is any function defined on an infinite subset of $\mathbb{N}$, taking values in X. It is easy to check that infinite subsets of $\mathbb{N}$ are characterized as follows.

Theorem 4.9 *Let N be a subset of $\mathbb{N}$. The following statements are equivalent:*

(a) N is an infinite set;
(b) for every $n \in \mathbb{N}$ there exists $p \in N$ such that $p \geq n$.

We will see later that (b) is actually the characterizing property of *nets*, a generalization of sequences.

Comparing sets by counting their elements obviously leads to a rather rough classification. However, this is the first appearance of the concept of *infinity*, which students consider from a philosophical viewpoint. We have proposed a standard approach to elementary cardinality of sets, and in particular we have avoided any explicit reference to the complicated issue of *choosing* elements from non-empty sets. This immediately leads to the Axiom of Choice and to the exhausting discussions about the necessity of using it.

Luckily, I have never found a student who needed an axiom to label the elements of a countable collection of countable sets, although such an operation requires some flavor of the Axiom of Choice. To clarify this point, we should always compare the sentences

1. A is a countable set;
2. let $\{s_1, s_2, s_3, \ldots\}$ be the elements of the countable set A.

The first statement is intrinsic, and we understand that an enumeration of the elements of A *exists*. The second statement already contains the *choice* of an enumeration of A, since the same countable set can be enumerated in infinitely many different ways. To summarize, the Axiom of Choice is not needed to define countable sets, but it comes into play as soon as we want to write down an enumeration of a countable set.

The Schröder-Bernstein Theorem is a useful result which can be proved in several ways. The first proof appears in [2] (but the author attributes it to G. Birkhoff and S. Mac Lane), while the second in based on the *fixed point* Lemma 4.1. I believe that both proofs are elegant and readable at an early stage.

References

1. S. Dolecki, F. Mynard, *Convergence Foundations of Topology* (World Scientific, 2016)
2. J.L. Kelley, *General Topology*. Graduate Texts in Mathematics, No. 27 (Springer, New York, 1975). Reprint of the 1955 edition [Van Nostrand, Toronto, Ont.]

Chapter 5
Distance, Topology and Sequences on the Set of Real Numbers

Abstract The set $\mathbb{R}$ has a rich algebraic structure. What is even more important for Analysis is that its structure of ordered field with the Completeness Axiom may be used to generate a topological environment.

We start with a fairly general definition that describes the possibility of measuring a distance.

Definition 5.1 Let X be a set. A distance on X is a function $d: X \times X \to \mathbb{R}$ with the following properties:

1. $d(x, y) \geq 0$ for each $(x, y) \in X \times X$; $d(x, y) = 0$ if and only if $x = y$.
2. $d(x, y) = d(y, x)$ for each $(x, y) \in X \times X$.
3. (triangle inequality) $d(x, y) \leq d(x, z) + d(z, y)$ for each x, y and z in X.

If a distance d is given on X, we say that (X, d) is metric space.

Definition 5.2 The standard (or Euclidean) metric on $\mathbb{R}$ is defined as $d(x, y) = |x - y|$ for each $(x, y) \in \mathbb{R} \times \mathbb{R}$.

Whenever a distance is available, we can introduce the idea of neighborhood. We will come back to this in greater generality; for the time being we stick to the particular case of the standard metric in $\mathbb{R}$.

Definition 5.3 An open interval is any set of the form $(a, b) = \{x \in \mathbb{R} \mid a < x < b\}$ for some real numbers $a < b$. If S is a subset of $\mathbb{R}$ and x_0 is a point, we say that x_0 is an interior point of S whenever there exists an open interval I such that $x_0 \in I \subset S$. A set S is called open, if each point of S is an interior point of S. A set S is called closed, if the complement $\mathbb{R} \setminus S$ is open.

We will also need *closed* intervals, i.e. set of the form $[a, b] = \{x \in \mathbb{R} \mid a \leq x \leq b\}$ for some real numbers $a \leq b$. Half-lines can be considered to be improper

© The Author(s), under exclusive license to Springer Nature Switzerland AG 2022
S. Secchi, *A Circle-Line Study of Mathematical Analysis*,
La Matematica per il 3+2 141, https://doi.org/10.1007/978-3-031-19738-3_5

47

intervals, for example

$$(a, +\infty) = \{x \in \mathbb{R} \mid a < x\}$$
$$(-\infty, b) = \{x \in \mathbb{R} \mid x < b\}$$
$$[a, +\infty) = \{x \in \mathbb{R} \mid a \leq x\}$$

and so on.

> The sentence "The set I is an interval" means that I is an interval of any kind: open, closed, half-open, a half-line or even the whole real line.

Definition 5.4 A neighborhood of a point x_0 in $\mathbb{R}$ is any set U such that an open interval (a, b) exists with $x_0 \in (a, b) \subset U$.

A neighborhood of x_0 is therefore any set that contains an open set that contains x_0.

Example 5.1 A neighborhood need not be open: $U = [0, 1]$ is a neighborhood of $x_0 = 1/2$, but U is not open.

Exercise 5.1 Prove that if x_0 is a positive real number, there exists a neighborhood of x_0 whose points are positive numbers. This harmless result will be used to prove some results about limits of sequences and functions.

Definition 5.5 Let S be a subset of $\mathbb{R}$. A point $x_0 \in \mathbb{R}$ is an accumulation (or limit, or cluster) point of S, if each neighborhood of x_0 contains a point of S different than x_0 itself (see Fig. 5.1).

Proposition 5.1 *A point x_0 is an accumulation point of S if and only if each neighborhood of x_0 contains infinitely many points of S.*

Proof We need to show that if a point x_0 is an accumulation point of S, then each neighborhood of x_0 contains infinitely many points of S. The other implication is clearly trivial. Suppose not: there exists a neighborhood U of x_0 that contains only finitely many elements of S, say $s_1, s_2, \ldots, s_k$. Pick a number δ such that

$$0 < \delta < \min\{d(x_0, s_j) \mid j = 1, 2, \ldots, k\}.$$

The set $(x_0 - \delta, x_0 + \delta)$ is an open neighborhood of x_0 that contains no points of S different than x_0, and this contradicts the assumption that x_0 is an accumulation point of S. $\square$

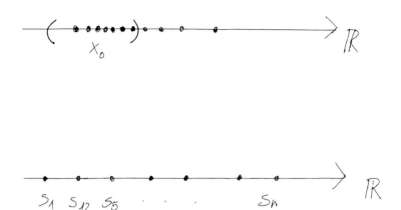

Fig. 5.1 Intuition of accumulation points

Definition 5.6 The set of accumulation points of $S \subset \mathbb{R}$ is denoted by S'.[1] The closure $\overline{S}$ of S is defined to be $S \cup S'$.

Example 5.2 Let $S = (0, 1)$. Each point of $[0, 1]$ is an accumulation point of S, for if $x_0 \in [0, 1]$, then each open interval centered at x_0 contains infinitely many points of S. Furthermore, no other point of $\mathbb{R}$ is an accumulation point of S. Indeed, if $x_0 \in (-\infty, 0) \cup (1, +\infty)$, it is easy to construct an open neighborhood of x_0 which does not intersect S. Suppose for instance that $1 < x_0 < +\infty$. Then the open interval $(x_0 - \frac{1}{2}(x_0 - 1), x_0 + \frac{1}{2}(x_0 - 1))$ is disjoint from S. A similar construction applies to points of $(-\infty, 0)$.

Theorem 5.1 *Suppose S is a subset of $\mathbb{R}$. Then*

(i) $\overline{S}$ is closed;
(ii) $S = \overline{S}$ if and only if S is closed
(iii) $\overline{S} \subset F$ for each closed subset F of $\mathbb{R}$ such that $S \subset F$.

Proof

(i) Let $x \in \mathbb{R}$, $x \notin \overline{S}$. Then x is neither a point of S nor an accumulation point of S. Therefore there exists a neighborhood of x that does not intersect S. This proves that $\complement \overline{S}$ is open, i.e. $\overline{S}$ is closed.
(ii) If $S = \overline{S}$, then S is closed by (i). On the other hand, if S is closed and x is an accumulation point of S, then $x \in S$. Indeed, supposing that $x \notin S$ would lead to a neighborhood U of x such that $x \in U \subset \complement S$ since $\complement S$ is open. This neighborhood would not contain any point of S at all, and x would not be an

---

[1] The notation $\mathcal{D}S$ is sometimes used instead of S'.

accumulation point of S. We have proved that $S' \subset S$ if S is closed, and thus $S = \overline{S}$.

(iii) If F is closed and $F \supset S$, then $F \supset F'$ and thus $F \supset S'$. Hence $F \supset \overline{S}$.

□

Theorem 5.2 *Let S be a subset of $\mathbb{R}$ that is bounded above. If $y = \sup S$, then $y \in \overline{S}$.*

Proof If $y \in S$, then $y \in \overline{S}$ by the very definition. Assume $y \notin S$, and we prove that y is an accumulation point of S. For each $\varepsilon > 0$ there exists a point $x \in S$ such that $y - \varepsilon < x < y$. Hence the arbitrary open neighborhood $(y - \varepsilon, y + \varepsilon)$ contains a point x of S different than y, This proves that y is an accumulation point of S.   □

A similar result holds true for the infimum of a subset bounded below.

5.1   Sequences and Limits

Recall that a sequence in a set X is any function whose domain is $\mathbb{N} \setminus F$ for some finite subset F of $\mathbb{N}$ and whose values lie in X (see Fig. 5.2).

Definition 5.7 Let $s = \{s_n\}_n$ be a sequence in $\mathbb{R}$. We say that s converges to a limit L, and we write

$$L = \lim_{n \to +\infty} s_n,$$

if for every neighborhood U of L there exists a positive integer n_0 such that $n > n_0$ implies $s_n \in U$.

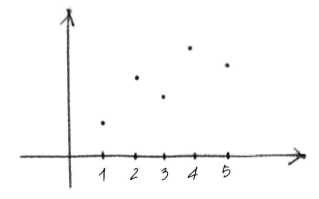

Fig. 5.2 The graph of a sequence of real numbers

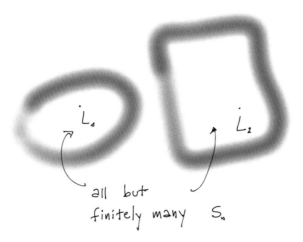

Fig. 5.3 Intuition about uniqueness of limits

Equivalently, s converges to L if and only if for every $\varepsilon > 0$ there exists a positive integer n_0 such that $n > n_0$ implies $d(s_n, L) < \varepsilon$, i.e. $|s_n - L| < \varepsilon$. This formulation is the most popular in undergraduate courses, but it does not work in general topological spaces. Notice that we will also write $s_n \to L$ as $n \to +\infty$ instead of $L = \lim_{n \to +\infty} s_n$.[2]

Exercise 5.2 Let $\{s_n\}_n$ be a sequence of real numbers. Prove that $s_n \to L$ as $n \to +\infty$ if and only if $|s_n - L| \to 0$ as $n \to +\infty$.

Exercise 5.3 Suppose that $\{s_n\}_n$ and $\{t_n\}_n$ are two sequences such that $s_n = t_n$ for every $n \in \mathbb{N}$ except for a finite number of values. Prove that $s_n \to L$ if and only if $t_n \to L$.

Theorem 5.3 (Uniqueness) *If $s_n \to L_1$ and $s_n \to L_2$ as $n \to +\infty$, then $L_1 = L_2$.*

Proof We will use the following fact: if L_1 and L_2 are distinct points on $\mathbb{R}$ there exist a neighborhood U_1 of L_1 and a neighborhood U_2 of L_2 such that $U_1 \cap U_2 = \emptyset$. Indeed, let $r = d(L_1, L_2) = |L_1 - L_2| > 0$. The sets $U_1 = (L_1 - r/2, L_1 + r/2)$ and $U_2 = (L_2 - r/2, L_2 + r/2)$ are clearly disjoint neighborhoods of L_1 and L_2, respectively.

We therefore choose U_1 and U_2 in such a way, and by assumption there exist two positive integers n_1 and n_2 such that $n > n_1$ implies $s_n \in U_1$, and $n > n_2$ implies $s_n \in U_2$. Therefore $n > \max\{n_1, n_2\}$ implies $s_n \in U_1 \cap U_2 = \emptyset$, a contradiction. We conclude that $L_1 = L_2$. Figure 5.3 should clarify our proof. $\square$

---

[2] Since we use s to name a sequence, it would be natural to write $s \to L$. This, albeit correct, might be easily confused with the identical symbol that describes the fact that a variable s tends to L independently, like in $\lim_{s \to L} f(s)$.

Theorem 5.4 (Squeezing Principle) *Let* $s = \{s_n\}_n$ *be a sequence in* $\mathbb{R}$. *The sequence* s *converges to the limit* L *if and only if* $d(s_n, L) = |s_n - L| \to 0$ *as* $n \to +\infty$. *In particular, if* $\{\varepsilon_n\}_n$ *is a sequence of positive numbers that converges to zero, and if* $d(s_n, L) < \varepsilon_n$ *for each* n, *then* $s_n \to L$ *as* $n \to +\infty$.

Proof If s converges to L, for every $\varepsilon > 0$ there exists a positive integer n_0 such that $n > n_0$ implies $d(s_n, L) < \varepsilon$. Hence $d(s_n, L) \to 0$ as $n \to +\infty$. Reversing this argument we prove the first statement. Suppose now that $\varepsilon_n \to 0$ as $n \to +\infty$, and that $d(s_n, L) < \varepsilon_n$ for each n. For every $\varepsilon > 0$ there exists a positive integer n_0 such that $n > n_0$ implies $0 < \varepsilon_n < \varepsilon$. Hence for any $n > n_0$ we get that $d(s_n, L) < \varepsilon$, or $s_n \to L$ as $n \to +\infty$. □

Theorem 5.5 *Any sequence that converges to a limit is bounded.*

Proof Let $s = \{s_n\}_n$ be a real sequence such that $\lim_{n \to +\infty} s_n = L$. Consider the neighborhood $(L - 1, L + 1)$ of L, and select a positive integer n_0 such that $n > n_0$ implies $s_n \in (L - 1, L + 1)$. Now define $\delta = \min\{d(L, s_j) \mid j = 1, 2, \ldots, n_0\}$. The union $U = (L - 1, L + 1) \cup (L - \delta, L + \delta)$ is an open neighborhood of L such that $s_n \in U$ for each n. We have proved that the range of the sequence s is contained in a bounded set, and the conclusion follows. □

If $\{s_n\}_n$ and $\{t_n\}_n$ are two sequences of real numbers, we can form their sum $\{s_n + t_n\}_n$ and their product $\{s_n t_n\}_n$. These algebraic operations are stable with respect to limits.

Theorem 5.6 *Suppose that* $s_n \to L$ *and* $t_n \to M$ *as* $n \to +\infty$. *Then, as* $n \to +\infty$,

1. $s_n + t_n \to L + M$
2. $s_n t_n \to LM$
3. $s_n / t_n \to L/M$ *whenever* $t_n \neq 0$ *and* $M \neq 0$.

Proof Let $\varepsilon > 0$. There exist positive integers n_1 and n_2 such that $n > n_1$ implies $|s_n - L| < \varepsilon$, and $n > n_2$ implies $|t_n - M| < \varepsilon$. If $n > \max\{n_1, n_2\}$, then

$$|(s_n + t_n) - (L + M)| = |(s_n - L) + (t_n - M)| \leq |s_n - L| + |t_n - M| < 2\varepsilon.$$

Since $\varepsilon > 0$ is arbitrary, this proves 1. Similarly,

$$|s_n t_n - LM| = |s_n t_n - s_n M + s_n M - LM| \leq |s_n(t_n - M)| + |M(s_n - L)|.$$

The sequence $\{s_n\}_n$ is bounded by Theorem 5.5, thus there exists $C > 0$ such that $|s_n| < C$ for each n. Therefore

$$|s_n t_n - LM| \leq |s_n(t_n - M)| + |M(s_n - L)| \leq C|t_n - M| + |M||s_n - L|$$
$$\leq C\varepsilon + |M|\varepsilon = (C + |M|)\,\varepsilon.$$

Since $C + |M|$ is independent of ε and of n, the proof of 2. is complete. To prove 3. we proceed as follows. First of all, since $M \neq 0$, there exists a positive integer n_3 such that $|t_n| > |M|/2 > 0$ for any $n > n_3$. This follows from the definition of limit with the neighborhood $(M - |M|/2, M + |M|/2)$. If $n > \max\{n_1, n_2, n_3\}$,

$$
\begin{aligned}
\left| \frac{s_n}{t_n} - \frac{L}{M} \right| = \left| \frac{s_n M - L t_n}{t_n M} \right| &= \left| \frac{s_n M - ML + ML - L t_n}{t_n M} \right| \\
&\leq \left| \frac{(s_n - L)M + L(M - t_n)}{t_n M} \right| \\
&\leq \frac{|s_n - L|}{|t_n|} + \frac{|L|}{|M|} \frac{|t_n - M|}{|t_n|} \\
&< \frac{2}{|M|} \varepsilon + \frac{|L|}{|M|} \frac{2}{|M|} \varepsilon
\end{aligned}
$$

and 3. follows again from the arbitrariness of $\varepsilon > 0$. $\qquad\square$

Theorem 5.7 (Monotone Sequences Have a Limit) *Let $\{s_n\}_n$ be a bounded increasing sequence of real numbers, Then $s_n \to \sup_k s_k$ as $n \to +\infty$.*

Proof Since $\{s_n\}_n$ is bounded, the quantity $L = \sup_k s_k$ is well defined in $\mathbb{R}$. If $\varepsilon > 0$, then there exists a term s_{n_0} such that $L - \varepsilon < s_{n_0} \leq L$. Since $\{s_n\}_n$ is increasing, $n > n_0$ implies $s_n > s_{n_0}$, so that $L - \varepsilon < s_n \leq L$. This proves that $s_n \to L$ as $n \to +\infty$. $\qquad\square$

Exercise 5.4 Let $\{s_n\}_n$ be a bounded decreasing sequence of real numbers. Then $s_n \to \inf_k s_k$ as $n \to +\infty$.

Monotone sequences in $\mathbb{R}$ provide the following important result.

Theorem 5.8 (Nested Intervals) *Suppose that for each positive integer n, $I_n = [a_n, b_n]$ is a closed interval, and suppose furthermore that $I_n \supset I_{n+1}$ for each n. Then $\bigcap_{n=1}^{\infty} I_n \neq \emptyset$.*

Proof The sequence $\{a_n\}_n$ is increasing and bounded above (by b_1, for instance). Hence it converges to $L = \sup_k a_k$. We claim that $L \in \bigcap_{n=1}^{\infty} I_n$. Indeed $L \geq a_n$ for each n, by definition of supremum. On the other hand, if n and p are positive integers, then $a_n \leq a_{n+p} \leq b_{n+p} \leq b_n$. Letting $p \to +\infty$ we get $a_n \leq L \leq b_n$, so that $L \in [a_n, b_n]$ for each n. The proof is complete. $\qquad\square$

Exercise 5.5 Show that there exists a sequence $\{I_n\}_n$ of nested intervals such that $\bigcap_{n=1}^{\infty} I_n$ contains infinitely many points.

Sometimes a more quantitative generalization is needed, as we are going to see.

Theorem 5.9 (Nested Intervals of Infinitesimal Length) *Let $\{I_n\}_n$ be a sequence of closed intervals of $\mathbb{R}$ such that (i) $I_{n+1} \subset I_n$ for each n, and (ii) for every $\varepsilon > 0$ there exists a positive integer n_ε such that the length of I_{n_ε} is smaller than ε. Then $\bigcap_{n=1}^{\infty} I_n$ is a singleton.*

Proof We have already proved that $\bigcap_{n \in \mathbb{N}} I_n \neq \emptyset$. Let us suppose that two distinct numbers $z < w$ belong to this intersection. We may choose $\varepsilon = w - z > 0$ in (ii) and obtain a contradiction. Hence $\bigcap_{n \in \mathbb{N}} I_n$ contains one and only one element. $\quad\square$

Example 5.3 We prove that $\mathbb{R}$ is uncountable as a consequence of Theorem 5.9. As already noticed before, we need to prove that a closed interval $[a, b]$ is uncountable. We suppose on the contrary that $[a, b]$ is countable. Of course $[a, b]$ contains infinitely many elements, so we may suppose that $\{x_n \mid n = 1, 2, 3, \ldots\}$ is an enumeration of $[a, b]$. We are going to construct a number $z \in [a, b]$ which is different than every term x_n of this enumeration. We divide $[a, b]$ into three intervals of equal length, and we choose one of them, called I_1, such that $x_1 \notin I_1$. If I_n has been chosen, we split it into three intervals of equal length, and we call I_{n+1} that interval which does not contain x_{n+1}. The length of I_n converges to zero as $n \to +\infty$, and by construction $I_{n+1} \subset I_n$ for each n. Hence there exists a unique real number z that belongs to every I_n. In particular $z \neq x_n$ for each n, since $z \in I_n$ but $x_n \notin I_n$. This contradiction proves the statement.

Theorem 5.10 *Let S be a subset of $\mathbb{R}$, and let x be an accumulation point of S. Then there exists a sequence $\{s_n\}_n$ of points of S such that $s_n \to x$ as $n \to +\infty$.*

Proof Consider the open neighborhood $U_1 = (x - 1, x + 1)$ of x, and select a point $s_1 \in S$ such that $s_1 \in U_1$. This is possible because x is an accumulation point of S. After point $s_2, s_3, \ldots, s_{n-1}$ are selected in S, choose s_n in S such that $s_n \in U_n = (x - 1/n, x + 1/n)$. Then $|s_n - x| < 1/n$ for each n, and therefore $s_n \to x$ as $n \to +\infty$. $\quad\square$

We record the following fact, which should be an easy exercise: a sequence $\{s_n\}_n$ converges to a limit L if and only if every subsequence of $\{s_n\}_n$ converges to L.

Definition 5.8 A set $K \subset \mathbb{R}$ is called sequentially compact, if every sequence in K has a subsequence that converges to a point of K.

Theorem 5.11 (Bolzano-Weierstrass) *Every bounded sequence in $\mathbb{R}$ contains a converging subsequence.*

Proof Let $s = \{s_n\}_n$ be a bounded sequence of real numbers. For some $M > 0$, this means that $s_n \in [-M, M]$ for each n. If the range of the sequence s is a finite set (in the sense that s takes on only a finite number of different values), then there exists infinitely many positive integers $n_1 < n_2 < n_3 < \ldots$ such that $s_{n_1} = s_{n_2} = s_{n_3} = \ldots$. Hence s contains a constant subsequence, which is clearly convergent to a point of K. We may now assume that the range of s is an infinite set.

We set $I_0 = [-M, M]$. We divide I_0 into two parts: at least one of them must contain infinitely many terms of the sequence, otherwise the range of s would be finite. Let us call I_1 this part, and we choose the smallest positive integer n_1 such that $s_{n-1} \in I_1$. Now we split I_1 into two parts as before, we call I_2 the part that contains infinitely many terms of the sequence s, and we choose the smallest positive integer $n_2 > n_1$ such that $s_{n_2} \in I_2$. Proceeding this way, we construct a subsequence $\{s_{n_k}\}_k$ of S.

By construction, $I_1 \supset I_2 \supset I_3 \supset \ldots$, and by the principle of nested intervals we know that there exists a point $x \in \bigcap_{k=1}^{\infty} I_k$. We claim that $s_{n_k} \to x$ as $k \to +\infty$. We fix a number $\varepsilon > 0$. By construction the length of the interval I_k is $m/2^{k-1}$. If the positive integer k is so large that $M/2^{k-1} < \varepsilon$, then the length of I_k is smaller than ε. Since s_{n_k} and x lie in I_k, then $d(s_{n_k}, x) < \varepsilon$, and the proof is complete. $\square$

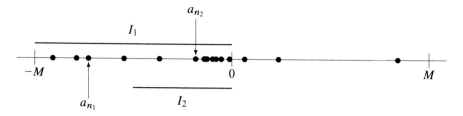

The Bolzano-Weierstrass Theorem can be proved along a different and delightful strategy. We now show that any real-valued sequence has a monotone subsequence. And since we already know that (bounded) monotone sequences converge…

Theorem 5.12 (Any Sequence Has a Monotone Subsequence) *Every sequence $\{s_n\}_n$ of real numbers has a monotone subsequence.*

Proof We provisionally agree that the n-th term of our sequence is dominant if $s_m < s_n$ for each $m > n$. There are now two cases. It might happen that there exist infinitely many dominant terms, and let $\{s_{n_k}\}_k$ be a subsequence consisting solely of dominant terms. Then $s_{n_{k+1}} < s_{n_k}$ for each k, and thus $\{s_{n_k}\}_k$ is monotonically increasing.

The second case is that only finitely many terms are dominant. In particular there exists a positive integer n_1 which is greater than any dominant term. There must exist $n_2 > n_1$ such that $s_{n_2} \geq s_{n_1}$. Suppose that $n_3, \ldots, n_{k-1}$ have been selected so that $n_3 < n_4 < \ldots < n_{k-1}$ and $s_{n_1} \leq s_{n_2} \leq s_{n_3} \leq \ldots \leq s_{n_{k-1}}$. As before, since only finitely many terms are dominant, there exists $n_k > n_{k-1}$ such that $s_{n_k} \geq s_{n_{k-1}}$. By induction we have constructed an increasing subsequence of $\{s_n\}_n$. $\square$

Remark 5.1 Theorem 5.12 is an ingenious trick that essentially reduces the theory of sequence in $\mathbb{R}$ to the theory of monotone sequences. Unfortunately this trick relies on the order property of the real numbers, and cannot be generalized to metric or topological spaces.

We now prove the basic characterization of compact subsets of the real line.

Theorem 5.13 *A set $K \subset \mathbb{R}$ is sequentially compact if and only if it is closed and bounded.*

Proof Suppose K is sequentially compact. If K is not bounded, then there exists $x_1 \in K$ such that $|x_1| > 1$. Likewise, there exists $x_2 \in K$ such that $|x_2| > 2$. In general, for each positive integer n there exists $x_n \in K$ such that $|x_n| > n$. From this we deduce that no converging subsequence of $\{x_n\}_n$ may exist. Indeed, any converging subsequence $\{x_{n_k}\}_k$ must be bounded by Theorem 5.5. But this contradicts the fact that $|x_{n_k}| > k$ for each k. Hence K is bounded.

To prove that K is closed, we show that K contains its accumulation points. So, let $\{x_n\}_n$ be a sequence in K that converges to a point x of $\mathbb{R}$. We have to show that $x \in K$. Since K is sequentially compact, there exists a subsequence $\{x_{n_k}\}_k$ that converges to a point $y \in K$. As we have remarked above, there results $y = x$ since $\{x_{n_k}\}_k$ is a subsequence of $\{x_n\}_n$. In particular $x \in K$, and K is a closed set.

Conversely, suppose that K is closed and bounded, and let $\{x_n\}_n$ be a sequence in K. Since K is bounded, by Bolzano-Weierstrass there exists a subsequence that converges to a point x of $\mathbb{R}$. Since K is closed, $x \in K$, and thus K is sequentially compact. The proof is complete. $\qquad\square$

The definition of limit (Definition 5.7) has a big weakness: we can check if a number is the limit of a sequence, but we need to have a good candidate at our disposal. In the framework of real numbers with the Euclidean metric, the existence of a limit can be ensured by a condition that only involves the terms of the sequence.

Definition 5.9 We say that a sequence $\{s_n\}_n$ of real numbers is a Cauchy sequence if and only if for every $\varepsilon > 0$ there exists a positive integer n_0 such that $n > n_0$, $m > n_0$ imply $d(s_n, s_m) < \varepsilon$.

Proposition 5.2 *Any converging sequence is a Cauchy sequence.*

Proof If $\{s_n\}_n$ is a converging sequence and L is its limit, then for every $\varepsilon > 0$ there exists a positive integer n_0 such that $n > n_0$ implies $d(s_n, L) < \varepsilon$. Hence $n > n_0$ and $m > n_0$ imply by the triangle inequality $d(s_n, s_m) \le d(s_n, L) + d(L, s_m) < 2\varepsilon$. $\qquad\square$

The surprising fact is that we can also reverse this statement: any Cauchy sequence of real numbers must have a limit. The proof is however more delicate.

Proposition 5.3 *Every Cauchy sequence is bounded.*

Proof Recalling the definition with $\varepsilon = 1$, we find a positive integer n_0 such that $n > n_0$ and $m > n_0$ imply $d(s_n, s_m) < 1$. In particular $|s_n| < |s_{n_0+1}| + 1$ for each $n > n_0$. It follows that

$$M = \max\{|s_1|, |s_2|, \ldots, |s_{n_0}|, |s_{n_0+1}| + 1\}$$

satisfies $s_n \in [-M, M]$ for each positive integer n. $\qquad\square$

Theorem 5.14 ($\mathbb{R}$ Is a Complete Metric Space) *A real-valued sequence converges if and only if it is a Cauchy sequence.*

Proof We have already proved one half of the statement. Suppose now that $\{s_n\}_n$ is a Cauchy sequence in $\mathbb{R}$. We know that it is bounded, and by Bolzano-Weierstrass it has a converging subsequence $\{s_{n_k}\}_k$: let L be its limit. We claim that the whole sequence $\{s_n\}_n$ converges to L. Indeed, let $\varepsilon > 0$, and choose a positive integer n_0 such that $n > n_0$ and $m > n_0$ imply $|s_n - s_m| < \varepsilon$. Since $s_{n_k} \to L$ as $k \to +\infty$, we may select an index k_0 such that

$$\left| s_{n_{k_0}} - L \right| < \varepsilon.$$

For each $n > n_0$ we have $|s_n - L| \leq |s_n - s_{n_{k_0}}| + |s_{n_{k_0}} - L| < 2\varepsilon$, and the proof is complete. $\qquad\square$

Exercise 5.6 Suppose that $\{s_n\}_n$ is a sequence of real numbers such that $|s_{n+1} - s_n| < 2^{-n}$ for each n. Prove that $\{s_n\}_n$ is a Cauchy sequence. Is this result true if we suppose that $|s_{n+1} - s_n| < 1/n$ for each n?

We complete this discussion with a few words about divergent sequences.

Definition 5.10 A sequence $\{s_n\}_n$ is divergent if it is not convergent.

This definition is more appreciated if one thinks that sequences may be more general than sequences of real numbers. In Calculus courses, the term *divergent* is usually associated to sequences that "have an infinite limit". When we think of sequences of terms in a general set, the meaning of infinity becomes hard to define, if not impossible at all.

Definition 5.11 Let $s = \{s_n\}_n$ be a sequence of real numbers. We say that s diverges to $+\infty$ if for every $\varepsilon > 0$ there exists a positive integer n_0 such that $n > n_0$ implies $s_n > 1/\varepsilon$. In this case we write $\lim_{n\to+\infty} s_n = +\infty$, or simply $s_n \to +\infty$ as $n \to +\infty$.

Similarly, we say that s diverges to $-\infty$ if for every $\varepsilon > 0$ there exists a positive integer n_0 such that $n > n_0$ implies $s_n < -1/\varepsilon$. In this case we write $\lim_{n\to+\infty} s_n = -\infty$, or simply $s_n \to -\infty$ as $n \to +\infty$.

We remark that the symbols $+\infty$ and $-\infty$ do not represent elements of a numeric set. We will see that we could extend the set $\mathbb{R}$ by adding them in such a way the previous definition becomes a particular case of a general definition of limit for sequences.

Theorem 5.15 *For a sequence $\{s_n\}_n$ of positive real numbers, we have*

$$\lim_{n\to+\infty} s_n = +\infty$$

if and only if

$$\lim_{n\to+\infty} \frac{1}{s_n} = 0.$$

Proof Let us suppose that our sequence diverges to $+\infty$. Let $\varepsilon > 0$ and $M = 1/\varepsilon$. By assumption there exists a positive integer N such that $n > N$ implies $s_n > M$. Therefore $n > N$ implies $\varepsilon > 1/s_n > 0$. This proves that $1/s_n$ converges to zero.

Conversely, let $M > 0$ and $\varepsilon = 1/M$. Since $1/s_n \to 0$, there exists a positive integer N such that $n > N$ implies $1/s_n < \varepsilon$, or $s_n > M$. This concludes the proof.   □

Exercise 5.7 Suppose that there exists a positive integer N_0 such that $s_n \le t_n$ for each $n > N_0$.

(a) Prove that if $\lim_{n \to +\infty} s_n = +\infty$, then $\lim_{n \to +\infty} t_n = +\infty$.
(b) Prove that if $\lim_{n \to +\infty} t_n = -\infty$, then $\lim_{n \to +\infty} s_n = -\infty$.
(c) Prove that if $\lim_{n \to +\infty} s_n$ and $\lim_{n \to +\infty} t_n$ exist, then $\lim_{n \to +\infty} s_n \le \lim_{n \to +\infty} t_n$.

Exercise 5.8 Suppose that $s_n \ne 0$ for each n, and that $L = \lim_{n \to +\infty} \left| \frac{s_{n+1}}{s_n} \right|$ exists.

(a) Show that if $L < 1$ then $\lim_{n \to +\infty} s_n = 0$. *Hint:* fix a number a such that $L < a < 1$, and obtain a positive integer N such that $|s_{n+1}| < a|s_n|$ for each $n > N$. Deduce that $|s_n| < a^{n-N}|s_N|$ for each $n > N$.
(b) Show that if $L > 1$, then $\lim_{n \to +\infty} |s_n| = +\infty$. *Hint:* apply (a) to the sequence $t_n = 1/|s_n|$.

5.2   A Few Fundamental Limits

We collect a few statements that follow from elementary estimates based on the Binomial Theorem. The reader is invited to appreciate the proofs.

Proposition 5.4 *If $p > 0$ is a real number, then*

$$\lim_{n \to +\infty} \frac{1}{n^p} = 0.$$

Proof Given $\varepsilon > 0$, just take $n > (1/\varepsilon)^{1/p}$. This is possible by the Archimedean property of $\mathbb{R}$.   □

Proposition 5.5 *If $p > 0$ is a real number, then*

$$\lim_{n \to +\infty} \sqrt[n]{p} = 1.$$

Proof If $p > 1$, put $x_n = \sqrt[n]{p} - 1$. Clearly $x_n > 0$ and Theorem 3.8 yields $1 + n x_n \le (1 + x_n)^n = p$. Hence $0 < x_n \le (p-1)/n$, and the conclusion follows by squeezing. If $p = 1$ the conclusion is trivial. If $0 < p < 1$, we set $q = 1/p > 1$ and the claim is reconducted to the previous case.   □

Proposition 5.6

$$\lim_{n \to +\infty} \sqrt[n]{n} = 1.$$

Proof Let $x_n = \sqrt[n]{n} - 1$, so that $x_n \geq 0$. By Theorem 3.8

$$n = (1 + x_n)^n \geq \frac{n(n-1)}{2} x_n^2.$$

Hence

$$0 \leq x_n \leq \sqrt{\frac{2}{n-1}}$$

for $n \geq 2$. We conclude by squeezing.   □

Proposition 5.7 *If $p > 0$ and α are real numbers, then*

$$\lim_{n \to +\infty} \frac{n^\alpha}{(1+p)^n} = 0.$$

Proof Fix a positive integer $k > \alpha$. For $n > 2k$

$$(1+p)^n > \binom{n}{k} p^k = \frac{n(n-1) \cdots (n-k+1)}{k!} p^k > \frac{n^k p^k}{2^k k!}.$$

Hence

$$0 < \frac{n^\alpha}{(1+p)^n} < \frac{2^k k!}{p^k} n^{\alpha - k}$$

for $n > 2k$. Since $\alpha - k < 0$, we conclude by squeezing and Proposition 5.4.   □

Proposition 5.8 *If x is a real number and $-1 < x < 1$, then*

$$\lim_{n \to +\infty} x^n = 0.$$

Proof Just take $\alpha = 0$ in Proposition 5.7.   □

5.3 Lower and Upper Limits

We now try to describe the loss of convergence for real sequences. The question is: why can a sequence be divergent?

Definition 5.12 Let $\{a_n\}_n$ be a sequence of real numbers. If it is not bounded above, we declare that

$$\limsup_{n \to +\infty} a_n = +\infty.$$

Similarly, if it is not bounded below, we declare that

$$\liminf_{n \to +\infty} a_n = -\infty.$$

We say that a number M is an eventual upper bound [resp. lower bound] for the sequence if there exists a positive integer ν such that $a_n \leq M$ [resp. $a_n \geq M$] for each $n \geq \nu$. The limsup of the sequence $\{a_n\}_n$ is the infimum of the set $\mathcal{M}$ of eventual upper bounds:

$$\limsup_{n \to +\infty} a_n = \inf \mathcal{M}.$$

In a similar fashion, the liminf of $\{a_n\}_n$ is the supremum of the set $\mathcal{N}$ of eventual lower bounds:

$$\liminf_{n \to +\infty} a_n = \sup \mathcal{N}.$$

Plainly $\liminf_{n \to +\infty} a_n \leq \limsup_{n \to +\infty} a_n$ in any case. The inequality can be strict: if $a_n = (-1)^n$, then $\liminf_{n \to +\infty} a_n = -1$ and $\limsup_{n \to +\infty} a_n = 1$.

Theorem 5.16 *The sequence $\{a_n\}_n$ converges to a limit L if and only if*

$$\liminf_{n \to +\infty} a_n = L = \limsup_{n \to +\infty} a_n.$$

Proof The cases $L \in \{-\infty, +\infty\}$ are clear by the initial definition. We now focus on the case $L \in \mathbb{R}$. Assume that $a_n \to L$: given $\varepsilon > 0$, there exists a positive integer ν such that $L - \varepsilon < a_n < L + \varepsilon$ for each $n > \nu$. Hence $L + \varepsilon \in \mathcal{M}$, and $\limsup_{n \to +\infty} a_n \leq L + \varepsilon$ by definition of infimum. Similarly $L - \varepsilon \leq \liminf_{n \to +\infty} a_n$. Since $\varepsilon > 0$ is arbitrary, this yields $\liminf_{n \to +\infty} a_n = \limsup_{n \to +\infty} a_n = L$.

Viceversa, assume that $\liminf_{n \to +\infty} a_n = L = \limsup_{n \to +\infty} a_n$. Fix any $\varepsilon > 0$. We know that $L + \varepsilon \in \mathcal{M}$ and $L - \varepsilon \in \mathcal{N}$. Hence there exists a positive integer ν such that for each $n > \nu$ we must have $L - \varepsilon \leq a_n \leq L + \varepsilon$. Therefore $a_n \to L$, and the proof is complete.   □

Exercise 5.9 Prove that the sequence $\{(-1)^n\}_n$ does not converge.

We provide a useful characterization of liminf and limsup. Sometimes this is taken as a definition.

Theorem 5.17 *For a bounded sequence* $\{a_n\}_n$,

$$\limsup_{n\to+\infty} a_n = \lim_{n\to+\infty} \sup_{k\ge n} a_k = \inf_{n\in\mathbb{N}} \sup_{k\ge n} a_k$$

$$\liminf_{n\to+\infty} a_n = \lim_{n\to+\infty} \inf_{k\ge n} a_k = \inf_{n\in\mathbb{N}} \inf_{k\ge n} a_k.$$

Proof The two statements are similar, and we prove the second one. We set $\lambda_n = \inf_{k\ge n} a_k$, and remark that $\lambda_{n+1} = \inf\{a_{n+1}, a_{n+2}, \ldots\} \ge \inf\{a_n, a_{n+1}, a_{n+2}, \ldots\} = \lambda_n$. The sequence $n \mapsto \lambda_n$ is thus increasing. Furthermore, $a_k \ge \lambda_n$ for each $k \ge n$, so that λ_n is an eventual lower bound. By definition, $\liminf_{n\to+\infty} a_n \ge \lambda_n$ and finally $\liminf_{n\to+\infty} a_n \ge \sup_n \lambda_n$. We need to prove the opposite inequality.

Fix any eventual lower bound ℓ: there exists a positive integer ν such that $a_k \ge \ell$ for each $k \ge \nu$. Hence $\ell \le \inf_{k\ge\nu} a_k = \lambda_\nu \le \sup_n \lambda_n$. The element $\ell \in \mathcal{N}$ is arbitrary, and so $\liminf_{n\to+\infty} a_n \le \sup_n \lambda_n$.   □

Theorem 5.18 (Monotonicity of liminf and limsup) *If* $a_n \le b_n$ *for each* n, *then* $\liminf_{n\to+\infty} a_n \le \liminf_{n\to+\infty} b_n$ *and* $\limsup_{n\to+\infty} a_n \le \liminf_{n\to+\infty} b_n$.

Proof If $\{a_n\}_n$ is unbounded below, or if $\{b_n\}_n$ is unbounded above, the conclusion is trivial. Let M be an eventual upper bound for $\{b_n\}_n$. There exists a positive integer ν such that $n \ge \nu$ implies $b_n \le M$. Then $a_n \le M$ for the same indices n, and thus $\limsup_{n\to+\infty} a_n \le M$. But M is arbitrary, so that $\limsup_{n\to+\infty} a_n \le \inf M = \limsup_{n\to+\infty} b_n$. The other statement is similar.   □

The following result is often used in the theory of numerical series.

Theorem 5.19 *Suppose* $\{a_n\}_n$ *is a sequence of real numbers. There results*

$$\liminf_{n\to+\infty} \left|\frac{a_{n+1}}{a_n}\right| \le \liminf_{n\to+\infty} |a_n|^{1/n} \le \limsup_{n\to+\infty} |a_n|^{1/n} \le \limsup_{n\to+\infty} \left|\frac{a_{n+1}}{a_n}\right|.$$

In particular, if $\lim_{n\to+\infty} \left|\frac{a_{n+1}}{a_n}\right|$ *exists and equals L, then also* $\lim_{n\to+\infty} |a_n|^{1/n}$ *exists and equals L.*

Proof We prove the last inequality, and leave the first as an exercise. The middle inequality is clear. Put $\alpha = \limsup_{n\to+\infty} |a_n|^{1/n}$ and $L = \limsup_{n\to+\infty} \left|\frac{a_{n+1}}{a_n}\right|$. If $L = +\infty$, the proof is complete, so we may assume that $L \in \mathbb{R}$. Furthermore, it suffices to show that $\alpha \le L + \varepsilon$ for each $\varepsilon > 0$. By definition,

$$L = \lim_{N\to+\infty} \sup\left\{\left|\frac{a_{n+1}}{a_n}\right| \,\middle|\, n > N\right\} < L + \varepsilon,$$

there exists a positive integer N such that

$$\sup\left\{\left|\frac{a_{n+1}}{a_n}\right|\ \middle|\ n \geq N\right\} < L + \varepsilon,$$

and thus

$$\left|\frac{a_{n+1}}{a_n}\right| < L + \varepsilon$$

for each $n \geq N$. For $n > N$ we can write

$$|a_n| = \left|\frac{a_n}{a_{n-1}}\right| \cdot \left|\frac{a_{n-1}}{a_{n-2}}\right| \cdots \left|\frac{a_{N+1}}{a_N}\right|.$$

Hence

$$|a_n| < (L + \varepsilon)^{n-N} |a_N|$$

for each $n > N$. It follows that for $n > N$ we have

$$|a_n|^{1/n} < (L + \varepsilon)^{\frac{n-N}{n}} |a_N|^{1/n},$$

and letting $n \to +\infty$ we see that $\alpha \leq L + \varepsilon$. $\square$

5.4 Problems

5.1 Let $\{s_n\}_n$ be a set of real numbers. The arithmetic mean σ_n are defined by

$$\sigma_n = \frac{s_0 + s_1 + \cdots + s_n}{n + 1}.$$

1. If $\lim_{n \to +\infty} s_n = s \in \mathbb{R}$, prove that $\lim_{n \to +\infty} \sigma_n = s$.
2. Construct a sequence $\{s_n\}_n$ which does not converge, but such that there results $\lim_{n \to +\infty} \sigma_n = 0$.
3. Define $a_n = s_n - s_{n-1}$ for each $n \geq 1$. Prove that

$$s_n - \sigma_n = \frac{1}{n+1} \sum_{k=1}^{n} k a_k.$$

Suppose that $\lim_{n \to +\infty} n a_n = 0$ and that $\{\sigma_n\}_n$ converges. Prove that $\{s_n\}_n$ converges.

4. Prove the same statement as in 3. under the weaker assumption that $\{na_n\}_n$ is bounded. As a hint, you may use the following approach. Suppose that $\lim_{n \to +\infty} \sigma_n = \sigma$ and that $|na_n| \leq M$ for each n. If $m < n$, then

$$s_n - \sigma_n = \frac{m+1}{n-m} (\sigma_n - \sigma_m) + \frac{1}{n-m} \sum_{j=m+1}^{n} (s_j - \sigma_j).$$

For each $j \in \{m+1, \ldots, n\}$,

$$|s_j - \sigma_j| \leq \frac{(n-j)M}{j+1} \leq \frac{(n-m-1)M}{m+2}.$$

For each $\varepsilon > 0$ and each positive integer n, let m be the positive integer such that

$$m \leq \frac{n-\varepsilon}{1+\varepsilon} < m+1.$$

Then $(m+1)/(n-m) \leq 1/\varepsilon$ and $|s_n - s_j| < M\varepsilon$. Hence

$$\limsup_{n \to +\infty} |s_n - \sigma| \leq M\varepsilon.$$

5.2 Let $b > 0$ be a given real number. Choose any real number $x_1 > \sqrt{b}$, and define recursively

$$x_{n+1} = \frac{1}{2} \left(x_n + \frac{b}{x_n} \right).$$

1. Prove that the sequence $\{x_n\}_n$ is decreasing, and that $\lim_{n \to +\infty} x_n = \sqrt{b}$.
2. Define $\varepsilon_n = x_n - \sqrt{b}$, and prove that

$$\varepsilon_{n+1} = \frac{\varepsilon_n^2}{2x_n} < \frac{\varepsilon_n^2}{2\sqrt{b}}.$$

Setting $\beta = 2\sqrt{b}$, deduce that

$$\varepsilon_{n+1} < \beta \left(\frac{\varepsilon_1}{\beta} \right)^{2^n}$$

for $n = 1, 2, 3, \ldots$

This problem describes a numerical algorithm for computing the square root of a given number.

5.3 Let $b > 1$ be a given real number. Choose any real number $x_1 > \sqrt{b}$, and define recursively

$$x_{n+1} = \frac{b + x_n}{1 + x_n} = x_n + \frac{b - x_n^2}{1 + x_n}.$$

1. Prove that $x_1 > x_3 > x_5 > \ldots$
2. Prove that $x_2 < x_4 < x_6 < \ldots$
3. Prove that $\lim_{n \to +\infty} x_n = \sqrt{b}$.

5.4 Compute the upper and the lower limit of the sequence $\{s_n\}_n$ defined recursively by

$$s_1 = 0$$
$$s_{2n} = \frac{s_{2n-1}}{2}$$
$$s_{2n+1} = \frac{1}{2} + s_{2n}.$$

5.5 Evaluate

$$\lim_{n \to +\infty} \sum_{k=1}^{n} \frac{1}{\sqrt{n^2 + k}}.$$

Hint: prove that the sum is smaller than 1 and larger than $n/\sqrt{n^2 + n}$.

5.6 We define the sequence of Fibonacci numbers recursively by $u_0 = 0$, $u_1 = 1$ and

$$u_{n+2} = u_n + u_{n+1}.$$

Set $x_n = u_{n+1}/u_n$ for each $n \in \mathbb{N}$.

(a) Prove that $x_1 < x_3 < x_5 < \ldots < x_6 < x_4 < x_2$.
(b) Prove that $\lim_{n \to +\infty} (x_{2n} - x_{2n-1}) = 0$.
(c) Compute $\lim_{n \to +\infty} x_n$.
(d) If α and β are the roots of the polynomial $x^2 = x + 1$ and if $w_n = a\alpha^n + b\beta^n$, prove that the sequence $\{w_n\}_n$ satisfies $w_{n+2} = w_{n+1} + w_n$ for each $n \in \mathbb{N}$.
(e) Deduce an explicit expression of u_n in terms of n.
(f) Use the result of (e) to solve (c).

5.5 Comments

Elementary Calculus can be introduced in a way that hides the topological (metric) structure of the set $\mathbb{R}$. This approach, in my opinion, is too radical: the passage from analysis in $\mathbb{R}$ to analysis in $\mathbb{R}^N$, $N \geq 2$, requires arguments that I prefer to introduce from the first time. The topology of the real line coincides with the order topology induced by the usual ordering $\leq$, but this is false in higher dimension. The fact that intervals are basic examples of: open sets, convex sets, connected sets, relatively compact sets should be seen both as a positive and a negative feature of $\mathbb{R}$.

The book [1] is a wonderful example of modern treatment of Calculus and Analysis based on topological tools. In my opinion, this book remains a masterwork in its field, although students may need to work hard before appreciating it.

Reference

1. W. Rudin, *Principles of Mathematical Analysis*. International Series in Pure and Applied Mathematics, 3rd edn. (McGraw-Hill Book Co., New York, 1976)

Chapter 6
Series

Abstract Series are just a special type of sequences. The main feature of numerical series is that they lead us to finding convergence theorems which do not involve the value of the limit.

If $a = \{a_n\}_n$ is a sequence of real numbers, we use the symbol

$$\sum_{n=p}^{q} a_n$$

to denote the finite sum $a_p + a_{p+1} + \cdots + a_{q-1} + a_q$. We use the sequence a to construct a new sequence $s = \{s_n\}_n$ by means of the formula

$$s_n = \sum_{k=1}^{n} a_k.$$

The sequence s is called the *sequence of partial sums* of a. It is customary to introduce a different notation for the sequence s:

$$s = \sum_{n=1}^{\infty} a_n.$$

In a really formal world, a series should be defined as an ordered couple (a, s) such that a is a sequence, s is a sequence, and $s_n = \sum_{k=1}^{n} a_k$ for each $n \in \mathbb{N}$.

Remark 6.1 The language about series is very unprecise. In a completely rigorous world, we should probably remove the word *series* and continue to use the word *sequence*, as in

consider the sequence $\left\{ \sum_{k=1}^{n} \frac{k}{k^2+1} \right\}_n$.

Furthermore, several mathematicians interpret $\sum_{n=1}^{\infty} a_n$ as $\lim_{N \to +\infty} \sum_{n=1}^{N} a_n$, which is either a real number of a symbol of infinity. Despite these difficulties, tradition rules, and in this chapter we will freely abuse of language and define a series with the symbol $\sum_n a_n$.

Definition 6.1 We say that the series $\sum_{n=1}^{\infty} a_n$ converges to s if $\lim_{n \to +\infty} s_n = s$. In this case, we will often say that s is the sum of the series.

Remark 6.2 It should be clear that sequences and series are the same object. Indeed, series are sequences by definition. Conversely, the sequence $\{a_n\}_n$ can be recovered from the sequence $\{s_n\}_n$ by writing $a_n = s_n - s_{n-1}$. Of course this logical equivalence is not a good reason to forget about numerical series at all.

We will often write $\sum_n a_n$ or even $\sum a_n$ to denote a series. We agree that the first index of the sum may also be different than 1, as in $\sum_{n=7}^{\infty} a_n$. Clearly, the convergence of a series does not depend on the first terms that we add or discard: remember that the character of a sequence is not altered by the modification of finitely many terms.

Example 6.1 Let us consider the series

$$\sum_{n=2}^{\infty} \frac{1}{n(n-1)}.$$

Since

$$\frac{1}{n(n-1)} = \frac{1}{n-1} - \frac{1}{n},$$

we see that

$$s_n = \sum_{k=2}^{n} \frac{1}{k(k-1)} = 1 - \frac{1}{n} \to 1$$

as $n \to +\infty$. Hence the series converges to the sum 1.

Example 6.2 The previous example can be easily generalized. Suppose that we are given the series

$$\sum_{n=1}^{\infty} (b_{n+1} - b_n),$$

where $\{b_n\}_n$ is a sequence such that $\lim_{n\to+\infty} b_n = b$. Then

$$\sum_{n=1}^{\infty} (b_n - b_{n+1}) = b_1 - b.$$

These are called *telescoping series*.

Since a closed formula for the partial sums of a sequence is usually unavailable, the whole theory of convergence must be based on some *indirect* approach. A very general one is the Cauchy characterization of convergence.

Theorem 6.1 (Cauchy for Series) *A series $\sum a_n$ converges if and only if for every $\varepsilon > 0$ there exists a positive integer N such that*

$$\left| \sum_{k=n}^{m} a_k \right| < \varepsilon$$

for any $m \geq n > N$.

Proof Since

$$\left| \sum_{k=n}^{m} a_k \right| = |s_m - s_{n-1}|,$$

the conclusion follows from Theorem 5.14.   □

Corollary 6.1 (Necessary Condition for Convergence) *If $\sum a_n$ converges, then $\lim_{n\to+\infty} a_n = 0$.*

Proof We take $m = n$ in the previous theorem.   □

Remark 6.3 We will see that this corollary cannot be reversed. For instance the harmonic series $\sum_{n=1}^{\infty} \frac{1}{n}$ diverges, although $1/n \to 0$ as $n \to +\infty$.

The necessary condition for convergence confirms an intuitive fact: you cannot sum infinitely many numbers and obtain a finite result, unless the numbers you add get smaller and smaller. As usual, intuitive results in mathematics are weak results.

Theorem 6.2 *Suppose that $a_n \geq 0$ for each n. The series $\sum a_n$ converges if and only if its partial sums form a bounded sequence.*

Proof For a series of non-negative terms, we clearly have

$$s_{n+1} = s_n + a_{n+1} \geq s_n$$

for every n. In other words, the sequence of partial sums is increasing. The conclusion follows from Theorem 5.7.   □

The most important test of convergence is based on comparison. We will see that actually all convergence tests are based on some comparison argument.

Theorem 6.3 (Comparison Test)

(a) *If $|a_n| \le c_n$ for $n \ge N_0$, where N_0 is some fixed positive integer, and if $\sum c_n$ converges, then $\sum a_n$ converges as well.*
(b) *If $a_n \ge d_n \ge 0$ for $n \ge N_0$, and if $\sum d_n$ diverges, then $\sum a_n$ diverges as well.*

Proof

(a) Given $\varepsilon > 0$, there exists a positive integer $n_0 \ge N_0$ such that $m \ge n > n_0$ implies $\sum_{k=n}^{m} c_k \le \varepsilon$. Hence $\left|\sum_{k=n}^{m} a_k\right| \le \sum_{k=n}^{m} |a_k| \le \sum_{k=n}^{m} c_k < \varepsilon$.
(b) If $\sum a_n$ converges, by (a) $\sum d_n$ converges. Contradiction.

                                                                                    □

An important corollary is described in the next result.

Theorem 6.4 (Asymptotic Comparison Test) *Let $\sum a_n$ and $\sum b_n$ be series of positive terms, and suppose that*

$$\lim_{n \to +\infty} \frac{a_n}{b_n} = 1.$$

The series $\sum a_n$ converges if and only if $\sum b_n$ converges.

Proof Indeed, there exists a positive integer N_0 such that $1/2 < a_n/b_n < 3/2$ for every $n > N_0$. Hence $\frac{b_n}{2} < a_n < \frac{3}{2}b_n$ for $n > N_0$. The conclusion follows from the Comparison test.   □

Example 6.3 The series

$$\sum_{n=1}^{\infty} \frac{1}{n^2}$$

converges. Indeed,

$$\frac{1}{n^2} \le \frac{1}{n(n-1)}$$

for $n = 2, 3, \ldots$ We conclude by comparison with Example 6.1.

Remark 6.4 If $a_n = 1/n^2$ and $b_n = \frac{1}{n(n-1)}$, we have $\lim_{n \to +\infty} a_n/b_n = 1$. This shows a typical application of the asymptotic comparison test to the series $\sum_{n=1}^{\infty} \frac{1}{n^2}$, which often requires less care in checking the validity of the comparison.

The triangle inequality always ensures that

$$\left| \sum_{k=n}^{m} a_k \right| \leq \sum_{k=n}^{m} |a_k|,$$ (6.1)

leading us to the following definition via the Cauchy condition for convergence.

Definition 6.2 (Absolute Convergence) We say that the series $\sum a_n$ converges absolutely, if the series $\sum |a_n|$ is convergent.

An easy but not trivial consequence of (6.1) is the next result.

Theorem 6.5 *Every absolutely convergent series is convergent.*

Proof Let $\sum a_n$ be an absolutely convergent series. By (6.1), the series $\sum a_n$ satisfies the Cauchy condition, and is therefore convergent. □

The converse is false, as Exercise 6.4 shows.

6.1 Convergence Tests for Positive Series

Theorem 6.2 says that series of positive terms are somehow easier to deal with, since no oscillation phenomenon can arise. In this section we develop several convergence tests for positive series, i.e. series of positive terms.

Important: Negative Series

Of course the very same tests can be applied to series of *negative* terms, just by changing signs to each term. For the sake of definiteness, we will always deal with positive series.

Let us start with a milestone of the theory.

Theorem 6.6 (Geometric Series) *If $0 \leq x < 1$, then*

$$\sum_{n=0}^{\infty} x^n = \frac{1}{1-x}.$$

If $x \geq 1$, the series $\sum_{n=0}^{\infty} x^n$ diverges.

Proof If $x = 1$, then $\sum_{k=0}^{n} 1^k = n + 1$, and the series diverges. Suppose $x \neq 1$, and compute

$$\sum_{k=0}^{n} x^k = \frac{1 - x^{n+1}}{1 - x}.$$

Indeed

$$(1 - x)(1 + x + x^2 + \cdots + x^n) = 1 + x - x + x^2 - x^2 + \cdots + x^n - x^n - x^{n+1}$$
$$= 1 - x^{n+1}.$$

The conclusion follows by letting $n \to +\infty$.   □

Exercise 6.1 Prove the identity

$$(1 - x)(1 + x + x^2 + \cdots + x^n) = 1 - x^{n+1}$$

by induction.

The following test is usually a difficult one for students. It states a rather surprising fact: under a monotonicity assumption, only those terms of a very particular subsequence decide whether a series converges.

Theorem 6.7 (Condensation Test) *Suppose that $a_1 \geq a_2 \geq a_3 \geq \ldots \geq 0$. The series $\sum_{n=1}^{\infty} a_n$ is convergent if and only if the series $\sum_{k=0}^{\infty} 2^k a_{2^k}$ is convergent.*

Proof It suffices to prove that the partial sums of the two series are simultaneously bounded from above. Set

$$s_n = a_1 + \cdots + a_n$$
$$t_k = 2^0 a_{2^0} + 2^1 a_{2^1} + \cdots + 2^k a_{2^k}.$$

We consider two cases. If $n < 2^k$, then

$$s_n \leq a_1 + (a_2 + a_3) + \cdots + (a_{2^k} + \cdots + a_{2^{k+1}-1})$$
$$\leq a_1 + 2a_2 + \cdots + 2^k a_{2^k}$$
$$= t_k$$

by the monotonicity of $\{a_n\}_n$. Notice that we have grouped terms in blocks that begin with a power of 2 and end one step before the subsequent power of 2. We deduce that $s_n \leq t_k$.

On the other hand, if $2^k < n$, we group terms in a different way:

$$s_n \geq a_1 + a_2 + (a_3 + a_4) + \cdots + (a_{2^{k-1}+1} + \cdots + a_{2^k})$$
$$\geq \frac{1}{2} a_1 + a_2 + 2a_4 + \cdots + 2^{k-1} a_{2^k}$$
$$= \frac{1}{2} t_k.$$

In this case, $t_k \leq 2s_n$. In any case the sequences $\{s_n\}_n$ and $\{t_k\}_k$ are both bounded or unbounded above, and the proof is complete.   □

Example 6.4 As a fundamental application, we consider the *generalized harmonic series*

$$\sum_{n=1}^{\infty} \frac{1}{n^p},$$

where p is a fixed real number. Clearly $p \leq 0$ implies divergence of the series, since the general term does not converge to zero. For $p > 0$ we use the condensation test, and look at the series

$$\sum_{k=0}^{\infty} 2^k \frac{1}{(2^k)^p} = \sum_{k=0}^{\infty} 2^{(1-p)k}.$$

This is a geometric series, and we know that the latter series converges if and only if $2^{1-p} < 1$, i.e. $p > 1$.

We propose the following tests for historical reasons. They are based on a comparison with a geometric series, and we will comment on the weakness of these tests after the proof.

Theorem 6.8 (Root and Ratio Tests) *The series $\sum a_n$*

(a) *converges, if* $\lim \sup_{n \to +\infty} \sqrt[n]{|a_n|} < 1$;
(b) *diverges, if* $\lim \sup_{n \to +\infty} \sqrt[n]{|a_n|} > 1$;
(c) *converges, if* $\lim \sup_{n \to +\infty} \left|\frac{a_{n+1}}{a_n}\right| < 1$;
(d) *diverges, if* $\left|\frac{a_{n+1}}{a_n}\right| \geq 1$ *for each* $n \geq n_0$, *where* n_0 *is some fixed positive integer.*

Proof Put $\alpha = \lim \sup_{n \to +\infty} \sqrt[n]{|a_n|}$. If $\alpha < 1$, we can choose β such that $\alpha < \beta < 1$, and a positive integer N such that $\sqrt[n]{|a_n|} < \beta$ for each $n \geq N$. Hence $n \geq N$ implies $|a_n| < \beta^n$. Since $\beta < 1$, the comparison test leads to (a).
   If $\alpha > 1$, then $\sqrt[n]{|a_n|} > 1$ for infinitely many indices n (otherwise 1 would be an eventual upper bound). This prevents a_n from converging to 0 as $n \to +\infty$, and the series $\sum a_n$ is divergent. This proves (b).
   Suppose that $\lim \sup_{n \to +\infty} \left|\frac{a_{n+1}}{a_n}\right| < 1$: we can find $\beta < 1$ and a positive integer N such that $\left|\frac{a_{n+1}}{a_n}\right| < \beta$ for each $n \geq N$. In particular

$$|a_{N+1}| < \beta|a_N|$$

$$|a_{N+2}| < \beta|a_{N+1}| < \beta^2|a_N|$$

$$\vdots$$

$$|a_{N+p}| < \beta^p|a_N|$$

for each positive integer p. Writing $n = N + p$ we discover that

$$|a_n| < |a_N|\beta^{-N} \cdot \beta^n$$

for each $n \geq N$. Again (c) follows from the comparison theorem. Finally, if $|a_{n+1}| \geq |a_n|$ for $n \geq n_0$, then the condition $a_n \to 0$ fails, and the series $\sum a_n$ is divergent.

$\square$

The root and the ration tests are popular but *weak*. We know that the series $\sum \frac{1}{n}$ diverges while $\sum \frac{1}{n^2}$ converges. The ratio and the root tests are both inconclusive, since the limsup equals 1.

Remark 6.5 It follows from Theorem 5.19 that the root test is stronger than the ratio test. In particular, if the root test is inconclusive, the ratio test must be inconclusive as well.

Example 6.5 Consider the series $\sum_n \frac{n}{n^2+3}$. If we put $a_n = \frac{n}{n^2+3}$, there results

$$\frac{a_{n+1}}{a_n} = \frac{n+1}{n} \frac{n^2+3}{n^2+2n+4}.$$

We deduce that $\lim_{n \to +\infty} |a_{n+1}/a_n| = 1$. Similarly, $\lim_{n \to +\infty} \sqrt[n]{|a_n|} = 1$. The root test and the ratio test are inconclusive, although the series is divergent by comparison:

$$a_n \geq \frac{n}{n^2+3n^2} = \frac{1}{4n}.$$

Once more, we remark that a clever direct comparison is often preferable to a standard test.

Exercise 6.2 Prove that if a series $\sum_n a_n$ of nonnegative numbers converges, then the series $\sum_n a_n^p$ converges for every real number $p > 1$. *Hint:* the inequality $a_n < 1$ must hold eventually.

Exercise 6.3 Prove that $\sum_n a_n$ and $\sum_n b_n$ are convergent series of nonnegative numbers, then the series $\sum_n \sqrt{a_n b_n}$ converges. *Hint:* prove that $\sqrt{a_n b_n} \leq a_n + b_n$.

6.2 Euler's Number as the Sum of a Series

The typical Calculus approach to the definition of the number e is via the "fundamental limit"

$$\lim_{n \to +\infty} \left(1 + \frac{1}{n}\right)^n.$$

Unfortunately the existence of this limit is not straightforward. In the next theorem we propose a different approach.

Theorem 6.9 (The Euler Number) *The series $\sum_{n=0}^{\infty} \frac{1}{n!}$ converges to a limit that is denoted by e and called the Euler number. Furthermore, $e = \lim_{n\to+\infty} \left(1 + \frac{1}{n}\right)^n$.*

Proof Recall that $0! = 1$ and, for any positive integer n, the factorial of n is defined as $n! = 1 \cdot 2 \cdots (n-1)n$. Since

$$s_n = 1 + \frac{1}{1} + \frac{1}{1 \cdot 2} + \frac{1}{1 \cdot 2 \cdot 3} + \cdots + \frac{1}{1 \cdot 2 \cdots n}$$
$$< 1 + 1 + \frac{1}{2} + \frac{1}{2^2} + \cdots + \frac{1}{2^{n-1}} < 3,$$

the series $\sum_{n=0}^{\infty} \frac{1}{n!}$ converges to a limit $e < 3$. To prove the second part, we introduce the sequences

$$s_n = \sum_{k=0}^{n} \frac{1}{k!}, \quad t_n = \left(1 + \frac{1}{n}\right)^n.$$

The binomial formula

$$(a+b)^n = \sum_{k=0}^{n} \binom{n}{k} a^{n-k} b^k = \sum_{k=0}^{n} \frac{n!}{k!(n-k)!} a^{n-k} b^k$$

yields

$$t_n = 1 + 1 + \frac{1}{2!}\left(1 - \frac{1}{n}\right) + \frac{1}{3!}\left(1 - \frac{1}{n}\right)\left(1 - \frac{2}{n}\right) + \cdots$$
$$+ \frac{1}{n!}\left(1 - \frac{1}{n}\right)\left(1 - \frac{2}{n}\right) \cdots \left(1 - \frac{n-1}{n}\right).$$

Then $t_n \le s_n$ and $\limsup_{n\to+\infty} t_n \le e$. If $n \ge m$,

$$t_n \ge 1 + 1 + \frac{1}{2!}\left(1 - \frac{1}{n}\right) + \cdots + \frac{1}{m!}\left(1 - \frac{1}{n}\right) \cdots \left(1 - \frac{m-1}{n}\right),$$

so that $s_m \le \liminf_{n\to+\infty} t_n$ for any m. Letting $m \to +\infty$, $e \le \liminf_{n\to+\infty} t_n$, and the proof is complete. $\quad\square$

The definition $e = \sum_{n=0}^{\infty} \frac{1}{n!}$ is rather flexible, and allows us to derive a theoretical property of the Euler number.

Theorem 6.10 *The number e is irrational.*

Proof We begin with an estimate of the convergence of the series $\sum 1/n!$ to e. Letting s_n denote the n-th partial sum of this series, we have

$$e - s_n = \frac{1}{(n+1)!} + \frac{1}{(n+2)!} + \frac{1}{(n+3)!} + \cdots$$

$$< \frac{1}{(n+1)!}\left(1 + \frac{1}{n+1} + \frac{1}{(n+1)^2} + \cdots\right)$$

$$= \frac{1}{n!n}.$$

Therefore $0 < e - s_n < \frac{1}{n!n}$ for each positive integer n. Now suppose that $e = p/q$ is a rational number, where p and q are positive integers. Then $0 < q!(e - s_q) < 1/q$. The number $q!e$ must be an integer, since e is rational. Also

$$q!s_q = q!\left(1 + 1 + \frac{1}{2!} + \cdots + \frac{1}{q}\right)$$

is an integer. Hence $q!(e - s_q)$ is an integer between 0 and 1: contradiction. The number e is therefore irrational. □

6.3 Alternating Series

The reader should suspect that a complete analysis of series whose terms do not have constant sign is out of reach. In this section we focus our attention on a particular class of series of variable sign. We begin with a general result which reminds us of the popular formula of integration by parts.

Proposition 6.1 (Summation by Parts) *Two sequences $\{a_n\}_n$ and $\{b_n\}_n$ are given. Put $A_{-1} = 0$ and $A_n = \sum_{k=0}^{n} a_k$ for $n \geq 0$. For each positive integers $p \leq q$ we have*

$$\sum_{n=p}^{p} a_n b_n = \sum_{n=p}^{q-1} A_n(b_n - b_{n+1}) + A_q b_q - A_{p-1} b_p.$$

Proof Since $a_n = A_n - A_{n-1}$, we write

$$\sum_{n=p}^{p} a_n b_n = \sum_{n=p}^{q} (A_n - A_{n-1})b_n = \sum_{n=p}^{q} A_n b_n - \sum_{n=p-1}^{q-1} A_n b_{n+1}.$$

The last difference is equal to $\sum_{n=p}^{q-1} A_n(b_n - b_{n+1}) + A_q b_q - A_{p-1} b_p$, and the proof is complete. ☐

Theorem 6.11 (Dirichlet's Test) *Suppose*

(a) *the partial sums A_n of $\sum a_n$ form a bounded sequence;*
(b) $b_0 \geq b_1 \geq b_2 \geq \ldots;$
(c) $\lim_{n \to +\infty} b_n = 0.$

Then the series $\sum a_n b_n$ is convergent.

Proof There exists $M > 0$ such that $|A_n| \leq M$ for each n. Let $\varepsilon > 0$, and pick a positive integer ν such that $b_\nu \leq \varepsilon/(2M)$. For $\nu \leq p \leq q$ we have by Proposition 6.1

$$\left| \sum_{n=p}^{q} a_n b_n \right| \leq \left| \sum_{n=p}^{q-1} A_n(b_n - b_{n+1}) + A_q b_q - A_{p-1} b_p \right|$$

$$\leq M \left| \sum_{n=p}^{q-1} (b_n - b_{n+1}) + b_q + b_p \right|$$

$$= 2M b_p \leq 2M b_\nu \leq \varepsilon.$$

The series $\sum a_n b_n$ converges by the Cauchy theorem. ☐

Choosing $a_n = (-1)^{n+1}$ and $b_n = |c_n|$ in the previous theorem yields a popular test for alternating series.

Theorem 6.12 (Leibnitz Theorem for Alternating Series) *Suppose that*

(a) $|c_1| \geq |c_2| \geq |c_3| \geq \ldots$
(b) $c_{2m-1} \geq 0,\ c_{2m} \leq 0\ for\ m = 1, 2, 3, \ldots$
(c) $\lim_{n \to +\infty} c_n = 0$

Then the series $\sum c_n$ is convergent.

Exercise 6.4 Prove that the series $\sum_{n=1}^{\infty} \frac{(-1)^n}{n}$ converges, but it does not converge absolutely. This fact often seems to be surprising, but we must remember that the factor $(-1)^n$ contributes to a huge balancing of the terms in the series.

6.3.1 Product of Series

Numerical series can be multiplied together. The definition is reminiscent of the product of two polynomials $p(x)$ and $q(x)$, in which terms are grouped according to the power of the unknown x.

Definition 6.3 (Cauchy Product of Two Series) The Cauchy product of the series $\sum a_n$ and $\sum b_n$ is the series $\sum c_n$ defined by

$$c_n = \sum_{k=0}^{n} a_k b_{n-k}.$$

Remark 6.6 Properly speaking, the Cauchy product of two series is a discrete *convolution* product. Since we do not assume the reader to be familiar with integral convolutions, we will not use this language in the book.

The convergence of a product of two series is a delicate issue. Consider for example the series

$$\sum_{n=0}^{\infty} \frac{(-1)^n}{\sqrt{n+1}} = 1 - \frac{1}{\sqrt{2}} + \frac{1}{\sqrt{3}} - \frac{1}{\sqrt{4}} + \cdots$$

Convergence follows from Theorem 6.12. Let us now multiply this series by itself, obtaining

$$\sum_{n=0}^{\infty} c_n = 1 - \left(\frac{1}{\sqrt{2}} + \frac{1}{\sqrt{2}} \right) + \left(\frac{1}{\sqrt{3}} + \frac{1}{\sqrt{2}\sqrt{2}} + \frac{1}{\sqrt{3}} \right) + \cdots$$

$$= \sum_{n=0}^{\infty} (-1)^n \sum_{k=0}^{n} \frac{1}{\sqrt{(n-k+1)(k+1)}}.$$

But

$$(n-k+1)(k+1) = \left(\frac{n}{2} + 1 \right)^2 - \left(\frac{n}{2} - k \right)^2 \le \left(\frac{n}{2} + 1 \right)^2,$$

and

$$|c_n| \ge \sum_{k=0}^{n} \frac{2}{n+2} = \frac{2(n+1)}{n+2}.$$

Since the necessary condition $c_n \to 0$ is violated, the series $\sum c_n$ must diverge. Here comes the basic convergence result about the product of convergent series.

Theorem 6.13 (Mertens) *Suppose that*

(a) $\sum_{n=0}^{\infty} a_n$ *converges absolutely*
(b) $\sum_{n=0}^{\infty} a_n = A$
(c) $\sum_{n=0}^{\infty} b_n = B$
(d) $c_n = \sum_{k=0}^{n} a_k b_{n-k}.$

Then $\sum_{n=0}^{\infty} c_n$ converges.

Proof We follow [1], and set

$$A_n = \sum_{k=0}^{n} a_k$$

$$B_n = \sum_{k=0}^{n} b_k$$

$$C_n = \sum_{k=0}^{n} c_k$$

$$\beta_n = B_n - B.$$

We compute

$$C_n = a_0 b_0 + (a_0 b_1 + a_1 b_0) + \cdots + (a_0 b_n + a_1 b_{n-1} + \cdots + a_n b_0)$$
$$= a_0 B_n + a_1 B_{n-1} + \cdots + a_n B_0$$
$$= a_0 (B + \beta_n) + \cdots + a_n (B + \beta_0)$$
$$= A_n B + a_0 \beta_n + a_1 \beta_{n-1} + \cdots + a_n \beta_0.$$

To conclude the proof, we must show that $\lim_{n \to +\infty} \gamma_n = 0$, where $\gamma_n = a_0 \beta_n + a_1 \beta_{n-1} + \cdots + a_n \beta_0$. Let $\alpha = \sum_{n=0}^{\infty} |a_n|$. Notice that this is the first time we invoke assumption (a). Given any $\varepsilon > 0$, we can choose a positive integer v such that $|\beta_n| \le \varepsilon$ for each $n \ge v$. Thus

$$|\gamma_n| \le |\beta_0 a_n + \cdots + \beta_v a_{n-v}| + |\beta_{v+1} a_{n-v-1} + \cdots + \beta_n a_0|$$
$$\le |\beta_0 a_n + \cdots + \beta_n a_{n-v}| + \varepsilon \alpha.$$

Since $\limsup_{n \to +\infty} (\beta_0 a_n + \cdots + \beta_n a_{n-v}) = 0$, we find $\limsup_{n \to +\infty} |\gamma_n| \le \varepsilon \alpha$, and the conclusion follows. $\square$

6.4   Problems

6.1 Decide whether the series

$$\sum_{n=1}^{\infty} \sin(\alpha) \sin(2\alpha) \cdots \sin(n\alpha)$$

is convergent, for any fixed value of $\alpha \in \mathbb{R}$.

6.2 Let $\{a_n\}_n$ be a sequence with the property that there exists a real number $h < 1$ such that $|a_{n+1} - a_n| \leq h|a_n - a_{n-1}|$ for each n. Prove that the sequence converges.

6.3 Using the previous problem, show that the sequence defined by choosing any two real numbers a_1 and a_2, and defining

$$a_{n+1} = \frac{a_{n-1} + a_n}{2}$$

converges. Compute its limit.

6.4 Let $\{a_n\}_n$ be a sequence of positive real numbers. Prove that the series $\sum_{n=1}^{\infty} a_n$ converges if and only if the series $\sum_{n=1}^{\infty} \frac{a_n}{1+a_n}$ converges.

6.5 Starting from

$$\frac{1}{1 - x} = \sum_{n=1}^{\infty} x^n$$

and using Cauchy products, prove that

$$\frac{1}{(1 - x)^2} = \sum_{n=1}^{\infty} nx^{n-1}$$

for each real number x with $|x| < 1$.

6.5   Comments

Once upon a time, the treatment of numerical series used to fill up long chapters in Calculus textbooks. As I have tried to show, the theory of series is indeed a long collection of sufficient conditions for the convergence of particular sequences of numbers. In recent years this awareness has become prevalent, and we no longer annoy our students with awful convergence tests. Last but not least, many of these tests are based on the algebraic properties of real numbers, and they do not extend to series of complex numbers, for instance.

Reference

1. W. Rudin, *Principles of Mathematical Analysis*. International Series in Pure and Applied Mathematics, 3rd edn. (McGraw-Hill Book Co., New York, 1976)

Chapter 7
Limits: From Sequences to Functions of a Real Variable

Abstract From a really abstract point of view, the whole theory of limits for functions of a real variable is an immediate consequence of the theory of limits for sequences. Furthermore, the definition of limit for a real-valued function can be easily reduced to a continuity request. In this chapter we introduce limits via sequential limits, and prove a standard characterization in terms of neighborhoods.

Definition 7.1 Let E be a nonempty subset of $\mathbb{R}$, and let p be an accumulation point of E. We say that a number $L \in \mathbb{R}$ is the limit of $f(x)$ as $x \to p$, if for every sequence $\{x_n\}_n$ of points of E such that

$$(x_n \to p) \wedge (\forall n)(x_n \neq p),$$

one has that $f(x_n) \to L$ as $n \to +\infty$. In this case we write

$$f(x) \to L \text{ as } x \to p,$$

or

$$\lim_{x \to p} f(x) = L.$$

Remark 7.1 Since $p \in E'$, at least one sequence of points x_n from E that converge to p while being always different than p exists. We point out that L has nothing to do with the value $f(p)$, and indeed p need not even belong to the domain E of f.[1]

Figure 7.1 shows a typical example of a function with a finite limit. We now introduce two more possible definitions of limits, and we finally prove that these definitions are indeed equivalent.

---

[1] We follow here the traditional definition of limit. In the recent French tradition, the condition $x_n \neq p$ is omitted. We will come back to this discussion later.

© The Author(s), under exclusive license to Springer Nature Switzerland AG 2022
S. Secchi, *A Circle-Line Study of Mathematical Analysis*,
La Matematica per il 3+2 141, https://doi.org/10.1007/978-3-031-19738-3_7

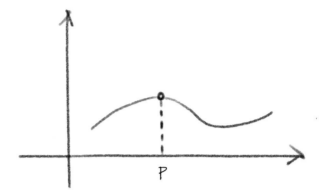

Fig. 7.1 A function with a limit

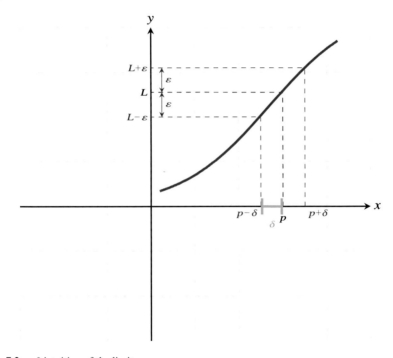

Fig. 7.2 ε-δ intuition of the limit

Definition 7.2 (ε-δ)   Let E be a nonempty subset of $\mathbb{R}$, and let p be an accumulation point of E. We say that a number $L \in \mathbb{R}$ is the limit of $f(x)$ as $x \to p$, if for every $\varepsilon > 0$ there exists $\delta > 0$ such that $x \in E$ and $0 < |x - p| < \delta$ imply $|f(x) - L| < \varepsilon$. See Fig. 7.2.

Definition 7.3 (Topological Limit)   Let E be a nonempty subset of $\mathbb{R}$, and let p be an accumulation point of E. We say that a number $L \in \mathbb{R}$ is the limit of $f(x)$ as

$x \rightarrow p$, if for every neighborhood V of L there exists a neighborhood U of p such that $f(U \setminus \{p\}) \subset V$.

Theorem 7.1 *Definitions 7.1, 7.2, and 7.3 are logically equivalent.*

Proof Since any neighborhood of a point x contains an interval of the form $(x - r, x + r)$ for some $r > 0$, it is clear that Definitions 7.2 and 7.3 are logically equivalent statements. We prove that Definitions 7.1 and 7.3 are equivalent. Suppose that $f(x) \rightarrow L$ as $x \rightarrow p$ in the sense of Definition 7.3, and let $\{x_n\}_n$ be a sequence of points from E such that $x_n \neq p$ for each n, and $x_n \rightarrow p$. If V is any neighborhood of L, we choose a neighborhood U of p as in Definition 7.3. There exists a positive integer n_0 such that $n > n_0$ implies $x_n \in U \setminus \{p\}$. Then $f(x_n) \in V$ for $n > n_0$, and this proves that $f(x_n) \rightarrow L$ as $n \rightarrow +\infty$.

Conversely, suppose that Definition 7.3 fails to hold. Then there exists some neighborhood V of L with the property that no neighborhood U of p satisfies the condition of Definition 7.3. Fixing such V, it follows that for each positive integer n there is some $x_n \in E$ such that $|x_n - p| < 1/n$, $x_n \neq p$, and $f(x_n) \notin V$. Thus $\{x_n\}_n$ is a sequence in $E \setminus \{p\}$ that converges to p, but the sequence $\{f(x_n)\}_n$ does not converge to L. Hence Definition 7.1 fails as well.   □

Remark 7.2 Definition 7.2 is for sure the most popular in Calculus courses. It has many computational advantages, but it cannot be extended to topological structures without a distance. We will come back to the theory of convergence in a later chapter.

Time has come to introduce another fundamental property of functions, which will produce important and useful consequences.

Definition 7.4 Let E be a nonempty subset of $\mathbb{R}$, and let $f : E \rightarrow \mathbb{R}$ be a function. For a point $p \in E$, we say that f is continuous at p if and only if for each neighborhood V of $f(p)$ there exists a neighborhood U of p such that $f(U) \subset V$. We say briefly that f is continuous on E if it is continuous at each point of E.

As a matter of facts, continuity and limits are essentially the same thing. Let us try to be more precise.

Theorem 7.2 *Let E be a nonempty subset of $\mathbb{R}$, let p be an accumulation point of E, and let $f : E \rightarrow \mathbb{R}$ be a function. The following statements are equivalent:*

(a) The limits $\lim_{x \rightarrow p} f(x)$ exists;
(b) there exists a number L such that the function defined on $E \cup \{p\}$ by

$$\tilde{f}(x) = \begin{cases} f(x) & \text{if } x \neq p \\ L & \text{if } x = p \end{cases}$$

is continuous at p.

Proof Suppose $L = \lim_{x\to p} f(x)$. Then the function $\tilde{f}$ is continuous at p, since $\lim_{x\to p} \tilde{f}(x) = \lim_{x\to p} f(x) = L = \tilde{f}(x)$. Conversely, suppose that L is chosen so that (b) holds. Then $\tilde{f}$ coincides with f on $E \setminus \{p\}$, and thus $\lim_{x\to p} f(x) = \lim_{x\to p} \tilde{f}(x)$. But $\tilde{f}$ is continuous at p, and we conclude that $\lim_{x\to p} f(x) = \tilde{f}(p) = L$. In particular $\lim_{x\to p} f(x)$ exists. $\qquad\square$

Remark 7.3 According to this result, we could consider the whole theory of limits as an application of the theory of continuous functions. We won't, both for pedagogical and theoretical reasons.

Sometimes[2] limits are just evaluation, as the next result shows.

Theorem 7.3 *Let E be a nonempty subset of $\mathbb{R}$, and let $f : E \to \mathbb{R}$ be a function. The function f is continuous at a point $p \in E$ if and only if either p is not an accumulation point of E, or p is an accumulation point of E and $\lim_{x\to p} f(x) = f(p)$.*

Proof The case in which p is not an accumulation point of E is clear. Indeed, in this case there is a neighborhood U of p such that no point of E other than p belongs to U. If V is any neighborhood of $f(p)$, then $f(U) = \{f(p)\} \subset V$, and f is continuous at p. We may therefore assume that p is an accumulation point of E. It suffices to compare Definitions 7.3 and 7.4. $\qquad\square$

We extend the definition of limit in the following way.

Definition 7.5

- Suppose that E contains a half-line $(a, +\infty)$ for some a. We say that the function $f : E \to \mathbb{R}$ converges to $L \in \mathbb{R}$ as $x \to +\infty$, if for any sequence $\{x_n\}_n$ of points in $(a, +\infty)$ such that $x_n \to +\infty$ as $n \to +\infty$, there results $f(x_n) \to L$ as $n \to +\infty$. In this case we write $f(x) \to L$ as $x \to +\infty$, or briefly $\lim_{x\to+\infty} f(x) = L$.
- Suppose that E contains a half-line $(-\infty, b)$ for some b. We say that the function $f : E \to \mathbb{R}$ converges to $L \in \mathbb{R}$ as $x \to -\infty$, if for any sequence $\{x_n\}_n$ of points in $(-\infty, b)$ such that $x_n \to -\infty$ as $n \to +\infty$, there results $f(x_n) \to L$ as $n \to +\infty$. In this case we write $f(x) \to L$ as $x \to -\infty$, or briefly $\lim_{x\to-\infty} f(x) = L$.
- Suppose that p is an accumulation point of E. We say that the function $f : E \to \mathbb{R}$ diverges to $+\infty$ as $x \to p$, if for every sequence $\{x_n\}_n$ of points in E such that $\forall n (x_n \neq p)$ and $x_n \to p$, there results $f(x_n) \to +\infty$ as $n \to +\infty$.
- Suppose that p is an accumulation point of E. We say that the function $f : E \to \mathbb{R}$ diverges to $-\infty$ as $x \to p$, if for every sequence $\{x_n\}_n$ of points in E such that $\forall n (x_n \neq p)$ and $x_n \to p$, there results $f(x_n) \to -\infty$ as $n \to +\infty$.

---

[2] Too many students tend to believe that this is indeed the general case. No, it isn't.

Similar definitions may be provided to describe the fact that $\lim_{x \to +\infty} f(x) = +\infty$ or $\lim_{x \to +\infty} f(x) = -\infty$ and so on. The details are left to the reader.

Dealing with several different definitions of limits may seem troublesome. An interesting way out consists in extending the set $\mathbb{R}$ so that it contains $+\infty$ and ∞.

Definition 7.6 We call the set $\mathbb{R}^* = \mathbb{R} \cup \{-\infty, +\infty\}$ the extended real line. A neighborhood of $-\infty$ is any set of the form $(-\infty, b)$ for some $b \in \mathbb{R}$. Analogously, a neighborhood of $+\infty$ is any set of the form $(a, +\infty)$ for some $a \in \mathbb{R}$. If $E \subset \mathbb{R}^*$ and $p \in \mathbb{R}^*$, we say that p is an accumulation point of E if any neighborhood of p contains a point of E, different than p itself.

The extended real line allows us to summarize a all the possible definitions of limit into a single topological definition.

Definition 7.7 (Limits in $\mathbb{R}^*$) Let $E \subset \mathbb{R}^*$, let $p \in \mathbb{R}^*$ an accumulation point of E, and let $L \in \mathbb{R}^*$. We say that $f(x)$ tends to L as $x \to p$, if for every neighborhood V of L in $\mathbb{R}^*$ there exists a neighborhood U of p in $\mathbb{R}^*$ such that $f(U \setminus \{p\}) \subset V$.

Arithmetic operations in the extended real line present some difficulties. Indeed, it is clear that $\lim_{x \to +\infty} \frac{x}{x} = 1$, while $\lim_{x \to +\infty} \frac{x^2}{x} = +\infty$ and $\lim_{x \to +\infty} \frac{x}{x^2} = 0$. There is no hope to define sums and products in $\mathbb{R}^*$ without any exceptional case. The algebra of limits is completely satisfactory only for finite limits.

Theorem 7.4 (Algebra of Finite Limits) *Retain the assumptions of Definition 7.7. If*

$$\lim_{x \to p} f(x) = L \in \mathbb{R}$$

$$\lim_{x \to p} g(x) = M \in \mathbb{R},$$

then

$$\lim_{x \to p} (f(x) + g(x)) = L + M$$

$$\lim_{x \to p} f(x)g(x) = LM,$$

and

$$\lim_{x \to p} \frac{f(x)}{g(x)} = \frac{L}{M}$$

provided that $M \neq 0$.

Proof The conclusion follows from the corresponding statements for sequences, see Theorem 5.6. $\square$

The circumstance that the algebra of limits cannot be completely extended to $\mathbb{R}^*$ should not be seen as a weakness. Limits are a topological object, and such is the set $\mathbb{R}^*$. On the contrary the algebra of limits describes an interplay of topology with the algebraic structure of real numbers. As a matter of facts, we cannot define algebraic operations with $\pm\infty$ without losing some properties. To rephrase this, there is no value of ∞/∞ or of $0 \cdot \infty$ that is compatible with our definition of limits.

7.1   Properties of Limits

Clearly enough, many properties of limits can be deduced from similar properties of limits for sequences. We just state a few important results that the reader may have studied in Calculus courses.

Theorem 7.5 (Limits and Order) *Let $E \subset \mathbb{R}$, $p \in \mathbb{R}^*$ be an accumulation point of E, and $L \in \mathbb{R}$.*

(a) *If $L > 0$, there exists a neighborhood U of p such that $f(x) > 0$ for each $x \in U \setminus \{p\}$.*

(b) *If $f(x) > 0$ for each $x \neq p$ that belongs to some neighborhood of p, then $L \geq 0$.*

Proof We first deal with the case $L \in \mathbb{R}$, and we use Definition 7.2. To prove (a), we select $\varepsilon = L/2$ so that a neighborhood U of p exists such that $L - L/2 < f(x) < L + L/2$ for each $x \in U \setminus \{p\}$. This shows that $f(x) > L/2 > 0$ for such values of x. If (b) were false, then $L < 0$. Applying (a) to $-f$ would yield a neighborhood of p in which $-f$ would be positive, i.e. f would be negative. This is a contradiction.

The case $L = +\infty$ is easier. Indeed, For each $M > 0$ there exists a neighborhood U of p such that $f(x) > M$ for each $x \in U \setminus \{p\}$. Conclusion (b) follows again by (a) as above. $\square$

Theorem 7.6 (Squeezing Property) *Let $p \in \mathbb{R}^*$ be an accumulation point of a set E. Suppose three functions f, g and σ are defined in some neighborhood of p, and that $|f(x) - g(x)| < \sigma(x)$ in that neighborhood. If $\sigma(x) \to 0$ and $g(x) \to L \in \mathbb{R}$ as $x \to p$, then also $f(x) \to L$ as $x \to p$.*

Proof Fix any $\varepsilon > 0$. By assumption there exists a neighborhood U of p such that $\sigma(x) < \varepsilon$ and $|g(x) - L| < \varepsilon$ provided that $x \in U \setminus \{p\}$. For these x's, $|f(x) - L| \leq |f(x) - g(x)| + |g(x) - L| < \sigma(x) + \varepsilon < 2\varepsilon$. $\square$

Remark 7.4 The condition $|f(x) - g(x)| < \sigma(x)$ is equivalent to $g(x) - \sigma(x) < f(x) < g(x) + \sigma(x)$, see Fig. 7.3. Under the assumptions of Theorem 7.6, $g(x) - \sigma(x)$ and $g(x) + \sigma(x)$ both converge to the finite limit L as $x \to p$, and this implies that $f(x) \to L$ as $x \to p$.

Fig. 7.3 Squeezing of functions

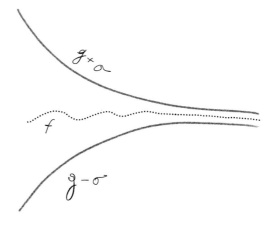

7.2   Local Equivalence of Functions

Since limits and continuity are related to the local behavior of functions, it seems rather natural to classify functions according to their asymptotic properties around a point

Definition 7.8 Let I be an interval, let $c \in \bar{I}$ or $c \in \{-\infty, +\infty\}$ (in case I is unbounded), and let f, g be two functions defined on I. We say that f and g are equivalent at c if and only if there exists a function $u : I \to \mathbb{R}$ such that $f(x) = u(x)g(x)$ for each $x \in I$ and $\lim_{x \to c} u(x) = 1$. In this case we write $f \sim g$ as $x \to c$.

Remark 7.5 If necessary, we will also write $f \sim g$ as $x \to c^-$ of $x \to c^+$, with an obvious meaning of symbols. We point out that the excluding the point c from the test of $f = ug$ is not particularly meaningful. Indeed, since $u(x) \to 1$ as $x \to c$, u can always be (re)defined at c so that $u(c) = 1$. This forces $f(c) = g(c)$, and therefore we may assume that f and g are defined at c without loss of generality.

Important: Simplified Definition

Several textbooks propose the following definition of local equivalence: $f \sim g$ as $x \to c$ if and only if

$$\lim_{x \to c} \frac{f(x)}{g(x)} = 1.$$

Unfortunately this becomes troublesome when $g = 0$ infinitely often around c, while there is no need to exclude this possibility.

Theorem 7.7 *Let $\mathcal{F}(c)$ be the set of all functions defined on I. Then $\sim$ is an equivalence relation on $\mathcal{F}(c)$.*

Proof For each $f \in \mathcal{F}(c)$, $f \sim f$ in a trivial way. Next, if $f \sim g$, there exists u such that $f = ug$ on I, with $u(x) \to 1$ as $x \to c$. In particular $u \neq 0$ near c, and thus $g = (1/u)f$ near c. Hence $g \sim f$. To conclude, if $f \sim g$ and $g \sim h$, we can find functions u and v such that $f = ug$ and $g = vh$ with $u(x) \to 1$, $v(x) \to 1$ as $x \to c$. Then $f = uvh$ and $u(x)v(x) \to 1$ as $x \to c$. Hence $f \sim h$.   □

Exercise 7.1 Suppose that $\lim_{x \to c} f(x) = \lim_{x \to c} g(x) = y_0 \neq 0$. Prove that $f \sim g$ as $x \to c$. Show with a counterexample that the condition $y_0 \neq 0$ is essential.

Local equivalence is just the first ingredient of local comparison.

Definition 7.9 Let I be an interval, let $c \in \bar{I}$ or $c \in \{-\infty, +\infty\}$ (in case I is unbounded), and let f, g be two functions defined on I.

(o) We say that $f = o(g)$ as $x \to c$ if and only if there exists a function σ such that $f(x) = \sigma(x)g(x)$ on I and $\lim_{x \to c} \sigma(x) = 0$. In this case we also say that f is negligible compared to g as $x \to c$.

(O) We say that $f = O(g)$ as $x \to c$ if and only if there exists a bounded function β such that $f(x) = \beta(x)g(x)$ on I.

Exercise 7.2 Suppose that g does not vanish in a neighborhood of c. Prove that $f = o(g)$ if and only if

$$\lim_{x \to c} \frac{f(x)}{g(x)} = 0.$$

Prove also that $f = O(g)$ if and only if $x \mapsto f(x)/g(x)$ remains bounded in a neighborhood of c.

We summarize the basic algebraic properties of local comparison in the following exercise. The proofs are simple consequences of the definition.

Definition 7.10 We say that a function g is bounded away from zero around a point x if and only if there exists a number $\varepsilon > 0$ such that $|g(x)| \geq \varepsilon$ for all x in a neighborhood of c.

Exercise 7.3 Prove that the following statements are true as $x \to c$.

1. If $f = o(h)$ and $g = o(h)$, then $f + g = o(h)$ and $fg = o(h)$.
2. If $f = o(h)$ and $\alpha \in \mathbb{R}$, then $\alpha f = o(h)$.
3. If $f = o(h)$ and g is bounded away from zero, then $f/g = o(h)$.
4. If $f = o(g)$ and $g = O(h)$, then $f = o(h)$.
5. If $f = o(1)$, then $1/(1 + f) = 1 - f + o(f)$.
6. If $f = O(h)$ and $g = O(h)$, then $f + g = O(h)$ and $fg = O(h)$.
7. If $f = O(h)$ and $\alpha \in \mathbb{R}$, then $\alpha f = O(h)$.
8. If $f = O(h)$ and g is bounded away from zero, then $f/g = O(h)$.
9. If $f = O(g)$ and $g = O(h)$, then $f = O(h)$.

10. If $f = o(1)$ and $g = O(h)$, then $fg = o(h)$.
11. If $f = O(1)$ and $g = O(h)$, then $fg = O(h)$.

Needless to say, the language of local comparison is a perfect dialect for speaking about limits.

Example 7.1

1. Since[3]

$$\lim_{x \to 0} \frac{e^x - 1}{x} = 1,$$

we can write $e^x = 1 + x + o(x)$ as $x \to 0$.
2. For a similar reason, $\sin x = x + o(x)$ as $x \to 0$, and $\cos x = 1 - (1/2)x^2 + o(x^2)$ as $x \to 0$.
3. If $f_1 = o(f)$ and $g_1 = o(g)$ as $x \to c$, then

$$\lim_{x \to c} \frac{f(x) + f_1(x)}{g(x) + g_1(x)} = \lim_{x \to c} \frac{f(x)}{g(x)}.$$

Indeed,

$$\lim_{x \to c} \frac{f(x) + f_1(x)}{g(x) + g_1(x)} = \lim_{x \to c} \frac{f(x)}{g(x)} \frac{1 + \frac{f_1(x)}{f(x)}}{1 + \frac{g_1(x)}{g(x)}}.$$

This is often called the *principle of negligible terms*. Its use in computing limits is ubiquitous.

7.3 Comments

We will discuss again the definition of limit in the chapter about topology. For the moment, I point out that our definition remains the most common in contemporary literature, although some alternatives actually exist. In particular, a few authors propose the following variant:

$\lim_{x \to p} f(x) = q$ if and only if for every $\varepsilon > 0$ there exists $\delta > 0$ such that $|x - p| < \delta$ implies $|f(x) - q| < \varepsilon$.

---

[3] We assume that the reader is familiar with a few limits that involve the elementary functions. Formal proofs will be given later on, when we discuss these functions from an advanced viewpoint.

The difference is that the case $x = p$ is not excluded. It is therefore clear that the condition $f(p) = q$ is a necessary condition for the existence of the limit with this definition. If you like this approach, you should always remember that limits are no longer independent of the value of the functions at the point p.

Chapter 8
Continuous Functions of a Real Variable

Abstract Calculus students tend to believe that continuous functions are those functions which "vary a little when the independent variable varies a little." In this chapter we define continuity in a rigorous way, and we invite the readers to convince themselves that the previous sentence is actually false.

First of all, we reformulate Definition 7.4 in a quantitative way.

Definition 8.1 Let E be a subset of $\mathbb{R}$, and let $f: E \to \mathbb{R}$ be a function. We say that f is continuous at the point $p \in E$ if for every $\varepsilon > 0$ there exists $\delta > 0$ such that $x \in E$ and $|x - p| < \delta$ imply $|f(x) - f(p)| < \varepsilon$. If f is continuous at any point of E, we say that f is continuous on E.

Two cases are possible. If the point p is also an accumulation point of E (hence $p \in E \cap E'$), then the previous definition merely says that $\lim_{x \to p} f(x) = f(p)$. The second case is $p \in E$ but not an accumulation point of E. This means that p is isolated, in the sense that there exists a neighborhood U of p such that $U \cap E = \{p\}$. In this situation, the continuity of f at p is always granted. Indeed, if $\varepsilon > 0$, we may choose $\delta > 0$ so small that the condition $|x - p| < \delta$ is satisfied only by $x = p$, and therefore $|f(x) - f(p)| = |f(p) - f(p)| = 0 < \varepsilon$.

Remark 8.1 Most Calculus books propose $\lim_{x \to p} f(x) = f(p)$ as the definition of continuity, but this is equivalent to ours only under the assumption that p is an accumulation point of the domain of f.

Recalling Theorem 7.1 and Definition 7.4, we may state

Theorem 8.1 *Let $f: E \to \mathbb{R}$. The function f is continuous at the point $p \in E$ if and only if one of the following conditions is met:*

(i) *for every $\varepsilon > 0$ there exists $\delta > 0$ such that $x \in E$ and $|x - p| < \delta$ imply $|f(x) - f(p)| < \varepsilon$;*
(ii) *For every neighborhood V of $f(p)$ there exists a neighborhood U of p such that $f(U \cap E) \subset V$;*

© The Author(s), under exclusive license to Springer Nature Switzerland AG 2022
S. Secchi, *A Circle-Line Study of Mathematical Analysis*,
La Matematica per il 3+2 141, https://doi.org/10.1007/978-3-031-19738-3_8

(iii) for every sequence $\{x_n\}$ of point from E that converges to p, there results $f(x_n) \to f(p)$ *as* $n \to +\infty$.

Corollary 8.1 *If there exists a sequence $\{x_n\}_n \subset E$ that converges to p, but $\{f(x_n)\}_n$ does not converge to $f(p)$, then f is discontinuous at p.*

The word discontinuity is used as the negation of continuity. In this book we will not enter into the troublesome classification of discontinuity points, since we believe that this is of little interest.

Exercise 8.1 Let $f: E \to \mathbb{R}$ be a function defined on $E \subset \mathbb{R}$. Prove that f is continuous at $x_0 \in E$ if and only if for every monotonic sequence $\{x_n\}_n$ converging to x_0, we have $f(x_n) \to f(x_0)$. *Hint:* use Theorem 5.12.

Example 8.1 Let $E = \mathbb{R}$, $p = 0$ and

$$f(x) = \begin{cases} \frac{x}{|x|} & \text{if } x \neq 0 \\ 0 & \text{if } x = 0. \end{cases}$$

Consider the two sequences defined by $x_n = 1/n$ and $y_n = -1/n$ for $n \in \mathbb{N}$. Then we form the sequence $\{z_n\}_n$ by the rule

$$x_1, y_1, x_2, y_2, \ldots, x_n, y_n, \ldots$$

Clearly $z_n \to 0$ as $n \to +\infty$, but the sequence $\{f(z_n)\}_n$ is

$$1, -1, 1, -1, \ldots$$

This sequence cannot converge to $f(0) = 0$, since $\liminf_{n \to +\infty} f(z_n) = -1 \neq 1 = \limsup_{n \to +\infty} f(z_n)$.

Example 8.2 Let us define f on $[0, 1) \cup (1, 2]$ by the following rule:

$$f(x) = \begin{cases} 0 & \text{if } 0 \leq x < 1 \\ 1 & \text{if } 1 < x \leq 2. \end{cases}$$

For each $x \in [0, 1) \cup (1, 2]$, the function f is continuous at x. However, if x is very close to 1 but smaller than 1, the value of $f(x)$ is zero. If x is very close to 1 but larger than 1, the value of $f(x)$ is 1. We are allowed to say that f changes *a lot* when x is varied *a little*! In other words, there is much more in the definition of continuity than the basic idea of "a little on the x-axis becomes a little on the y axis."

The algebraic rules for computing (finite) limits immediately implies the following result.

Theorem 8.2 *Suppose that* $f : E \to \mathbb{R}$ *and* $g : E \to \mathbb{R}$ *are continuous at* $p \in E$. *Then*

(i) $x \mapsto kf(x)$ *is continuous at* p *for every* $k \in \mathbb{R}$;
(ii) $x \mapsto f(x) + g(x)$ *is continuous at* p;
(iii) $x \mapsto f(x)g(x)$ *is continuous at* p;
(iv) $x \mapsto f(x)/g(x)$ *is continuous at* p, *provided that the quotient is defined.*

As a consequence, any polynomial function $x \mapsto a_0 + a_1 x + a_2 x^2 + \cdots + a_n x^n$ is continuous on $\mathbb{R}$, as the sum of continuous functions.

Example 8.3 We consider the function $h(x) = [x]$, where $[x]$ denotes the largest integer $n \in \mathbb{Z}$ such that $n \leq x$. Given a point $p \in \mathbb{Z}$, we consider the sequence $x_n = p - 1/n$. Clearly $x_n \to p$, but $h(x_n) \to p - 1$, which is different than $h(p) = p$. Hence h is discontinuous at all integer points. On the contrary, h is continuous at all non-integers. Indeed, let us fix $p \notin \mathbb{Z}$. Given $\varepsilon > 0$, we want to find a neighborhood U of p such that $h(U) \subset (h(p) - \varepsilon, h(p) + \varepsilon)$. Since p is not an integer, there exists an integer n such that $n < p < n + 1$. Let us take $\delta = \min\{p - n, (n + 1) - p\}$: it follows from the definition of h that $h(x) = h(p)$ for every $x \in (p - \delta, p + \delta)$. Thus we certainly have $h(x) \in (h(p) - \varepsilon, h(p) + \varepsilon)$ for every $x \in (p - \delta, p + \delta)$.

Since elementary functions (polynomials, sine, cosine, exponentials and logarithms) are continuous functions in their natural domains of definition, no arithmetic operation on them can produce discontinuous functions. We now observe that even their composition must be continuous.

Theorem 8.3 (Continuity of Composite Functions) *Given* $f : E \to \mathbb{R}$ *and* $g : F \to \mathbb{R}$, *assume that the range* $f(E)$ *is contained in* F, *so that the composition* $g \circ f$ *is defined on* E. *If* f *is continuous at a point* $p \in E$ *and if* g *is continuous at the point* $f(p) \in F$, *then* $g \circ f$ *is continuous at* p.

Proof We use Theorem 8.1. Let $\{x_n\}_n$ be a sequence in E that converges to p. By assumption $f(x_n) \to f(p)$, and so $g(f(x_n)) \to g(f(p))$ since g is continuous at $f(p)$ and $f(x_n) \to f(p)$. This shows that $g \circ f(x_n) \to g \circ f(p)$, namely that $g \circ f$ is continuous at p. $\qquad\square$

The previous theorem is indeed a particular case of the following formula about change of variables in limits.

Theorem 8.4 (Changing Variables in Limits) *Let* D *and* E *be subsets of* $\mathbb{R}^* = \mathbb{R} \cup \{-\infty\} \cup \{+\infty\}$, *let* c *and* p *be accumulation points of* D *and* E *respectively.* *Let* $f : D \to \mathbb{R}^*$ *be a function, and let* $\varphi : E \to D$ *be a bijective function such that*

$$\lim_{t \to p} \varphi(t) = c, \quad \lim_{x \to c} \varphi^{-1}(x) = p.$$

Under these assumptions, $\lim_{x \to c} f(x)$ exists in $\mathbb{R}^*$ if and only if $\lim_{t \to p} f \circ \varphi(t)$ exists in $\mathbb{R}^*$, and in such a case these limits coincide.

Proof We refer to Fig. 8.1. Let us assume that $L = \lim_{x \to c} f(x)$ exists in $\mathbb{R}^*$. Let V be a neighborhood of L in $\mathbb{R}^*$, and let U be a neighborhood of c in $\mathbb{R}^*$ such that $f(U \cap D \setminus \{c\}) \subset V$. Furthermore, given the neighborhood U of c there exists a neighborhood W of p in $\mathbb{R}^*$ such that $\varphi(W \cap E \setminus \{p\}) \subset U$. The crucial remark is now that φ is bijective: there exists one and only one element $a \in E$ such that $\varphi(a) = c$. In case $a \neq p$, we may take a smaller neighborhood W so that $a \notin W$. In this way we are sure that $\varphi(W \cap E \setminus \{p\}) \subset U \setminus \{c\}$. As a consequence $f(\varphi(t)) \in V$ whenever $t \in W \cap E \setminus \{p\}$. This shows that $L = \lim_{t \to p} f(\varphi(t))$.

To prove the converse, we simply apply the result to the function $f \circ \varphi$ and φ^{-1}, and we conclude because $(f \circ \varphi) \circ \varphi^{-1} = f$. $\qquad\qquad\qquad\qquad\qquad\qquad\square$

The rough interpretation of Theorem 8.4 is that $x = \varphi(t)$; hence $x \to c$ if and only if $t \to p$, and there is no difference between the variable x and the variable t when limits are considered.

Example 8.4 The following limits are equivalent:

$$\lim_{x \to 0} \frac{e^x - 1}{x} = 1$$

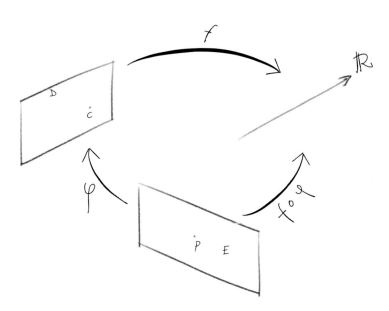

Fig. 8.1 Change of variable in the limit

and

$$\lim_{x \to 0} \frac{\log(x+1)}{x} = 1.$$

Indeed, setting $e^t - 1 = x$, we get $t = \log(x+1)$ and $x \to 0$ if and only if $t \to 0$.

The following example introduces a quantitative approach to continuity of real-valued functions.

Example 8.5 Let $f : \mathbb{R} \to \mathbb{R}$ be a function. For each $x \in \mathbb{R}$, we define

$$\omega_f(x) = \inf_{\delta > 0} \sup \{|f(y) - f(z)| \mid \{y, z\} \subset (x_0 - \delta, x_0 + \delta)\}.$$

We claim that the set $\{x \in \mathbb{R} \mid \omega_f(x) < \varepsilon\}$ is open for each $\varepsilon > 0$. Indeed, we suppose that $\omega_f(x_0) < \varepsilon$, and we must prove that the same inequality holds in an open interval containing x_0. By definition, there exists δ_0 such that

$$\sup \{|f(y) - f(z)| \mid \{y, z\} \subset (x_0 - \delta, x_0 + \delta)\} < \varepsilon.$$

For every $x \in (x_0 - \delta/2, x_0 + \delta/2)$ we have $(x_0 - \delta/2, x + \delta/2) \subset (x_0 - \delta, x_0 + \delta)$, and therefore

$$\omega_f(x) \leq \sup \{|f(y) - f(z)| \mid y, z \in x - \delta/2, x + \delta/2)\}$$
$$\leq \sup \{|f(y) - f(z)| \mid y, z \in (x_0 - \delta, x_0 + \delta)\}$$
$$< \varepsilon.$$

Exercise 8.2 Prove that a function f is continuous at a point x_0 if and only if $\omega_f(x_0) = 0$.

8.1 Continuity and Compactness

The most interesting properties of continuous functions are related to the compactness of their domains.

Theorem 8.5 (Preservation of Compactness) *Let $f : E \to \mathbb{R}$ be a continuous function, and suppose that K is a (sequentially) compact subset of A. Then $f(K)$ is a (sequentially) compact subset of $\mathbb{R}$.*

Proof Let $\{y_k\}_k$ be a sequence in the range $f(K)$. Hence for each positive integer k, there exists an element $x_k \in K$ such that $f(x_k) = y_k$. The sequence $\{x_k\}_k$ possesses a converging subsequence $\{x_{n_k}\}_k$. Call x its limit. By continuity of f at x, there results $f(x) = \lim_{k \to +\infty} f(x_{n_k})$. Hence $y_{n_k} \to f(x)$ as $k \to +\infty$. □

Preservation of compactness under continuous functions ensures the existence of maxima and minima. The following is a fundamental result of Real Analysis.

Theorem 8.6 (Weierstrass) *If $f : K \to \mathbb{R}$ is a continuous function on a (sequentially) compact subset K of $\mathbb{R}$, then f attains a maximum and a minimum value. In particular, there exist points x_0 and x_1 in K such that $f(x_0) \le f(x) \le f(x_1)$ for every $x \in K$.*

Proof Since the range $f(K)$ is (sequentially) compact, we can set $\alpha = \sup f(K)$ and deduce that $\alpha \in f(K)$. Indeed $f(K)$ is a closed set, and α is an accumulation point of $f(K)$ (by Theorem 5.2). This shows that f attains a maximum value. In a similar fashion we can show that f attains a minimum value.   □

Continuity is essentially a *local* property, since a function is continuous on a set if and only if it is continuous at each point of the set. We introduce now a genuinely *global* definition which extends continuity in a proper way.

Definition 8.2 Let E be a subset of $\mathbb{R}$, $f : E \to \mathbb{R}$ a function. We say that f is uniformly continuous on E if for every $\varepsilon > 0$ there exists $\delta > 0$ such that $x \in E$, $y \in E$ and $|x - y| < \delta$ imply $|f(x) - f(y)| < \varepsilon$.

It is worth noticing that uniform continuity implies at once continuity on E: just fix $y \in E$ and use the definition. On the other hand, the following example shows that the two definitions remain distinct.

Example 8.6 The function $g : x \mapsto x^2$ is not uniformly continuous on $\mathbb{R}$. Indeed, we can consider the two sequences $x_n = n$ and $y_n = n + 1/n$. Clearly $\lim_{n \to +\infty} x_n = \lim_{n \to +\infty} y_n = +\infty$, while $\lim_{n \to +\infty} |x_n - y_n| = 0$. Now,

$$|f(x_n) - f(y_n)| = \left| n^2 - \left(n + \frac{1}{n} \right)^2 \right|$$

$$= \left| 2 + \frac{1}{n^2} \right|.$$

Hence $\lim_{n \to +\infty} |f(x_n) - f(y_n)| = 2$, against the definition of uniform continuity.

The previous argument can be turned into a general test for uniform continuity.

Theorem 8.7 *A function $f : E \to \mathbb{R}$ fails to be uniformly continuous on E if and only if there exist a positive number $\varepsilon_0 > 0$ and two sequences $\{x_n\}_n$, $\{y_n\}_n$ in E that satisfy $|x_n - y_n| \to 0$, $|f(x_n) - f(y_n)| \ge \varepsilon_0$.*

Proof If ε_0 and the two sequences exist, it is immediate to check that f cannot be uniformly continuous. Actually, no matter how close x_n and y_n are, their images $f(x_n)$ and $f(y_n)$ remain far away from each other.

Conversely, we suppose that f fails to be uniformly continuous on E. Negating the definition, we discover that there exists ε_0 such that for every $\delta > 0$ there must exist points x_δ and y_δ in E satisfying $|x_\delta - y_\delta| < \delta$, but $|f(x_\delta) - f(y_\delta)| \ge \varepsilon_0$.

Now $\delta > 0$ is a free variable, and we can select $\delta = \delta_n = 1, 2, 3, \ldots$. Hence two sequences $\{x_{\delta_n}\}_n$ and $\{y_{\delta_n}\}_n$ are born such that $|x_{\delta_n} - y_{\delta_n}| < \delta_n$, but $|f(x_{\delta_n}) - f(y_{\delta_n})| \geq \varepsilon_0$. □

Again compactness does the magic: uniform continuity boils down to continuity on (sequentially) compact sets.

Theorem 8.8 *A function that is continuous on a (sequentially) compact subset K of $\mathbb{R}$ is uniformly continuous.*

Proof Let f be a continuous function on K. We argue by contradiction, assuming that f fails to be uniformly continuous on K. Then there exist $\varepsilon_0 > 0$ and two sequences $\{x_n\}_n$, $\{y_n\}_n$ in E that satisfy $|x_n - y_n| \to 0$, $|f(x_n) - f(y_n)| \geq \varepsilon_0$. Compactness comes into play: a subsequence $\{x_{n_k}\}_k$ of $\{x_n\}_n$ converges to a point $x \in K$. Then also y_{n_k} converges to the same limit x, since $|x_{n_k} - y_{n_k}| \to 0$ (write $|y_{n_k} - x| = |y_{n_k} - x_{n_k} + x_{n_k} - x|$ and use the triangle inequality). Since f is continuous at x, we know that $f(x) = \lim_{k \to +\infty} f(x_{n_k}) = \lim_{k \to +\infty} f(y_{n_k})$. This contradicts the condition $|f(x_{n_k}) - f(y_{n_k})| \geq \varepsilon_0$. □

Exercise 8.3 Prove that uniformly continuous functions map Cauchy sequences into Cauchy sequences. More precisely, if f is uniformly continuous on a set S and if $\{x_n\}_n$ is a Cauchy sequence in S, then $\{f(x_n)\}_n$ is a Cauchy sequence. *Hint:* just apply the definition of uniform continuity.

Theorem 8.9 *A real-valued function f on (a, b) is uniformly continuous on (a, b) if and only if it can be extended to a continuous function $\tilde{f}$ on $[a, b]$.*

Proof If f can be extended, then $\tilde{f}$ is automatically uniformly continuous on the compact set $[a, b]$, and thus f is uniformly continuous on (a, b) as a restriction. Conversely, we need to define $\tilde{f}(a)$ and $\tilde{f}(b)$ in such a way that the extended function is continuous. We construct the value $\tilde{f}(a)$, since the other case is similar. Let $\{x_n\}_n$ any sequence in (a, b) such that $x_n \to a$. The sequence $\{x_n\}_n$ is a Cauchy sequence, and therefore $\{f(x_n)\}_n$ is a Cauchy sequence by the previous exercise. Thus $\lim_{n \to +\infty} f(x_n)$ exists: we set

$$\tilde{f}(a) = \lim_{n \to +\infty} f(x_n).$$

Now, the value of $\tilde{f}(a)$ depends on the *choice* of the sequence $\{x_n\}_n$. But this is not the case. Indeed, let $\{y_n\}_n$ be another sequence in (a, b) such that $y_n \to a$. Then the sequence $\{u_n\}_n$ defined by

$$\{x_1, y_1, x_2, y_2, \ldots, x_n, y_n, \ldots\}$$

is a sequence in (a, b) which converges to a. We deduce that $\lim_{n \to +\infty} f(u_n)$ exists, and since $\{f(x_n)\}_n$ and $\{f(y_n)\}_n$ are subsequences of $\{f(u_n)\}_n$, the limits $\lim_{n \to +\infty} f(x_n)$ and $\lim_{n \to +\infty} f(y_n)$ must be equal. We have thus proved that the definition of $\tilde{f}(a)$ is unambiguous, and plainly $\tilde{f}$ is continuous at a. □

8.2   Intermediate Value Property

The next results are of topological nature, and in a perfect mathematical world they should be obtained from the properties of *connected sets*. We present them in this chapter, and we invite the interested reader to think back of them after studying the chapter on General Topology.

Theorem 8.10 *Let $f : [a, b] \to \mathbb{R}$ be a continuous function. If L is a real number with either $f(a) < L < f(b)$ or $f(a) > L > f(b)$, then there exists a point $c \in (a, b)$ such that $f(c) = L$.*

Proof By considering $\tilde{f} : x \mapsto f(x) - L$ instead of f, we can always assume that $L = 0$. We prove the Theorem under the assumption $f(a) < 0 < f(b)$, the other case being similar. Therefore we look for a point $c \in (a, b)$ such that $f(c) = 0$. Let us introduce the set

$$K = \{x \in [a, b] \mid f(x) \le 0\}.$$

The number b is an upper bound for K, and $K \ne \emptyset$ because $a \in K$. Then $c = \sup K$ exists in $\mathbb{R}$. Let us prove that $f(c) = 0$. If not, either $f(c) > 0$ of $f(c) < 0$. Suppose $f(c) > 0$: by continuity, $f > 0$ in a neighborhood $(c - \delta, c + \delta)$ of c. In particular $K \subset [a, c - \delta]$, against the definition of c as the supremum of K. By the same token, one excludes the case $f(c) < 0$, and the conclusion follows.   □

Exercise 8.4   In this exercise we propose a second proof of Theorem 8.10. Suppose $L = 0$ and $f(a) < 0 < f(b)$. Define $I_0 = [a, b]$ and consider the mid-point $z = (a + b)/2$. If $f(z) \ge 0$, set $a_1 = a$ and $b_1 = z$. If $f(z) < 0$, set $a_1 = z$ and $b_1 = b$. This gives rise to an interval $I_1 \subset I_0$. Use this scheme and the principle of infinitesimal nested intervals to prove Theorem 8.10.

The previous result suggests the following question: if a function attains every value between its infimum and its supremum, is it a continuous function?

Definition 8.3   A function f has the intermediate value property on an interval $[a, b]$ if for all $x < y$ in $[a, b]$, and all L between $f(x)$ and $f(y)$, it is possible to find a point $c \in (x, y)$ such that $f(c) = L$.

We have seen in Theorem 8.10 that any continuous function has the intermediate value property. it is a common mistake to assume that the intermediate value property is a characterization of continuous functions.

Example 8.7   Define $g : (-\infty, +\infty) \to \mathbb{R}$ by

$$g(x) = \begin{cases} \sin(1/x) & \text{if } x \ne 0 \\ 0 & \text{otherwise.} \end{cases}$$

For each $L \in [-1, 1]$, the equation $\sin z = L$ possesses infinitely many solutions $z \neq 0$, and therefore g has the intermediate value property. However g is not continuous at 0 (why?).

Theorem 8.11 *If f is increasing on $[a, b]$ and satisfies the intermediate value property, then f is continuous on $[a, b]$. The same results holds for a decreasing function.*

Proof We consider a point $c \in (a, b)$: the continuity of f at a and b is left as an exercise. By monotonicity,

$$f(c-) = \lim_{x \to c^-} f(x) = \sup \{f(x) \mid a \leq x < c\}$$

$$f(c+) = \lim_{x \to c^+} f(x) = \inf \{f(x) \mid c < x \leq b\},$$

and both limits are finite. Again by monotonicity, $f(c-) \leq f(c+)$. We need to prove that actually $f(c-) = f(c+)$. We suppose on the contrary that $f(c-) < f(c+)$, and pick a number L such that $f(c-) < L < f(c+)$. We have now two cases: (i) if $f(c) \neq L$, we reach a contradiction with the intermediate value property. If (ii) $f(c) = L$, we select $L' \neq L$ and fall into case (i). The proof is complete. □

8.3 Continuous Invertible Functions

A word of warning: the results of this section are typical of functions of a single variable. We will see in the chapter on General Topology that the continuity of an inverse function is not for free. Here the algebraic properties of the real line add a remarkable richness.

Example 8.8 Let A be a subset of $\mathbb{R}$, and let f be an injective continuous function from A to $\mathbb{R}$. We show that f^{-1} need not be continuous. Consider $A = [0, 1] \cup [2, 3]$ and

$$f(x) = \begin{cases} x & \text{if } 1 \leq x \leq 2 \\ x - 1 & \text{if } 2 \leq x \leq 3. \end{cases}$$

Clearly f is continuous on A, but its inverse f^{-1} is defined on $[0, 2]$ by

$$f^{-1}(y) = \begin{cases} y & \text{if } 0 \leq y \leq 1 \\ y + 1 & \text{if } 1 \leq y \leq 2. \end{cases}$$

The point $y = 1$ is a discontinuity point.

Theorem 8.12 *If f is a continuous, invertible function on an interval I, then f is strictly monotonic.*

Proof Let $x_0 \in I$ be a fixed point. Suppose that $f(\bar{x}) > f(x_0)$ for some $\bar{x} > x_0$. If a point $x' > x_0$ exists such that $f(x') \le f(x_0)$, then Theorem 8.10 yields a point ξ between x_0 and $\bar{x}$ with $f(\xi) = f(x_0)$. This is impossible, since f is invertible. We have proved that $x \mapsto f(x) - f(x_0)$ keeps the same sign in $I \cap [x_0, +\infty)$.

Let us consider any two points a and b of I, with $a < b$ and $f(a) < f(b)$. We claim that f is monotonically increasing on I. Let $x_1 < x_2$ be two points of I, and let $b_1 = \max\{b, x_2\}$. Since $f(b) > f(a)$, we see from the previous argument that $f(x) > f(a)$ for every $x > a$. In particular $f(b_1) > f(a)$. By the same token, $f(x) < f(b_1)$ for every $x < b_1$, hence $f(x_1) < f(b_1)$. Finally, $f(x_1) < f(x)$ for every $x > x_1$, and in particular $f(x_1) < f(x_2)$. The claim is proved.

If $f(a) > f(b)$, in a similar way we prove that f is monotonically decreasing on I. The proof is complete. $\square$

Theorem 8.13 *Let g be a function defined on an interval J, and monotonic on J. The function g is continuous on J if and only if the range $g(J)$ is an interval.*

Proof Without loss of generality we assume that g is increasing on J. If g is continuous, then $g(J)$ is an interval by Theorem 8.10.

We suppose now that $g(J)$ is an interval. If g is discontinuous at some point $x_0 \in J$, then

$$l = \lim_{x \to x_0^-} g(x) < \lim_{x \to x_0^+} g(x) = L.$$

For any $x < x_0$ we have $g(x) < l$, while for any $x > x_0$ we have $g(x) > L$. Among all points of (l, L), at most one can belong to $g(J)$, a set that contains both points smaller than l and points larger than L. This contradicts the assumption that $g(J)$ is an interval. The proof is complete. $\square$

Theorem 8.14 (Continuity of the Inverse Function) *If a function f is continuous on an interval I and invertible, then f^{-1} is continuous on $f(I)$.*

Proof By Theorem 8.12 f is strictly monotonic on I. By Theorem 8.10 $J = f(I)$ is an interval. The function f^{-1} is defined on J and is monotonic on J. Its range is I. By Theorem 8.13 f^{-1} is therefore continuous. $\square$

Example 8.9 The previous results justify several elementary statements. For instance, the continuity of the exponential function $x \mapsto e^x$ is equivalent to the continuity of the logarithm function $x \mapsto \log x$, since they both map an interval into an interval.

Important: Warning

We will not spend too much time in proving the continuity of elementary functions. Most traditional proofs are actually *circular*, the reason being the lack of a rigorous definition of the functions themselves. Consider the following sketch of a proof that $x \mapsto e^x$ is continuous at $x = 0$. For each $\varepsilon > 0$, the inequalities

$$1 - \varepsilon < e^x < 1 + \varepsilon$$

is equivalent to

$$\log (1 - \varepsilon) < x < \log (1 + \varepsilon),$$

at least when $0 < \varepsilon < 1$. Hence (?) the continuity is proved.

Unfortunately we have tacitly used several properties of the exponential without any authorization. Monotonicity is the most evident, but monotonicity is indeed the main reason why the exponential function is continuous! In a future chapter we will define several elementary functions in such a way that their properties can be proved.

8.4   Problems

8.1 Let y be a point of $\mathbb{R}$ and let f be a function defined on an interval containing y. We define

$$\limsup_{x \to y} f(x) = \inf \{\sup \{f(x) \mid 0 < |x - y| < \delta\} \mid \delta > 0\}$$

$$\liminf_{x \to y} f(x) = \sup \{\inf \{f(x) \mid 0 < |x - y| < \delta\} \mid \delta > 0.\}$$

Prove the following statements, and conjecture similar statements for lim inf.

(a) $\limsup_{x \to y} f(x) \le A$ if and only if for every $\varepsilon > 0$ there exists $\delta > 0$ such that $0 < |x - y| < \delta$ implies $f(x) \le A + \varepsilon$.
(b) $\limsup_{x \to y} f(x) \ge A$ if and only if for every $\varepsilon > 0$ and for every $\delta > 0$ there exists a point x such that $0 < |x - y| < \delta$ and $f(x) \ge A - \varepsilon$.
(c) $\liminf_{x \to y} f(x) \le \limsup_{x \to y} f(x)$ with equality if and only if $\lim_{x \to y} f(x)$ exists.
(d) If $\limsup_{x \to y} f(x) = A$ and if $\{x_n\}_n$ converges to y, then $\limsup_{n \to +\infty} f(x_n) = A$.
(e) If $\limsup_{x \to y} f(x) = A$, then there exists a sequence $\{x_n\}_n$ such that $x_n \to y$ and $A = \lim_{n \to +\infty} f(x_n) = A$.

8.2 A function $f: \mathbb{R} \to \mathbb{R}$ is homogeneous of degree one if $f(\lambda x) = \lambda f(x)$ for each $x \in \mathbb{R}$ and each $\lambda \in \mathbb{R}$. Prove that f is continuous.

8.3 Let $\{a_n\}_n$ and $\{b_n\}_n$ be two sequences of real numbers. Suppose that each $a_n \geq 0$ and that $\sum_n a_n$ converges to a finite sum a. A function f_n is defined for each n by

$$f_n(x) = \begin{cases} 0 & \text{if } x < b_n \\ a_n & \text{if } x \geq b_n. \end{cases}$$

Let $f(x) = \sum_n f_n(x)$ for all x. prove that

1. f is non-decreasing.
2. f is discontinuous on the set $A = \{b_n \mid n \in \mathbb{N}\}$.
3. f is continuous on $\mathbb{R} \setminus A$.

8.4 Let f and g be real-valued functions, uniformly continuous on a set A.

1. Prove that $f + g$ is uniformly continuous on A.
2. Prove that $f \circ g$ is uniformly continuous on $g(A) \cap A$.
3. Prove that fg is uniformly continuous if $A = [a, b]$. Give a counterexample if A is not a compact interval.

8.5 Let A be a non-empty subset of $\mathbb{R}$. A function $f_A : \mathbb{R} \to \mathbb{R}$ is defined by $f_A(x) = \inf\{|x - a| \mid a \in A\}$. Prove that f_A is uniformly continuous on A.

8.6 Let K be a compact subset of $\mathbb{R}$, and let $f : K \to \mathbb{R}$ be a continuous function. Prove that for each $\varepsilon > 0$ there exists $M \in \mathbb{R}$ such that $|f(x) - f(y)| \leq M|x - y| + \varepsilon$ for each $x, y \in K$.

8.7 A function $f : [0, 1] \to \mathbb{R}$ is upper semicontinuous if given $x \in [0, 1]$ and $\varepsilon > 0$ there exists $\delta > 0$ such that $|y - x| < \delta$ implies $f(y) < f(x) + \varepsilon$. Prove that an upper semicontinuous function on $[0, 1]$ is bounded above and attains its maximum value at some point of $[0, 1]$.

Chapter 9
Derivatives and Differentiability

Abstract Derivatives are usually introduced by fully exploiting the possibility of *dividing* real numbers. We propose an approach that can be extended almost literally to function defined on general normed spaces.

Definition 9.1 Let $f\colon (a, b) \to \mathbb{R}$ be a function, and let $x_0 \in (a, b)$ be a distinguished point. We say that f is differentiable at x_0 if a real number A exists with the property that

$$f(x) = f(x_0) + A(x - x_0) + o(|x - x_0|) \quad \text{as } x \to x_0. \tag{9.1}$$

The number A is called the *derivative* of f at x_0, and is denoted by any of the symbols

$$f'(x_0), \quad Df(x_0), \quad \mathrm{d}f(x_0), \quad \dot{f}(x_0).$$

The reader should recall that (9.1) is an equivalent way of requiring

$$\lim_{x \to x_0} \frac{f(x) - f(x_0) - A(x - x_0)}{|x - x_0|} = 0.$$

Remark 9.1 The assumption that f be defined on an open interval (a, b) is essentially for definiteness. Equation (9.1) shows that x_0 must be an accumulation point of the domain of f, but it should also belong to it. It would be possible to differentiate functions defined on a closed interval $[a, b]$, for instance, but at the end-points the derivative would lose several properties. For this reason we define the derivative of a function only at interior points of its domain.

Exercise 9.1 Suppose that f is differentiable at x_0. We want to prove that the number A in (9.1) is uniquely determined. For the sake of contradiction, assume that

$$f(x) = f(x_0) + A(x - x_0) + o(|x - x_0|)$$
$$f(x) = f(x_0) + B(x - x_0) + o(|x - x_0|)$$

as $x \to x_0$. Deduce that $A - B = o(1)$ as $x \to x_0$, and conclude that $A = B$.

If (9.1) holds, then

$$\lim_{x \to x_0} \frac{f(x) - f(x_0)}{x - x_0} = A. \tag{9.2}$$

On the other hand, if

$$\lim_{x \to x_0} \frac{f(x) - f(x_0)}{x - x_0} = A,$$

then

$$\frac{f(x) - f(x_0)}{x - x_0} = A + o(1) \quad \text{as } x \to x_0,$$

or $f(x) - f(x_0) = A(x - x_0) + o(|x - x_0|)$ as $x \to x_0$. We have proved

Theorem 9.1 *For a function $f : (a, b) \to \mathbb{R}$ the following conditions are equivalent:*

(i) f is differentiable at $x_0 \in (a, b)$ and $f'(x_0) = A$;
(ii) the limit $\lim_{x \to x_0} \frac{f(x) - f(x_0)}{x - x_0}$ exists as a real number and is equal to A.

Remark 9.2 Calculus books usually propose the derivative as the limit of the incremental ratio, namely (9.2). Our Definition 9.1 can be formally generalized to the case in which the function f is defined on a normed vector space, like $\mathbb{R}^n$ for $n \geq 2$. Equivalent definitions may be used as they are needed: we will see that (9.1) is the most convenient characterization of the derivative for proving the chain rule.

We record a third definition of the derivative in terms of continuous functions.

Theorem 9.2 *A function $f : (a, b) \to \mathbb{R}$ is differentiable at $x_0 \in (a, b)$ if and only if there exists a continuous function $\omega \colon (a, b) \to \mathbb{R}$ such that*

$$f(x) = f(x_0) + \omega(x)(x - x_0) \quad \text{for every } x \in (a, b). \tag{9.3}$$

In this case, $f'(x_0) = \omega(x_0)$.

Proof Condition (9.3) simply means that the function

$$x \mapsto \frac{f(x) - f(x_0)}{x - x_0}, \quad x \neq x_0$$

can be extended at $x = x_0$ continuously. Of course this is true if and only if the limit

$$\lim_{x \to x_0} \frac{f(x) - f(x_0)}{x - x_0}$$

exists as a real number. This means that f is differentiable at x_0, and by (9.3) we must have $\omega(x_0) = \lim_{x \to x_0} \frac{f(x) - f(x_0)}{x - x_0}$. $\qquad\square$

Corollary 9.1 *If a function is differentiable at a point, then it is continuous at that point.*

Proof This is immediate from (9.3). $\qquad\square$

Exercise 9.2 Prove the previous Corollary by using each of the equivalent definitions of the derivative.

9.1 Rules of Differentiation, or the Algebra of Calculus

If two functions f and g are defined on a neighborhood of a point x_0, we can define pointwise the functions $f + g$ and $f \cdot g$: indeed $x \mapsto f(x) + g(x)$ and $x \mapsto f(x)g(x)$ are well defined in a neighborhood of x_0. If $g \neq 0$ in a neighborhood of x_0, then the quotient $x \mapsto f(x)/g(x)$ is also defined.

Theorem 9.3 (Differentiation Rules) *Suppose that f and g are defined on a neighborhood (a, b) of the point x_0. Then*

(i) the function $f + g$ is differentiable at x_0, and $(f + g)'(x_0) = f'(x_0) + g'(x_0)$;
(ii) the function $f \cdot g$ is differentiable at x_0, and

$$(f \cdot g)'(x_0) = f'(x_0)g(x_0) + f(x_0)g'(x_0);$$

(iii) if $g(x_0) \neq 0$, then the function f/g is differentiable at x_0, and

$$(f/g)'(x_0) = \frac{f'(x_0)g(x_0) - f(x_0)g'(x_0)}{g(x_0)^2}.$$

Proof The proof of (i) is left as an easy exercise. A standard proof of (ii) is as follows:

$$\frac{f(x)g(x) - f(x_0)g(x_0)}{x - x_0} = \frac{f(x)g(x) - f(x_0)g(x) + f(x_0)g(x) - f(x_0)g(x_0)}{x - x_0}$$

$$= \frac{f(x) - f(x_0)}{x - x_0}g(x) + f(x_0)\frac{g(x) - g(x_0)}{x - x_0}$$

$$\to f'(x_0)g(x_0) + f(x_0)g'(x_0).$$

Another proof, based instead on Definition 9.1, starts from the assumptions $f(x) = f(x_0) + f'(x_0)(x - x_0) + o(1)$, $g(x) = g(x_0) + g'(x_0)(x - x_0) + o(1)$ and then

$$f(x)g(x) = \left[f(x_0) + f'(x_0)(x - x_0) + o(1)\right]\left[g(x_0) + g'(x_0)(x - x_0) + o(1)\right]$$
$$= f(x_0)g(x_0) + f'(x_0)g(x_0)(x - x_0)$$
$$+ f(x_0)g'(x_0)(x - x_0) + [\ldots]o(1),$$

where $[\ldots]$ contains all the terms that are multiplied by some $o(1)$ in the algebraic expansion. It follows that the linearization of fg at x_0 (exists and) is $f'(x_0)g(x_0) + f(x_0)g'(x_0)$.

The proof of (iii) is more traditional. First of all, since differentiability implies continuity, the condition $g(x_0) \neq 0$ implies that $g \neq 0$ in a neighborhood of x_0. Then we construct

$$\frac{\frac{f(x)}{g(x)} - \frac{f(x_0)}{g(x_0)}}{x - x_0} = \frac{1}{x - x_0} \frac{f(x)g(x_0) - f(x_0)g(x)}{g(x)g(x_0)}$$
$$= \frac{1}{x - x_0} \frac{f(x)g(x_0) - f(x_0)g(x_0) + f(x_0)g(x_0) - f(x_0)g(x)}{g(x)g(x_0)}$$
$$\to \frac{f'(x_0)g(x_0) - f(x_0)g'(x_0)}{g(x_0)^2}.$$

$\square$

Remark 9.3 Formula (iii) is not easily proved by means of Definition 9.1. The trouble is that it is not trivial to extract a linearization formula from the quotient

$$\frac{f(x_0) + f'(x_0)(x - x_0) + o(1)}{g(x_0) + g'(x_0)(x - x_0) + o(1)}.$$

Exercise 9.3 Try to deduce formula (iii) from

$$\frac{f(x_0) + f'(x_0)(x - x_0) + o(1)}{g(x_0) + g'(x_0)(x - x_0) + o(1)}.$$

As a hint, you may begin with the identity

$$\frac{1}{1 - z} = 1 + z + z^2 + z^3 + \cdots = 1 + z + O(z^2).$$

Theorem 9.4 (Chain Rule) *Let* $f : (a, b) \to \mathbb{R}$ *be differentiable at a point* $x_0 \in (a, b)$, *and let* g *be a function defined on a neighborhood of the range* $f((a, b))$. *If* g *is differentiable at the point* $f(x_0)$, *then* $g \circ f$ *is differentiable at* x_0, *and*

$$(g \circ f)'(x_0) = g'(f(x_0))f'(x_0). \tag{9.4}$$

Proof From Definition 9.1, we know that there exists function σ and τ such that $\sigma(x) = o(1)$ as $x \to x_0$, $\tau(y) = o(1)$ as $y \to f(x_0)$, and

$$f(x) = f(x_0) + f'(x_0)(x - x_0) + \sigma(x)|(-x_0)$$
$$g(y) = g(f(x_0)) + g'(f(x_0))(y - f(x_0) + \tau(y)(y - f(x_0)).$$

Then

$$g \circ f(x) = g(f(x_0)) + g'(f(x_0))(f(x) - f(x_0)) + \tau(f(x)(f(x) - f(x_0))$$
$$= g(f(x_0)) + g'(f(x_0))(f'(x_0) + \sigma(x)|x - x_0|)$$
$$+ \tau(f(x)(f(x) - f(x_0))$$
$$= g(f(x_0)) + g'(f(x_0))(f'(x_0) + \sigma(x)|x - x_0|)$$
$$+ \tau(f(x))(f'(x_0)(x - x_0) + \sigma(x)(x - x_0))$$
$$= g(f(x_0)) + g'(f(x_0) f'(x_0)(x - x_0) +$$
$$+ ([\ldots]\sigma(x) + [\ldots]\tau(f(x))) (x - x_0).$$

As $x \to x_0$, it is immediate to check that $[\ldots]\sigma(x) + [\ldots]\tau(f(x)) \to 0$, and the conclusion follows.   □

Important: Warning

The Calculus "proof" of the Chain Rule goes as follows:

$$\frac{g(f(x)) - g(f(x_0))}{x - x_0} = \frac{g(f(x)) - g(f(x_0))}{f(x) - f(x_0)} \frac{f(x) - f(x_0)}{x - x_0}$$
$$\to g'(f(x_0)) f'(x_0).$$

There is a subtle flaw in this computation, since division by $f(x) - f(x_0)$ is legitimate only if $f(x) \neq f(x_0)$ in a neighborhood of x_0. Unfortunately the assumptions of the Theorem do not ensure that this additional condition is satisfied by f. There is a way out, but we do not emphasize this approach, since it cannot be generalized to higher dimension.

Exercise 9.4 Provide a proof of the Chain Rule according to Theorem 9.2.

Theorem 9.5 (Differentiation of the Inverse Function) *Suppose that f is an invertible function on an interval (a, b). If f is differentiable at a point $x_0 \in (a, b)$ and $f'(x_0) \neq 0$, then f^{-1} is differentiable at $y_0 = f(x_0)$. Moreover,*

$$Df^{-1}(y_0) = \frac{1}{f'(x_0)}.$$

Fig. 9.1 Differentiating the
inverse function

Proof Since f is continuous and invertible on an interval, its inverse function f^{-1}
is continuous on the range of f. Then

$$\lim_{y \to y_0} \frac{f^{-1}(y) - f^{-1}(y_0)}{y - y_0} = \lim_{x \to x_0} \frac{1}{\frac{f(x)-f(x_0)}{x-x_0}}$$

$$= \frac{1}{f'(x_0)}.$$

□

The necessity of all the assumptions should be clear from Fig. 9.1.

Example 9.1

1. The function f defined by

$$f(x) = \begin{cases} x \sin \frac{1}{x} & \text{if } x \neq 0 \\ 0 & \text{otherwise} \end{cases}$$

is differentiable at any $x \neq 0$: indeed

$$f'(x) = \sin \frac{1}{x} - \frac{1}{x} \cos \frac{1}{x}.$$

However

$$\frac{f(x) - f(0)}{x - 0} = \sin \frac{1}{x}$$

does not converge as $x \to 0$. Hence f is not differentiable at $x = 0$.
2. The function f defined by

$$f(x) = \begin{cases} x^2 \sin \frac{1}{x} & \text{if } x \neq 0 \\ 0 & \text{otherwise} \end{cases}$$

is differentiable at $x = 0$, since

$$\frac{f(x) - f(0)}{x - 0} = x \sin \frac{1}{x},$$

and $0 \le |x \sin(1/x)| \le x$ for every x. Therefore $f'(0) = 0$.

Exercise 9.5 Suppose that $f : (a, b) \to \mathbb{R}$ is differentiable at a point a. Show that

$$f'(a) = \lim_{h \to 0} \frac{f(a + h) - f(a - h)}{2h}.$$

Provide an example of a function such that the limit on the right-hand side exists, but the function is not differentiable at a.

9.2 Mean Value Theorems

The derivative is a local object that can provide *global* properties of functions. This is essentially the basis of mean value theorems.

Theorem 9.6 (Fermat) *Let f be a function defined on the interval $[a, b]$. If f has a local maximum or a local minimum at a point $x_0 \in (a, b)$, and if $f'(x_0)$ exists, then $f'(x_0) = 0$.*

Proof Suppose that x_0 is a local maximum of f, so that there exists $\delta > 0$ such that $a < x_0 - \delta < x_0 < x_0 + \delta < b$. If $x_0 - \delta < x, x_0$, then

$$\frac{f(x) - f(x_0)}{x - x_0} \ge 0,$$

since f attains a local maximum at x_0. Letting $x \to x_0$ in the last inequality, we get $f'(x_0) \ge 0$. Similarly, if $x_0 < x < x_0 + \delta$, then

$$\frac{f(x) - f(x_0)}{x - x_0} \le 0,$$

and letting $x \to x_0$ we get $f'(x_0) \le 0$. Necessarily $f'(x_0) = 0$. The proof for a local minimum reduces to the previous one by considering $-f$ instead of f. $\quad\square$

Theorem 9.7 (Cauchy) *Let f and g be two functions defined on $[a.b]$, which are differentiable on (a, b) and continuous on $[a, b]$. Then there exists a point $c \in (a, b)$ such that*

$$[f(b) - f(a)]g'(c) = [g(b) - g(a)]f'(c).$$

Proof Let us introduce the function h defined on $[a, b]$ by

$$h(x) = [f(b) - f(a)]g'(x) - [g(b) - g(a)]f'(x).$$

obviously h is continuous and differentiable on (a, b), and $h(a) = h(b)$. It remains to prove that the derivative of h vanishes somewhere inside (a, b). If h turns out to be constant on $[a, b]$, then the proof is complete.

If $h(x) > h(a)$ for some $x \in (a, b)$, then h must attain a global maximum inside (a, b), and at this point h' vanishes by Theorem 9.6.

If $h(x) < h(a)$ for some $x \in (a, b)$, then h must attain a global minimum inside (a, b), and at this point h' vanishes again by Theorem 9.6.   □

The simple choice $g = $ id is surprisingly important: see Fig. 9.2.

Theorem 9.8 (Lagrange) *Let f be a function defined on $[a.b]$, which is differentiable on (a, b) and continuous on $[a, b]$. Then there exists a point $c \in (a, b)$ such that*

$$[f(b) - f(a)]g'(c) = (b - a)f'(c).$$

Corollary 9.2 (Monotonicity) *Suppose f is differentiable on (a, b).*

(a) If $f' \geq 0$ on (a, b), then f is monotonically increasing on (a, b).
(b) If $f' \leq 0$ on (a, b), then f is monotonically decreasing on (a, b).
(c) if $f' = 0$ identically on (a, b), then f is constant on (a, b).

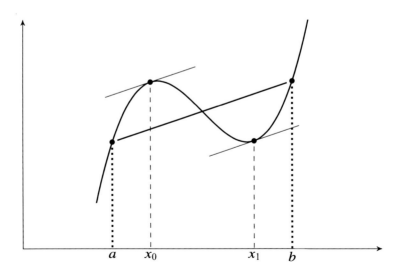

Fig. 9.2 Lagrange's theorem

Proof For any points x_1 and x_2 in (a, b), Theorem 9.8 yields $f(x_2) - f(x_1) = (x_2 - x_1) f'(c)$ for some c between x_1 and x_2. It is now immediate to conclude according to the sign of f'.   □

Exercise 9.6 Suppose that f is a differentiable function such that $f'(x) = f(x)$ for each $x \in \mathbb{R}$. If $f(0) = 1$, prove that $f(x) = e^x$ for each $x \in \mathbb{R}$.

Exercise 9.7 Let f be a differentiable function on $\mathbb{R}$ with

$$L = \sup \left\{ |f'(x)| \mid x \in \mathbb{R} \right\} < 1.$$

(a) Fix any $s_0 \in \mathbb{R}$, and define $s_n = f(s_{n-1})$ for each $n = 1, 2, \ldots$ Prove that the sequence $\{s_n\}_n$ is convergent. *Hint:* show that $\{s_n\}_n$ is a Cauchy sequence.

(b) Prove the Banach-Caccioppoli Fixed Point Theorem: there exists a point $x \in \mathbb{R}$ such that $f(x) = x$.

Mean value theorems are typically used in Calculus to derive criteria for the existence of limits. The well-known result which goes under the name of De l'Hospital is the most celebrated one.[1] We follow [2] for the proof.

Theorem 9.9 (De l'Hospital) *Suppose f and g are differentiable on (a, b), and $g'(x) \neq 0$ for all $x \in (a, b)$, where $-\infty \leq a < b \leq +\infty$. Suppose*

$$\lim_{x \to a} \frac{f'(x)}{g'(x)} = A, \tag{9.5}$$

where $A \in \tilde{\mathbb{R}}$. If either

$$\lim_{x \to a} f(x) = \lim_{x \to a} g(x) = 0 \tag{9.6}$$

or

$$\lim_{x \to a} g(x) = +\infty, \tag{9.7}$$

then

$$\lim_{x \to a} \frac{f(x)}{g(x)} = A. \tag{9.8}$$

An analogous statement holds as $x \to b$.

We remark that A may be infinite.

---

[1] We write De l'Hospital since this is the ancient and original name. Nowadays it is customary to write De l'Hôpital.

Proof Let us start with the case $-\infty \le A < +\infty$. Let $q > A$ be any real number, and choose r such that $A < r < q$. By (9.5) there exists a point $c \in (a, b)$ such that $a < x < c$ implies

$$\frac{f'(x)}{g'(x)} < r. \tag{9.9}$$

If $a < x < y < b$, Theorem 9.7 yields a point $t \in (x, y)$ such that

$$\frac{f(x) - f(y)}{g(x) - g(y)} = \frac{f'(t)}{g'(t)} < r. \tag{9.10}$$

Suppose that (9.6) holds. When $x \to a$ in (9.10) we see that $a < y < c$ implies

$$\frac{f(y)}{g(y)} \le r < q \tag{9.11}$$

Suppose now that (9.7) holds. We fix y in (9.10) and select $c_1 \in (a, y)$ such that $g(x) > g(y)$ and $g(x) > 0$ for every $x \in (a, c_1)$. Then it follows from (9.10) that

$$\frac{f(x)}{g(x)} < r - r\frac{g(y)}{g(x)} + \frac{f(y)}{g(x)} \tag{9.12}$$

for every $x \in (a, c_1)$. Letting $x \to a$ in (9.12) we see that the right-hand side of (9.12) converges to r, and therefore there exists a point $c_2 \in (a, c_1)$ such that

$$\frac{f(x)}{g(x)} \le r < q \tag{9.13}$$

for every $x \in (a, c_2)$. Equations (9.11) and (9.13) imply that $f(x)/g(x) < q$ for every $x \in (a, c_2)$.

If $-\infty < A \le +\infty$, a completely similar argument shows that, given any $p < A$, there exists a point c_3 such that $p < f(x)/g(x)$ for every $x \in (a, c_3)$. Since p and q are arbitrary, we have proved that $f(x)/g(x) \to A$ as $x \to a$. □

Exercise 9.8 For every $x \in \mathbb{R}$, consider the functions

$$f(x) = x + \sin x \cos x$$
$$g(x) = e^{\sin x} f(x).$$

(a) Show that $\lim_{x \to +\infty} f(x) = \lim_{x \to +\infty} g(x) = +\infty$.
(b) Show that $f'(x) = 2 \cos^2 x$ and $g'(x) = e^{\sin x} \cos x (2 \cos x + f(x))$.
(c) Show that $f'(x)/g'(x) = \frac{2e^{-\sin x} \cos x}{2 \cos x + f(x)}$ if $\cos x \ne 0$ and $x > 3$.

(c) Show that $\lim_{x \to +\infty} \frac{2e^{-\sin x}\cos x}{2\cos x + f(x)} = 0$ and yet $\lim_{x \to +\infty} f(x)/g(x)$ does not exist.

This exercise shows that the assumption "$g'(x) \neq 0$" is necessary in Theorem 9.9.

Remark 9.4 Calculus books often write that De l'Hospital's theorem is a tool for the analysis of indeterminate forms $[0/0]$ and $[\infty/\infty]$. As we have seen, no condition on f is needed when $g(x) \to +\infty$ as $x \to a$.

Exercise 9.9 Suppose that $f(x) + f'(x) \to L$ as $x \to +\infty$. Prove that $f(x) \to L$ as $x \to a$ and $f'(x) \to 0$ as $x \to +\infty$. *Hint:* write $f(x) = e^x f(x)/e^x$, and remark that $e^x \to +\infty$ as $x \to +\infty$. De l'Hospital's theorem applies even if we have no information about $e^x f(x)$ as $x \to +\infty$.

9.3   The Intermediate Property for Derivatives

Although a differentiable function may have a discontinuous derivative, it is interesting that derivatives always have the intermediate value property.

Theorem 9.10 (Darboux) *Suppose that f is a real differentiable function on $[a, b]$, and suppose that $f'(a) < \lambda < f'(b)$. Then there exists a point $\xi \in (a, b)$ such that $f'(\xi) = \lambda$.*

Proof We define $g(x) = f(x) - \lambda x$. By assumption $g'(a+) < 0$ and $g'(b-) > 0$. It follows that $g(t_1) < g(a)$ and $g(t_2) < g(b)$ for some t_1, t_2 in (a, b). As a consequence, the function g must attain its minimum at some point $\xi \in (a, b)$. We already know that $g'(\xi) = 0$, and thus $f'(\xi) = \lambda$.   □

Example 9.2 Define the polynomial $P(x) = (x^2 - 1)^2$. Let $f : [0, 1] \to \mathbb{R}$ be the function such that $f(0) = 0$ and

$$f(x) = \frac{1}{n^{3/2}} P\left(2n(n+1)x - 2n - 1\right) \quad \text{if } \frac{1}{n+1} \leq x \leq \frac{1}{n}.$$

Clearly f is a differentiable function, but f' is not continuous, and event not bounded on $[0, 1]$. Indeed $f'_+(0) = 0$, but $f(b_n) \to +\infty$ at the points

$$b_n = \frac{4n + 1}{4n(n+1)} \to 0.$$

Nevertheless, Theorem 9.10 applies, and f' attains every positive value γ on every interval $[0, b_n]$.

9.4   Derivatives at End-Points

The idea of linearization is fruitful at *inner* points of the domain of definition: the function can be identified, at an infinitesimal scale, with a linear function. It is nonetheless convenient, from time to time, to extend the definition of derivative at end-points.

Definition 9.2 Let $f: [a, b] \to \mathbb{R}$ be a function. We say that f is differentiable at a if the limits

$$f'(a+) = \lim_{x \to x_0^+} \frac{f(x) - f(x_0)}{x - x_0}$$

exists as a real number. In this case, we call $f'(a+)$ the right-derivative of f at a. Similarly we define the left derivative of f at b.

Unfortunately, several fundamental results of differential calculus do not extend to end-point derivatives. As an example, we propose the function $f: [a, b] \to \mathbb{R}$ such that $f(x) = mx + q$, where $m \neq 0$ and q are real numbers. It is easy to check that a and b are global extremum points of f (their nature depends on the sign of m), but $f'(a+) = m = f'(b-)$ are different than zero. In other words, Fermat's Theorem does not hold at end-points.

9.5   Derivatives of Derivatives

A nice feature of derivatives in one variable is that we can easily differentiate derivatives. We will see that this requires much more attention in higher dimension, since the derivative is no longer a real number.

Definition 9.3 Suppose that a function f is defined on an interval (a, b), and that the derivative f' of f exists at every point of (a, b). Hence the function $f': (a, b) \to \mathbb{R}$ is defined in such a way that $f': x \mapsto f'(x)$ for every $x \in (a, b)$. We say that f is twice differentiable at $x \in (a, b)$ if f' is differentiable at x. In this case we denote by $f''(x)$ or $D^2 f(x)$ or $d^2 f(x)$ the derivative of f' at x, and call it the second derivative of f at x.

More generally, if f is differentiable n times at every point of (a, b), we say that f is differentiable $(n + 1)$-times at $x \in (a, b)$ if the function $f^{(n)}$ is differentiable at x. In this case we denote by $f^{(n+1)}(x)$, or $D^{n+1} f(x)$ or $d^{n+1} f(x)$ the derivative of $f^{(n)}$ at x, and call it the derivative of f of order $n + 1$ at x.

Definition 9.4 (Regularity Classes) Let $f: (a, b) \to \mathbb{R}$ be a function. We write $f \in C^0(a, b)$ if f is continuous on (a, b). For $n \in \mathbb{N}$, we write $f \in C^n(a, b)$ if the derivatives f', f'', $\ldots$, $f^{(n-1)}$ exist on (a, b), and if f^n exists and is a continuous function on (a, b). We formally write $f \in C^\infty(a, b)$ to mean that

$f \in \bigcap_{n=0}^{\infty} C^n(a, b)$. Hence, a function is of class C^∞ if and only if it can be differentiated as many times as we please.

We now attach a very specific polynomial to every function with a high degree of differentiability.

Definition 9.5 (Taylor Polynomial) Let $f : (a, b) \to \mathbb{R}$ be n-times differentiable at a point $x_0 \in (a, b)$. The Taylor polynomial of degree n at x_0 is defined to be

$$P(n, x_0; x) = \sum_{k=0}^{n} \frac{f^{(k)}(x_0)}{k!}(x - x_0)^k$$

$$= f(x_0) + f'(x_0)(x - x_0) + \frac{f''(x_0)}{2!}(x - x_0)^2 + \cdots$$

$$\cdots + \frac{f^{(n)}(x_0)}{n!}(x - x_0)^n. \tag{9.14}$$

Taylor polynomials express the local behavior of functions, and generalize the concept of linear approximation which was introduced in the definition of the first derivative.

Theorem 9.11 (Local Polynomial Approximation) *Let* $f : (a, b) \to \mathbb{R}$ *be n-times differentiable at a point* $x_0 \in (a, b)$, *and let* $P(n, x_0; \cdot)$ *be its Taylor polynomial of degree n. Then*

$$f(x) = P(n, x_0; x) + (x - x_0)^n o(1) \quad as \ x \to x_0. \tag{9.15}$$

If, in addition, $f^{(n+1)}(x_0)$ *exists, then*

$$\lim_{x \to x_0} \frac{\zeta(x)}{x - x_0} = \frac{f^{(n+1)}(x_0)}{(n+1)!}.$$

Proof We consider the function

$$\zeta(x) = \frac{f(x) - P(n, x_0; x)}{(x - x_0)^n}, \tag{9.16}$$

defined for $x \neq x_0$. It is easy to check that all the derivatives of order $1 \le j \le n - 1$ of the numerator of ζ vanish at x_0. We apply Theorem 9.9 $n - 1$ times to (9.16), to get

$$\lim_{x \to x_0} \zeta(x) = \lim_{x \to x_0} \frac{f^{(n-1)}(x) - f^{(n-1)}(x_0) - (x - x_0)f^{(n)}(x_0)}{n!(x - x_0)}, \tag{9.17}$$

provided the last limit exists. By definition,

$$\lim_{x \to x_0} \frac{f^{(n-1)}(x) - f^{(n-1)}(x_0)}{x - x_0} = f^{(n)}(x_0),$$

thus from (9.17) we deduce $\lim_{x \to x_0} \zeta(x) = 0$. We can define $\zeta(x_0) = 0$, so that ζ becomes a continuous function on (a, b). By applying again Theorem 9.9 n times, we finally see that

$$\lim_{x \to x_0} \frac{\zeta(x)}{x - x_0} = \frac{1}{(n+1)!} \lim_{x \to x_0} \frac{f^{(n)}(x) - f^{(n)}(x_0)}{(x - x_0)} = \frac{f^{(n+1)}(x_0)}{(n+1)!}.$$

□

Theorem 9.12 (Lagrange Remainder) *Let* $f \colon (a, b) \to \mathbb{R}$ *be* $n + 1$ *times differentiable on* (a, b), *and let* $x_0 \in (a, b)$. *For each* $x \in (a, b)$, $x \neq x_0$, *there exists a point* ξ *between* x_0 *and* x *such that*

$$f(x) = P(n, x_0; x) + \frac{f^{(n+1)}(\xi)}{(n+1)!}(x - x_0)^{n+1}.$$

Proof Suppose without loss of generality that $x_0 < x$. We define $F \colon [x_0, x] \to \mathbb{R}$,

$$F(t) = f(x) - f(y) - f'(t)(x - t) - \frac{f''(t)}{2!}(x - t)^2 - \cdots - \frac{f^{(n)}(t)}{n!}(x - t)^n.$$

Then

$$F'(t) = -\frac{f^{(n+1)}(t)}{n!}(x - t)^n.$$

Next we introduce the function $G \colon [x_0, x] \to \mathbb{R}$,

$$G(t) = \frac{(x - t)^{n+1}}{(n+1)!}.$$

We have $F(x) = G(x) = 0$, and $F'(t)/G'(t) = f^{(n+1)}(t)$. We now apply Theorem 9.7 to F and G on $[x_0, x]$, and find a point ξ between x_0 and x such that

$$\frac{F(x_0)}{G(x_0)} = \frac{F(x) - \Gamma(x_0)}{G(x) - G(x_0)} = \frac{F'(\xi)}{G'(\xi)} = f^{(n+1)}(\xi).$$

□

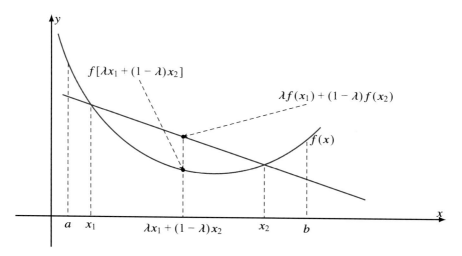

Fig. 9.3 A convex function

9.6 Convexity

Convexity is a fundamental property in mathematical analysis. Most Calculus books propose the definition of convexity as a property of the second derivative. We are going to see that this is just the top of the iceberg.

Definition 9.6 Let I be an interval[2] (open, closed, bounded or unbounded), and let $f : I \to \mathbb{R}$ be a function. We say that f is a convex function on I, if for each $x_1 \in I$, $x_2 \in I$, and for all real numbers $\lambda \geq 0$, $\mu \geq 0$ such that $\lambda + \mu = 1$, there results

$$f(\lambda x_1 + \mu x_2) \leq \lambda f(x_1) + \mu f(x_2). \qquad (9.18)$$

The function f is concave on I if and only if the function $-f$ is convex on I.

Exercise 9.10 Prove that a function f is convex on an interval I if and only if for every $x_1 \in I$, $x_2 \in I$ and $\lambda \in [0, 1]$ there results

$$f(\lambda x_1 + (1 - \lambda)x_2) \leq \lambda f(x_1) + (1 - \lambda)f(x_2).$$

See Fig. 9.3.

Remark 9.5 The crucial fact is that $\lambda x_1 + \mu x_2$ must be an element of I, as soon as x_1 and x_2 belong to I and $\lambda + \mu = 1$. It is easy to check that this is actually correct, since I is an interval.

---

[2] Hence I is characterized by the following property: if $x \in I$, $y \in I$ and $x < z < y$, then $z \in I$.

Let us manipulate the *convexity inequality* (9.18). Let x be any point between x_1 and x_2. We set $x = \lambda x_1 + \mu x_2$. Since $\lambda + \mu = 1$, we see that

$$\lambda = \frac{x_2 - x}{x_2 - x_1}, \qquad \mu = \frac{x - x_1}{x_2 - x_1}.$$

Hence (9.18) is equivalent to

$$f(x) \le \frac{x_2 - x}{x_2 - x_1} f(x_1) + \frac{x - x_1}{x_2 - x_1} f(x_2). \tag{9.19}$$

By symmetry, we can always suppose that $x_2 > x_1$. Then we get

$$(x_2 - x_1) f(x) \le (x_2 - x) f(x_1) + (x - x_1) f(x_2). \tag{9.20}$$

Writing $x_2 - x = (x_2 - x_1) - (x - x_1)$ in (9.20), we find

$$\frac{f(x) - f(x_1)}{x - x_1} \le \frac{f(x_2) - f(x_1)}{x_2 - x_1}. \tag{9.21}$$

Writing $x_2 - x_1 = (x_2 - x) + (x - x_1)$ in (9.20) we find

$$\frac{f(x_1) - f(x)}{x_1 - x} \le \frac{f(x_2) - f(x)}{x_2 - x}. \tag{9.22}$$

Comparing (9.18), (9.21) and (9.22), we have proved

Theorem 9.13 *The function f is convex on the interval I if and only if, for every $x_0 \in I$, the map*

$$x \mapsto \frac{f(x) - f(x_0)}{x - x_0}$$

is (defined for $x \ne x_0$ and) monotonically increasing.

We thus see that convexity is just another way of stating that the incremental ratio is an increasing function. Since monotone functions always have one-sided limits, we deduce

Corollary 9.3 *A convex function f defined on an interval (a, b) is left- and right-differentiable at every $x_0 \in (a, b)$. Moreover $f'(x_0-) \le f'(x_0+)$.*

If we remember (9.22) and let $x \to x_1$ and then $x \to x_2$, we see that

$$f'(x_1+) \le \frac{f(x_2) - f(x_1)}{x_2 - x_1} \le f'(x_2-). \tag{9.23}$$

We can finally relate convexity to derivatives.

Theorem 9.14 *Let f be a differentiable function in the interval [a, b]. A necessary and sufficient condition for f to be convex is that f' be monotonically increasing.*

Proof If f is convex, then (9.23) implies $f'(x_1) \le f'(x_2)$, so that f' is increasing. In the other direction, we remark that convexity is equivalent to (9.22) for all points x_1, x and x_2 such that $x_1 < x < x_2$. If f' is increasing, then there exists points ξ_1 and ξ_2 such that $x_1 < \xi_1 < x < \xi_2 < x_2$ and

$$\frac{f(x_1) - f(x)}{x_1 - x} = f'(\xi_1), \qquad \frac{f(x_2) - f(x)}{x_2 - x} = f'(\xi_2).$$

Since $f'(x_1) \le f'(x_2)$, the conclusion follows. □

In particular, it is indeed true that convex functions are those functions whose second derivative is positive, but there is no need to restrict our definitions to twice differentiable functions.

Corollary 9.4 *Let f be twice differentiable in [a, b]. The function f is convex if and only if $f''(x) \ge 0$ for every x.*

Proof This follows immediately from the characterization of increasing functions in terms of the first derivative, see Corollary 9.2. □

Example 9.3 Prove that the function $x \mapsto |x|$ is convex on $I = \mathbb{R}$. Of course this conclusion would be meaningless if we had defined convex functions through the sign of the second derivative.

9.7 Problems

9.1 Suppose that f is differentiable at the point x_0. Prove that

$$\lim_{n \to +\infty} n \left[f\left(x_0 + \frac{\alpha}{n}\right) - f\left(x_0 - \frac{\beta}{n}\right) \right] = (\alpha + \beta) f'(x_0).$$

Give an example to show that the existence of the previous limit does not imply the differentiability of f at x_0.

9.2 Suppose that f is differentiable at the point x_0. let $\{h_n\}_n$ and $\{k_n\}_n$ be two nonincreasing sequences which converge to x_0. Prove that

$$\lim_{n \to +\infty} \frac{f(x_0 + h_n) - f(x_0 - k_n)}{h_n + k_n} = f'(x_0).$$

Give an example to show that the existence of the previous limit does not imply the differentiability of f at x_0.

9.3 Suppose that f' is continuous on an interval $[a, b]$. Prove that for every $\varepsilon > 0$ there exists $\delta > 0$ such that

$$\left| \frac{f(t) - f(x)}{t - x} - f'(x) \right| < \varepsilon$$

whenever $0 < |t - x| < \delta$ and $x \in [a, b], t \in [a, b]$.

9.4 Let f be differentiable on $[a, b]$. Suppose that $0 < m \le f'(x) \le M$ for each $x \in [a, b]$, and that $f(a) < 0 < f(b)$. Given $x_1 \in [a, b]$, define a sequence $\{x_n\}_n$ by

$$x_{n+1} = x_n - \frac{f(x_n)}{M}$$

for $n = 1, 2, 3, \ldots$ Prove that $\{x_n\}_n$ converges to a limit x_0 such that $f(x_0) = 0$. Furthermore, prove that

$$|x_{n+1} - x_n| \le \frac{f(x_1)}{m} \left(1 - \frac{m}{M} \right)^n.$$

9.5 Suppose f is a real-valued function defined on the half-line $(a, +\infty)$. Suppose that f is twice differentiable on $(a, +\infty)$, and define

$$M_0 = \sup_{x > a} |f(x)|$$

$$M_1 = \sup_{x > a} |f'(x)|$$

$$M_2 = \sup_{x > a} |f''(x)|.$$

Prove that $M_1^2 \le M_0 M_2$ as follows: for each $h > 0$ deduce from Taylor's expansion that there exists $\xi \in (x, x + 2h)$ such that

$$f'(x) = \frac{f(x + 2h) - f(x)}{2h} - hf'(\xi).$$

Therefore

$$|f'(x)| \le hM_1 + \frac{M_0}{h}.$$

Now optimize the right-hand side with respect to $h > 0$.

9.6 If f is twice differentiable on $(0, +\infty)$, f'' is bounded and $\lim_{x \to +\infty} f(x) = 0$, prove that $\lim_{x \to +\infty} f'(x) = 0$. *Hint:* consider the limit $a \to +\infty$ in the previous problem.

9.7 (a) Prove that for each $x > 0$, $x \neq 1$, we have

$$\frac{x - 1}{x} < \log x < x - 1.$$

(b) For each $j \in \mathbb{N}$, $j > 1$, prove that

$$\log \frac{j + 1}{j} < \frac{1}{j} < \log \frac{j}{j - 1}.$$

(c) For each $n \in \mathbb{N}$, $k \in \mathbb{N}$, $n > 1$, prove that

$$\log \left(k + \frac{1}{n} \right) < \sum_{j=n}^{kn} \frac{1}{j} < \log \left(k + \frac{k}{n - 1} \right)$$

and

$$\lim_{n \to +\infty} \sum_{j=n}^{kn} \frac{i}{j} = \log k.$$

(d) Deduce from (c) and from the identity

$$\sum_{j=1}^{2n} \frac{(-1)^{j+1}}{j} = \sum_{j=1}^{2n} \frac{1}{j} - 2 \sum_{k=1}^{n} \frac{1}{2k}$$

that

$$\sum_{j=1}^{\infty} \frac{(-1)^{j+1}}{j} = \log 2.$$

9.8 If $x > 1$, $x \neq e$, prove that there exists one and only one number $f(x) > 0$ such that $f(x) \neq x$ and

$$x^{f(x)} = (f(x))^x.$$

Hint: $x^y = y^x$ if and only if $\frac{\log x}{x} = \frac{\log y}{y}$.

9.8   Comments

The standard definition of derivative as the limit of the incremental ratio should not be considered as the one used by mathematicians from the beginning of Calculus. They would rather use a principle of disappearing quantities which roughly correspond to an expansion of functions at first order as in

$$f(t + \Delta t) = f(t) + f'(t)\Delta t + f''(t)(\Delta t)^2 \approx f(t) + f'(t)\Delta t.$$

In this sense, the one-dimensional derivative has progressively lost its definition as a linearization procedure in favor of an iconic limit:

$$f'(t) = \lim_{\Delta t \to 0} \frac{f(t + \Delta t) - f(t)}{\Delta t}.$$

There are several good reasons to define the derivative as a linearization, and the most important one is that the derivative of a function of several variables is not a number.

The theory of convex functions is a long but elementary exercise, in the case of functions of a single real variable. The topic becomes much more exciting in higher dimensions, where intervals must be replaced by convex sets and a new fact comes into play: it is possible to draw conclusion about a function on a convex set by assuming a property of that function on every straight line. The interested reader may start from [1].

References

1. R.T. Rockafellar, *Convex Analysis* (Princeton University Press, 1970)
2. W. Rudin, *Principles of Mathematical Analysis*. International Series in Pure and Applied Mathematics, 3rd edn. (McGraw-Hill Book Co., New York, 1976)

Chapter 10
Riemann's Integral

Abstract The basic theory of definite integration can be named after Bernhard Riemann. In this chapter we will propose a rigorous introduction to it. Although we have in mind Rudin's lucid chapter in his *Principles of Mathematical Analysis*, we prefer to avoid the additional complication of the Stieltjes generalization. By the way, a later chapter will show that the much more flexible integral of Lebesgue can be presented without too much effort.

10.1 Partitions and the Riemann Integral

We will systematically consider *bounded* functions defined on *bounded* intervals.

This is by far the worst weakness of the Riemann integral. The rough idea is to construct finite sums of values of a function at suitable points, and then pass to the limit is a suitable sense. The next definition gives a name to the selection of the suitable points.

Definition 10.1 A partition P of a closed and bounded interval $[a, b]$ is a finite set of points $x_0, x_1, \ldots, x_n$ such that $a = x_0 < x_1 < x_2 < \ldots < x_{n-1} < x_n = b$. We write $\Delta x_i = x_i - x_{i-1}$ for $i = 1, \ldots, n$.

Let $f : [a, b] \to \mathbb{R}$ be a bounded function. To any partition P of $[a, b]$ we attach the quantities

$$M_i = \sup \{ f(x) \mid \Delta x_{i-1} \leq x \leq x_i \}$$
$$m_i = \inf \{ f(x) \mid \Delta x_{i-1} \leq x \leq x_i \}$$

La Matematica per il 3+2 141, https://doi.org/10.1007/978-3-031-19738-3_10

Fig. 10.1 Riemann sums

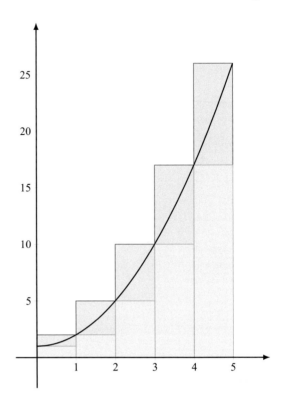

$$U(f, P) = \sum_{i=1}^{n} M_i \Delta x_i$$

$$L(f, P) = \sum_{i=1}^{n} m_i \Delta x_i;$$

see Fig. 10.1.

Definition 10.2 The upper and the lower Riemann integral of f on $[a, b]$ are defined respectively by

$$\overline{\int_a^b} f \, dx = \inf U(f, P), \quad \underline{\int_a^b} f \, dx = \sup L(f, P)$$

where inf and sup are taken over all partitions P of $[a, b]$.

Remark 10.1 Since $m_i \leq M_i$, we see that $L(f, P) \leq U(f, P)$ for every partition P. Since f is bounded, say $m \leq f \leq M$ on $[a, b]$, then $m(b - a) \leq L(f, P) \leq U(f, P) \leq M(b - a)$. As a consequence, the upper and the lower integrals of f are finite. Of course this is the reason why boundedness cannot be dispensed with.

Definition 10.3 A bounded function f is R-integrable on $[a, b]$ if and only if

$$\underline{\int_a^b} f \, dx = \overline{\int_a^b} f \, dx.$$

In this case the common value of the upper and the lower integral is denoted by $\int_a^b f \, dx$, and it is called the Riemann integral of f on $[a, b]$.

Remark 10.2 There are good reasons to criticize the symbol associated to the Riemann integral. Indeed, the letter x in $\int_a^b f \, dx$ is completely useless. The symbol $\int_a^b f$ would be a natural choice, but we prefer to use the traditional notation.

Unlike the derivative, the Riemann integral is a *global* object, in the sense that it involves the values of the function f on the whole interval $[a, b]$. It is a matter of fact that Definition 10.3 is not particularly concrete. We need to deploy some general condition for integrability.

Definition 10.4 A partition P^* is a refinement of a partition P of $[a, b]$, if $P \subset P^*$. If P_1 and P_2 are two partitions of $[a, b]$, their union $P^* = P_1 \cup P_2$ is called the common refinement of P_1 and P_2.

Exercise 10.1 Prove that the common refinement of two partitions P_1 and P_2 is the smallest partition which refines both P_1 and P_2.

Theorem 10.1 *If P^* is a refinement of P, then*

$$L(f, P) \leq L(f, P^*), \qquad U(f, P^*) \leq U(f, P).$$

Proof We suppose first that P^* contains just one point more that P. If this point is x^*, for some index i we must have $x_{i-1} < x^* < x_i$. Let

$$w_1 = \inf\{f(x) \mid x_{i-1} \leq x \leq x^*\} \tag{10.1}$$

$$w_2 = \inf\{f(x) \mid x^* \leq x \leq x_i\}. \tag{10.2}$$

By set inclusion, $m_i \leq w_1$ and $m_i \leq w_2$, hence

$$L(f, P^*) - L(f, P) = w_1[x^* - x_{i-1}] + w_2[x_i - x^*] - m_i[x_i - x_{i-1}]$$
$$= (w_1 - m_i)[x^* - x_{i-1}] + (w_2 - m_i)[x_i - x^*] \geq 0.$$

In the general case, a finite number of points are added to those of P. We just repeat the same construction for each of them, and we conclude. The proof of the inequality $U(f, P^*) \leq U(f, P)$ is similar, and we omit the details. $\qquad \square$

Corollary 10.1 $\underline{\int_a^b} f \, dx \leq \overline{\int_a^b} f \, dx.$

Proof Let P_1 and P_2 be arbitrary partitions of $[a, b]$, and let P^* their common refinement. As we have just seen,

$$L(f, P_1) \le L(f, P^*) \le U(f, P^*) \le U(f, P_2).$$

Taking the supremum over P_1, we see that $\underline{\int}_a^b f \, dx \le U(f, P_2)$. Taking now the infimum over P_2 we conclude the proof.   □

Theorem 10.2 (Integrability Condition) *A bounded function f is R-integrable on $[a, b]$ if and only if for every $\varepsilon > 0$ there exists a partition P_ε of $[a, b]$ such that*

$$U(f, P_\varepsilon) - L(f, P_\varepsilon) < \varepsilon. \tag{10.3}$$

Proof If P is any partition, then

$$L(f, P) \le \underline{\int}_a^b f \, dx \le \overline{\int}_a^b f \, dx \le U(f, P).$$

Then (10.3) implies $0 \le \overline{\int}_a^b f \, dx - \underline{\int}_a^b f \, dx < \varepsilon$, and $\overline{\int}_a^b f \, dx = \underline{\int}_a^b f \, dx$ because $\varepsilon > 0$ is arbitrary.

Conversely, let f be integrable, and let $\varepsilon > 0$ be fixed. There exist two partitions P_1 and P_2 such that

$$U(f, P_2) < \int_a^b f \, dx + \frac{\varepsilon}{2}, \quad L(f, P_1) > \int_a^b f \, dx - \frac{\varepsilon}{2}.$$

Choosing P_ε as the common refinement of P_1 and P_2, then

$$U(f, P_\varepsilon) \le U(f, P_2) < \int_a^b f \, dx + \frac{\varepsilon}{2} < L(f, P_1) + \varepsilon \le L(f, P_\varepsilon) + \varepsilon,$$

and (10.3) is proved.   □

Exercise 10.2

(a) Prove that a bounded function f is integrable on $[a, b]$ if and only if there exists a sequence $\{P_n\}_n$ of partitions of $[a, b]$ such that

$$\lim_{n \to +\infty} (U(f, P_n) - L(f, P_n)) = 0,$$

and in this case $\int_a^b f \, dx = \lim_{n \to +\infty} U(f, P_n) = \lim_{n \to +\infty} L(f, P_n)$. *Hint:* apply (10.3) with $\varepsilon = \varepsilon_n \to 0$ as $n \to +\infty$.

(b) For each n, let P_n be the partition of $[0, 1]$ into n equally spaced points. Find a closed formula for $U(f, P_n)$ and $L(f, P_n)$ in case $f(x) = x$. *Hint:* first prove by induction that $1 + 2 + \ldots + n = n(n + 1)/2$.

(c) Deduce that $f(x) = x$ is integrable on $[0, 1]$, and compute $\int_0^1 x \, dx$.

We can now relate our definition of the Riemann integral to the usual approach of Calculus via Riemann sums. In undergraduate introductions to the integral, it is customary to select arbitrary points between two consecutive nodes of a partition, so that sums like

$$\sum_{i=1}^{n} f(t_i) \Delta x_i$$

are considered for points $t_i \in [x_{i-1}, x_i]$.

Theorem 10.3

(i) *If (10.3) holds for a partition P and a value of $\varepsilon > 0$, then (10.3) holds for any refinement of P, with the same value of ε.*

(ii) *If (10.3) holds for a partition $P = \{x_0, x_1, \ldots, x_n\}$ and if s_i and t_i are points belonging to $[x_{i-1}, x_i]$, then*

$$\sum_{i=1}^{n} |f(s_i) - f(t_i)| \, \Delta x_i < \varepsilon.$$

(iii) *If f is R-integrable on $[a, b]$ and the assumptions of (ii) hold, then*

$$\left| \sum_{i=1}^{n} f(t_i) \Delta x_i - \int_a^b f \, dx \right| < \varepsilon.$$

Proof Part (i) follows immediately from Theorem 10.1. Let us prove part (ii). Of course $f(s_i)$ and $f(t_i)$ lie in $[m_i, M_i]$, so that $|f(s_i) - f(t_i)| \leq M_i - m_i$. This yields

$$\sum_{i=1}^{n} |f(s_i) - f(t_i)| \Delta x_i \leq U(f, P) - L(f, P) < \varepsilon.$$

Part (iii) follows from the relations

$$L(f, P) \leq \sum_{i=1}^{n} f(t_i) \Delta x_i \leq U(f, P)$$

$$L(f, P) \leq \int_a^b f \, dx \leq U(f, P).$$

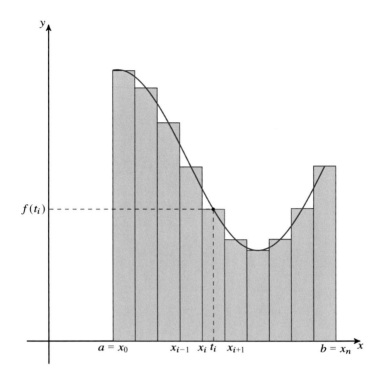

Fig. 10.2 Intuition of Theorem 10.3

□

Remark 10.3 Part (iii) actually says that the Riemann integral can be approximated by a Riemann sum as soon as it exists, see Fig. 10.2.

10.2 Integrable Functions as Elements of a Vector Space

Functions can be added pointwise, and pointwise multiplied by constants. This fact induces a vector space structure to the class of R-integrable functions, as the next result states.

Theorem 10.4

(a) *If f_1 and f_2 are R-integrable on $[a, b]$ and if c is a real number, then $f_1 + f_2$ and cf_1 are R-integrable; moreover*

$$\int_a^b (f_1 + f_2)\, dx = \int_a^b f_1\, dx + \int_a^b f_2\, dx,$$

$$\int_a^b cf_1\, dx = c \int_a^b f_1\, dx.$$

(b) *If $f_1 \leq f_2$ on $[a, b]$, then $\int_a^b f_1\, dx \leq \int_a^b f_2\, dx$.*

(c) *If f is R-integrable on $[a, b]$ and if $a < c < b$, then f is R-integrable on $[a, c]$ and on $[c, b]$, and $\int_a^b f\, dx = \int_a^c f\, dx + \int_c^b f\, dx$.*

Proof For $f = f_1 + f_2$ and any partition P of $[a, b]$, we notice that

$$L(f_1, P) + L(f_2, P) \leq L(f, P) \leq U(f, P) \leq U(f_1, P) + U(f_2, P).$$

Let $\varepsilon > 0$, and choose partitions P_1, P_2 so that

$$U(f_1, P_1) - L(f_2, P_2) < \varepsilon,$$

and

$$U(f_2, P_2) - L(f_2, P_2) < \varepsilon.$$

Call P the common refinement of P_1 and P_2. Then $U(f, P) - L(f, P) < 2\varepsilon$. Furthermore

$$U(f_j, P) < \int_a^b f_j\, dx + \varepsilon, \quad j = 1, 2$$

Hence

$$\int_a^b f\, dx \leq U(f, P) < \int_a^b f_1\, dx + \int_a^b f_2\, dx + 2\varepsilon,$$

and

$$\int_a^b f\, dx \leq \int_a^b f_1\, dx + \int_a^b f_2\, dx$$

follows from the arbitrariness of ε. Replacing f_1 and f_2 with $-f_1$ and $-f_2$, we see that $\int_a^b f\, dx \geq \int_a^b f_1\, dx + \int_a^b f_2\, dx$. The claim about cf is easier, and we leave it as an exercise. We have thus proved (a).

To prove (b) we notice that, for every partition P of $[a, b]$,

$$L(f_1, P) \leq L(f_2, P), \quad U(f_1, P) \leq U(f_2, P).$$

Hence $\int_a^b f_1 \, dx \leq U(f_2, P)$, and thus $\int_a^b f_1 \, dx \leq \int_a^b f_2 \, dx$.

To prove (c), write $f = g + h$ in such a way that $g = 0$ on $[c, b]$, $h = 0$ on $[a, c]$. It is clear that g and h are R-integrable on $[a, b]$, and $\int_a^b g \, dx = \int_a^c f \, dx$, while $\int_a^b h \, dx = \int_c^b f \, dx$. The conclusion follows from (a).   □

Exercise 10.3 Suppose that $f(x) > 0$ for each $x \in [a, b]$. If f is integrable on $[a, b]$, prove that $\int_a^b f \, dx > 0$. *Hint:* recall that $L(f, P) \leq \int_a^b f \, dx \leq U(f, P)$ for each partition P of $[a, b]$. What is the sign of $L(f, P)$?

10.3 Classes of Integrable Functions

Example 10.1 Dirichlet's function is defined on $[0, 1]$ by

$$h(x) = \begin{cases} 0 & \text{if } x \in [0, 1] \cap \mathbb{Q} \\ 1 & \text{otherwise.} \end{cases}$$

Let P be any partition of $[0, 1]$. Since $\mathbb{Q}$ is dense in $\mathbb{R}$, between two consecutive nodes x_{i-1} and x_i of the partition there are infinitely many rational points, and infinitely many irrational points. If follows that $U(f, P) = 1$ and $L(f, P) = 0$. This clearly shows that h is not R-integrable on $[0, 1]$.

The previous example suggests the following question: are there general properties that imply the R-integrability of functions?

Theorem 10.5 (Monotonic Functions Are R-integrable) *Let f be a bounded monotonic function on $[a, b]$. Then f is R-integrable on $[a, b]$.*

Proof For the sake of definiteness, we assume that f is increasing. Pick any $\varepsilon > 0$, and split $[a, b]$ into a number n of equal parts so that $(b - a)/n < \varepsilon/[f(b) - f(a)]$. This gives rise to a partition P of $[a, b]$. Monotonicity implies that $m_i = f(x_{i-1})$, $M_i = f(x_i)$, so that

$$U(f, P) - L(f, P) = \sum_{i=1}^n [f(x_i) - f(x_{i-1})]\Delta x_i = \frac{b - a}{n}[f(b) - f(a)] < \varepsilon.$$

The conclusion follows from Theorem 10.2.   □

Theorem 10.6 (Continuous Functions Are R-integrable) *If f is continuous on $[a, b]$, then f is R-integrable on $[a, b]$.*

Proof Pick any $\varepsilon > 0$. By Theorem 8.8 f is uniformly continuous. There exists $\delta > 0$ such that $|f(x) - f(t)| < \varepsilon/(b-1)$ whenever $|x - t| < \delta$. Let P a partition of $[a, b]$ such that $\Delta x_i < \delta$ for every i. Clearly $M_i - m_i < \varepsilon/(b-a)$ for every i, and thus

$$U(f, P) - L(f, P) = \sum_{i=1}^{n} [M_i - m_i] \Delta x_i \leq \frac{\varepsilon}{b-a}(b-a) = \varepsilon.$$

The conclusion follows from Theorem 10.2. □

Remark 10.4 The passage from pointwise continuity to uniform continuity should not be a surprise: integration is not a local property, and we cannot just sum up infinitely many local inequalities to get a global inequality.

Too many discontinuity points can prevent integrability, as we have seen in Example 10.1. However, discontinuous functions may well be integrable.

Theorem 10.7 *Suppose f is a bounded function on $[a, b]$ that possesses finitely many discontinuity points. Then f is R-integrable.*

Proof Pick any $\varepsilon > 0$, and write $M = \sup_{x \in [a,b]} |f(x)|$. By assumption, the set E of points at which f is not continuous is a finite set. Let us cover E by finitely many intervals $[u_j, v_j]$ such that $\sum_j |v_j - u_j| < \varepsilon$. Without loss of generality, we can choose $[u_j, v_i]$ such that each points of E is an interior point of some $[u_j, v_j]$.

By removing the open sets (u_j, v_j) from $[a, b]$ we get a compact set K, on which f is uniformly continuous: there exists $\delta > 0$ such that $|f(s) - f(t)| < \varepsilon$ whenever $s \in K$, $t \in K$, $|s - t| < \delta$. Now we construct a partition $P = \{x_0, \dots, x_n\}$ as follows: each u_j is in P. Each v_j is in P. No point of any segment (u_j, v_j) is in P. If x_{i-1} is not one of the u_j, then $\Delta x_i < \delta$.

Since $M_i - m_i \leq 2M$ for every j, and $M_i - m_i \leq \varepsilon$ unless x_{i-1} is one of the u_j, we see that

$$U(f, P) - L(f, P) \leq (b-a)\varepsilon + 2M\varepsilon.$$

The conclusion follows again from Theorem 10.2. □

Despite the technicalities of the proof, we strongly urge the reader to understand the general idea: we remove small neighborhoods of each point of discontinuity, obtaining a compact subset K. We know that f is integrable on K by continuity. A finite number of small intervals remains, but we have chosen the size of these intervals so small that their contribution to the Riemann sum is arbitrarily small.

Theorem 10.8 (Integrability of a Composite Function) *Suppose that f is R-integrable on $[a, b]$, $m \leq f \leq M$ on $[a, b]$, ϕ is continuous on $[m, M]$, and $h(x) = \phi(f(x))$ on $[a, b]$. Then h is R-integrable on $[a, b]$.*

Proof Let $\varepsilon > 0$. Since $[m, M]$ is a compact set, ϕ is uniformly continuous. Thus we get $\delta > 0$ such that $\delta < \varepsilon$ and $|\phi(s) - \phi(t)| < \varepsilon$ whenever $|s - t| \leq \delta$ in $[m, M]$.

Select a partition $P = \{x_0, \ldots, x_n\}$ of $[a, b]$ such that

$$U(f, P) - L(f, P) < \delta^2. \tag{10.4}$$

We use the symbols m_i and M_i as before, and we denote

$$m_i^* = \inf\{h(x) \mid x_{i-1} \leq x \leq x_i\},$$
$$M_i^* = \sup\{h(x) \mid x_{i-1} \leq x \leq x_i\}.$$

The idea is to split the partition in two parts: a part on which $M_i - m_i < \delta$, and the remaining part. Precisely, for $i = 0, \ldots, n$ we say that $i \in A$ if $M_i - m_i < \delta$, and $i \in B$ if $M_i - m_i \geq \delta$. For $i \in A$ we have $M_i^* - m_i^* \leq \varepsilon$. For $i \in B$, we have $M_i^* - m_i^* \leq 2K$, where $K = \sup_{x \in [a,b]} |\phi(x)|$. By (10.4) we have

$$\delta \sum_{i \in B} \Delta x_i \leq \sum_{i \in B} (M_i - m_i) \Delta x_i \leq \delta^2.$$

Hence

$$U(h, P) - L(h, P) = \sum_{i \in A} (M_i^* - m_i^*) \Delta x_i + \sum_{i \in B} (M_i^* - m_i^*) \Delta x_i$$
$$\leq (b - a)\varepsilon + 2K\delta < \varepsilon(b - a + 2K).$$

Since $\varepsilon > 0$ is arbitrary, the proof is complete.   □

Corollary 10.2 *If f and g are two R-integrable functions on $[a, b]$, then*

(a) the product fg is R-integrable;
(b) $|f|$ is R-integrable, and $\left| \int_a^b f \, dx \right| \leq \int_a^b |f| \, dx$.

Proof We take $\phi(t) = t^2$ in Theorem 10.8. This yields that f^2 is R-integrable, and the identity

$$4fg = (f + g)^2 - (f - g)^2$$

shows (a).

Taking $\phi(t) = |t|$ in Theorem 10.8 shows that $|f|$ is R-integrable. Let $c \in \{-1, +1\}$ be chosen so that $c \int_a^b f \, dx \geq 0$. Then

$$\left| \int_a^b f \, dx \right| = c \int_a^b f \, dx = \int_a^b cf \, dx \leq \int_a^b |f| \, dx.$$

This completes the proof of (b).   □

Exercise 10.4 Let $f : E \to \mathbb{R}$ be a bounded function on a set E. We define

$$f^+ = \max\{f, 0\}, \qquad f^- = \max\{-f, 0\},$$

$$M = \sup\{f(x) \mid x \in E\}$$
$$m = \inf\{f(x) \mid x \in E\}$$
$$M' = \sup\{|f(x)| \mid x \in E\}$$
$$m' = \inf\{|f(x)| \mid x \in E\}.$$

(i) Prove that $f = f^+ - f^-$, $|f| = f^+ + f^-$ on E. Deduce that $2f^- = |f| - f$.
(ii) Show that $2\sup\{f^-(x) \mid x \in E\} = M' - m$ and that $2\inf\{f^-(x) \mid x \in E\} = m' - M$. In particular $M - m \geq M' - m'$.
(iii) Use (ii) to prove (b) of Corollary 10.2.

Exercise 10.5 Suppose f and g are continuous functions on $[a, b]$ such that $\int_a^b f \, dx = \int_a^b g \, dx$. Prove that there exists a point $x \in [a, b]$ such that $f(x) = g(x)$.

Theorem 10.9 (Change of Variable) *Let φ be a strictly increasing functions from an interval $[A, B]$ onto an interval $[a, b]$. Let f be R-integrable on $[a, b]$, and suppose that φ' is R-integrable on $[A, B]$. Defining g on $[A, B]$ by $g(y) = f(\varphi(y))$, the function $g\varphi'$ is R-integrable on $[A, B]$, and $\int_A^B g\varphi' \, dy = \int_a^b f \, dx$.*

Proof Let $\varepsilon > 0$, and pick a partition $Q = \{y_0, \ldots, y_n\}$ of $[A, B]$ such that $U(\varphi', Q) - L(\varphi', Q) < \varepsilon$. The mean value theorem furnishes points $t_i \in [y_{i-1}, y_i]$ such that $\varphi(y_i) - \varphi(y_{i-1}) = \varphi'(t_i) \Delta y_i$ for every $i = 1, \ldots, n$.

We observe that any partition $Q = \{y_0, \ldots, y_n\}$ of $[A, B]$ corresponds to a partition $P = \{x_0, \ldots, x_n\}$ of $[a, b]$ such that $x_i = \varphi(y_i)$. Furthermore the values taken by f on $[x_{i-1}, x_i]$ are the same as the values taken by g on $[y_{i-1}, y_i]$. In particular

$$U(f, Q) = \sum_{i=1}^{n} (\sup\{g(y) \mid y_{i-1} \leq y \leq y_i\}) [\varphi(y_i) - \varphi(y_{i-1})] \qquad (10.5)$$

$$L(f, Q) = \sum_{i=1}^{n} (\inf\{g(y) \mid y_{i-1} \leq y \leq y_i\}) [\varphi(y_i) - \varphi(y_{i-1})]. \qquad (10.6)$$

Let $s_i \in [y_{i-1}, y_i]$, and observe that

$$\sum_{i=1}^{n} |\varphi'(s_i) - \varphi'(t_i)| \Delta y_i < \varepsilon$$

by Theorem 10.3. If $M = \sup |f|$, it follows from

$$\sum_{i=1}^{n} g(s_i)[\varphi(y_i) - \varphi(y_{i-1})] = \sum_{i=1}^{n} g(s_i)\varphi'(t_i)\Delta y_i$$

that

$$\left| \sum_{i=1}^{n} g(s_i)[\varphi(y_i) - \varphi(y_{i-1})] - \sum_{i=1}^{n} g(s_i)\varphi'(s_i)\Delta y_i \right| \leq M\varepsilon.$$

In particular we get

$$\sum_{i=1}^{n} g(s_i)[\varphi(y_i) - \varphi(y_{i-1})] \leq U(g\varphi', P) + M\varepsilon$$

for all choices of $s_i \in [y_{i-1}, y_i]$. In a similar way we also see that

$$U(g\varphi', P) \leq \sum_{i=1}^{n} g(s_i)[\varphi(y_i) - \varphi(y_{i-1})] + M\varepsilon$$

for all choices of $s_i \in [y_{i-1}, y_i]$. Comparing with (10.5) we conclude that

$$\left| U(f, Q) - U(g\varphi', P) \right| \leq M\varepsilon.$$

Analogously we see that $\left| L(f, Q) - L(g\varphi', P) \right| \leq M\varepsilon$, and the conclusion follows.
□

10.4   Antiderivatives and the Fundamental Theorem

How do we actually compute Riemann integrals? Despite all the definitions and all the results we have proved, there is just one universal approach: we need to find an *antiderivative*.

Definition 10.5 A function F is an antiderivative of a function f on the interval $[a, b]$, if F is differentiable on $[a, b]$, and $F' = f$ at every point.

It is customary to collect under the symbol $\int f$ (or $\int f(x)\,dx$) the set of all antiderivatives of f (on an interval that is not specified in the notation). The symbol $D^{-1}f$ would probably be a better choice, but it is not customary in the literature.

Remark 10.5 Although legitimate, we do not define antiderivatives of a function on more general sets like the union of disjoint intervals. The main reason is that the description of the set of all antiderivatives becomes less explicit, and could induce mistakes in applications.

Theorem 10.10 *Two antiderivatives of the same function differ by a constant on any interval.*

Proof Let F_1 and F_2 be two antiderivatives of a function f on the interval $[a, b]$. Since $(F_1 - F_2)' = F_1' - F_2' = f - f = 0$ on $[a, b]$, the function $F_1 - F_2$ is constant on $[a, b]$. The conclusion follows. □

Example 10.2 For any choice of the real numbers c_1 and c_2, the function

$$f(x) = \begin{cases} c_1 & \text{if } 0 \le x \le 1 \\ c_2 & \text{if } 2 \le x \le 3 \end{cases}$$

is an antiderivative of the zero function on $[0, 1] \cup [2, 3]$. This shows that Theorem 10.10 does not hold on domains different than intervals.

Theorem 10.11 (Existence of Antiderivatives) *Let f be R-integrable on $[a, b]$. For $a \le x \le b$ we define*

$$F(x) = \int_a^x f(t)\, dt.$$

The function F is uniformly continuous on $[a, b]$. If f is continuous at a point $x_0 \in [a, b]$, then F is also differentiable at x_0, and there results $F'(x_0) = f(x_0)$.

Proof As an integrable function, f must be bounded. So we can put $M = \sup |f|$. If $a \le x < y \le b$, then

$$|F(x) - F(y)| \le \left| \int_x^y f(t)\, dt \right| \le M(y - x).$$

Hence for every $\varepsilon > 0$ we see that $|y - x| < \varepsilon/M$ implies $|F(x) - F(y)| < \varepsilon$.

Now suppose that f is continuous at x_0. Given $\varepsilon > 0$ there exists $\delta > 0$ such that $|x - x_0| < \delta$ implies $|f(x) - f(x_0)| < \varepsilon$. As a consequence, if $x_0 - \delta < s \le x_0 \le t < x_0 + \delta$ and $a \le s < t \le b$, then

$$\left| \frac{F(t) - F(s)}{t - s} - f(x_0) \right| = \left| \frac{1}{t - s} \int_s^t [f(u) - f(x_0)]\, du \right| < \varepsilon.$$

It follows that $F'(x_0) = f(x_0)$. □

Theorem 10.12 (The Fundamental Theorem of Calculus) *If f is R-integrable on $[a, b]$, and if there exists a differentiable function F on $[a, b]$ such that $F' = f$,* *then*

$$\int_a^b f \, dx = F(b) - F(a).$$

Proof Fix any $\varepsilon > 0$. A partition $P = \{x_0, \dots, x_n\}$ of $[a, b]$ exists such that $U(f, P) - L(f, P) < \varepsilon$. The mean value theorem yields points $t_i \in [x_{i-1}, x_i]$ such that $F(x_i) - F(x_{i-1}) = f(t_i)\Delta x_i$. It follows that $\sum_{i=1}^n f(t_i)\Delta x_i = F(b) - F(a)$. Theorem 10.3 now ensures that

$$\left| F(b) - F(a) - \int_a^b f \, dx \right| < \varepsilon.$$

The arbitrariness of $\varepsilon > 0$ implies that $F(b) - F(a) = \int_a^b f \, dx$.   □

Example 10.3 Following Archimedes, we can easily claim that $\int_{-1}^1 x^2 \, dx = 2/3$, since $F(x) = \frac{1}{3}x^3$ is an antiderivative of $f(x) = x^2$ on $[-1, 1]$.

Theorem 10.13 (Integration by Parts) *Suppose that F and G are differentiable functions on $[a, b]$, and $F' = f$, $G' = g$ are R-integrable. Then*

$$\int_a^b F(x)g(x) \, dx = F(b)G(b) - F(a)G(a) - \int_a^b f(x)G(x) \, dx.$$

Proof Let $H = FG$, so that Theorem 10.12 yields $\int_a^b H'(x) \, dx = H(b) - H(a)$. But $H' = fG + Fg$, and we conclude.   □

Remark 10.6 It is customary to present integration by parts in a more symmetric manner:

$$\int_a^b uv' \, dx = u(b)v(b) - u(a)v(a) - \int_a^b u'v \, dx.$$

Exercise 10.6

(a) Choose $u(x) = \arctan x$ and $v(x) = x^2/2$ to compute the integral $\int_0^1 x \arctan x \, dx$.
(b) Choose now $u(x) = \arctan x$ and $v(x) = (x^2 + 1)/2$ and repeat the computations.

Remark 10.7 Theorem 10.12 provides an easier proof of Theorem 10.9. Indeed one observes that, if $F' = f$, then $(F \circ \varphi)' = (f \circ \varphi)\varphi'$. Hence

$$\int_A^B f(\varphi(y))\varphi'(y) \, dy = F(\varphi(B)) - F(\varphi(A)) = F(b) - F(a) = \int_a^b f \, dx.$$

However this approach requires that an antiderivative F of f exists, while Theorem 10.9 does not need this assumption.

10.5 Problems

10.1 Assume that f is a bounded function defined on $[a, b]$. If $x \mapsto f(x)^2$ is R-integrable on $[a, b]$, can we deduce that f is R-integrable?

10.2 Let $f : (a, b] \to \mathbb{R}$ be a function such that f is R-integrable on each interval $[c, b]$ with $a < c < b$. Define

$$\int_a^b f \, dx = \lim_{c \to a+} \int_c^b f \, dx,$$

provided that this limit exists as a real number.

1. If f is R-integrable on $[a, b]$, prove that this definition coincides with the definition of the Riemann integral of f.
2. Construct a function f such that the previous limit exists, but the same limit for $|f|$ does not exist.

This exercise suggests a reasonable definition for the *improper* Riemann integral of an unbounded function f on a bounded interval.

10.3 Let f be R-integrable on every interval $[a, b]$ with $b > a$. Define

$$\int_a^{+\infty} f \, dx = \lim_{b \to +\infty} \int_a^b f \, dx,$$

provided that this limit exists as a real number. Suppose that f is monotonically decreasing on $[1, +\infty)$ and that $f(x) \geq 0$ for each $x \geq 1$. Prove that the series $\sum_{n=1}^{+\infty} f(n)$ converges if and only if $\int_1^{+\infty} f \, dx$ exists as a finite number. This is sometimes called the *test of the improper integral* for numerical series.

10.4 Let p and q be two positive real numbers such that $\frac{1}{p} + \frac{1}{q} = 1$. Prove the following statements.

1. If $u \geq 0$ and $v \geq 0$, then

$$uv \leq \frac{u^p}{p} + \frac{v^q}{q}.$$

Furthermore the equality sign holds if and only if $u^p = v^q$.

2. If f and g are R-integrable, $f \geq 0$ and $g \geq 0$, then

$$\int_a^b fg\,\mathrm{d}x \leq \left(\int_a^b f^p\mathrm{d}x\right)^{\frac{1}{p}}\left(\int_a^b g^q\mathrm{d}x\right)^{\frac{1}{q}}.$$

3. Deduce that if f and g are R-integrable functions, then

$$\left|\int_a^b fg\,\mathrm{d}x\right| \leq \left(\int_a^b |f|^p\mathrm{d}x\right)^{\frac{1}{p}}\left(\int_a^b |g|^q\mathrm{d}x\right)^{\frac{1}{q}}.$$

This is called the *Hölder inequality* with conjugate exponents p and q. The case $p = q = 2$ is usually called the Cauchy-Schwarz inequality.

10.6   Comments

My personal position is that a young (and also an old) mathematician should have a very clear view of Riemann's integral for functions of one variable, while a rough idea of its extension to functions of several variables is more than enough. The construction I have proposed in this chapter is standard, and can be found on many textbooks.

Riemann integration theory is weak and easy in one dimension, but it becomes weak and tricky in two or more dimensions. For this reason I always recommend colleagues and students to replace Riemann's integral in $\mathbb{R}^n$ with the (concrete) Lebesgue integral as soon as possible.

Chapter 11
Elementary Functions

Abstract Most of us make use of *elementary* functions in a formal way. We graph exponentials, logarithms, sines, cosines, we differentiate and integrate them. But Calculus does not teach us an acceptable definition of these functions. We accept their existence, and we keep using their properties. In this chapter we offer a more advanced description of the most important functions, and show that their definition is indeed far from being *elementary*.

11.1 Sequences and Series of Functions

A particular type of sequence is of fundamental importance for defining the elementary functions in a rigorous way: sequences (and series) of *functions*. Although these are nothing else than sequences in a *set* of functions, we follow here a classical approach which does not lean on General Topology.

Suppose that E is a set and that for every positive integer n we have a function f_n defined on E. We can say that $\{f_n\}_n$ is a sequence of functions on E.[1] When we speak of sequences, sooner or later we speak of limits.

Definition 11.1 Let $\{f_n\}_n$ be a sequence of functions on a set[2] E into $\mathbb{R}$. We say that this sequence converges pointwise to a function f, if the numerical sequence $\{f_n(x)\}_n$ converges for every $x \in E$. The function $f : E \to \mathbb{R}$ defined by

$$f(x) = \lim_{n \to +\infty} f_n(x)$$

is the (pointwise) limit of the sequence $\{f_n\}_n$.

---

[1] It should be remarked that E does not depend on the index n. From a theoretical view point we could consider sequences of functions $f_n : E_n \to \mathbb{R}$ in which every term is defined on a set E_n. For our purposes such a generality can be troublesome.

[2] The nature of E is not particularly relevant. In many concrete cases, E is a subset of $\mathbb{R}$ of $\mathbb{C}$.

In a similar way—and indeed in an equivalent way—we can define *series* of functions. Indeed we can consider the numerical series $\sum_n f_n(x)$ for every $x \in E$: if this series converges, the (pointwise) limit of the series $\sum_n f_n$ is $f(x) = \sum_n f_n(x)$.

Example 11.1 For $m = 1, 2, 3, \ldots$ we define

$$f_m(x) = \lim_{n \to +\infty} (\cos(m!\pi x))^{2n}.$$

When $m!x \in \mathbb{Z}$, we have $f_m(x) = 1$. Otherwise we have $f_m(x) = 0$. Let $f(x) = \lim_{m \to +\infty} f_m(x)$.

If x is irrational, then $f_m(x) = 0$ for every m, so $f(x) = 0$. For rational values of $x = p/q$, we see that $m!x$ is an integer for every $m \geq q$, and therefore $f(x) = 1$. To summarize,

$$f(x) = \begin{cases} 0 & \text{if } x \text{ is irrational} \\ 1 & \text{if } x \text{ is rational.} \end{cases}$$

This shows that the pointwise limit of very smooth functions may well be a very irregular function.

Exercise 11.1 Consider

$$f_n(x) = \frac{x^2}{(1 + x^2)^n}$$

for $x \in \mathbb{R}$ and $n = 0, 1, 2, 3, \ldots$ Let $f(x) = \sum_{n=0}^{\infty} f_n(x)$. Show that f is defined for every real x, and that

$$f(x) = \begin{cases} 0 & \text{if } x = 0 \\ 1 + x^2 & \text{if } x \neq 0. \end{cases}$$

Deduce that the pointwise limit of a sequence of continuous functions is in general a discontinuous function.

Example 11.2 Let $f_n : [0, 1] \to \mathbb{R}$ be defined by $f_n(x) = n^2 x(1 - x^2)^n$. Since $f_n(0) = 0$, we have that

$$f(0) = \lim_{n \to +\infty} f_n(0) = 0.$$

For $0 < x \leq 1$, we trivially have $\lim_{n \to +\infty} n^2 x (1 - x^2)^n = 0$. By the Fundamental Theorem of Calculus,

$$\int_0^1 x(1 - x^2)^n \, dx = \frac{1}{2n + 2},$$

so that

$$\lim_{n\to+\infty} \int_0^1 f_n(x)\,dx = \lim_{n\to\infty} \frac{n^2}{2n+2} = +\infty.$$

In particular $0 = \int_0^1 f(x)\,dx \neq \lim_{n\to+\infty} \int_0^1 f_n(x)\,dx$.

Exercise 11.2 Consider instead $f_n(x) = nx(1 - x^2)^n$ for $x \in [0, 1]$. Show that $\lim_{n\to+\infty} \int_0^1 f_n(x)\,dx = 1/2$, while $f(x) = \lim_{n\to+\infty} f_n(x) = 0$ for every $x \in [0, 1]$.

To summarize: the pointwise limit of a sequence of function does not preserve continuity, differentiability, integrability. The natural question is whether we can replace our pointwise convergence with another convergence which preserves these properties.

11.2 Uniform Convergence

Definition 11.2 A sequence $\{f_n\}_n$ of functions defined on a set E converges uniformly to a limit f on E, if and only if for every $\varepsilon > 0$ there exists a positive integer N such that $x \in E$ and $n \geq N$ imply

$$|f_n(x) - f(x)| < \varepsilon.$$

When dealing with series of functions, we say that $\sum_n f_n$ converges uniformly on E if the sequence $\{s_n\}_n$ of partial sums defined by $s_n(x) = \sum_{j=1}^{n} f_j(x)$ converges uniformly (to some limit).

Remark 11.1 It is easy to check that $\{f_n\}_n$ converges uniformly to f on E if and only if

$$\lim_{n\to+\infty} \sup\{|f_n(x) - f(x)| \mid x \in E\} = 0.$$

The quantity $\sup\{|f_n(x) - f(x)| \mid x \in E\}$ is often denoted by $\|f_n - f\|_{\infty, E}$.

Theorem 11.1 (Cauchy Criterion for Uniform Convergence) *A sequence $\{f_n\}_n$ of functions defined on a set E converges uniformly if and only if the Cauchy condition holds: for every $\varepsilon > 0$ there exists a positive integer N such that $m \geq N$, $n \geq N$, $x \in E$ imply*

$$|f_n(x) - f_m(x)| < \varepsilon.$$

Proof Suppose that $\{f_n\}_n$ converges uniformly on E to a limit f, and let $\varepsilon > 0$ be fixed. By Definition 11.2 there exists a positive integer N such that $n \geq N$ and $x \in E$ imply $|f_n(x) - f(x)| < \varepsilon/2$. Thus

$$|f_n(x) - f_m(x)| \leq |f_n(x) - f(x)| + |f_m(x) - f(x)| < \varepsilon$$

for every $n \geq N, m \geq N, x \in E$.

On the contrary, suppose that the Cauchy condition holds. For every $x \in E$, the numerical sequence $\{f_n(x)\}_n$ is then a Cauchy sequence in $\mathbb{R}$, so it converges to some limit that we call $f(x)$. We need to prove that this convergence is uniform on E. Let $\varepsilon > 0$ be given, and choose a positive integer N such that $|f_n(x) - f_m(x)| < \varepsilon$ for every $m \geq N, n \geq N, x \in E$. Letting $m \to +\infty$, since $f_m(x) \to f(x)$, we deduce that

$$|f_n(x) - f(x)| \leq \varepsilon$$

for every $n \geq N$ and $x \in E$. Hence $\{f_n\}_n$ converges uniformly on E to the function f. □

The Cauchy condition immediately implies a useful test for the uniform convergence of series of functions.

Theorem 11.2 (Weierstrass' M-Test for Series) *Suppose that $\{f_n\}_n$ is a sequence of functions on a set E, and suppose that $|f_n(x)| \leq M_n$ for every $x \in E$ and every $n \in \mathbb{N}$. If $\sum_n M_n$ converges, then $\sum_n f_n$ converges uniformly on E.*

Proof For any $\varepsilon > 0$, let N be an integer such that $n \geq N, m \geq N$ and $m \geq n$ imply $\sum_{j=n}^m M_j < \varepsilon$. For every $x \in E$ we see that

$$\left| \sum_{j=n}^m f_j(x) \right| \leq \sum_{j=n}^m M_j \leq \varepsilon,$$

and the conclusion follows from Theorem 11.1. □

We will now present a few statements which relate uniform convergence with limits, derivatives and integrals.

Theorem 11.3 (Uniform Convergence and Continuity) *Suppose that $f_n \to f$ uniformly on a set E in a metric space. Let x be an accumulation point of E, and suppose that*

$$\lim_{t \to x} f_n(t) = A_n$$

for n = 1, 2, 3, . . . Then $\{A_n\}_n$ converges, and

$$\lim_{t \to x} f_n(t) = \lim_{n \to +\infty} A_n.$$

Proof Fix $\varepsilon > 0$. By uniform convergence there exists N such that $n \geq N, m \geq N$, $t \in E$ imply

$$|f_n(t) - f_m(t)| < \varepsilon.$$

As $t \to x$, $|A_n - A_m| \leq \varepsilon$. Hence $\{A_n\}_n$ is a Cauchy sequence in $\mathbb{R}$, and it converges to a limit A. Now

$$|f(t) - A| \leq |f(t) - f_n(t)| + |f_n(t) - A_n| + |A_n - A|.$$

We choose $n_0 \geq N$ so that $|f(t) - f_{n_0}(t)| < \varepsilon/3$ for all $t \in E$, and $|A_{n_0} - A| < \varepsilon/3$. With this n_0 we choose a neighborhood V of x such that $t \in V \cap E$, $t \neq x$ imply $|f_{n_0}(t) - A_n| < \varepsilon/3$. It follows that $|f(t) - A| < \varepsilon$ for every $t \in V \cap E, t \neq x$. The proof is complete. $\qquad\square$

Corollary 11.1 *If $\{f_n\}_n$ is a sequence of continuous functions on E that converges uniformly to a limit f, then f is a continuous function.*

Theorem 11.4 (Uniform Convergence and Differentiation) *Suppose that $\{f_n\}_n$ is a sequence of functions, differentiable on $[a, b]$ and such that there exists a point $x_0 \in [a, b]$ such that $\{f_n(x_0)\}_n$ converges. If $\{f_n'\}_n$ converges uniformly on $[a, b]$, then $\{f_n\}_n$ converges uniformly on $[a, b]$ to a limit f, and*

$$f'(x) = \lim_{n \to +\infty} f_n'(x)$$

for every $x \in [a, b]$.

In other words, uniform convergence of the derivatives a pointwise convergence of the functions at *some* point x_0 imply uniform convergence of the functions.

Proof We fix $\varepsilon > 0$. By assumption there exists a positive integer N such that

$$|f_n(x_0) - f_m(x_0)| < \frac{\varepsilon}{2}$$

and

$$|f_n'(t) - f_m'(t)| < \frac{\varepsilon}{2(b - a)} \tag{11.1}$$

for every $n \geq N, m \geq N$. Let us apply the mean value theorem to $f_n - f_m$: Eq. (11.1) yields

$$|f_n(x) - f_m(x) - (f_n(t) - f_m(t))| \leq \frac{|x - t|\varepsilon}{2(b-a)} \leq \frac{\varepsilon}{2} \qquad (11.2)$$

for every $x \in [a, b], t \in [a, b], n \geq N, m \geq N$. Splitting

$$|f_n(x) - f_m(x)| \leq |f_n(x) - f_m(x) - f_n(x_0) + f_m(x_0)| + |f_n(x_0) - f_m(x_0)|$$

we deduce that $x \in [a, b], n \geq N, m \geq N$ imply

$$|f_n(x) - f_m(x)| \leq \varepsilon.$$

This proves that $\{f_n\}_n$ converges uniformly to a limit that we call f. We fix a point $x \in [a, b]$ and introduce the incremental ratios

$$\phi_n(t) = \frac{f_n(t) - f_n(x)}{t - x}, \qquad \phi(t) = \frac{f(t) - f(x)}{t - x}$$

for any $t \in [a, b] \setminus \{x_0\}$. Clearly $\lim_{t \to x} \phi_n(t) = f_n'(x)$. From (11.2) we see that $n \geq N, m \geq N$ imply

$$|\phi_n(t) - \phi_m(t)| \leq \frac{\varepsilon}{2(b-a)}.$$

Hence $\{\phi_n\}_n$ converges uniformly on $[a, b] \setminus \{x\}$. But $\{f_n\}_n$ converges to f, hence $\lim_{n \to +\infty} \phi_n(t) = \phi(t)$ uniformly on $[a, b] \setminus \{x\}$. We conclude from Theorem 11.3 that $\lim_{t \to x} \phi(t) = \lim_{n \to +\infty} f_n'(x)$. The proof is complete.   □

Theorem 11.5 (Uniform Convergence and Integration) *Suppose that each f_n is R-integrable on $[a, b]$, and suppose that $f_n \to f$ uniformly on $[a, b]$. Then f is R-integrable on $[a, b]$ and $\int_a^b f \, dx = \lim_{n \to +\infty} \int_a^b f_n \, dx$.*

Proof Let

$$\varepsilon_n = \sup\{|f_n(x) - f(x)| \mid x \in [a, b]\}.$$

Hence $f_n - \varepsilon_n \leq f \leq f_n + \varepsilon_n$, and this implies

$$\int_a^b (f_n - \varepsilon_n) \, dx \leq \underline{\int_a^b} f \, dx \leq \overline{\int_a^b} f \, dx \leq \int_a^b (f_n + \varepsilon_n) \, dx.$$

Hence

$$0 \le \overline{\int_a^b} f \, dx - \underline{\int_a^b} f \, dx \le 2\varepsilon_n (b-a).$$

Since $\varepsilon_n \to 0$ as $n \to +\infty$, we conclude that $\underline{\int_a^b} f \, dx = \overline{\int_a^b} f \, dx$, and f is R-integrable. As before,

$$\left| \int_a^b f \, dx - \int_a^b f_n \, dx \right| = \left| \int_a^b (f - f_n) \, dx \right| \le \varepsilon_n (b-a),$$

and it follows that $\int_a^b f_n \, dx \to \int_a^b f \, dx$ as $n \to +\infty$.   □

Remark 11.2 The rigidity of the Riemann integral under passage to the limit is one of the reasons why it has been superseded by more flexible integrals. We will meet Lebesgue's generalization in Chap. 15.

11.3   The Exponential Function

It has been said that the exponential function is the most important function in Mathematical Analysis. We propose a definition which entails a lot of useful properties.

Definition 11.3 For each $z \in \mathbb{C}$, we define

$$\exp z = \sum_{n=0}^{\infty} \frac{z^n}{n!}.$$

By the Ratio Test, the series converges absolutely. The function

$$\exp \colon \mathbb{C} \to \mathbb{C}$$

is the exponential function.[3]

Proposition 11.1

1. For every $z \in \mathbb{C}$, $w \in \mathbb{C}$, there results $\exp(z+w) = \exp z \cdot \exp w$.
2. $\exp 0 = 1$ *and* $\exp 1 = e$.
3. $\exp z \ne 0$ *for every* $z \in \mathbb{C}$.

---

[3] We refrain from writing e^z instead of $\exp z$. This is only a pedagogical choice, since we want to prevent the reader from believing that all properties of this function are trivial since we are deling with an ordinary power.

4. $\exp(-z) = 1/\exp z$ *for every $z \in \mathbb{C}$.*
5. exp *is a continuous function.*

Proof We form the Cauchy product according to Definition 6.3:

$$
\exp z \cdot \exp w = \sum_{n=0}^{\infty} \sum_{k=0}^{n} \frac{z^k}{k!} \frac{w^{n-k}}{(n-k)!}
$$

$$
= \sum_{n=0}^{\infty} \frac{1}{n!} \sum_{k=0}^{n} \frac{n!}{k!(n-k)!} z^k w^{n-k}
$$

$$
= \sum_{n=0}^{\infty} \frac{1}{n!} (z+w)^n = \exp(z+w).
$$

This proves 1. Part 2 is clear from the definitions. Using 1 and 2 we see that $\exp(-z)\exp z = 1$ for every $z \in \mathbb{C}$, and both 3 and 4 follow. To prove 5, we fix a point $z \in \mathbb{C}$ and $\varepsilon > 0$. Let

$$
\delta = \min\left\{1, \frac{\varepsilon}{2|\exp z|}\right\}.
$$

Hence $h \in \mathbb{C}$ and $|h| < \delta$ imply

$$
|\exp(z+h) - \exp z| = |\exp z| \cdot |\exp h - 1|
$$

$$
\leq \frac{\varepsilon}{2\delta} \left| \sum_{n=1}^{\infty} \frac{h^n}{n!} \right| \leq \frac{\varepsilon}{2\delta} \sum_{n=1}^{\infty} \frac{|h|^n}{n!}
$$

$$
< \frac{\varepsilon}{2} \sum_{n=1}^{\infty} \frac{1}{n!} = \frac{\varepsilon}{2}(e-1) < \varepsilon.
$$

The proof of 5 is complete.   □

Theorem 11.6 *There results*

1. $\exp x > 0$ *for every $x \in \mathbb{R}$.*
2. exp *is strictly increasing on $\mathbb{R}$.*
3. $\lim_{x \to +\infty} \exp x = +\infty$.
4. $\lim_{x \to -\infty} \exp x = 0$.
5. $\exp(\mathbb{R}) = (0, +\infty)$.
6. $\lim_{x \to +\infty} x^{-n} \exp x = +\infty$ *for every integer $n \geq 0$ and every $x \in \mathbb{R}$.*

Proof We first notice that $\exp(\mathbb{R}) \subset \mathbb{R}$. Since $x > 0$ implies $\exp x > 1 + x$, 1 follows for $x \geq 0$. If $x < 0$, $\exp(-x) = 1/\exp x$, and 1 follows also in this case. To prove 2, we fix $x \in \mathbb{R}$ and $h > 0$. Then $\exp(x+h) = \exp x \cdot \exp h > \exp x$ since $\exp h > 1$. Similarly, $\lim_{x \to +\infty} \exp x \geq \lim_{x \to +\infty} (1+x) = +\infty$, and 3 follows.

Recalling again that

$$\lim_{x \to -\infty} \exp x = \lim_{x \to +\infty} \exp(-x) = \lim_{x \to +\infty} \frac{1}{\exp x},$$

we see that 4 follows from 3. The proof of 6 goes as follows: $x > 0$ implies

$$x^{-n} \exp x > x^{-n} \frac{x^{n+1}}{(n+1)!} = \frac{x}{(n+1)!}$$

To conclude, 5 follows from 1, 3, 4 and the fact that exp is injective on $\mathbb{R}$.   □

Theorem 11.7 *The restriction of* exp *to* $\mathbb{R}$ *is differentiable at every point, and there results* $\exp' = \exp$.

Proof Just differentiate the series of function that defines the exponential, and use Theorem 11.4.   □

One an exponential function has been introduced, the logarithm comes into play as its inverse.

Definition 11.4 The (real) logarithm is defined as

$$\log = \left(\exp_{|\mathbb{R}} \right)^{-1}.$$

As a function, $\log \colon (0, +\infty) \to \mathbb{R}$.

Theorem 11.8 *There results*

1. $\log((0, +\infty)) = \mathbb{R}$.
2. $\log$ *is strictly increasing on* $(0, +\infty)$.
3. $\log$ *is continuous on* $(0, +\infty)$.
4. $\lim_{x \to +\infty} \log x = +\infty$.
5. $\lim_{x \to 0+} \log x = -\infty$.
6. $\log 1 = 0$ *and* $\log e = 1$.
7. $\log(ab) = \log a + \log b$ *for every* $a > 0, b > 0$.
8. $\log(a^n) = n \log a$ *for every* $a > 0$ *and every* $n \in \mathbb{Z}$.
9. $\log(a^{1/n}) = (1/n) \log a$ *for every* $a > 0$ *and* $n \in \mathbb{N}$.
10. $\lim_{x \to +\infty} \frac{\log x}{\sqrt[n]{x}} = 0$ *for every* $n \in \mathbb{N}$.

Proof Properties from 1 to 6 follow from the analogous properties of the exponential. For $a > 0$ and $b > 0$ we have

$$\exp(\log a + \log b) = \exp(\log a) \cdot \exp \log b = ab,$$

and this proves 7. Properties 8 and 9 are left as a simple exercise about induction. To prove 10 we fix $\varepsilon > 0$ and $n \in \mathbb{N}$. By Theorem 11.6 we can choose $\alpha > 1$ such that $y > \alpha$ implies $y^{-n} \exp y > \varepsilon^{-n}$. By property 4 there exists $\beta > 1$ such that

$x > \beta$ implies $\log x > \alpha$. Therefore $x > \beta$ implies

$$(\log x)^{-n} x > \varepsilon^{-n},$$

or

$$0 < \frac{\log x}{\sqrt[n]{x}} < \varepsilon.$$

□

Theorem 11.9 *The function* log *is differentiable at every point of* $(0, +\infty)$, *and there results* $(\log)'(y) = 1/y$ *for every* $y > 0$.

Proof Exercise on the derivative of the inverse function!   □

Remark 11.3 Another approach is possible, which avoids the use of series. The first idea is to define a function log: $(0, +\infty) \to \mathbb{R}$ by

$$\log x = \int_1^x \frac{dt}{t}.$$

Here we are using the convention $\int_a^b = -\int_b^a$. The main properties of the real logarithm follows at once, in particular the fact that log has a continuous inverse. We call the inverse the real exponential function.

11.4   Sine and Cosine

Definition 11.5 The functions sin and cos are defined on $\mathbb{C}$ by

$$\sin z = \sum_{n=0}^{\infty} (-1)^n \frac{z^{2n+1}}{(2n+1)!}$$

$$\cos z = \sum_{n=0}^{\infty} (-1)^n \frac{z^{2n}}{(2n)!}$$

for every $z \in \mathbb{C}$. The convention $0^0 = 1$ is used in these formulas.

Theorem 11.10 (Euler) *For every* $z \in \mathbb{C}$ *we have*

1. $\exp(iz) = \cos z + i \sin z$.
2. $\exp(-iz) = \cos z - i \sin z$.
3. $\sin z = \frac{\exp(iz)-\exp(-iz)}{2i}$.
4. $\cos z = \frac{\exp(iz)+\exp(-iz)}{2}$.

Proof Observe that $(-1)^n z^{2n} = (iz)^{2n}$ and $i(-1)^n z^{2n+1} = (iz)^{2n+1}$. Then 1 follows from the definition of sin and cos. Property 2 follows by replacing z by $-z$ in 1. If we solve the system

$$\begin{cases} \exp(iz) = \cos z + i \sin z \\ \exp(-iz) = \cos z - i \sin z \end{cases}$$

with respect to $\sin z$ and $\cos z$, we find 3 and 4. $\square$

The next result may also be taken as a *definition* of π.

Theorem 11.11 *There exists one and only one real number π such that*

(i) $\pi > 0$;
(ii) $\cos(\pi/2) = 0$;
(iii) $0 < x < \pi/2$ *implies* $\cos x > 0$.

Furthermore, $\pi < 4$.

Proof If π and π' are different numbers which satisfy (i)–(iii) and $\pi < \pi'$, then $\cos(\pi/2) > 0$. This contradiction shows the uniqueness of π.

To prove the existence, we reason as follows. For every $n > 1$ we have

$$\frac{2^{2n}}{(2n)!} > \frac{2^{2n+2}}{(2n+2)!},$$

hence

$$\cos 2 = 1 - \frac{2^2}{2!} + \frac{2^4}{4!} - \frac{2^6}{6!} + \cdots$$

$$= -1 + \frac{16}{24} - \cdots < -1 + \frac{2}{3} < 0.$$

Since $x \mapsto \cos x$ is a continuous real-valued function on $[0, 2]$ and $\cos 0 = 1 > 0$, it follows that the set

$$A = \{x \in [0, 2] \mid \cos x = 0\}$$

is non-empty. We define $\pi/2 = \inf A$. As an accumulation point of the closed set A, we have $\pi/2 \in A$, and $\pi/2 > 0$ since $\cos 0 = 1$. As a consequence π satisfies (i) and (ii). Suppose (iii) was false. Then another application of the Intermediate Value Theorem would produce an element of A which would be smaller than $\pi/2$. Finally, since $\cos 2 < -1 + 2/3$, we find $\pi/2 < 2$, and the proof is complete. $\square$

From Euler's Theorem we can recover all the properties of the trigonometric functions. The interested reader can expand the details and prove the main results. It is however remarkable that it would be impossible to postpone the use of these functions until they can be rigorously defined. Calculus is built around elementary functions, but it does not provide sufficient tools to define them without any reference to geometric or intuitive facts.

Remark 11.4 The approach to trigonometric functions without power series is slightly involved. A possible approach is to begin with arctan: $\mathbb{R} \to (-\pi/2, \pi/2)$ in terms of a definite R-integral:

$$\arctan x = \int_0^x \frac{dt}{1+t^2}.$$

This function is invertible, so that tan is defined on $(-\pi/2, \pi/2)$ as $\arctan^{-1}$. Then sin and cos are recovered as suggested by the identity $\sin^2 + \cos^2 = 1$, i.e. $\tan^2 + 1 = 1/\cos^2$.

We end this section with a discussion about periodic functions.

Definition 11.6 A function $f : \mathbb{R} \to \mathbb{R}$ is periodic if and only if there exists a real number $T \neq 0$ such that $f(x + T) = f(x)$ for every $x \in \mathbb{R}$.

Exercise 11.3 Let $P = \{T \in \mathbb{R} \setminus \{0\} \mid T \text{ is a period of } f\}$ be the set of all periods of a given function f. Prove that the sum and the difference of two elements of P are elements of P. We can summarize this by saying that P is an additive subgroup of $\mathbb{R}$.

We now investigate additive subgroups of $\mathbb{R}$.

Theorem 11.12 *If H is an additive subgroup of $\mathbb{R}$, then either $\overline{H} = \mathbb{R}$ or there exists $T^* \neq 0$ such that $H = \{mT^* \mid m \in \mathbb{Z}\}$.[4]*

Proof Since H is an additive subgroup, $H \cap [0, +\infty) \neq \emptyset$. We define

$$\eta = \inf(H \cap [0, +\infty)).$$

Two cases are possible. If $\eta > 0$, then we pick $h \in H$ and $m \in \mathbb{Z}$ such that

$$m\eta \leq |h| < (m+1)\eta.$$

Since $|h| - m\eta \in H$ and $0 \leq |h| - m\eta < (m+1)\eta - m\eta = \eta$, the definition of η implies $|h| - m\eta = 0$, or $h = \pm m\eta$. Hence H consists of all integer multiples of η.

If $\eta = 0$, we must prove that H is dense in $\mathbb{R}$. Let $r \in \mathbb{R}$ and let $\varepsilon > 0$. Since $\eta = 0$, there exists $h_\varepsilon \in H \cap [0, \varepsilon]$. We may suppose $r \geq 0$, the case $r < 0$ being similar. By the Archimedean property, there exists $k \in \mathbb{N}$ such that $kh \leq r < (k+1)h$. Since $hk \in H$ and $0 \leq r - kh < (k+1)h - kh = h \leq \varepsilon$, we see that $|r - kh| \leq \varepsilon$, which shows that H is dense in $\mathbb{R}$. The proof is complete. □

---

[4] Algebraists say that H is a cyclic group.

Theorem 11.13 *Let* $f: \mathbb{R} \to \mathbb{R}$ *be a periodic continuous function. If* f *is non-constant, then there exists a smallest positive period of* f.

Proof Let P be the set of all periods of f. We are going to rule out the possibility that P is dense in $\mathbb{R}$. Indeed, if this were true, then f would be constant on the dense subset P, and thus f would be globally constant by continuity. Since this contradicts the assumption, we conclude that all periods are integer multiples of some period $T^* \neq 0$. Since $-T^*$ is also a period, we may assume that $T^* > 0$, and by Theorem 11.12 T^* is the smallest positive period of f. The proof is complete.   □

Important: Periodic Functions Without a Smallest Period

Many Calculus students believe that any periodic function possesses *the* period, i.e. a unique number like π for the sine or cosine. Although some textbooks restrict the definition of periodicity to functions which do have a smallest period, in this book we will always think of constant functions as periodic functions. We invite the reader to elaborate on the function

$$f(x) = \begin{cases} 0 & \text{if } x \text{ is rational} \\ 1 & \text{if } x \text{ is irrational.} \end{cases}$$

It is clear that no smallest positive period exists, although f is surely periodic in the sense of Definition 11.6.

11.5 Polynomial Approximation

Polynomials are the most elementary functions of mathematical analysis. They are built on arithmetic operations, and they turn out to be a flexible class of infinitely differentiable functions. Of course not all functions are polynomial.

Exercise 11.4 Prove rigorously that not all functions are polynomials. *Hint:* if P is a non-constant polynomial, either $\lim_{x \to \pm\infty} |P(x)| = +\infty$.

Nevertheless, polynomials do *approximate* continuous functions in a strong way.

Theorem 11.14 (Weierstrass Approximation Theorem) *If* $f: [a, b] \to \mathbb{R}$ *is a continuous function and* $\varepsilon > 0$, *there exists a polynomial* P *such that* $\sup_{x \in [a,b]} |f(x) - P(x)| < \varepsilon$.

Proof We will prove an equivalent statement: there exists a sequence $\{P_n\}_n$ of polynomials such that $P_n \to f$ uniformly on $[a, b]$. Considering an affine change of variable, we may assume that $[a, b] = [0, 1]$. Furthermore, we may also assume that $f(0) = f(1) = 0$. Indeed, the function f may be replaced by the function

$$x \mapsto f(x) - f(0) - x(f(1) - f(0)).$$

This function differs from f by a polynomial (of degree ≤ 1), so that a uniform approximation of this function by means of polynomials implies that f is approximated by polynomials.

Finally, for technical reasons, we define $f(x) = 0$ for each $x \in \mathbb{R} \setminus [0, 1]$. Hence f is defined on the whole real line. For each $n = 1, 2, 3, \ldots$ define

$$c_n = \frac{1}{\int_{-1}^{1}(1 - x^2)^n \, dx}$$

and

$$Q_n(x) = c_n(1 - x^2)^n.$$

Trivially, $\int_{-1}^{1} Q_n(x) \, dx = 1$ for each n. Furthermore,

$$\int_{-1}^{1}(1 - x^2)^n \, dx = 2 \int_{0}^{1}(1 - x^2)^n \, dx$$

$$\geq 2 \int_{0}^{1/\sqrt{n}}(1 - x^2)^n \, dx$$

$$\geq 2 \int_{0}^{1/\sqrt{n}}(1 - nx^2) \, dx$$

$$= \frac{4}{3\sqrt{n}} > \frac{1}{\sqrt{n}}.$$

As a consequence, $c_n < \sqrt{n}$ for each n. Now, fix any $\delta > 0$, and observe that $\delta \leq |x| \leq 1$ implies

$$Q_n(x) \leq \sqrt{n}(1 - \delta^2)^n.$$

The right-hand side converges to zero, hence Q_n converges to zero uniformly in the region $\delta \leq |x| \leq 1$, i.e. away from zero.

We introduce the sequence of functions

$$P_n(x) = \int_{-1}^{1} f(x + t) Q_n(t) \, dt,$$

defined for $0 \le x \le 1$. A change of variable shows that

$$P_n(x) = \int_0^1 f(t)Q_n(t-x)\,dt,$$

which is a polynomial function. We claim that $\{P_n\}_n$ is the approximating sequence of polynomials we are looking for. Indeed, given $\varepsilon > 0$ we choose $\delta > 0$ so that $|y - x| < \delta$ implies $|f(y) - f(x)| < \varepsilon/2$. Here we are exploiting the uniform continuity of f on $[0, 1]$. Call $M = \sup\{|f(x)| \mid 0 \le x \le 1\}$. Recalling that $Q_n \ge 0$ we see that for each $0 \le x \le 1$ we have

$$|P_n(x) - f(x)| = \left| \int_{-1}^1 (f(x+t) - f(x))\, Q_n(t)\, dt \right|$$

$$\le \int_{-1}^1 |f(x+t) - f(x)|\, Q_n(t)\, dt$$

$$\le 2M \int_{-1}^{-\delta} Q_n(t)\, dt + \frac{\varepsilon}{2} \int_{-\delta}^{\delta} Q_n(t)\, dt + 2M \int_{\delta}^1 Q_n(t)\, dt$$

$$\le 4M \sqrt{n}(1 - \delta^2)^n + \frac{\varepsilon}{2}$$

$$< \varepsilon$$

provided that n is big enough.   □

Example 11.3 Consider the function $f : [-1, 1] \to \mathbb{R}$ defined by $f(x) = |x|$. By the previous result, a sequence $\{\tilde{P}_n\}_n$ of polynomials exists which converges uniformly to f on $[-1, 1]$. Setting $P_n(x) = \tilde{P}_n(x) - \tilde{P}_n(0)$, we see that $\{P_n\}_n$ is a sequence of polynomials that converges to f on $[-1, 1]$ and such that $P_n(0) = 0$ for every n. Notice that f is not differentiable at $x = 0$: in some sense, the sequence $\{P_n\}$ is a smooth uniform approximation of f.

11.6   A Continuous Non-differentiable Function

Every student learns that a continuous function may fail to be differentiable at all points, and the simplest example is usually the absolute value $x \mapsto |x|$. Karl Weierstrass proved a much stronger result about a continuous function exists which is *nowhere* differentiable. Intuitively, such a function cannot be represented by a simple formula. In this section we propose a reasonable construction.

Theorem 11.15 *There exists a continuous function on the real line $\mathbb{R}$ which is nowhere differentiable .*

Proof Let us start with $\varphi(x) = |x|$, for each $x \in [-1, 1]$. Then we extend it by periodicity to $\mathbb{R}$, i.e. $\varphi(x + 2) = \varphi(x)$ for each $x \in \mathbb{R}$. It is clear that

$$|\varphi(s) - \varphi(t)| \le |s - t| \tag{11.3}$$

for each s, t in $\mathbb{R}$. The function $\varphi \colon \mathbb{R} \to \mathbb{R}$ is uniformly continuous.

Now we define

$$f(x) = \sum_{n=0}^{\infty} \left(\frac{3}{4}\right)^n \varphi\left(4^n x\right).$$

The M-test 11.2 shows that f is defined by a series that converges uniformly on $\mathbb{R}$, so that f is a continuous function. We claim that f is differentiable at *no* point of $\mathbb{R}$. To prove this claim, we pick any $x \in \mathbb{R}$ and any positive integer m. Let

$$\delta_m = \pm \frac{1}{2} \cdot 4^{-m},$$

where the sign is chosen so that the interval $[4^m x, 4^m (x + \delta_m)]$ contains no integer. Since $4^m |\delta_m| = 1/2$, this is indeed possible. Consider now the incremental ratio

$$\gamma_n = \frac{\varphi(4^n (x + \delta_m)) - \varphi(4^n x)}{\delta_m}.$$

If $n > m$, the number $4^n \delta_m$ is an even integer, so that $\gamma_n = 0$. If $0 \le n \le m$, it follows from (11.3) that $|\gamma_n| \le 4^n$. Recalling that $|\gamma_m| = 4^m$, we see that

$$\left| \frac{f(x + \delta_m) - f(x)}{\delta_m} \right| = \left| \sum_{n=0}^{\infty} \left(\frac{3}{4}\right)^n \gamma_n \right|$$

$$\ge 3^m - \sum_{n=0}^{m-1} 3^n$$

$$= \frac{1}{2} \left(3^m + 1\right).$$

Letting $m \to +\infty$ we get $\delta_m \to 0$. Hence f is not differentiable at x, and the proof is complete. $\square$

Remark 11.5 Such a function is a typical example of a *fractal* curve, whose graph is essentially impossible to sketch. Our function f is based on the function φ, which is already irregular at countably many points. However, Weierstrass constructed a more complicated example of the form

$$f(x) = \sum_{n=0}^{\infty} a^n \cos\left(b^n \pi x\right),$$

where $0 < a < 1$ and b is a positive odd integer such that

$$ab > 1 + \frac{3}{2}\pi.$$

The function f is then a *trigonometric* series, and each term of the infinite sum is a smooth function. Once more we see that a limit of regular functions may be a very irregular function.

11.7 Asymptotic Estimates for the Factorial Function

Asymptotic estimates are a major tool in several fields of mathematics. We want to present a couple of results which describe the behavior of the factorial $n!$ as n gets larger and larger. Since the factorial has been introduced as a discrete function, our road is not really straight.

Definition 11.7 (Double Factorial) We define inductively on $n \in \mathbb{N}$,

$$(-1)!! = 1, \quad 0!! = 1, \quad (n+1)!! = (n+1) \cdot (n-1)!!$$

Exercise 11.5 Prove that $n!!$ is the product of all odd numbers $m \le n$ when n is odd, and it is the product of all even numbers $m \le n$ when n is even.

Theorem 11.16 (Wallis Integrals) *If $n \in \mathbb{N}$, then*

$$\int_0^{\pi/2} (\sin x)^{2n+1} \, dx = \frac{(2n)!!}{(2n+1)!!} \tag{11.4}$$

$$\int_0^{\pi/2} (\sin x)^{2n} \, dx = \frac{\pi}{2} \frac{(2n-1)!!}{(2n)!!} \tag{11.5}$$

$$\int_0^{\pi/2} (\sin x)^{2n+1} \, dx \le \int_0^{\pi/2} (\sin x)^{2n} \, dx \le \int_0^{\pi/2} (\sin x)^{2n-1} \, dx. \tag{11.6}$$

Proof For every $x \in [0, \pi/2]$ we have $0 \le \sin x \le 1$. Hence, for any such x, the sequence $n \mapsto (\sin x)^n$ is decreasing. In particular (11.6) follows at once from the monotonicity properties of the Riemann integral. We prove (11.4) and (11.5) by induction on n. They clearly hold true when $n = 0$, and we integrate by parts as follows:

$$\int_0^{\pi/2} (\sin x)^m \, dx = \int_0^{\pi/2} (\sin x)^{m-2} (1 - \cos^2 x) \, dx$$

$$= \int_0^{\pi/2} (\sin x)^{m-2} \, dx - \int_0^{\pi/2} \cos x \cdot (\sin x)^{m-2} x \cos x \, dx$$

$$= \int_0^{\pi/2} (\sin x)^{m-2} \, dx - \frac{1}{m-1} \int_0^{\pi/2} (\sin x)^m \, dx,$$

deducing the identity

$$\int_0^{\pi/2} (\sin x)^m \, dx = \frac{m-1}{m} \int_0^{\pi/2} (\sin x)^{m-2} \, dx$$

for every $m \geq 2$. The induction step is now easy. $\square$

Theorem 11.17 (Wallis Formulas) *As $n \to +\infty$,*

$$\frac{(2n)!!}{(2n-1)!!} \sim \sqrt{n\pi} \tag{11.7}$$

$$\binom{2n}{n} \sim \frac{2^{2n}}{\sqrt{n\pi}} \tag{11.8}$$

$$\frac{2 \cdot 2 \cdot 4 \cdot 4 \cdot 6 \cdot 6 \cdot 8 \cdot 8 \cdots 2n \cdot 2n}{1 \cdot 3 \cdot 3 \cdot 5 \cdot 5 \cdot 7 \cdot 7 \cdots (2n-1) \cdot (2n-1)} = \frac{\pi}{2} + o(1). \tag{11.9}$$

Proof We set

$$q = \frac{\sqrt{n\pi}}{\frac{(2n)!!}{(2n-1)!!}}$$

and we see from (11.4) and (11.5) that

$$q = \frac{2n}{\sqrt{n\pi}} \int_0^{\pi/2} (\sin x)^{2n} \, dx \geq \frac{2n}{\sqrt{n\pi}} \int_0^{\pi/2} (\sin x)^{2n+1} \, dx$$

$$= \frac{2n}{\sqrt{n\pi}} \frac{(2n)!!}{(2n-1)!!} \frac{1}{2n+1} = \frac{1}{q} \frac{2n}{2n+1}$$

and

$$q = \frac{2n}{\sqrt{n\pi}} \int_0^{\pi/2} (\sin x)^{2n} \, dx \leq \frac{2n}{\sqrt{n\pi}} \int_0^{\pi/2} (\sin x)^{2n-1} \, dx$$

$$= \frac{2n}{\sqrt{n\pi}} \frac{(2n)!!}{(2n-1)!!} \frac{1}{2n} = \frac{1}{q}.$$

We have thus proved that

$$\frac{2n}{2n+1} \leq \left(\frac{\sqrt{n\pi}}{\frac{(2n)!!}{(2n-1)!!}} \right)^2 \leq 1$$

for every $n \geq 1$. Now (11.7) follows easily. Since

$$\binom{2n}{n} = \frac{(2n)!!}{(n!)^2} = \frac{2^{2n}(2n-1)!!}{(2n)!!} \sim \frac{2^{2n}}{\sqrt{n\pi}},$$

also (11.8) follows. To conclude, we observe that

$$\frac{2 \cdot 2 \cdots (2n) \cdot (2n)}{1 \cdot 3 \cdots (2n-1) \cdot (2n-1)} \sim \frac{(2n)!!(2n)!!}{(2n-1)!!(2n-1)!!(2n+1)} \sim \frac{n\pi}{2n+1} \sim \frac{\pi}{2}.$$

$\square$

Theorem 11.18 (Stirling) *As* $n \to +\infty$,

$$n! \sim n^n e^{-n} \sqrt{2n\pi}.$$

Proof We define the sequence

$$x_n = n! e^n n^{-n-1/2}.$$

A direct calculation shows that

$$\frac{x_{n+1}}{x_n} = e^{-\left(n+\frac{1}{2}\right) \log \frac{n+1}{n} + 1}.$$

By taking logarithms we see that

$$\log x_{n+1} - \log x_n = \log \frac{x_{n+1}}{x_n} = -\left(n + \frac{1}{2}\right) \log \frac{n+1}{n} + 1 \sim -\frac{1}{12n^2}$$

as $n \to +\infty$. As a consequence, the series $\sum_n \log x_{n+1} - \log x_n$ converges, and

$$k = \lim_{n \to +\infty} \log x_n$$

exists in $\mathbb{R}$. Thus $n! \sim n^n e^{-n} \sqrt{n} e^k$. Inserting this into (11.8) we see that

$$\frac{2^{2n}}{\sqrt{n\pi}} \sim \binom{2n}{n} = \frac{(2n)!}{(n!)^2} \sim \frac{(2n)^{2n} e^{-2n} \sqrt{2n} e^k}{n^{2n} e^{-2n} n e^{2k}} = \frac{2^{2n} \sqrt{2}}{\sqrt{n} e^k} = \frac{2^{2n}}{\sqrt{n\pi}} \frac{\sqrt{2\pi}}{e^k},$$

and necessarily $e^k = \sqrt{2\pi}$. The proof is complete.   □

11.8   Problems

11.1 Consider the series of functions

$$f(x) = \sum_{n=1}^{\infty} \frac{1}{1 + n^2 x}.$$

For what values of x does the series converge absolutely? In what intervals does the series converge uniformly? If the series converges, is f a continuous function?

11.2 For each $n = 1, 2, 3, \ldots$ let

$$f_n(x) = \begin{cases} 0 & \text{if } x < \frac{1}{n+1} \\ \sin^2 \frac{\pi}{x} & \text{if } \frac{1}{n+1} \leq x \leq \frac{1}{n} \\ 0 & \text{if } x > \frac{1}{n}. \end{cases}$$

Prove that the sequence $\{f_n\}_n$ converges pointwise to a continuous function, but the converges is not uniform.

11.3 For each real number x, let $\{x\} = x - [x]$, where $[x]$ denotes the integer part of x. Let

$$f(x) = \sum_{n=1}^{\infty} \frac{\{nx\}}{n^2}.$$

Find all discontinuity points of f, and prove that these points form a dense, countable subset of $\mathbb{R}$. Prove also that f is R-integrable on each bounded interval.

11.4 Let f be a continuous function on $[0, 1]$, and suppose that

$$\int_0^1 f(x)x^n \, dx = 0 \quad \text{for } n = 0, 1, 2, 3, \dots$$

Prove that $f(x) = 0$ for each $x \in [0, 1]$.

11.5 Suppose that a sequence $\{f_n\}_n$ converges pointwise to f on a compact set K, and suppose moreover that $f_n(x) \le f_{n+1}(x)$ for each $x \in K$ and each $n \in \mathbb{N}$.

1. By setting $g = f - f_n$, reduce to the case of a sequence of function which converges pointwise to zero in a decreasing way.
2. Assume that f and each f_n are continuous functions on K. Fix any $\varepsilon > 0$ and define for each n the set $K_n = \{x \in K \mid g_n(x) \ge \varepsilon\}$. Prove that

$$K_1 \supset K_2 \supset K_3 \supset \dots,$$

and conclude that $f_n \to f$ uniformly on K.

Part II
Second Half of the Journey

Chapter 12
Return to Set Theory

Abstract This chapter introduces the second part of the book. Why should we bother about set theory, again? And what is *axiomatic* set theory? In some sense, learning mathematics has a privileged direction: we take something for granted, and then we proceed. Mathematical analysts are usually satisfied with a good knowledge of *naïve* set theory, since this is all they need to do their job. Going backwards is another story. We are always worried by *primitive* knowledge, in the sense of something that we agree to know before we start out journey. So, what is a set? Why do textbooks begin with the definition of a function as a black box that turns an element (of a set) into another element (of another set)? And, after all: is this the only way to begin?

The roots of mathematics are close to philosophy, in the sense that a beginning must exist. If nothing exists, how can we exist? We only (!) have to choose where to start from.

It is a general agreement that set theory is indeed the first mathematical chapter in the big book of all Mathematics. The rigorous construction of sets, functions and all that has been a long and recent process. If we agree that mathematics should set *axioms* and deduce *theorems*, we should isolate the axioms of set theory.

Nowadays the most popular axiomatization of set theory is **ZF**, or Zermelo-Fraenkel. Another axiomatization is due to Bernays, Gödel and Von Neumann. It is not our purpose to discuss the pros and cons of each theory, since this is a deep aspect that soon involves mathematical logic. Since this is a book in mathematical analysis, we prefer to present an overview of two less known theory of sets. The first one is due to John Kelley, see [3]. The second one is actually a variation on Kelley's theme, due to J. D. Monk. These theories share the simple approach of considering only *classes*, an undefined category that contains *sets*. The idea of *elements* is not a different object: everything is a class. For the reader's convenience, a short account of **ZF** is also provided.

© The Author(s), under exclusive license to Springer Nature Switzerland AG 2022
S. Secchi, *A Circle-Line Study of Mathematical Analysis*,
La Matematica per il 3+2 141, https://doi.org/10.1007/978-3-031-19738-3_12

Morse-Kelley theory is suitable for mind shape of a mathematical analyst, because it provides a straightforward escape from one of the worst paradoxes of mathematics:

No matter what you suppose to know about sets, the set of all sets cannot be a set.

We will come back to this paradox later in the chapter.

Remark 12.1 From a more advanced viewpoint, Morse-Kelley axiomatization has been superseded by the NBG axiomatization of Von Neumann, Bernays and Gödel. The classification axiom scheme that we will introduce below is the most flexible feature of Morse-Kelley, although it is so general that experts of mathematical logic prefer to replace it with something weaker but easier from their perspective. This is however an issue that does not bother us.

12.1   Kelley's System of Axioms

We will see that the word "class" does not appear in any axiom of Kelley's theory: instead of assuming the existence of sets, Kelley assumes the existence of classes, and sets are just special classes.

The natural question now is: why don't we simply agree that "set = class"? Of course this is a matter of language, but we always keep in mind that the set of all sets is not a set.

Important: Variables

We stick here to Kelley's original habit of using variables in a broad sense, so that any object should be considered as a class unless otherwise stated. Another popular approach is to reserve lower case letters to sets, and upper case letters to classes.

Axiom of Extent For each x and each y, it is true that $x = y$ if and only if for each z, $z \in x$ if and only if $z \in y$.

Two classes are equal if and only if each element of each is a member of the other. The reader will notice that this is the usual definition of equal sets. And here is the rigorous definition of a set, at last! Sets are just elements of some class.

Definition 12.1 A class x is a set if and only if there exists a class y such that $x \in y$.

Remark 12.2 Maybe some reader will remember a common approach to naïve set theory: whenever we name a set, we must agree that it is a subset of some larger set. This is a naïve response to the paradox of the set of all sets: since we agree to work with subsets of a fixed large set, we will never have to deal with the set of all sets. This is a useful agreement, but it does not face the deep meaning of the paradox.

Now that we have sets, we want to describe the use of the classifier $\{\ldots \mid \ldots\}$. The are two blanks: the first blank for a variable, and a second blank for a formula.

> **Axiom of the Classifier** For each u and y, $u \in \{x \mid x \in y\}$ if and only if u is a set and $u \in y$.

This axiom is of fundamental importance: it states that whenever we write $u \in \ldots$, we immediately understand that u is a set. More generally, each statement of the form

$$u \in \{x \mid \ldots x \ldots\}$$

is considered to be an axiom with the meaning that u is a set and $\ldots u$. Here

$$\ldots x \ldots$$

is a formula and

$$\ldots u \ldots$$

is the formula which is obtained from it by replacing each occurrence of x with u. For example, $u \in \{x \mid x \in y$ and $z \in x\}$ if and only if u is a set, $u \in y$ and $z \in u$.

The axiom scheme that we have just introduced obviously corresponds to the familiar way of constructing sets by specifying properties that characterize its elements. However, we added the requirement "u is a set". This is surely unnatural, but again: $\{x \mid x$ is a set$\}$ would coincide with the set of all sets, and this is not a set. With our axiom scheme, $u \in \{x \mid x$ is a set$\}$ just means that u is a set, and no paradox arises.

Formulae We agree that:

(a) The result of replacing α and β by variables is, for each of the following, a formula: $\alpha = \beta$ and $\alpha \in \beta$.
(b) The result of replacing α and β by variables and A and B by formulae is, for each of the following, a formula:

    1. if A then B,
    2. A if and only if B,
    3. it is false that A,
    4. A and B,
    5. A or B,

6. for every α, A
7. for some α, A
8. $\beta \in \{\alpha \mid A\}$
9. $\{\alpha \mid A\} \in \beta$
10. $\{\alpha \mid A\} \in \{\beta \mid B\}$

(c) Formulae are constructed recursively, beginning with the primitive formulae of (a) and then proceeding via the constructions allowed by (b).

Classification Axiom Scheme An axiom results if in the following α and β are replaced by variables, A by a formula $\mathcal{A}$ and B by the formula obtained from $\mathcal{A}$ by replacing each occurrence of the variable that replaced α by the variable that replaced β:

For each β, $\beta \in \{\alpha \mid A\}$ if and only if β is a set and B.

Important: Braces

In old books a peculiar use of braces was common. For instance, one would write $M = \{m\}$ to mean the "typical" element of the class/set M is m. In modern mathematics and according to our use of the classifier $\{\ldots \mid \ldots\}$, $M = \{m\}$ means $\{x \mid x = m\}$, so that the class/set M contains exactly one element, i.e. m. Please avoid old-fashioned notation if you do not want to produce tragic results.

Luckily enough, our axioms already allow us to introduce new definitions and prove some theorems. Let us see some of them.

Definition 12.2 (Union of Classes) $x \cup y = \{z \mid z \in x \text{ or } z \in y\}$.

Definition 12.3 (Intersection of Classes) $x \cap y = \{z \mid z \in x \text{ and } z \in y\}$.

Theorem 12.1 *For each z, $z \in x \cup y$ if and only if $z \in x$ or $z \in y$. For each z, $z \in x \cap y$ if and only if $z \in x$ and $z \in y$.*

Proof It follows from the classification axiom that $z \in x \cup y$ if and only if z is a set, and $z \in x$ or $z \in y$ Recalling Definition 12.1, $z \in x$ or $z \in y$ and z is a set if and only if $z \in x$ or $z \in y$. By the same token one proves the second statement about the intersection. $\square$

Remark 12.3 Some reader may think that we are playing a strange game. This is more or less the case. We should always remember that mathematics is concerned with proving theorems from axioms. In this particular case, it would be much stranger if our theorems were unexpected ones! We are trying to define rigorously what we handle every day, this is the point.

Theorem 12.2 *For each x, $x \cup x = x$, and $x \cap x = x$.*

Proof Every element of x is a member of $x \cup x$, by definition of the union. Conversely, if $z \in x \cup x$, then z is a set and $z \in x$ or $z \in x$. Hence $z \in x$. The second statement is left as an exercise. □

Exercise 12.1 Prove the theorem $x \cap x = x$ by mimicking the previous proof.

We collect now three statements that follow directly from the properties of logical quantifiers "and", "or".

Theorem 12.3 *For each x, y and z,*

1. $x \cup y = y \cup x$
2. $x \cap y = y \cap x$
3. $(x \cup y) \cup z = x \cup (y \cup z)$
4. $(x \cap y) \cap z = x \cap (y \cap z)$
5. $x \cap (y \cup z) = (x \cap y) \cup (x \cap z)$
6. $x \cup (y \cap z) = (x \cup y) \cap (x \cup z)$.

Exercise 12.2 Prove the previous Theorem, by showing that any two classes separated by the symbol $=$ have the same members.

Definition 12.4 For each x and y, $x \notin y$ if and only if it is false that $x \in y$.

Definition 12.5 For each x, $\complement x = \{y \mid y \notin x\}$. The class $\complement x$ is the complement of the class x.

Theorem 12.4 *For each x, $\complement(\complement x) = x$.*

Proof By Definition 12.5, $\complement(\complement x)$ is the class

$$\{y \mid y \notin \complement x\} = \{y \mid \text{it is false that } y \in \complement x\} = \{y \mid y \in x\}.$$

  □

Theorem 12.5 (De Morgan Laws) *For each x and y, $\complement(x \cup y) = \complement x \cap \complement y$, and $\complement(x \cap y) = \complement x \cup \complement y$.*

Proof The second statement is left as an exercise. Let us see why the previous statement is true. For each z, $z \in \complement(x \cup y)$ if and only if z is a set and it is false that $z \in x \cup y$. Recalling Theorem 12.1, $z \in x \cup y$ if and only if $z \in x$ or $z \in y$. Consequently, $z \in \complement(x \cup y)$ if and only if z is a set and $z \notin x$ and $z \notin y$. This means exactly that $z \in \complement x$ and $z \in \complement y$. By Theorem 12.1, $z \in \complement x$ and $z \in \complement y$ if and only if $z \in \complement x \cap \complement y$. We conclude by the axiom of extent. □

Exercise 12.3 Prove that $\complement(x \cap y) = \complement x \cup \complement y$ by mimicking the previous proof.

Definition 12.6 For each x and y, $x \setminus y = x \cap \complement y$.

Remark 12.4 We use the modern symbol \ throughout the book. Other symbols are found in the literature, like $x - y$ or $x \sim y$. Kelley systematically writes $\sim x$ instead of $\complement x$.

Remark 12.5 In naïve set theory, the class $\complement x$ is usually undefined as a primitive object. The reason is that all sets must be subsets of some universe U. Hence $\complement x$ must reduce to $U \setminus x$. In other words, only the *relative* complement of a set must be defined in naïve set theory.

Definition 12.7 (Empty Class) $\emptyset = \{x \mid x \neq x\}$.

Remark 12.6 The empty class is denoted by 0 in [3]. We prefer to use the dedicated symbol $\emptyset$, since 0 is used with a lot of different meanings in mathematics. Another popular notation is $\{\}$.

Theorem 12.6 *For each x, $x \notin \emptyset$.*

Proof By definition equality is reflexive, in the sense that for each x, $x = x$. Hence it is false that $x \neq x$, and therefore $\emptyset$ cannot contain any element.   □

Exercise 12.4 Prove that for each x, $\emptyset \cup x = x$ and $\emptyset \cap x = \emptyset$.

Definition 12.8 (Universal Class) $\mathcal{U} = \{x \mid x = x\}$. The class $\mathcal{U}$ is called the universe.

Important: Warning

Beware! $\mathcal{U}$ is not a set! Look ahead in this chapter.

Theorem 12.7 *For each x, $x \in \mathcal{U}$ if and only if x is a set.*

Proof $x \in \mathcal{U}$ if and only if x is a set and $x = x$, hence if and only if x is a set.   □

We may now say that $\mathcal{U}$ is the class that contains every set, although $\mathcal{U}$ is not a set itself. This is essentially the reason why you are reading this chapter.

Exercise 12.5 Prove that for each x, $x \cup \mathcal{U} = \mathcal{U}$ and $x \cap \mathcal{U} = x$. Roughly speaking, $\mathcal{U}$ is so large that we cannot add anything that is not already an element, and intersecting with $\mathcal{U}$ is equivalent to doing nothing.

Theorem 12.8 $\complement \emptyset = \mathcal{U}$ *and* $\complement \mathcal{U} = \emptyset$.

Proof Indeed $z \in \complement \emptyset$ if and only if z is a set and it is false that $z \in \emptyset$. By Theorem 12.6, we conclude that $z \in \complement \emptyset$ if and only if z is a set, namely if and only if $z \in \mathcal{U}$. The proof of the second equality is similar.   □

Definition 12.9 The intersection the members of x is

$$\bigcap x = \{z \mid (\forall y)(y \in x \Rightarrow z \in y)\}.$$

Definition 12.10 The union of the members of x is

$$\bigcup x = \{z \mid (\exists y)(z \in y \text{ and } y \in x)\}.$$

Remark 12.7 We are using here an intrinsic notation. Most readers are probably familiar with the bound variable notation for a "set of sets", as we have seen in (2.1) and (2.2).

Important: Sets or Elements?

In my experience, the hardest step of learning axiomatic Set Theory consists in getting rid of the naïve idea that an element cannot be a set. Consider the following identity:

$$\bigcup \{a, b, c\} = a \cup b \cup c.$$

This is actually correct, since $x \in \bigcup \{a, b, c\}$ if and only if $x \in a$ or $x \in b$ or $x \in c$. However, at the beginning, we tend to believe that the equality is non-sense, since the right-hand side looks like a union of elements, which is naïvely undefined (what is $0 \cup 1$, if 0 and 1 represent the usual natural number we first met at school?). Well, this is precisely how and when *abstraction* is needed. To be honest, in naïve set theory nobody ever writes $\bigcup \{a, b, c\}$, and the issue disappears. But a, b and c may very well be sets (or classes)!

The popular belief that lower-case letters are elements and upper-case letters are sets/classes is, in this perspective, tragic. It should be encouraged only if a basic knowledge of Set Theory is sufficient.

Theorem 12.9 $\bigcap \emptyset = \mathcal{U}$ *and* $\bigcup \emptyset = \emptyset$.

Proof For each x, $x \in \bigcap \emptyset$ if and only if x is a set and x belongs to each member of $\emptyset$. We already know that $\emptyset$ contains no member at all, hence $x \in \bigcap \emptyset$ if and only if x is a set, i.e. $x \in \mathcal{U}$.

Similarly, for each x, $x \in \bigcup \emptyset$ if and only if x is a set and x belongs to some member of $\emptyset$. Since $\emptyset$ has no member, such an x cannot exist. Hence $\bigcup \emptyset$ contains no member. □

Definition 12.11 For each x and y, $x \subset y$ if and only if for each z, if $z \in x$ then $z \in y$. In this case we say that x is a subclass of y, or that the class x is contained in the class y, or that the class y contains the class x.

It should be noticed that $x \subset y$ does not exclude that $x = y$. We will not use $x \subset y$ in the sense of "every element of x is an element of y and $x \neq y$".

Remark 12.8 It is tempting to confuse $\in$ and $\subset$. The language does not help, since $x \in y$ is often read "y contains x". However it would be definitely wrong to use $\subset$ as a replacement of $\in$.

Exercise 12.6 Prove that $\emptyset \subset \emptyset$, but $\emptyset \notin \emptyset$.

We collect some basic properties of subclasses, whose proofs are a straightforward application of the definition.

Theorem 12.10 *For each x, y and z,*

1. $\emptyset \subset x$ *and* $x \subset \mathcal{U}$.
2. $x = y$ *if and only if* $x \subset y$ *and* $y \subset x$.
3. *If* $x \subset y$ *and* $y \subset z$, *then* $x \subset z$.
4. $x \subset y$ *if and only if* $x \cup y = y$.
5. $x \subset y$ *if and only if* $x \cap y = x$.
6. *If* $x \subset y$, *then* $\bigcup x \subset \bigcup y$ *and* $\bigcap y \subset \bigcap x$.
7. *If* $x \in y$, *then* $x \subset \bigcup y$ *and* $\bigcap y \subset x$.

Exercise 12.7 Prove the previous Theorem. When union and intersection of classes are involved, it could be helpful to temporarily switch to a bound variable notation, e.g. $\bigcup_{\alpha \in A} x_\alpha$ instead of $\bigcup x$.

Question Do sets exist?

Axiom of Subsets If x is a set, there is a set y such that for each z, if $z \subset x$, then $z \in y$.

In words, given a set x there exists a set y such that any subclass of x is a member of y. Let us see a useful consequence of this axiom.

Theorem 12.11 *If x is a set and* $z \subset x$, *then z is a set.*

Proof If x is a set, there is a set y such that $z \subset x$ implies $z \in y$. Hence z is a set. □

Exercise 12.8 Read carefully the previous proof, and notice that the axiom of subsets was not used in its full strength. *Hint:* by definition, if y is a class and $z \in y$, then z is a set. Did we use every property of the classes in the axiom of subsets?

Theorem 12.12 $\emptyset = \bigcap \mathcal{U}$ *and* $\mathcal{U} = \bigcup \mathcal{U}$.

Proof If $x \in \bigcap \mathcal{U}$, then x is a set and since $\emptyset \subset x$, it follows that $\emptyset$ is also a set. Then $\emptyset \in \mathcal{U}$ and each member of $\bigcap \mathcal{U}$ belongs to $\emptyset$. It now follows that $\bigcap \mathcal{U}$ has no member.

To prove the second statement, Theorem 12.10 implies that $\bigcup \mathcal{U} \subset \mathcal{U}$. On the other hand, if $x \in \mathcal{U}$ (i.e. if x is a set) by the axiom of subsets there exists a set y

such that, if $z \subset x$, then $z \in y$. In particular $x \in y$, and since $y \in \mathcal{U}$ it follows that $x \in \bigcup \mathcal{U}$. Consequently $\mathcal{U} \subset \bigcup \mathcal{U}$, and the proof is complete.   □

Theorem 12.13 (A Set Exists) *If $x \neq \emptyset$, then $\bigcap x$ is a set.*

Proof If $x \neq \emptyset$, then there exists y such that $y \in x$. Then y is a set, and since $\bigcap x \subset y$, by the axiom of subsets it follows that $\bigcap x$ is also a set.   □

Definition 12.12 (Power Set) For each x,

$$2^x = \{y \mid y \subset x\}.$$

Theorem 12.14 $\mathcal{U} = 2^{\mathcal{U}}$.

Proof Every element of $2^{\mathcal{U}}$ is a set and therefore belongs to $\mathcal{U}$. Conversely, every member of $\mathcal{U}$ is a set and is contained in $\mathcal{U}$, so that it belongs to $2^{\mathcal{U}}$.   □

Theorem 12.15 *If x is a set, then 2^x is a set, and for each y, $y \subset x$ if and only if $y \in 2^x$.*

Exercise 12.9 Prove the previous Theorem, by using the fact that $\mathcal{U} = 2^{\mathcal{U}}$.

We are now ready for the *Russel paradox*, which has been haunting us so far.

Example 12.1 Let $R = \{x \mid x \notin x\}$. By the classification axiom, $R \in R$ if and only if $R \notin R$ and R is a set. Therefore R is not a set.

Theorem 12.16 $\mathcal{U}$ *is not a set.*

Proof Otherwise $R \subset \mathcal{U}$, and R would be a set by Theorem 12.11.   □

Definition 12.13 (Singleton) For each x, $\{x\} = \{z \mid \text{if } x \in \mathcal{U}, \text{ then } z = x\}$.

Theorem 12.17 *If x is a set, for each y, $y \in \{x\}$ if and only if $y = x$.*

Exercise 12.10 Prove the previous Theorem.

Theorem 12.18 *1. If x is a set, then $\{x\}$ is a set.*
2. For each x, $\{x\} = \mathcal{U}$ if and only if x is not a set.

Proof If x is a set, then $\{x\} \subset 2^x$ and 2^x is a set. Hence $\{x\}$ is a set. This proves 1. Let us prove 2. If x is a set, then $\{x\}$ is a set and therefore is not equal to $\mathcal{U}$ (since $\mathcal{U}$ is *not* a set). If x is not a set, then $x \notin \mathcal{U}$ and $x \in \{x\}$ by definition.   □

Axiom of Finite Union If x is a set and y is a set, then $x \cup y$ is a set.

Definition 12.14 (Unordered Pair) For each x and y,

$$\{x, y\} = \{x\} \cup \{y\}.$$

Remark 12.9 The symbol $\{xy\}$ has some advantages over $\{x, y\}$. However we prefer the standard notation with a comma which cannot be confused with "the singleton of the product xy", as soon as a product of x and y is defined and denoted by xy.

If x and y are sets, by the axiom of union also $\{x, y\}$ is a set. In general, however, $\{x, y\}$ is a class.

Definition 12.15 (Ordered Pair) For each x and y,

$$(x, y) = \{\{x\}, \{x, y\}\}$$

is the ordered pair of x and y.

Theorem 12.19 (Equality of Ordered Pairs) *If x is a set, y is a set and $(x, y) = (u, v)$, then $x = u$ and $y = v$.*

Proof Since x and y are sets, so is (x, y) as the unordered pair of two sets. If $x = y$, then $(x, x) = \{\{x\}\}$, and also $u = v$. If $x \neq y$, then $(x, y) = (u, v)$ implies $\{x\} = \{u\}$ and hence $x = u$. This in turn implies $\{x, y\} = \{x, v\}$, and thus $y = v$.   □

Theorem 12.19 contains the essential information about ordered couples. For the sake of completeness we record now a deeper survey of both unordered and ordered couples. We omit their proofs, since we will not refer to them in the sequel.

Theorem 12.20

1. *If x is a set and y is a set, then $\{x, y\}$ is a set and $z \in \{x, y\}$ if and only if $z = x$ or $z = y$. Furthermore, $\{x, y\} = \mathcal{U}$ if and only if x is not a set or y is not a set.*
2. *For each x and y, (x, y) is a set if and only if x is a set and y is a set. If (x, y) is not a set, then $(x, y) = \mathcal{U}$.*

Theorem 12.21 *For all x and y, there results*

1. $\bigcap\bigcap(x, y) = x$;
2. $\bigcap\bigcap\{(x, y)\}^{-1} = y$.

Proof Indeed statement 1. follows from

$$\bigcap\bigcap(x, y) = \bigcap\bigcap\{\{x\}, \{x, y\}\} = \bigcap\{x\} = x.$$

Statement 2. follows from 1. and the fact that $\bigcap\{(x, y)\}^{-1} = (y, x)$.   □

We are therefore led to the following definition, which is mainly interesting in the case of an ordered pair.

Definition 12.16 For each x,

$$\pi_1 x = \bigcap\bigcap x$$

$$\pi_2 x = \bigcap\bigcap\{x\}^{-1}.$$

We call $\pi_1 x$ the first coordinate of x, and $\pi_2 x$ the second coordinate of x.

We now want to define functions. As we saw in Chap. 1, a function is a particular type of relation, and relations were defined as sets of ordered pairs. It is by now clear that we can formulate the following definition.

Definition 12.17 A class r is a relation if and only if for each member z of r there exist x and y such that $z = (x, y)$.

Remark 12.10 We state clearly that a relation is *not* a *set* of ordered pairs. It is a *class* of ordered pairs.

Definition 12.18 The composition $r \circ s$ of the relations r and s is

$$r \circ s = \left\{ u \;\middle|\; \begin{array}{l} \text{for some } x, \text{ some } y \text{ and some } z, \; u = \\ (x, z), \; (x, y) \in s \text{ and } (y, z) \in r \end{array} \right\}$$

Remark 12.11 To save space, we will abbreviate $\{u \mid$ for some x, some z, $u = (x, z)$ and $\ldots\}$ with $\{(x, z) \mid \ldots\}$. In particular, $r \circ s = \{(x, z) \mid$ for some y, $(x, y) \in s$ and $(y, z) \in r\}$.

Definition 12.19 For each relations r, $r^{-1} = \{(x, y) \mid (y, x) \in r\}$. The relation r^{-1} is the inverse relation of r.

Exercise 12.11 Prove that, for each relation r and each relation s, $(r^{-1})^{-1} = r$, $(r \circ s)^{-1} = s^{-1} \circ r^{-1}$.

Definition 12.20 A relation f is a function if and only if for each x, each y and each z, if $(x, y) \in f$ and $(x, z) \in f$, then $y = z$.

Two functions can always be composed as relations. But they can also be composed as functions.

Theorem 12.22 *If f is a function and g is a function, then $g \circ f$ is a function.*

Proof Assume that $(x, y) \in g \circ f$, $(x, z) \in g \circ f$. There exists y_1 such that $(x, y_1) \in f$ and $(y_1, y) \in g$. Similarly there exists y_2 such that $(x, y_2) \in f$ and $(y_2, z) \in g$. Since f is a function, $y_1 = y_2$. For the same reason, $y = z$, and the proof is complete. $\square$

Definition 12.21 For each f,

$$\mathrm{dom}\, f = \{x \mid (\exists y)(x, y) \in f\}$$

and

$$\operatorname{ran} f = \{y \mid (\exists x)(x, y) \in f\}.$$

The class dom f is the domain of f, the class ran f is the range of f.

Exercise 12.12 Prove that dom $\mathcal{U} = \mathcal{U} = \operatorname{ran} \mathcal{U}$. *Hint:* If $x \in \mathcal{U}$, then x is a set and both $(x, \emptyset)$, $(\emptyset, x)$ belong to $\mathcal{U}$.

Definition 12.22 For each f and each x, we define

$$f(x) = \bigcap \{y \mid (x, y) \in f\}.$$

Hence $z \in f(x)$ if and only if z belongs to the second coordinate of each member of f whose first coordinate is x. The class $f(x)$ is the value of f at x.

Remark 12.12

$$\bigcap \{y \mid (x, y) \in f\} = \{z \mid \text{for each } w, \text{ if } w \in \{y \mid (x, y) \in f\} \text{ then } z \in w\},$$

and this is—in general—different than

$$\{y \mid \exists z (z \in x) \wedge (y = f(z))\} = \bigcup \{\pi_2 z \mid \pi_1 z \in x\}.$$

Remark 12.13 The previous Definition is usually a source of bad nightmares. It looks totally different than naïve definition of $f(x)$ as the unique y such that $f(x) = y$. Nightmares disappear as soon as we realize that MK theory does not distinguish between *sets* and *points*. By the way, a point is just another name for an element of some class, and therefore a point is just another name for a set. But there is also another reason behind Definition 12.22: the image of x is always defined, no matter if x is an element of dom f, and no matter if f is actually a function. Consider for example

$$f = \{(1, 1), (1, 2), (3, 4), (9, 9)\}.$$

Clearly f is not a function, and yet $f(3) = 4$, $f(9) = 9$ are defined. But what is the image of 1? In the naïve sense, 1 this question is meaningless, since f is not a function. In MK theory $f(1)$ must belong to both $\{1\}$ and $\{2\}$, i.e. $f(1) = \emptyset$. Similarly, $f(0) = \emptyset$, since 0 is not an element of the domain of f. Needless to say, nobody really needs to define the image of a point outside the domain, in everyday mathematics. But Axiomatic Set Theory exists because we need a coherent deductive theory that does not lean on intuition to prove or disprove statements about sets.

Theorem 12.23 *If $x \notin \operatorname{dom} f$, then $f(x) = \mathcal{U}$. If $x \in \operatorname{dom} f$, then $f(x) \in \mathcal{U}$.*

Proof In the first case, $\{y \mid (x, y) \in f\} = \emptyset$, and the intersection of the empty class is $\mathcal{U}$. In the second case, $\{y \mid (x, y) \in f\} \neq \emptyset$, and Theorem 12.13 implies that $\bigcap \{y \mid (x, y) \in f\}$ is a set. $\square$

Theorem 12.24 *If f is a function, then $f = \{(x, y) \mid y = f(x)\}$.*

Exercise 12.13 Prove the previous Theorem.

Theorem 12.25 (Equality of Functions) *For each function f and each function g, $f = g$ if and only if $f(x) = g(x)$ for each x.*

Proof Indeed $f = g$ if and only if $\{(x, y) \mid y = f(x)\} = \{(x, y) \mid y = g(x)\}$ and this happens if and only if $f(x) = g(x)$ for each x. $\square$

We are in a position to enlarge our collection of axioms.

Axiom of Substitution If f is a function and dom f is a set, then ran f is a set.

Axiom of Amalgamation If x is a set then $\bigcup x$ is a set.

Remark 12.14 We should compare Theorem 12.13 with the axiom of amalgamation.

Definition 12.23 For each x and each y,

$$x \times y = \{(u, v) \mid u \in x, \ v \in y\}.$$

The class $x \times y$ is the cartesian product of the classes x and y.

Theorem 12.26

(a) If u and y are sets, then $\{u\} \times y$ is a set.
(b) If x and y are sets, then $x \times y$ is a set.

Proof Consider the function $\{(w, z) \mid w \in y$ and $z = (u, w)\}$. Its domain is y and its range is $\{u\} \times y$. By the axiom of substitution, $\{u\} \times y$ is a set. This proves (a).

To prove (b), let f be the function $f = \{(u, z) \mid u \in x$ and $z = \{u\} \times y\}$. By the axiom of substitution, ran f is a set. But ran $f = \{z \mid \exists u(u \in x) \wedge (z = \{u\} \times y)\}$. Since $x \times y = \bigcup$ ran f, the axiom of amalgamation implies the result. $\square$

Theorem 12.27 *If f is a function and dom f is a set, then f is a set.*

Proof Indeed $f \subset \operatorname{dom} f \times \operatorname{ran} f$. Since the right-hand class is a set, also f is a set. ∎

Definition 12.24 For each x and each y,

$$y^x = \{f \mid f \text{ is a function, } \operatorname{dom} f = x \text{ and } \operatorname{ran} f \subset y\}.$$

Theorem 12.28 *If x and y are sets, then y^x is a set.*

Proof For each $f \in y^x$ we have $f \subset x \times y$, and $x \times y$ is a set. Hence $f \in 2^{x \times y}$, and $2^{x \times y}$ is a set. We conclude that $y^x \subset 2^{x \times y}$, and the axiom of subsets implies that y^x is a set. ∎

A remarkable fact is that we a function is *not* a triple (X, Y, f) in which X is the domain, Y is the codomain, and f is the law. However, it is convenient to introduce some terminology that may be recall the Calculus approach.

Definition 12.25

(i) f is on x if and only if f is a function and $\operatorname{dom} f = x$.
(ii) f is to y if and only if f is a function and $\operatorname{ran} f \subset y$.
(iii) f is onto y if and only if f is a function and $\operatorname{ran} f = y$.

The last axiom of Kelley's set theory is a returning object of contemporary mathematics.

Definition 12.26 (Choice Function) A class c is a choice function if and only if c is a function and $c(x) \in x$ for each $x \in \operatorname{dom} c$.

Remark 12.15 The existence of a choice function ensures the possibility of simultaneously select a member from each set that belongs to the domain of the choice function. This looks like an almost trivial fact, but it cannot be deduced from the previous axioms.

Axiom of Choice There exists a choice function c whose domain is $\mathcal{U} \setminus \emptyset$.

Remark 12.16 Equivalently, the axiom of choice gives us a function c such that $c(x) \in x$ for each non-empty set x.

The axiom of choice is not necessary to develop all modern mathematics. It is however essential to prove a few results that are of fundamental importance in mathematical analysis. We present now a list of axioms which are equivalent to the axiom of choice, although we will not prove these equivalences in this book. We will follow [2], but a richer statement is also presented in [3].

Well-Ordering Principle Every set can be well-ordered. More precisely, any non-empty subset has a smallest element.

Definition 12.27 A subset C of a partially ordered set $(P, <)$ is chain in P if and only if C is totally ordered by $<$. An element $u \in P$ is an upper bound of C if and only if $c \leq u$ for every $c \in C$. Finally, an element $a \in P$ is a maximal element if there is no $x \in P$ such that $a < x$.

Zorn's Lemma 1 Let $(P, <)$ be a non-empty partially ordered set. If every chain in P has an upper bound, then P has a maximal element.

Definition 12.28 A collection $\mathcal{F}$ of sets has finite character if and only if the following condition holds: for every X, $X \in \mathcal{F}$ if and only if every finite subset of X belongs to $\mathcal{F}$.

Tukey's Lemma 2 If a non-empty collection $\mathcal{F}$ of sets has finite character, then $\mathcal{F}$ has a maximal element with respect to the inclusion $\subset$.

Theorem 12.29 *The following statements are equivalent:*

(i) the Axiom of Choice;
(ii) the Well-ordering Principle;
(iii) Zorn's Lemma;
(iv) Tukey's Lemma.

Proof Suppose (i), and let S be any set. Let F be a choice function on the family of all non-empty subsets of S. Now we define $a_0 = F(S)$, $a_\xi = F(S \setminus \{a_\eta \mid \eta < \xi\})$. The construction stops as soon as we exhaust all elements of S. Hence (ii) holds.

Now assume that (ii) holds, and let $(P, <)$ be a non-empty partially ordered set. Assume that every chain of P has an upper bound. To construct a maximal element, we start from the assumption that P can be well-ordered, so that there is an enumeration

$$P = \{p_0, p_1, \ldots, p_\xi, \ldots\} \quad \xi < \alpha$$

for some ordinal number α. We set $c_0 = p_0$ and $c_\xi = p_\gamma$, where γ is the smallest ordinal such that p_γ is an upper bound of the chain $C = \{c_\eta \mid \eta < \xi\}$ and $p_\gamma \notin C$. We remark that $\{c_\eta \mid \eta < \xi\}$ is always a chain, and that p_γ exists unless $c_{\xi-1}$ is a maximal element of P. In the end, the construction must stop, and we obtain a maximal element of P. This proves (iii).

Suppose now that (iii) holds. We consider a non-empty family $\mathcal{F}$ of sets and we assume that $\mathcal{F}$ has finite character. Clearly $\mathcal{F}$ is partially ordered by inclusion. If C is a chain in $\mathcal{F}$ and if $A = \{X \mid X \in C\}$, then every finite subset of A belongs to $\mathcal{F}$ and therefore $A \in \mathcal{F}$. It follows at once that A is an upper bound of C. We may apply Zorn's Lemma and obtain a maximal element of the collection $\mathcal{F}$. This proves the validity of (iv).

Finally, assume that (iv) holds, and let $\mathcal{F}$ be a collection of non-empty sets. We need to construct a choice function on $\mathcal{F}$. To this aim we consider the collection

$$\mathcal{G} = \{f \mid f \text{ is a choice function on some } \mathcal{E} \subset \mathcal{F}\}.$$

Since a subset of a choice function is a choice function, it follows that G has finite character. By assumption G possesses a maximal element F. By maximality, the domain of F is $\mathcal{F}$, and the proof is complete.   □

12.2   From Sets to $\mathbb{N}$

Our next task is to construct the (positive) integers from set theory. As we will see, this is somehow a painful task, while the subsequent steps—from integers to rationals, and from rationals to real numbers—are rather standard.

Axiom of Regularity For each x, if $x \neq \emptyset$, there is $y \in x$ such that $x \cap y = \emptyset$.

Remark 12.17 Using Kelley's words, the axiom of regularity avoids the possibility that there exist a class z whose member exist by "taking in each other's laundry, in the sense that every member of z consists of members of z."

Axiom of Infinity For some y, y is a set, $\emptyset \in y$, and $x \cup \{x\} \in y$ whenever $x \in y$.

Remark 12.18 An immediate consequence of this axiom is that $\emptyset$ is actually a set, since it is an element of the set y. Notice that $\emptyset$ was merely a class, before adding the axiom of infinity.

Theorem 12.30 *For each x, $x \notin x$.*

Proof If not, $x \in x$, x is a non-empty set and is the only member of $\{x\}$. By the axiom of regularity there exists y in $\{x\}$ such that $y \cap \{x\} = \emptyset$, and necessarily $y = x$. But then $y \in y \cap \{x\}$, against $y \cap \{x\} = \emptyset$.   □

Theorem 12.31 *The statement*

$$\forall x \forall y (x \in y \wedge y \in x)$$

is false.

Proof If $x \in y$ and $y \in x$, then both x and y are sets and are the only members of $\{z \mid z = x \text{ or } z = y\}$. To this class we apply the axiom of regularity, and we reach a contradiction exactly as in the proof of Theorem 12.30.   □

Good. Let us now sketch the basic idea to construct (positive) integers from set theory. We want to start from $0 = \emptyset$, then define $1 = 0 \cup \{0\}$, $2 = 1 \cup \{1\}$, $3 = 2 \cup \{2\}$, and so on. Of course this is not a full definition, since the previous "and so on" requires explanation. We must now provide a rigorous definition of the *ordinals*.

Definition 12.29 (The ∈-Relation) $E = \{(x, y) \mid x \in y\}$.

Theorem 12.32 E *is not a set.*

Proof Assume that $E \in \mathcal{U}$, then $\{E\} \in \mathcal{U}$ and $(E, \{E\}) \in E$. We recall that $(x, y) = \{\{x\}, \{x, y\}\}$ and, if (x, y) is a set, $z \in (x, y)$ if and only if $z = \{x\}$ or $z = \{x, y\}$. We deduce that $E \in \{E\} \in \{\{E\}, \{E, \{E\}\}\} \in E$. Let us summarize: we know that $a \in b \in c \in a$. A contradiction now arises if we apply the axiom of regularity to the class $\{z \mid z = a \text{ or } z = b \text{ or } z = c\}$. Hence E cannot be a set. □

Definition 12.30 (Complete Classes) For each x, x is complete if and only if for each $y \in x$, $y \subset x$.

Remark 12.19 We prefer to use the word "complete" instead of Kelley's "full". If it is true that "complete" refers to several mathematical properties, we believe that it is impossible to systematically select different words.

Definition 12.31 (Ordinals) A class x is an ordinal if and only if x is complete and the following condition holds:

(C) when u and v belong to x, either $(u, v) \in E$, or $(v, u) \in E$, or $u = v$.

Definition 12.32 $R = \{x \mid x \text{ is an ordinal}\}$. A class x is an ordinal number if and only if $x \in R$.

In particular, an ordinal number is a set, being a member of R.

Definition 12.33 (Successor) For each x, $x + 1 = x \cup \{x\}$.

To move from ordinals to integers, we need to explore relation E more deeply.

Definition 12.34 We say that E^{-1} well-orders x if and only if condition (C) is satisfied, and for each $y \subset x$, $y \neq \emptyset$, there exists $z \in y$ such that for each $w \in y$, it is false that $(w, z) \in E^{-1}$.

Definition 12.35 (Positive Integers) A class x is an integer if and only if x is an ordinal and E^{-1} well-orders x.

Definition 12.36 $\mathbb{N} = \{x \mid x \text{ is an integer}\}$.

Remark 12.20 In axiomatic set theory, the letter ω is used instead of $\mathbb{N}$. We believe that $\mathbb{N}$ is a better choice from the viewpoint of a mathematical analyst.

The next result contains the celebrated *Peano axioms* of positive integers. A mathematical analyst need not know its proof, which therefore we omit.

Theorem 12.33 (Peano Axioms)

1. *If $x \in \mathbb{N}$, then $x + 1 \in \mathbb{N}$.*
2. *$\emptyset \in \mathbb{N}$, and if $x \in \mathbb{N}$, then $\emptyset \neq x + 1$.*
3. *If x and y are members of $\mathbb{N}$ and $x + 1 = y + 1$, then $x = y$.*
4. *If $x \subset \mathbb{N}$, $0 \in x$ and $u + 1 \in x$ whenever $u \in x$, then $x = \mathbb{N}$.*

Theorem 12.34 $\mathbb{N} \in R$. *In particular, $\mathbb{N}$ is a set.*

Proof By the axiom of infinity, there exists a set y such that $\emptyset \in y$ and $x + 1 \in y$ whenever $x \in y$. By Theorem 12.33, statement 4, $\mathbb{N} \cap y = \mathbb{N}$ and hence $\mathbb{N}$ is a set because $\mathbb{N} \subset y$. Since $\mathbb{N}$ consists of ordinal numbers, Definition 12.32 is satisfied by $\mathbb{N}$, and therefore $\mathbb{N} \in R$.   □

Definition 12.37 (Ordering) For each x and y, $x < y$ if and only if $x \in y$, i.e. $(x, y) \in E$. Similarly, $x \leq y$ if and only if $x \in y$ or $x = y$.

With this definition in mind, we see that positive integers are ordered in a way that is compatible with the usual naïve definition of inequality between numbers.

12.3   A Summary of Kelley's Axioms

For the reader's convenience, we summarize here Kelley's axioms.

Axiom of Extent For each x and each y, it is true that $x = y$ if and only if for each z, $z \in x$ if and only if $z \in y$.

Axiom of the Classifier For each u and y, $u \in \{x \mid x \in y\}$ if and only if u is a set and $u \in y$.

Classification Axiom Scheme An axiom results if in the following α and β are replaced by variables, A by a formula $\mathcal{A}$ and B by the formula obtained from $\mathcal{A}$ by replacing each occurrence of the variable that replaced α by the variable that replaced β:

For each β, $\beta \in \{\alpha \mid A\}$ if and only if β is a set and B.

Axiom of Subsets If x is a set, there is a set y such that for each z, if $z \subset x$, then $z \in y$.

Axiom of Union If x is a set and y is a set, then $x \cup y$ is a set.

Axiom of Substitution If f is a function and dom f is a set, then ran f is a set.

Axiom of Amalgamation If x is a set then $\bigcup x$ is a set.

Axiom of Choice There exists a choice function c whose domain is $\mathcal{U} \setminus \emptyset$.

12.4 Set Theory According to J.D. Monk

As we said at the beginning of this chapter, we want to introduce a variation of Kelley's construction. Although essentially equivalent, Monk's revisited Set Theory is based on a stronger Axiom of Choice, and on a more restrictive definition of relations and functions.

Definition 12.38 The primitive notions are those of classes and membership. We will use capital letters like A, $B,\ldots$, X, Y, Z for classes, and the usual symbol $\in$ for membership. The negation of $\in$ will be denoted by $\notin$.

Axiom 1 (Extensionality Axiom) $\forall A \, \forall B \, (\forall C (C \in A \iff C \in B) \implies A = B)$.

By this Axiom, two classes A and B are the same class if and only if they share the same members.

Definition 12.39 (Sets) A class A is a set if and only if there exists a class B such that $A \in B$. A is a proper class if and only if A is not a set.

As a rule, sets will be denoted by lower case letters: $a, b, \ldots, x, y, z$.

Exercise 12.14 Prove that $\forall a \, \exists B \, (a \in B)$.

Exercise 12.15 Prove that $\forall x (x \in A \iff x \in B) \implies A = B$.

Definition 12.40 The expressions $A = A$, $A = B$, $A = C, \ldots$, $B = A$, $B = B$, $B = C, \ldots C = A, C = B, C = C, \ldots$ are all set-theoretical formulas, as are $A \in A$, $A \in B, A \in C, \ldots, B \in A, B \in B, B \in C, \ldots C \in A, C \in B, C \in C, \ldots$

If φ and ψ are set-theoretical formulas, so are $\neg \varphi$, $\varphi \vee \psi$, $\varphi \wedge \psi$, $\varphi \implies \psi$, $\varphi \iff \psi$, $\exists A \varphi$, $\exists B \varphi, \ldots$, $\forall A \varphi$, $\forall B \varphi$, $\ldots$ Set-theoretical formulas can only be obtained by finitely many applications of the processes just mentioned.

> **Axiom 2 (Class-Building Axiom)** *If $\varphi(X)$ is a set-theoretical formula not involving the letter A, then the following is an axiom:*
>
> $$\exists A \forall X (X \in A \iff X \text{ is a set and } \varphi(X)).$$
>
> *Similarly, if $\varphi(X)$ is a set-theoretical formula not involving the letter B, then the following is an axiom:*
>
> $$\exists B \forall X (X \in B \iff X \text{ is a set and } \varphi(X)),$$
>
> *and so on for other letters. Letters different than X may also be used.*

This axiom allows us to define classes by specifying the properties which each member must satisfy.

Example 12.2 If $\varphi(X)$ is $X \in X$, the class-building axiom allows us to construct the class of all sets.

Definition 12.41 For any set-theoretical formula $\varphi(X)$ not involving A, let

$$\{X \mid \varphi(X)\}$$

be the unique class A such that

$$\forall X (X \in A \iff X \text{ is a set and } \varphi(X)).$$

As before, letters other than A and X may be used.

Definition 12.42 $A \subset B$ if and only if $\forall C \, (C \in A \implies C \in B)$. We say that A is a subclass of B. If A is a set, then A is a subset of B.

Theorem 12.35 $A \subset B$ *if and only if* $\forall x (x \in A \iff x \in B)$.

Proof If $A \subset B$, certainly the conclusion is true. Conversely, assume that $\forall x\,(x \in A \iff x \in B)$. If C is an arbitrary class such that $C \in A$, then C is a set and $C \in B$. Since C is arbitrary, $\forall X\,(X \in A \iff X \in B)$, and thus $A \subset B$. ☐

Axiom 3 (Power-Set Axiom) $\forall a \exists b \forall C (C \subset a \implies C \in b)$.

Axiom 4 (Pairing Axiom) $\forall a \forall b \exists c (a \in c \land b \in c)$

Beware that the set c need not contain *only* a and b.

Axiom 5 (Union Axiom) $\forall a \exists b \forall C (C \in a \implies C \subset b)$.

This axiom becomes clearer if we imagine that a is a family of sets, so that there exists another set b which contains every member of a,

Definition 12.43 (Empty Class) $\emptyset = \{x \mid x \neq x\}$.

Theorem 12.36 *For all X, $X \notin \emptyset$.*

Proof If $X \in \emptyset$, then X is a set and $X \neq X$, a contradiction. ☐

Definition 12.44 The intersection of two classes A and B is defined as

$$A \cap B = \{x \mid x \in A \land x \in B\}.$$

The following axiom is usually considered as an useless one. However classes are so large that counterintuitive possibilities may arise, such as the existence of a class A with $A \in A$.

Axiom 6 (Regularity Axiom) $\forall A (A \neq \emptyset \implies \exists X (X \in A \land X \cap A = \emptyset))$.

Definition 12.45 The successor of a class A is the class

$$S(A) = \{x \mid x \in A \lor x = A\}.$$

Remark 12.21 We cannot write $S(A) = A \cup \{A\}$, since singletons have not been introduced yet.

Theorem 12.37 *If A is a proper class, then $S(A) = A$.*

Proof Suppose that X is any class. If $X \in A$, then X is a set, and $X \in A$ or $X = A$. Hence $X \in S(A)$. If $X \in S(A)$, then $X \in A$ or $X = A$, and X is a set. But A is not a set, hence $X = A$. By the axiom of extensionality, $A = S(A)$.   □

Axiom 7 (Axiom of Infinity) $\exists a(\emptyset \in a \wedge \forall X(X \in a \implies S(X) \in a))$.

So a set a exists such that $\emptyset \in A$, $S(\emptyset) \in a$, $S(S(\emptyset)) \in a$, and so on.

Definition 12.46 (Unordered Pairs) The unordered pair[1] of two classes A and B is

$$\{A, B\} = \{x \mid x = A \vee x = B\}.$$

Theorem 12.38 *If a and b are sets, so is $\{a, b\}$.*

Proof By the pairing axiom, a set c exists such that $a \in c$ and $b \in c$. Thus $\{a, b\} \subset c$. By the power-set axiom, a set d exists such that $\forall X(X \subset c \implies X \in d)$. In particular we have $\{a, b\} \in d$, hence $\{a, b\}$ is a set.   □

Definition 12.47 The singleton A is $\{A\} = \{A, A\}$.

The following result describes the equality of unordered pairs.

Theorem 12.39 *If $\{a, b\} = \{c, d\}$, then $a = c$ and $b = d$, or $a = d$ and $b = c$.*

Proof Clearly $a \in \{a, b\}$. By assumption $a \in \{c, d\}$, so that $a = c$ or $a = d$. These cases are symmetric, hence we may assume that $a = c$. Similarly $b \in \{a, b\}$, and $b = c$ or $b = d$. If $b = d$, the proof is complete. If $b = c$, then $a = b = c$.

Again, $d \in \{a, b\}$, and by the same token $d = a$ or $d = b$. Thus $a = b = c = d$, and the desired conclusion has been reached.   □

We now propose a familiar definition of ordered pair.

Definition 12.48 The ordered pairs of two classes A and B is

$$(A, B) = \{\{A\}, \{A, B\}\}.$$

By imitating Theorem 12.38, one can prove that the ordered pair of two sets is a set.

---

[1] The old-fashioned term *doubleton* has been used as well.

Theorem 12.40 *If* $(a, b) = (c, d)$ *then* $a = c$ *and* $b = d$.

Proof Since $(a, b) = \{\{a\}, \{a, b\}\}$ and $(c, d) = \{\{c\}, \{c, d\}\}$ and $\{a\}, \{a, b\}, \{c\}$ and $\{c, d\}$ are all sets, by Theorem 12.39 two cases may happen. In the first case, $\{a\} = \{c\}$ and $\{a, b\} = \{c, d\}$. Now $a \in \{a\}$, thus $a \in \{c\}$ and $a = c$. It remains to show that $b = d$. We apply Theorem 12.39 again and we conclude that either $a = c$ and $b = d$, or $a = d$ and $b = c$. In both cases $a = c$ and $b = d$.

In the second case, $\{a\} = \{c, d\}$ and $\{a, b\} = \{c\}$. Now $c \in \{c, d\}$, so that $c \in \{a\}$ and consequently $c = a$. Similarly, $d = a$ and $b = c$, so that $a = c$ and $b = d$. The proof is complete.   □

Let us turn to relations.

Definition 12.49 A class R is a relation if and only if

$$\forall A \, (A \in R \implies \exists c \exists d (A = (c, d))).$$

The domain of a relation R is the class

$$\text{Dom } R = \{x \mid \exists y ((x, y) \in R)\}.$$

The range of R is the class

$$\text{Ran } R = \{y \mid \exists x ((x, y) \in R)\}.$$

Definition 12.50 A class F is a function if and only if F is a relation and

$$\forall x \forall y \forall z ((x, y) \in F \wedge (x, z) \in F \implies y = z).$$

As usual, a relation is just a any class of ordered pairs. A function is just a "rule" which assigns to each set of its domain a unique set of its codomain.

Axiom 8 (Axiom of Substitution) *If* F *is a function and* $\text{Dom } F$ *is a set, then* $\text{Ran } F$ *is a set.*

Monk's final axiom is a very strong form of the Axiom of Choice. It is indeed stronger than the usual one.

Axiom 9 (Relational Axiom of Choice) *If* R *is a relation, there exists a function* F *such that* $F \subset R$ *and* $\text{Dom } F = \text{Dom } R$.

With these axioms one can construct the familiar boolean algebra of sets and of relations, with minor differences with respect to our previous discussion. Monk's Theory of Sets is actually equivalent to Kelley's one. The strength of the Relational Axiom of Choice may be used to prove stronger results directly in Monk's theory, but most mathematicians do not usually see any concrete difference.

12.5   ZF Axioms

As we have stated at the beginning, the most popular axiomatization of Set Theory goes under the name of Zermelo and Fraenkel. Although we have preferred another approach, we believe it may be useful for the reader to have at least an account of **ZF**. It should be noticed that **ZF** requires both sets and elements.

Axiom 10 (Extensionality Axiom) *Two sets are equal if and only if they have the same elements. In symbols:*

$$\forall A \forall B (A = B \iff \forall x (x \in A \iff x \in B))$$

Axiom 11 (Empty Set) *There exists a set with no elements. In symbols:*

$$\exists A \forall x (x \notin A)$$

Axiom 12 (Subset Axiom) *Let $\varphi(x)$ be a formula. For every set A there exists a set S that consists of all the elements $x \in A$ such that $\varphi(x)$ holds. In symbols,*

$$\forall A \exists S \forall x (x \in S \iff (x \in A \wedge \varphi(x)))$$

Axiom 13 (Pairing Axiom) *For every u and v there exists a set that consists of just u and v. In symbols:*

$$\forall u \forall v \exists A \forall x (x \in A \iff (x = u \vee x = v))$$

Axiom 14 (Union Axiom) *For every set $\mathcal{F}$ there exists a set U that consists of all the elements that belong to at least one set in $\mathcal{F}$. In symbols,*

$$\forall \mathcal{F} \exists U \forall x (x \in U \iff \exists C (x \in C \land C \in \mathcal{F}))$$

Axiom 15 (Power Set Axiom) *For every set A there exists a set P that consists of all the sets that are subsets of A. In symbols:*

$$\forall A \exists P \forall x (x \in P \iff \forall y (y \in x \implies y \in A))$$

Axiom 16 (Infinity Axiom) *There exists a set I that contains the empty set as an element and whenever $x \in I$, then $x \cup \{x\} \in I$. In symbols:*

$$\exists I (\emptyset \in I \land \forall x (x \in I \implies x \cup \{x\} \in I))$$

Axiom 17 (Replacement Axiom) *Let $\psi(x, y)$ be a formula. For every set A, if for each $x \in A$ there exists a unique y such that $\psi(x, y)$, then there exists a set S that consists of all the elements y such that $\psi(x, y)$ for some $x \in A$. In symbols:*

$$\forall A ((\forall x \in A) \exists! y \, \psi(x, y) \implies \exists S \forall y (y \in S \iff (\exists x \in A) \psi(x, y)))$$

Axiom 18 (Regularity Axiom) *Every non-empty set A has an element that is disjoint from A. In symbols:*

$$\forall A (A \neq \emptyset \implies \exists x (x \in A \land x \cap A = \emptyset))$$

12.6   From ℕ to ℤ

Once we have a rigorous foundation of positive integers, the construction of *relative integers* is standard. Let us set $X = \mathbb{N} \times \mathbb{N}$, the cartesian product of the positive integers with themselves.

Definition 12.51 For each (m, n) and (m', n') in X, we say that $(m, n) \sim (m', n')$ if and only if $m + n' = m' + n$.

Remark 12.22 The intuition behind the previous definition is that we want to identify the "numbers" $m - n$ and $m' - n'$. Actually, -3 is both equal to $2 - 5$ and $7 - 10$. However this intuition cannot be a rigorous definition o relative integers, since $m - n$ is not—in general—a positive integer. In other words, we know what we want, but we cannot just take it for granted that what we want does exist.

Exercise 12.16

1. Prove that $\sim$ is an equivalence relation.
2. Prove that $(m, n) \sim (m', n')$ if and only if $(m + p, n + p) \sim (m' + p, n' + p)$ for each $p \in \mathbb{N}$.
3. Prove that $m = n$ if $(m, p) \sim (n, p)$ or if $(p, m) \sim (p, n)$.

Definition 12.52 (Addition) For each (m, n) and (m', n') in X, we define

$$(m, n) + (m', n') = (m + m', n + n').$$

Definition 12.53 (The Relative Integers) The symbol $\mathbb{Z}$ denotes the set $X/\sim$, i.e. the set of equivalence classes determined by $\sim$. The equivalence class of the element (m, n) is denoted by $[(m, n)]$.

Definition 12.54 $0 = [(1, 1)]$.

Definition 12.55 For each $a \in \mathbb{Z}$ and $b \in \mathbb{Z}$, we define $a + b$ as follows. We write $a = [(m, n)]$ and $b = [(p, q)]$ for some (m, n) and (p, q) in X. Then

$$a + b = [(m + p, n + q)].$$

Exercise 12.17 Prove that the sum $a + b$ is independent of the particular ordered couples (m, n) and (p, q) that describe the classes a and b. More precisely, show that $(m + p, n + q) \sim (m' + p', n' + p')$ whenever $(m, n) \sim (m', n')$ and $(p, q) \sim (p', q')$.

Definition 12.56 (Opposite) For each $a \in \mathbb{Z}$, we define $-a$ as the unique element $x \in \mathbb{Z}$ such that $a + x = 0$.

Exercise 12.18 Prove that the previous definition is consistent, in the sense that $-a$ does exist for each $a \in \mathbb{Z}$.

Definition 12.57 (Difference) For each a and b in $\mathbb{Z}$, we define $a - b = a + (-b)$.

We can now *embed* $\mathbb{N}$ into $\mathbb{Z}$ as follows.

Definition 12.58 $f_+ : \mathbb{N} \to \mathbb{Z}$ is defined by $f_+(n) = [(n+1, 1)]$. We define $\mathbb{Z}_+ =$ ran f_+.

Exercise 12.19 Prove that $[(n+1, 1)] = [(n'+1, 1)]$ implies $n = n'$. Deduce that f_+ is injective, and therefore its restriction $f_+ : \mathbb{N} \to \mathbb{Z}_+$ is a bijection.

By the previous exercise, for each $a \in \mathbb{Z}_+$ there exists a unique $n \in \mathbb{N}$ such that $f_+(n) = a$. We write $n = f_+^{-1}(a)$.

Definition 12.59 Let $s : \mathbb{N} \to \mathbb{N}$ the standard successor function defined by $s(n) = n + 1$. We define $s' : \mathbb{Z}_+ \to \mathbb{Z}_+$ by $s'(a) = f_+(s(f_+^{-1}(a)))$.

To conclude, we define the product of relative integers.

Definition 12.60 An operation $*$ is defined on X by

$$(m, n) * (p, q) = (mp + nq, mq + np).$$

Exercise 12.20 Prove that $(m, n) \sim (m', n')$ and $(p, q) \sim (p', q')$ imply $(m, n) * (p, q) \sim (m', n') * (p', q')$.

By the previous exercise, the product on X can be extended to a product on $\mathbb{Z}$.

Definition 12.61 For each $a = [(m, n)]$ and $b = [(p, q)]$ in $\mathbb{Z}$, we define

$$ab = a * b = [(mp + nq, mq + np)].$$

Remark 12.23 In general this product does not possess a inverse element, in the sense that, given a and b in $\mathbb{Z}$, $a \neq [(0, 0)]$, the equation $a * x = b$ is not solvable in $\mathbb{Z}$.

12.7 From $\mathbb{Z}$ to $\mathbb{Q}$

The next extension of our number sets will be the set of rational numbers. While everybody thinks of a rational number as a *fraction* like p/q with $p \in \mathbb{Z}$ and $q \in \mathbb{N} \setminus \{0\}$, this approach is purely naïve. Indeed we did not—and could not—define a *division* operation in $\mathbb{Z}$. Again, the road goes through equivalence classes.

Definition 12.62 $V = \{(r, s) \in \mathbb{Z} \times \mathbb{Z} \mid s \neq 0\}$. We define an equivalence relation $\sim$ on V by setting $(r, s) \sim (t, u)$ if and only if $ru = st$. Notice that we are writing rs instead of $r * s$, and so on. The equivalence class of the element (r, s) will be denoted by $[(r, s)]$.

Definition 12.63 (Rational Numbers) A rational number is any element of $V / \sim$. If $[(r, s)] \in V / \sim$, we simply write r/s. The set $\mathbb{Q}$ is the set whose members are all the rational numbers.

Definition 12.64 $\iota: \mathbb{Z} \to \mathbb{Q}$ is defined by $\iota(r) = [(r, 1)]$ for each r.

Theorem 12.41 ι *is an injective function.*

Proof Assume that $\iota(x) = \iota(y)$ for some x, y in $\mathbb{Z}$. Then $[(x, 1)] = [(y, 1)]$, i.e. $(x, 1) \sim (y, 1)$, and this means that $x * 1 = y * 1$, hence $x = y$.   □

Definition 12.65 (Sum of Rational Numbers) For each $[(r, s)]$ and $[(t, u)]$ in $\mathbb{Q}$, we define

$$[(r, s)] + [(t, u)] = [(ru + st, su)].$$

Exercise 12.21 Prove that the previous definition is consistent: if $(r', s') \sim (r, s)$ and if $(t', u') \sim (t, u)$, then $(ru + st, su) \sim (r'u' + s't', s'u')$.

Definition 12.66 (Multiplication of Rational Numbers) For each $[(r, s)]$ and $[(t, u)]$ in $\mathbb{Q}$, we define

$$[(r, s)] * [(t, u)] = [(rt, su)].$$

Exercise 12.22 Prove that the previous definition is consistent.

Remark 12.24 It must be observed that our operations of sum and multiplication is compatible with the corresponding operation of $\mathbb{Z}$. More precisely,

$$\iota(x) + \iota(y) = [(x, 1)] + [(y, 1)]$$
$$= [(x * 1 + 1 * y, 1 * 1)] = [(x + y, 1)]$$
$$= \iota(x + y)$$

and

$$\iota(x) * \iota(y) = [(x, 1)] * [(y, 1)]$$
$$= [(xy, 1)]$$
$$= \iota(x * y)$$

Exercise 12.23

1. Prove that $[(1, 1)]$ is neutral for multiplication: for each $[(r, s)]$ there results $[(r, s)] * [(1, 1)] = [(r, s)]$.
2. Prove that for each $[(r, s)] \neq [(0, 1)]$, there exists a unique $[(x, y)]$ such that $[(r, s)] * [(x, y)] = [(1, 1)]$. *Hint:* a good candidate is $x/y = s/r$.

It is now easy to prove the following result.

Theorem 12.42 *The set* $\mathbb{Q}$ *is a field under the operations of sum and multiplication.*

Remark 12.25 Every rational number can be uniquely written as r/s with $s \geq 1$. Indeed, if $s \leq -1$, we recall that $[(r, s)] = [(-r, -s)]$. In the rest of this section, we will systematically assume that rational numbers are presented in this form.

Definition 12.67 (Ordering) A relation $<$ is defined on $\mathbb{Q}$ by $[(r, s)] < [(t, u)]$ if and only if $ru < st$.

Remark 12.26 If $r/s = r'/s'$ and $t/u = t'/u'$, then $rs' = r's$ and $tu' = t'u$. Therefore $[(r, s)] < [(t, u)]$ if and only if $ru < st$ if and only if $rus'u' < sts'u'$ if and only if $r'suu' < ss't'u$ if and only if $r'u' < s't'$ if and only if $[(r', s')] < [(t', u')]$. We deduce that the ordering definition is consistent in $\mathbb{Q}$.

12.8 From ℚ to ℝ

Our last step in set and number theory is a *concrete* construction of real numbers. We have introduced $\mathbb{R}$ as a totally ordered field which satisfies Dedekind's axiom. But this definition remains somehow vague until we prove that such a numerical structure exists. To this aim several approaches have been proposed. A classical one is via *Dedekind cuts*, see [4].

We present a different approach that lies on a more analytical construction, due to Cantor.

Definition 12.68 A sequence $x = \{x_n\}_n$ of rational numbers is a Cauchy sequence of rational numbers if for each rational number $a > 0$, there exists $N \in \mathbb{N}$ such that $|x_m - x_n| < a$ for each $m \geq N$ and each $n \geq N$. We denote by C the set of all Cauchy sequences of rational numbers.

Definition 12.69 A sequence $x = \{x_n\}_n$ of rational numbers is a zero sequence of rational numbers if and only if for each rational number $a > 0$ there exists $N \in \mathbb{N}$ such that $|x_n| < a$ for each $n \geq N$. We denote by $\mathcal{Z}$ the set of all zero sequences of rational numbers.

Exercise 12.24 Prove that $\mathcal{Z} \subset C$.

Definition 12.70 Each rational number q induces a constant sequence $\bar{q} : \mathbb{N} \to \mathbb{Q}$ by setting $\bar{q}_n = q$ for each n. We denote by Q the set of all constant sequences of rational numbers.

Definition 12.71 If $\{x_n\}_n$ and $\{y_n\}_n$ are sequences in $\mathbb{Q}$, we define their sum as $n \mapsto x_n + y_n$ and their product as $n \mapsto x_n y_n$.

Theorem 12.43 *Let $\{x_n\}_n$ and $\{y_n\}_n$ be sequences in $\mathbb{Q}$.*

1. *If $\{x_n\}_n \in C$ and $\{x_n\}_n \in C$ then their product is a member of C.*
2. *If $\{x_n\}_n \in \mathcal{Z}$ and $\{x_n\}_n \in \mathcal{Z}$ then their product is a member of $\mathcal{Z}$.*
3. *If $\{x_n\}_n \in C$ and $\{x_n\}_n \in \mathcal{Z}$ then their product is a member of $\mathcal{Z}$.*

Proof To prove 1, we first remark that $\{x_n\}_n$ and $\{y_n\}_n$ are bounded sequences as in Proposition 5.3. Then we merely use the triangle inequality:

$$|x_n y_n - x_m y_m| \leq |(x_n - x_m)y_n| + |x_m(y_n - y_m)| = |x_n - x_m||y_n| + |x_m||y_n - y_m|.$$

Since the two sequences and in C and bounded, the right-hand side of the last inequality can be made smaller than any given positive rational number by taking m and n sufficiently large. In a similar way we prove 2 and 3. $\qquad\square$

Definition 12.72 An equivalence relation on C is defined as follows: for each $x \in C$ and each $y \in C$, $x \sim y$ if and only if their difference $x - y$ is an element of $\mathcal{Z}$. We denote by $[x]$ the equivalence class of the sequence $x \in C$.

Definition 12.73 A real number is any element of $C/\sim$, i.e. any equivalence class of a sequence in C. We denote by $\mathbb{R}$ the set of all real numbers.

Let us stop for a moment: what are we doing? Well, the intuition is that a real number is the "limit" of a sequence of rational numbers. But this would not be a good definition, since—of course—there are sequences of rational numbers that do not converge to a rational number. We are forced to take limits in $\mathbb{Q}$, but we need to go outside $\mathbb{Q}$! The only way out is to remove the requirement that our sequences of rational numbers have a limit, and this is done by using the Cauchy condition. Then we identify rational sequences that "converge" to the same number, in the sense that their difference converges to zero. This is the equivalence relation $\sim$.

In this way we have avoided any use of the term *limit*, which would be somehow misleading at this stage. But we have preserved the original intuition of "adding to $\mathbb{Q}$ the limits of rational sequences that satisfy the Cauchy property."

Theorem 12.44 *Let x, x', y and y' be elements of C. If $x \sim x'$ and $y \sim y'$, then $x + y \sim x' + y'$ and $xy \sim x'y'$.*

Proof Let $x' = x + p$ and $y' = y + q$ for some p and q in $\mathcal{Z}$. Then $(x' + y') - (x + y) = p + q \in \mathcal{Z}$ by Theorem 12.43. Also $x'y' - xy = py + xq + pq \in \mathcal{Z}$ by Theorem 12.43. The conclusion follows. $\qquad\square$

Definition 12.74 A sequence $x: \mathbb{N} \to \mathbb{Q}$ is eventually positive if there exists $N \in \mathbb{N}$ such that $x_n > 0$ for each $n \geq N$.

Definition 12.75 A real number $[x]$ is positive if each $y \in [x]$ is an eventually positive sequence of rational numbers. The subset of positive real numbers is denoted by $\mathbb{R}^+$.

Remark 12.27 Notice that $\mathbb{R}^+$ is what we usually denote by $(0, +\infty)$. The real number 0 does not belong to $\mathbb{R}^+$. Indeed $y \in [0]$ if and only if $y \in \mathcal{Z}$. The constant rational sequence $\bar{0}$ belongs to $\mathcal{Z}$, but it is not eventually positive.

The following theorem shows that $\mathbb{R}$ is indeed a totally ordered field in the sense of Chap. 2. Its proof is not difficult, but it is somehow boring. We leave the details as an exercise for the interested reader.

Theorem 12.45 *The set $\mathbb{R}$ with the operations defined by*

$$[x] + [y] = [x + y], \qquad [x] * [y] = [x * y]$$

for each $[x]$ and each $[y]$ in $\mathbb{R}$, is a field. It is also totally ordered by the relation $[x] < [y]$ if and only if $y - x$ is a positive real number in the sense of Definition 12.75.

Luckily enough, $\mathbb{Q}$ may be identified with a subset of $\mathbb{R}$. Indeed, let $\varphi \colon \mathbb{Q} \to \mathbb{R}$ be the function defined by $\varphi(p) = [\bar{p}]$, where $\bar{p}$ is the constant sequence $n \mapsto p$. It can be easily proved that φ is injective, so that $\varphi \colon \mathbb{Q} \to \varphi(\mathbb{Q})$ is a bijection. Furthermore $\varphi(p + q) = \varphi(p) + \varphi(q)$ and $\varphi(pq) = \varphi(p)\varphi(q)$.

The last—and the most intriguing—property of $\mathbb{R}$ is clearly the fact that $\mathbb{R}$ has the least upper bound property: if a subset E of $\mathbb{R}$ has an upper bound, then there exists the least upper bound of E. The proof of this fact requires some work.

Theorem 12.46 *If $r = [x]$ is a real number and if there exists $N \in \mathbb{N}$ such that $x_N \geq 0$, then $r \geq 0$.*

Proof Indeed, either $r = 0$, or r is positive, or $-r$ is positive. The last possibility is ruled out by assumption, since every element of the class r would be eventually positive. But we know that $-x \in r$ and $-x$ is not eventually positive. We deduce that either $r = 0$ or r is a positive real number. $\square$

Theorem 12.47 *If r is a positive real number, then there is $p \in \mathbb{Q}$ such that $0 < p < r$.*

Proof Indeed, let $r = [x]$. By assumption x is eventually positive, and it follows easily that for some $a \in \mathbb{Q}$, $a > 0$ and some $N \in \mathbb{N}$, either $x_n > a$ or $x_n < -a$ for each $n \geq N$. Since x is eventually positive, the second alternative cannot hold. Hence $x_n > a$ whenever $n \geq N$. By Theorem 12.46 $0 < a \leq r$. Let $p = a/2$, then $0 < a/2 < a \leq r$, and the proof is complete. $\square$

Theorem 12.48 (Density or Rational Numbers) *For each real numbers r, s with $r < s$, there exists $q \in \mathbb{Q}$ such that $r < q < s$.*

Proof By assumption, $0 < s - r$. By the previous results, we can choose $p \in \mathbb{Q}$ such that $0 < p < s - r$. Let x and y be sequences in the classes r and s, respectively. Now $0 < (s - r) - p$ implies that the sequence $n \mapsto y_n - x_n - p$ is eventually positive. We choose $M \in \mathbb{N}$ with $y_m > x_m + p$ for each $m \geq M$. Since x is a Cauchy sequence, there exists $N \geq M$, $N \in \mathbb{N}$, such that $|x_N - n_n| \leq p/4$ for each $n \geq N$.

Let $q_1 = x_N + p/4$ and $q_2 = x_N + 3p/4$. For each $n \geq N$, $q_1 - x_n = p/4 + x_N - x_n \geq p/4 + |x_N - x_n| \geq 0$ implies $r \leq q_1$, and similarly $q_2 \leq s$. The rational number $q = (q_1 + q_2)/2$ satisfies $r < q < s$. $\square$

Theorem 12.49 *Let A be a nonempty subset of $\mathbb{R}$. If A has an upper bound, then a number $p \in \mathbb{Q}$ exists such that p is not an upper bound of A but $p + 1$ is an upper bound of A.*

Proof Pick $a \in A$. Surely $r = a - 1$ is not an upper bound of A, since $r < a$. Let $s < r$, $s \in \mathbb{Q}$. A fortiori s is not an upper bound of A. The sequence $n \mapsto s_n = s + n$ is unbounded from above. Hence there exists $n \in \mathbb{N}$ such that $s + n$ is an upper bound of A. By induction, we know that there exists a smallest positive integer m such that $q = s + m$ is an upper bound of A. Then $p = q - 1 = s + m - 1$ is not an upper bound of A.   □

Theorem 12.50 *Let A be a nonempty subset of $\mathbb{R}$. If A has an upper bound, then there exist two sequences of rational numbers $\{p_n\}_n$ and $\{q_n\}_n$ such that, for each $n \in \mathbb{N}$, p_n is not an upper bound of A and q_n is an upper bound of A. Furthermore, $q_n - p_n = 2^{1-n}$.*

Proof Let p and q be as in Theorem 12.49. We construct the two sequences inductively. Let $p_1 = p$ and $q_1 = q$. Suppose that p_n and q_n have been chosen, and let $s_n = (p_n + q_n)/2$. If s_n is not an upper bound of A, then we set $p_{n+1} = s_n$, $q_{n+1} = q_n$. Otherwise we set $p_{n+1} = p_n$ and $q_{n+1} = s_n$. It follows that $q_{n+1} - p_{n+1} = 2^{-n}$, and the proof is complete.   □

Theorem 12.51 *The sequences $\{p_n\}_n$ and $\{q_n\}_n$ are equivalent Cauchy sequences.*

Proof By construction, $p_n \leq p_{n+1} \leq q_{n+1} \leq q_n$ for each n. It follows that $p_n \leq p_{n+k} \leq q_{n+k} \leq q_n$ for each k. Thus, if $m \geq n$, $|p_n - p_m| \leq |p_n - q_n| \leq 2^{1-n}$. Since $\{2^{1-n}\}_n$ is a zero sequence, we see that $\{p_n\}_n$ is a Cauchy sequence. By the same token, $\{q_n\}_n$ is a Cauchy sequence. Since $q_n - p_n = 2^{1-n}$, $\{p_n\}_n \sim \{q_n\}_n$.   □

Theorem 12.52 (Existence of the Least Upper Bound) *Let A be a non-empty subset of $\mathbb{R}$. If A has an upper bound, then there exists in $\mathbb{R}$ a least upper bound of A.*

Proof We begin with the sequences $\{p_n\}_n$ and $\{q_n\}_n$ constructed above. Since they are equivalent Cauchy sequences, they represent the same real number r. We claim that $r = \sup A$.

Suppose r is not an upper bound of A. Then there exists $a \in A$ such that $r < a$. By Theorem 12.48 we can select a rational number q with $r < q < a$. The sequence $n \mapsto q - q_n$ represents the number $q - a \in \mathbb{R}$. Hence this sequence must be eventually positive., and there exists $N \in \mathbb{N}$ such that $q > q_n$ for each $n \geq N$. This implies $q_n < q < a$ for each $n \geq N$, so that q_n is not an upper bound of A, against the property of q_n. This contradiction proves that r is an upper bound of A.

To see that $r = \sup A$, we suppose that A possesses an upper bound $s \in \mathbb{R}$ with $s < r$. Again, we find $p \in \mathbb{Q}$ such that $s < p < r$, and $n \mapsto p_n - p$ is eventually positive. As above, $p_n > p > s$ for large $n \in \mathbb{N}$. This implies that p_n is an upper bound of A, against the main property of p_n. This contradiction shows that no real number less than r can be an upper bound of A, i.e. $r = \sup A$.   □

12.9 About the Uniqueness of $\mathbb{R}$

One of the most *folkloristic* statements of Mathematical Analysis says that the system of real numbers is unique, up to isomorphisms. However, the vast majority of the textbooks that I have used in my life omit a rigorous proof of this (true) fact. Following the survey [1], we will learn that all models (i.e. constructions) of $\mathbb{R}$ are indeed equivalent up to renaming objects.

Definition 12.76 Let $\mathbb{K} = (\mathbb{K}, +, \cdot, \leq)$ be a totally ordered field, in the sense described in Chap. 3. For any non-empty subsets X and Y of $\mathbb{K}$, we write $X \leq Y$ to mean that

$$\forall x \forall y (x \in X \wedge y \in Y \implies x \leq y).$$

Definition 12.77 (Dedekind Completeness) The field $\mathbb{K}$ is Dedekind complete if and only if for every non-empty subsets X and Y of $\mathbb{K}$ such that $X \leq Y$, there exists $z \in K$ with the property that

$$\forall x \forall y (x \in X \wedge y \in Y \implies x \leq z < y)$$

The element z separates X and Y, in the sense of the order relation $\leq$.

As a first step, we prove that Dedekind completeness is equivalent to the property of the least upper bound.

Theorem 12.53 *Let $\mathbb{K}$ be a totally ordered field. The following properties are equivalent:*

(i) *$\mathbb{K}$ is Dedekind complete;*
(ii) *every non-empty subset X of $\mathbb{K}$ which is bounded from above possesses a least upper bound in $\mathbb{K}$;*
(iii) *every non-empty subset Y of $\mathbb{K}$ which is bounded from below possesses a least upper bound in $\mathbb{K}$;*

Proof Suppose (i) holds true, and let X be a non-empty subset of $\mathbb{K}$ which is bounded from above. The set $\mathcal{U}$ of all upper bounds of X is non-empty, and trivially $X \leq \mathcal{U}$. By (i), there exists $z \in \mathbb{K}$ such that $X \leq \{z\} \leq \mathcal{U}$. Since $z \in \mathcal{U}$, we conclude that $z = \min \mathcal{U}$, which implies that z is the least upper bound of X. Hence (ii) is proved.

Suppose now that (ii) holds, and let X, Y be two non-empty subsets of $\mathbb{K}$ such that $X \leq Y$. Since every element of Y is an upper bound of X, the set $\mathcal{U}$ of all upper bound of X is non-empty. By (ii), there exists $z \in \mathbb{K}$ such that $z = \sup X = \min \mathcal{U}$. But $Y \subset \mathcal{U}$, hence $z = \min Y$ and $\{z\} \leq Y$. On the other hand, $X \leq \{z\}$ because $z \in \mathcal{U}$, and thus (i) is proved.

The proof that (i) is equivalent to (iii) is similar, and it is left as an exercise. $\square$

Here comes the most technical part: we will be somehow sketchy, and refer to [1] for the details. First of all, any ordered field contains an isomorphic copy of the rational numbers.

Theorem 12.54 *Let* $\mathbb{K}$ *be an ordered field. Then there exists a subfield* $\mathbb{Q}(\mathbb{K})$ *of* $\mathbb{K}$ *isomorphic to* $\mathbb{Q}$.

Proof We temporarily denote by $1_{\mathbb{K}}$ the unit element (with respect to the multiplication) of $\mathbb{K}$. First we embed $\mathbb{N}$ into $\mathbb{K}$ as follows:

$$\varphi : n \in \mathbb{N} \mapsto 1_{\mathbb{K}} + \cdots + 1_{\mathbb{K}},$$

where the right-hand side contains n terms. It is easy to check that $\varphi : \mathbb{N} \to \mathbb{K}$ is an injective homomorphism which preserves the order.

Then we define $\psi : \mathbb{Z} \to \mathbb{K}$ as follows:

$$\psi(n) = \begin{cases} \varphi(n) & \text{if } n > 0 \\ 0 & \text{if } n = 0 \\ (-1_{\mathbb{K}})\varphi(-n) & \text{if } n < 0. \end{cases}$$

Finally, the required copy of the rational numbers is the range of the function $e : \mathbb{Q} \to \mathbb{K}$ defined as follows: for every $r = m/n \in \mathbb{Q}$ such that $m \in \mathbb{Z}, \in \mathbb{N}$, $n > 0$ and either $m = 0$ or m and n are coprime, we let

$$e(r) = \psi(m)\,(\varphi(n))^{-1}.$$

It can be verified that e is an injective homomorphism which preserves the order relation. $\square$

It follows from the previous result that we may systematically write $\mathbb{Q}$ to denote the copy of the standard rational numbers embedded into a given ordered field.[2]

With a minor modification of the proof of Theorem 3.14, we can see that the Archimedean property of $\mathbb{K}$ implies the density of $\mathbb{Q}$ in $\mathbb{K}$.

Theorem 12.55 *Let* $\mathbb{K}$ *be a totally ordered field with the Archimedean property. Let* $\varphi : \mathbb{K} \to \mathbb{K}$ *be an increasing function, in the sense that*

$$\forall x \forall y (x \in \mathbb{K} \wedge y \in \mathbb{K} \wedge x < y \implies \varphi(x) < \varphi(y)).$$

If $\varphi(x) = x$ *for every* $x \in \mathbb{Q}$, *then* $\varphi(x) = x$ *for every* $x \in \mathbb{K}$.

---

[2] More precisely, we identify $\mathbb{Q}(\mathbb{K})$ and $e(\mathbb{Q})$.

Proof Pick $x \in \mathbb{K}$ and $\varepsilon > 0$. Since $\mathbb{Q}$ is dense in $\mathbb{K}$, there exist numbers $p \in \mathbb{Q}$ and $q \in \mathbb{Q}$ such that

$$x - \varepsilon < p < x < q < x + \varepsilon.$$

The assumptions imply that

$$\varphi(p) = p < \varphi(x) < q = \varphi(q),$$

which in turn imply $x - \varepsilon < \varphi(x) < x + \varepsilon$. Since ε does not depend on x, we conclude that $\varphi(x) = x$, and the proof is complete.   □

We are ready to learn that Dedekind complete totally ordered fields are unique up to monotonic isomorphisms. Although the meaning of the terms could be already clear, we formalize some definitions.

Definition 12.78 Let $\mathbb{K}_1 = (\mathbb{K}_1, +_1, \cdot_1, \leq_1)$ and $\mathbb{K}_2 = (\mathbb{K}_2, +_2, \cdot_2, \leq_2)$ be ordered fields. A function $\varphi \colon \mathbb{K}_1 \to \mathbb{K}_2$ is an increasing isomorphism if and only if φ is bijective and

$$\forall x \forall y (x \in \mathbb{K}_1 \wedge y \in \mathbb{K}_2 \wedge x \leq_1 y \implies \varphi(x) \leq_2 \varphi(y)).$$

Theorem 12.56 (Uniqueness up to Isomorphisms) *For every Dedekind complete totally ordered fields $\mathbb{K}_1$ and $\mathbb{K}_2$ there exists an increasing isomorphism $\varphi \colon \mathbb{K}_1 \to \mathbb{K}_2$.*

Proof Let $\mathbb{K}_1 = (\mathbb{K}_1, +_1, \cdot_1, \leq_1)$ and $\mathbb{K}_2 = (\mathbb{K}_2, +_2, \cdot_2, \leq_2)$ be Dedekind complete totally ordered fields. To ease notation, we will omit the subscripts 1 and 2 in the rest of the proof.

For every $x \in \mathbb{K}_1$ we consider

$$L_x = \{q \in \mathbb{Q} \mid q \leq x\}.$$

It follows that L_x is bounded from above in $\mathbb{Q}$. Understanding the identification of the rational numbers in $\mathbb{K}_1$ and $\mathbb{K}_2$, we deduce that L_x is bounded from above also in $\mathbb{K}_2$. The Dedekind completeness of $\mathbb{K}_2$ allows us to define $\varphi \colon \mathbb{K}_1 \to \mathbb{K}_2$ by declaring that $\varphi(x)$ is the least upper bound in $\mathbb{K}_2$ of L_x.

If $x < y$ in $\mathbb{K}_1$, then $L_x \subset L_y$ by density of $\mathbb{Q}$ in $\mathbb{K}_1$, so that φ is increasing. Furthermore, if $x \in \mathbb{Q}$, then x is trivially the maximum of L_x in $\mathbb{K}_2$, and then $\varphi(x) = x$.

By the same token, we can construct an increasing function $\psi \colon \mathbb{K}_2 \to \mathbb{K}_1$ such that ψ is the identity on $\mathbb{Q}$. If we consider the functions $\psi \circ \varphi$ and $\varphi \circ \psi$, we may apply Theorem 12.55 and conclude that these two compositions coincide with the identities of $\mathbb{K}_1$ and $\mathbb{K}_2$, respectively. In conclusion, φ turns out to be an order-preserving bijection.

To complete the proof, we must ensure that φ also preserves the algebraic operations., namely

(a) $\varphi(x + y) = \varphi(x) + \varphi(y)$;
(b) $\varphi(xy) = \varphi(x)\varphi(y)$

for every $x \in \mathbb{K}_1$, $y \in \mathbb{K}_1$. We will exploit the density of the rational numbers: more precisely, we claim that, given $x \in \mathbb{K}_1$, $y \in \mathbb{K}_2$ and $\varepsilon > 0$, there results

$$\varphi(x) + \varphi(y) - \varepsilon < \varphi(x + y) < \varphi(x) + \varphi(y) + \varepsilon, \tag{12.1}$$

$$\varphi(x)\varphi(y) - \varepsilon < \varphi(xy) < \varphi(x)\varphi(y) + \varepsilon. \tag{12.2}$$

Fix rational numbers p_1, q_1, p_2 and q_2 such that

$$x - \frac{\varepsilon}{2} < p_1 < x < p_2 < x + \frac{\varepsilon}{2}, \quad y - \frac{\varepsilon}{2} < q_1 < y < q_2 < x + \frac{\varepsilon}{2}.$$

It follows from the properties of φ that

$$\varphi(x) < \varphi\left(p_1 + \frac{\varepsilon}{2}\right),$$

or

$$\varphi(x) - \frac{\varepsilon}{2} < p_1.$$

Similarly it follows that

$$\varphi(x) - \frac{\varepsilon}{2} < p_1$$

$$p_2 < \varphi(x) + \frac{\varepsilon}{2}$$

$$\varphi(y) - \frac{\varepsilon}{2} < q_1$$

$$q_2 < \varphi(y) + \frac{\varepsilon}{2}.$$

Putting these inequalities together, we deduce that

$$\varphi(x) + \varphi(y) - \varepsilon < p_1 + q_1, \quad p_2 + q_2 < \varphi(x) + \varphi(y) + \varepsilon.$$

But $p_1 + q_1 < x + y < p_2 + q_2$, hence $p_1 + q_1 < \varphi(x + y) < p_2 + q_2$, and the proof of (12.1) follows.

We prove (12.2) under the additional assumptions that $0 \leq x$ and $0 \leq y$ in $\mathbb{K}_1$. The general case follows by replacing x and y with $-x$ and $-y$.

Let $r \in \mathbb{Q}$ be such that $r > 0$ and

$$r < x, \quad r < y, \quad r < \frac{\varepsilon}{2(x+y)}, \quad r^2 < \frac{\varepsilon}{2}.$$

Fix rational numbers p_1, q_1, p_2 and q_2 such that

$$x - r < p_1 < x < p_2 < x + r, \quad y - r < q_1 < y < q_2 < x + r.$$

It follows that $p_1 q_1 < xy < p_2 q_2$, $p_2 - r < x < p_1 + r$, $q_2 - r < y < q_1 + r$. The properties of φ imply that $p_1 q_1 < \varphi(xy) < p_2 q_2$, and $p_2 - r < \varphi(x) < p_1 + r$, $q_2 - r < \varphi(y) < q_1 + r$. Then

$$(p_2 - r)(q_2 - r) < \varphi(x)\varphi(y) < (p_1 + r)(q_1 + r).$$

This implies that

$$\varphi(xy) - \varphi(x)\varphi(y) < p_2 q_2 - (p_2 - r)(q_1 + r) = r(p_2 + q_2 - r)$$
$$< r(x + y + r)$$
$$\varphi(x)\varphi(y) - \varphi(x)\varphi(y) > p_1 q_1 - (p_1 + r)(q_1 + r) = -r(p_1 + q_1 + r)$$
$$> -r(x + y + r).$$

We conclude that

$$|\varphi(xy) - \varphi(x)\varphi(y)| < r(x + y) + r^2 < \varepsilon,$$

and the proof of (12.2) is complete.   □

At the end of this journey, we must observe that sentences like

$\mathbb{R}$ is the unique totally ordered field with the least upper bound property

remain logically ambiguous. Theorem 12.56 ensures that we can speak of real numbers only up to *relabeling* elements in an increasing manner. Since Analysts need to make computations, what really matters to their eyes are *models* of $\mathbb{R}$. We have seen Dedekind cuts and equivalence classes of rational Cauchy sequences, but other modes are possible. In everyday life, most mathematicians simply *use* the formal properties of the real numbers to carry on.

References

1. G. Devillanova, G. Molica Bisci, The fabuloud destiny of Richard Dedekind. Atti Accad. Peloritana Pericolanti **99**(S1), A18 (2021)
2. T.J. Jech, *The Axiom of Choice* (North-Holland Publishing Company, 1973)

3. J.L. Kelley, *General Topology*. Graduate Texts in Mathematics, No. 27 (Springer, New York, 1975). Reprint of the 1955 edition [Van Nostrand, Toronto, Ont.]
4. W. Rudin, *Principles of Mathematical Analysis*. International Series in Pure and Applied Mathematics, 3rd edn. (McGraw-Hill Book Co., New York, 1976)

Chapter 13
Neighbors Again: Topological Spaces

Abstract The theory of Calculus can be successfully developed in the restricted framework of the real line. We have seen that $\mathbb{R}$ enjoys several properties that are related to the concept of *distance* between two numbers. In this chapter we propose a robust introduction to General Topology, which is the branch of mathematics that deals with the idea of proximity.

> I am happy to say that the disease of axiomatic topology has been almost totally cured. Right now I don't care a bit whether every β-capsule of type Δ is also a T-spot of the second kind.
>
> (Edwin Hewitt)

13.1 Topological Spaces

Definition 13.1 A topology is a family τ of sets which satisfies two conditions:

(τ_1) the intersection of any two members of τ is a member of τ;
(τ_2) the union of the members of any subfamily of τ is a member of τ.

A subset U of X is open if and only if $U \in \tau$. The set

$$X = \bigcup \tau = \bigcup \{U \mid U \in \tau\}$$

is usually called the space of the topology τ, and τ is a topology for X. The pair (X, τ) is a topological space.

Remark 13.1 Our conditions imply that $X = \bigcup \tau$ is necessarily a member of τ, since τ is a subfamily of itself and every member of τ is a subset of X. Similarly, the empty set $\emptyset$ is a member of τ, since it is the union of all the elements of the empty subfamily of τ.

© The Author(s), under exclusive license to Springer Nature Switzerland AG 2022
S. Secchi, *A Circle-Line Study of Mathematical Analysis*,
La Matematica per il 3+2 141, https://doi.org/10.1007/978-3-031-19738-3_13

Important: About Language

We admit that our language here may sound peculiar. Most textbooks of General Topology begin with a set X and define a topology τ as a family of subsets of X which contains X, $\emptyset$, arbitrary unions and finite intersections of its members. The two approaches are indeed equivalent, since X is always the union of all its open subsets.

Example 13.1 For a given set X, the topology $\tau = \{X, \emptyset\}$ is called the indiscrete topology. This topology possesses only two open sets.

Example 13.2 For a given set X, the topology $\tau = \{U \mid U \subset X\}$ is called the discrete topology. Every set is open in this topology.

Exercise 13.1

1. Let $X = \{1, 2, 3\}$ and $\tau = \{\emptyset, \{1, 2, 3\}, \{1, 2\}, \{3\}\}$. Show that τ is a topology on X.
2. Let $X = \{1, 2, 3\}$ and $\tau = \{\emptyset, \{1, 2, 3\}, \{1, 2\}, \{1, 3\}\}$. Show that τ is not a topology on X.
3. Let $X = \{1, 2\}$. Write down all sets of subsets of X, and decide which are topologies on X.
4. Let τ consist of $\emptyset$, $\mathbb{R}$, and all sets of the form $[a, +\infty), a \in \mathbb{R}$. Show that τ is not a topology on $\mathbb{R}$.

Exercise 13.2 Show that $\tau = \{S \subset \mathbb{R} \mid \mathbb{R} \setminus S$ is a finite set$\} \cup \{\emptyset\}$ is a topology on $\mathbb{R}$. This is the topology of finite complements.

Definition 13.2 Let τ_1 and τ_2 be two topologies for a set X. If $\tau_1 \subset \tau_2$, we say that τ_1 is smaller than τ_2.

Remark 13.2 If τ_1 is smaller than τ_2, many mathematicians say that τ_1 is stronger than τ_2, or equivalently that τ_2 is weaker than τ_1. This is somehow counterintuitive, since a smallness is seldom associated to strongness. For this reason we will try to use the set-theoretic language of smaller and larger in the rest of this chapter.

Exercise 13.3 Let X be a set, and let τ be a topology for X. Show that any open set for τ is also an open set for the discrete topology, and that any open set for the indiscrete topology is also an open set for τ. Deduce that the indiscrete and the discrete topology for a set X are respectively the largest and the smallest topology for X.

Example 13.3 Let us go back to the real line. The usual topology of $\mathbb{R}$ is the family of all subsets U of $\mathbb{R}$ such that for any $x \in U$ there exists an open interval (a, b) such that $(a, b) \subset U$.

Exercise 13.4 Prove that the usual topology of $\mathbb{R}$ is indeed a topology in the sense of our definition. In particular, notice that any open interval is an open set for the usual topology.

Exercise 13.5 We define a family σ of subsets of $\mathbb{R}$ as follows: $W \in \sigma$ if either $W = \emptyset$, or $\mathbb{R} \setminus W$ is a finite set. Prove that σ is a topology on $\mathbb{R}$, and that this topology is smaller that the usual topology of $\mathbb{R}$.

Definition 13.3 A set U of a topological space (X, τ) is a neighborhood of a point x if and only if U contains an open set to which x belongs.

Remark 13.3 A neighborhood need not be an open set. The reader should be aware that several mathematicians say that U is a neighborhood of x if and only if $x \in U$ *and U is open*. The two definitions are not equivalent, of course.

Example 13.4 In $\mathbb{R}$ with the usual topology, $(0, 1)$ is clearly a neighborhood of $1/2$. But also $[0, 1)$ is a neighborhood of $1/2$. On the contrary, $(0, 1)$ is not a neighborhood of 0, and $[0, 1)$ is not a neighborhood of 0.

Exercise 13.6 If $x \notin U$, can U be a neighborhood of x? If $x \in U$, is U necessarily a neighborhood of x? Discuss.

The whole topology of a space can be described in terms of its neighborhoods. The next result characterizes open sets.

Theorem 13.1 *A set is open if and only if it contains a neighborhood of each of its points.*

Proof Let A a subset of a topological space. If A is open, then $A \subset A$, so that it contains a neighborhood of each of its points. Conversely, suppose that A contains a neighborhood of each of its points. The union U of all open subsets of A is an open set, according to the definition of a topology. Then each $x \in A$ belongs to some open subset of A, and hence $x \in U$. Hence $A = U$, and A is open. □

Definition 13.4 The neighborhood system of a point is the family of all neighborhoods of that point.

Definition 13.5 (Closed Sets) In a topological space (X, τ), a subset A is closed if and only if $X \setminus A$ is an open set.

Exercise 13.7 Recall that $X \setminus (X \setminus A) = A$. Deduce that the complement of a closed set is an open set. In conclusion, a subset is open if and only if its relative complement is closed.

Exercise 13.8 Luckily enough, any closed interval $[a, b]$ is a closed set of $\mathbb{R}$ with its usual topology. Prove this.

Exercise 13.9 What are the closed sets for the discrete and the indiscrete topologies on a give set X?

Exercise 13.10 Using De Morgan's laws from Set Theory, prove that the union of any two closed subsets is a closed subset. Furthermore, the intersection of any subfamily of closed sets is a closed set.

Example 13.5 The topology of upper semicontinuity on $\mathbb{R}$ is the topology whose open sets are all the subsets of the form $(-\infty, a)$, where $a \in \mathbb{R}$.

Example 13.6 The cofinite topology on a set X is the topology whose closed sets are those subsets C such that either $X = C$ or C is finite.

Definition 13.6 (Accumulation Points) A point x is an accumulation point of a subset A of a topological space (X, τ) if and only if every neighborhood of x contains points of A different than x. The set A' of all the accumulation points of A is called the derived set of A.[1]

Theorem 13.2 *A subset is closed if and only if it contains its derived set.*

Proof Indeed, if A is a subset, each neighborhood of a point x intersects A if and only if either $x \in A$ or x is an accumulation point of A.   □

Theorem 13.3 *If A is a subset, $A \cup A'$ is a closed set.*

Proof If x is neither a point of A nor an accumulation point of A, then there exists an open neighborhood U of x which does not intersect A. But U is a neighborhood of each of its points, and none of these are accumulation points of A. This proves that the union $A \cup A'$ is the complement of an open set.   □

Definition 13.7 (Closure) The closure $\overline{A}$ of a subset A of a topological space (X, τ) is the intersection of all closed sets that contain A. Hence $\overline{A}$ is the smallest[2] closed set containing A.

Theorem 13.4 *The closure of any set if the union of the set and of its derived set.*

Proof Every accumulation point of a set A in an accumulation point of each set containing A, and is therefore a member of each closed set containing A. Hence $\overline{A}$ contains both A and A'. Conversely, $A \cup A'$ is closed by Theorem 13.3, and therefore it contains $\overline{A}$.   □

Exercise 13.11 Let $A = (-1, 3) \cup [5, 6)$ be a subset of $\mathbb{R}$ with the usual topology. What is $\overline{A}$?

Definition 13.8 (Dense Subsets) A set A is dense in a topological space X if and only if $\overline{A} = X$.

Example 13.7 $\mathbb{Q}$ is dense in $\mathbb{R}$, since for any real numbers $x < y$ there exists a rational number r such that $x < r < y$.

---

[1] Sometimes denoted by $\mathcal{D}A$ or DA.

[2] In the sense of inclusion of sets.

Definition 13.9 (Interior Points) A point x of a subset A of a topological space is an interior point of A if and only if A is a neighborhood of x. The set A° of all interior points of A is called the interior of A.

We collect several properties of the interior in a single statement.

Theorem 13.5 *Let A be a subset of a topological space X. Then the interior of A is open, and it is the largest open subset of A. A set A is open if and only if $A = A^\circ$. The set of all points that are not accumulation points of $X \setminus A$ is A°. The closure of $X \setminus A$ is $X \setminus A^\circ$.*

Proof If a point x belongs to A°, then x belongs to some open subset U of A. Every element of U is also an element of A°, and consequently A° contains a neighborhood of each of its points, and it is therefore open. Now, if V in an open subset of A and $y \in V$, then A is a neighborhood of y, so $y \in A^\circ$. This shows that A° contains each open subset of A and it is the largest open subset of A. If A is open, it is clearly equal to the largest open subset of A. Hence A is open if and only if $A + A^\circ$.

If x is a point of A which is not an accumulation point of $X \setminus A$, then there exists a neighborhood U of x which does not intersect $X \setminus A$ and is therefore a subset of A. Then A° is a neighborhood of x and $x \in A^\circ$. Conversely, A° is a neighborhood of each of its points and A° does not intersect $X \setminus A$. Hence there is no point A° which is an accumulation point of $X \setminus A$.

To conclude, since A° consists of the points of A which are not accumulation points of $X \setminus A$, the complement $X \setminus A^\circ$ consists of the points which are either points of $X \setminus A$ or accumulation points of $X \setminus A$; In other words, A° is the complement of $\overline{X \setminus A}$. □

Definition 13.10 (Boundary of a Subset) The boundary of a subset A of a topological space X is the set of all points x such that every neighborhood of x intersects both A and $X \setminus A$. The boundary of A is denoted by ∂A.

Exercise 13.12 Prove that ∂A is precisely the set of all points that are interior to neither A nor $X \setminus A$.

Exercise 13.13 Prove that $\partial A = \partial(X \setminus A)$.

Example 13.8 If X has the indiscrete topology and if $A \neq X$, $A \neq \emptyset$, then $\partial A = X$. If X has the discrete topology, the boundary of every subset is empty.

Exercise 13.14 Let $[a, b]$ a proper interval in $\mathbb{R}$ with the usual topology. Check that $\partial[a, b] = \{a, b\}$.

Exercise 13.15 Prove that the boundary of both $\mathbb{Q}$ and $\mathbb{R} \setminus \mathbb{Q}$ is $\mathbb{R}$. *Hint:* remember that between any two distinct real numbers there is a rational number.

13.2 The Special Case of $\mathbb{R}^N$

In the first part of the book we have extensively described the real line $\mathbb{R}$ and its properties. We have seen that a distance function exists on it: $d(x, y) = |x - y|$. In terms of this distance we have defined open sets, accumulation points, and so forth. Our aim now is to introduce finitely many additional *dimensions* to the space $\mathbb{R} = \mathbb{R}^1$, a process that is needed is almost every application of modern mathematics to the real world.

Definition 13.11 Let $N \geq 1$ be an integer. The (vector) space $\mathbb{R}^N$ is the set of all N-tuples $x = (x_1, \ldots, x_N)$ of real numbers, together with the following operations: a sum defined by

$$x + y = (x_1 + y_1, \ldots, x_N + y_N)$$

for each $x = (x_1, \ldots, x_N)$ and $y = (y_1, \ldots, y_N)$, and a multiplication by a scalar defined by

$$\lambda x = (\lambda x_1, \ldots, \lambda x_N)$$

for each $x = (x_1, \ldots, x_N)$ and $\lambda \in \mathbb{R}$.

Remark 13.4 It is sometimes desirable to distinguish $\mathbb{R}^N$ as a set and $\mathbb{R}^N$ as a vector space. For instance, the set $\mathbb{C}$ of complex numbers is built on the *set* $\mathbb{R}^2$, but no multiplication by a real scalar is necessary at the beginning. More correctly, our vector space $\mathbb{R}^N$ should be denoted by $(\mathbb{R}^N, +, \cdot)$, but we will not be so pedantic in the sequel.

Here is how we introduce a topology on $\mathbb{R}^N$.

Definition 13.12 The usual distance on $\mathbb{R}^N$ is defined by

$$d(x, y) = \sqrt{\sum_{j=1}^{N} |x_j - y_j|^2}$$

for each $x = (x_1, \ldots, x_N)$ and $y = (y_1, \ldots, y_N)$. This distance is often called the Euclidean distance. The open ball centered at some $x \in \mathbb{R}^N$ with radius $r > 0$ is

$$B(x, r) = B_r(x) = \left\{ y \in \mathbb{R}^N \,\middle|\, d(x, y) < r \right\};$$

see Fig. 13.1. A subset A of $\mathbb{R}^N$ is open if for each $x \in A$ there exists $r > 0$ such that $B_r(x) \subset A$; see Fig. 13.2.

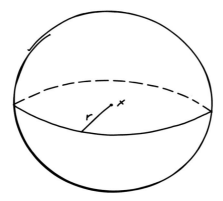

Fig. 13.1 The ball $B(x, r)$

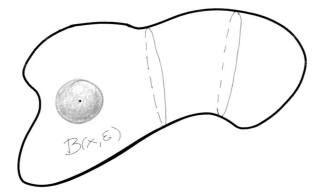

Fig. 13.2 An open subset

Remark 13.5 From the topological viewpoint, $\mathbb{R}^N$ is a metric space. But the vector space structure is often so important that mathematicians prefer to see $\mathbb{R}^N$ as a normed vector space. For this reason the usual distance $d(x, y)$ is often denoted by $\|x - y\|$. This is indeed the standard notation for any distance induced by a norm $\|\cdot\|$. We will come back to normed vector spaces in the chapter about differentiation in abstract spaces.

Once we can recognize open sets in $\mathbb{R}^N$, we can speak of closed sets, accumulation points, closure, interior, and so on. If not explicitly stated, we will always endow $\mathbb{R}^N$ with the usual distance and the usual topology induced by this distance.

13.3   Bases and Subbases

As a rule, in order to define a topology it is not strictly necessary to specify the family of all open subsets. Exactly as a maximal set of linearly independent vectors spans the whole vector space (via finite linear combinations), a maximal family of subsets generates all open sets.

Definition 13.13 (Base)   A family $\mathcal{B}$ is a base for a topology τ if and only if $\mathcal{B} \subset \tau$ and for each point x of the space, and each neighborhood U of x, there exists a member V of $\mathcal{B}$ such that $x \in V \subset U$.

Example 13.9   The collection of all open intervals is a base for the usual topology of $\mathbb{R}$. Notice that this is essentially a restatement of the definition of open sets in $\mathbb{R}$.

Example 13.10   We consider the family of all intervals $[a, b)$ in $\mathbb{R}$. As a and b range over $\mathbb{R}$, this family if the base for a topological space called the Sorgenfrey line.

Exercise 13.16   Prove that the topology of the Sorgenfrey line is larger than the usual topology of $\mathbb{R}$. *Hint:* just observe that $(a, b) = \bigcup_{c>a}[c, b)$.

Thus a base for a topology must be a family of open subsets. Can we characterize those families of open subsets that are a base for the topology? The answer is contained in the next theorem.

Theorem 13.6   *A family $\mathcal{B}$ of sets is a base for some topology for the set $X = \bigcup\{B \mid B \in \mathcal{B}\}$ if and only if for every two elements U and V of $\mathcal{B}$ and each point $x \in U \cap V$, there exists $W \in \mathcal{B}$ such that $x \in W \subset U \cap V$.*

Proof   Suppose first that $\mathcal{B}$ is a base for some topology, U and V are members of $\mathcal{B}$, and $x \in U \cap V$. Since $U \cap V$ is open, some $W \in \mathcal{B}$ contains x and is contained in $U \cap V$. Conversely, let τ be the union of all members of $\mathcal{B}$. Clearly any union of members of τ is a union of members of $\mathcal{B}$, hence it is a member of τ. We only need to show that the intersection of any two members U, V of τ is a member of τ. But if $x \in U \cap V$, we can choose U_1 and V_1 in $\mathcal{B}$ and select $W \in \mathcal{B}$ such that $x \in U_1 \subset U$ and $x \in V_1 \subset V$. Hence there exists $W \in \mathcal{B}$ such that $x \in W \subset U_1 \cap V_1 \subset U \cap V$. This shows that $U \cap V$ is the union of members of $\mathcal{B}$, and τ is a topology.   □

Definition 13.14 (Subbase)   A family $\mathcal{S}$ of sets is a subbase for a topology τ if and only if the family of all finite intersections of members of $\mathcal{S}$ is a base for τ.

Example 13.11   Since $(a, b) = (-\infty, b) \cap (a, +\infty)$, the family of all half-lines is a subbase for the usual topology of $\mathbb{R}$.

Subbases are a very general and flexible tool for topologizing sets.

Theorem 13.7   *Let X be a non-empty set, and let $\Sigma = \{A_\alpha \mid \alpha \in \mathcal{A}\}$ be any collection of subsets of X. There exists a unique topology $\tau(\Sigma)$ on X which contains Σ. This topology is the smallest topology with this property, and it can be described as follows: it consists of $\emptyset$, X, all finite intersections of elements of Σ, and all unions of these finite intersections.*

We say that $\tau(\Sigma)$ is the topology on X generated by the collection Σ.

Proof It is obvious that 2^X (the collection of all subsets of X) is a topology on X containing Σ. Therefore the intersection $\tau(\Sigma)$ of all topologies on X which contain Σ is non-empty. It is a simple exercise to check that $\tau(\Sigma)$ is a topology on X containing Σ, and it is both unique and the smallest one with such a property. To prove the stated description of open sets, we notice that $\tau(\Sigma)$ contains Σ, and thus it must contain all the sets listed in the statement of the theorem. On the other hand, since unions distribute over intersections, the sets listed actually form a topology on X containing Σ, and therefore such a topology must contain $\tau(\Sigma)$. The proof is complete.   □

13.4 Subspaces

Imagine we have a topological space (X, τ), and that Y is a subset of X. A natural question at this point is whether Y can be endowed with some topology related to τ. We answer this question in the following definition.

Definition 13.15 (Relative Topology) Let (X, τ) be a topological space, and let Y be a subset of X. The induced (or relative) topology on Y is the family of all intersections of members of τ with Y, and is denoted by $\tau|Y$. Explicitly,

$$\tau|Y = \{V \cap Y \mid V \in \tau\}.$$

We say that $(Y, \tau|Y)$ is a subspace of (X, τ).

Example 13.12 Let $Y = [0, 1]$. Each set of the form $(a, b) \cap [0, 1]$ is an open set in the induced topology on Y. Observe in particular that $[0, 1/2)$ is open in the relative topology of Y, but not in the usual topology of $\mathbb{R}$.

This examples shows a general fact: if $A \subset Y \subset X$, the set A may be open in the relative topology of Y, but this does not imply that A is open as a subset of X. Trivially, Y is always open — and closed — in the relative topology of Y.

Exercise 13.17 Describe the topology induced by the indiscrete and the discrete topologies on a generic subset.

Theorem 13.8 *Let (X, τ) be a topological space, and let Y be a subset of X. If A is a subset of Y, then:*

 (i) *the set A is closed in $\tau|Y$ if and only if it is the intersection of Y and a set closed in τ.*
 (ii) *A point $y \in Y$ is an accumulation point of A in $\tau|Y$ if and only if it is an accumulation point of A in τ.*
 (iii) *The closure of A in $\tau|Y$ is the intersection of Y and the closure of A in τ.*

Proof The set A is closed in Y if and only if $Y \setminus A$ has the form $V \cap X$ for some $V \in \tau$. This is true if and only if $A = (X \setminus V) \cap Y$, and (i) is proved. Statement (ii) follows directly from the definitions of relative topology and accumulation point. Finally, the closure of A in $\tau | Y$ is the union of A and the set of accumulation points of A in $\tau | Y$. Then (iii) follows from (ii).   □

Exercise 13.18 Prove that if Y is open in X, each open set of $(Y, \tau | Y)$ is also open in X. Is this true for closed sets, provided that Y is closed in X?

As a matter of fact, the information that a subset A is open in the relative topology of Y tells very little about the properties of A as a subset of X.

Example 13.13 Suppose that $X = Y \cup Z$, and $A \subset X$ is such that $A \cap Y$ is open in Y, $A \cap Z$ is open in Z. Although we might expect that A be open in X, this is false. Just take $A = Y$, $Z = X \setminus Y$: clearly $Y \cap Y$ and $Y \cap Z$ are open in Y and Z respectively, but Y may well be any subset of X.

13.5 Connected Spaces

Connected topological spaces are, from an analyst's viewpoint, the straightforward generalization of *intervals* in the real line. Recall that a (nonempty) set A in $\mathbb{R}$ is an interval if and only if, for each $x < y$ in A, each number z with $x < z < y$ is a member of A. Roughly speaking, intervals cannot have holes.

Definition 13.16 (Connected Spaces) A topological space X is connected if and only if the only subsets of X which are both open and closed are X and $\emptyset$. Disconnected is the logical negation of connected. A set $Y \subset X$ is connected if and only if Y is connected in the induced topology.

Exercise 13.19 Prove that a topological space X is connected if and only if X is not the union of two disjoint open subsets. Equivalently, X is connected if and only if X is not the union of two disjoint closed subsets. See Fig. 13.3.

This is by far the most elegant definition of connectedness. However an equivalent definition can be provided in terms of *separation*.

Definition 13.17 Two subsets A and B are separated in a topological space X if and only if $\overline{A} \cap B = \emptyset$ and $A \cap \overline{B} = \emptyset$.

Theorem 13.9 *Let X be a topological space. A subspace Y of X is connected if and only if Y is not the union of two non-empty separated sets.*

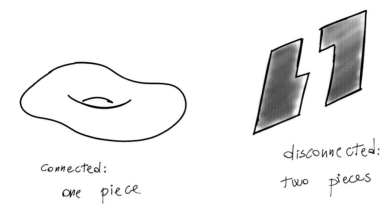

connected:

one   piece

disconnected:

two   pieces

Fig. 13.3 Connected vs. disconnected

Proof Suppose that Y is disconnected, so that $Y = A \cup B$, and A, B are non-empty open subsets of Y such that $A \cap B = \emptyset$. Clearly A and B are separated in X, since

$$A \cap \mathrm{Cl}(B, X) = (A \cap Y) \cap \mathrm{Cl}(B, X)$$
$$= H \cap (Y \cap \mathrm{Cl}(B, X))$$
$$= A \cap \mathrm{Cl}(B, Y) = \emptyset,$$

and similarly for $\mathrm{Cl}(A, X) \cap B$.[3] Conversely, if A and B are separated in X and $Y = A \cup B$, then

$$\mathrm{Cl}(A, Y) = Y \cap \mathrm{Cl}(A, X) = (A \cup B) \cap \mathrm{Cl}(A, X)$$
$$= (A \cap \mathrm{Cl}(A, X)) \cup (B \cap \mathrm{Cl}(A, X))$$
$$= A,$$

so that A is closed in Y. A similar computation proves that B is closed in Y.   □

The topology of the real line is intimately related to the order properties, and this characterizes connected subsets.

Theorem 13.10 *A subset A of $\mathbb{R}$ (with the usual topology) is connected if and only if it has the following property: if $x \in A$, $y \in A$ and $x < z < y$, then $z \in A$.*

---

[3] We have denoted $\mathrm{Cl}(A, X)$ the closure of A in X, since we need to distinguish the closure in X and the closure in Y.

Proof Suppose there exist $x \in A$, $y \in A$ and $z \in (x, y)$ with $z \notin A$. Then $A = U_z \cup V_z$ with

$$U_z = A \cap (-\infty, z), \quad V_z = A \cap (z, +\infty).$$

Clearly U_z and V_z are nonempty. Since $U_z \subset (-\infty, z)$ and $V_z \subset (z, +\infty)$, they are separated. Hence A is the union of two nonempty separated sets, and is not connected.

Conversely, suppose that A is not connected, and has the form $A = U \cup V$ for some nonempty separated sets U and V. Let $x \in U$, $y \in V$, assuming $x < y$ for the sake of definiteness. We define $z = \sup (U \cap [x, y])$. As a supremum, we know that z is in the closure of U, so that $z \notin V$ because U and V are separated. In particular, $x \le z < y$. If $z \notin U$, it follows that $x < z < y$ and $z \notin A$. If $z \in U$, then $z \notin \overline{V}$, so that there exists z_1 such that $z < z_1 < y$ and $z_1 \notin V$. Again $x < z_1 < y$ and $z_1 \notin A$. □

Important: Beware!

The last result shows that the only connected subsets of the real line (with the usual topology) are the intervals, of any kind. This is no longer true if we replace $\mathbb{R}$ with $\mathbb{Q}$ with the induced topology. Indeed, for any irrational number a, the two sets $\{x \in \mathbb{Q} \mid x < a\}$ and $\{x \in \mathbb{Q} \mid x > a\}$ are separated in this topology.

Remark 13.6 In a geometric language, we have proved the connected subsets of $\mathbb{R}$ coincide with the *convex* subsets of $\mathbb{R}$. Indeed $A \subset \mathbb{R}$ is convex if and only if

$$\lambda x + (1 - \lambda)y \in A$$

for every $x \in A$, $y \in A$, $\lambda \in [0, 1]$.

Exercise 13.20 Prove the last statement of the warning above.

Connectedness is stable under closure and unions.

Theorem 13.11

(a) *The closure of a connected set is connected.*
(b) *If $\mathcal{A}$ is a family of connected sets of a topological space, and if no two members of $\mathcal{A}$ are separated, then $\bigcup \mathcal{A}$ is connected.*

Proof Suppose that Y is a connected subset of a topological space X, and that $\overline{Y} = A \cup B$ for some sets A and B that are both open and closed in $\overline{Y}$. Then each of $A \cap Y$ and $B \cap Y$ is open and closed in Y, and since Y is connected, one of these two sets must be empty. Suppose $B \cap Y$ is empty. Then Y is a subset of A and $\overline{Y}$ is a subset of A because A is closed in $\overline{Y}$. Hence $B = \emptyset$, and it follows that $\overline{Y}$ is connected. This proves (a).

To prove (b), we denote by C the union of the members of $\mathcal{A}$. Suppose that D is both open and closed in C. For each $A \in \mathcal{A}$, we have that $A \cap D$ is open and closed in A, and since A is connected we conclude that either $A \subset D$ or $A \subset C \setminus D$. We claim that either each member of $\mathcal{A}$ is contained in D and $C \setminus D = \emptyset$, or it is contained in $C \setminus D$ and $D = \emptyset$. Indeed, if A and B are members of $\mathcal{A}$, it is impossible that $A \subset D$ and $B \subset C \setminus D$, for in this case A and B would be separated as subsets of the separated sets D and $C \setminus D$. This contradicts the assumption, and proves the claim and the proof.   □

Definition 13.18 (Connected Components) A (connected) component of a topological space is a maximal connected subset. More precisely, it is a connected subset which is properly contained in no other connected subset.

Example 13.14 If a topological space is connected, then it is the only connected component. Indeed, if A is a connected subset, then A is contained in the connected topological space, and A cannot be maximal.

Example 13.15 In a discrete topological space, each connected component is a singleton.

Theorem 13.12 *Each connected subset of a topological space is contained in a connected component. Each component is closed. If A and B are distinct connected components of a space, then A and B are separated.*

Proof Let A be a nonempty connected subsets of a space X, and let C be the union of all the connected sets containing A. As a consequence of Theorem 13.11, C is connected. If D contains C and D is connected, then $D \subset C$, so that $D = C$. Hence C is a maximal connected subset, i.e. a connected component of the space. Each component C is connected by definition, so that its closure is connected by Theorem 13.11. Therefore $C = \overline{C}$, and C is closed. Finally, if A and B are disjoint components but they are not separated, then $A \cup B$ is a connected subset, a contradiction to the maximality of A and B.   □

Definition 13.19 (Totally Disconnected Spaces) A topological space X is totally disconnected if and only if the connected component containing any point $x \in X$ coincides with $\{x\}$.

The intuition behind connectedness is often expressed by saying that any two points of the space can be *connected* by a path. Although it can be proved the this is *not* an equivalent definition, it is nonetheless an *interesting* definition.

Definition 13.20 (Arc-Wise Connected Spaces) A topological space X is arc-wise connected if and only if for each points x_0 and x_1 in X there exists a continuous map $\alpha : [0, 1] \to X$ such that $\alpha(0) = x_0$ and $\alpha(1) = x_1$. See Fig. 13.4.

Fig. 13.4 A path joining two points

Theorem 13.13 *An arc-wise connected space is always connected.*

Proof Let X be a connected space, and suppose that $X = A \cup B$ for some open sets A and B. We will prove that $A \cap B \neq \emptyset$. Pick any two points $x_0 \in A$ and $x_1 \in B$, and consider a path α that joins them as in the previous definition. The subsets $\alpha^{-1}(A)$ and $\alpha^{-1}(B)$ are open in $[0, 1]$, non-empty, and their union is $[0, 1]$. Since $[0, 1]$ is connected, we must have $\alpha^{-1}(A) \cap \alpha^{-1}(B) \neq \emptyset$. But then $A \cap B \neq \emptyset$.   □

Example 13.16 The space $\mathbb{R}^N$ is arc-wise connected: if x_0 and x_1 are two points, the path $\alpha(t) = t x_0 + (1 - t) x_1$ is a continuous map that joins them. Using polar coordinates, it is easy to check that the unit sphere $\mathbb{S}^{N-1} = \{x \in \mathbb{R}^N \mid \|x\| = 1\}$ is arc-wise connected (by a geodesic arc).

13.6 Nets and Convergence

So far we have described topologies in terms of their open sets. This is by far the most common approach to General Topology. But there are other viewpoints which can be even more useful to mathematical analysts.

If we think of our previous chapters, we immediately see that elementary real analysis in one variable leans on a single idea: that of *limits*. In any reasonable generalization, limits should therefore play a crucial role. And indeed limits can be defined as soon as a topology exists on a set. However, in this section we want to define a broader definition that can include sequences, functions, Riemann sums, and much more.

Recall that a sequence is just a function defined on the complement of a finite subset N of $\mathbb{N}$. With an abuse of notation, and since our interest is towards the theory of convergence, we will assume that sequences are functions defined on the whole set $\mathbb{N}$ of natural numbers. Indeed, any two sequences which differ for finitely many terms have the same character.

The value of a sequence S at $n \in \mathbb{N}$ is denoted either by the functional symbol $S(n)$ of by the traditional symbol S_n. We say that a sequence S is in a set A if and only if $S_n \in A$ for each n.

Definition 13.21 A sequence S is eventually in a set A if and only if there exists an integer m such that $S_n \in A$ for each $n \geq m$.

Example 13.17 A sequence S of real numbers converges to a limit $L \in \mathbb{R}$ if and only if S_n is eventually in each neighborhood of L.

The definition of subsequence is usually hard to grasp at first sight. Since we are going to introduce a more general definition, it is advisable to look back to subsequences. Suppose a sequence S is given, and that we want to select an integer N_i for each integer i, such that $S_{N_i} = S(N(i))$ converges. A good candidate could be $N_i = 1$ for each i, since $S(N(i)) = S(1)$ for each i, and convergence would be ensured. But this is not an interesting candidate, and some additional condition on $i \mapsto N(i)$ should be imposed.

The usual condition, which we actually used in a previous chapter, is that $i < j$ imply $N(i) < N(j)$, i.e. a strict monotonicity condition. If this condition is satisfied, we then say that $\{S_{N_i}\}_i$ is a subsequence of S. But is this really *the* good condition? Well, not really. What we actually need is that N_i becomes large as i becomes large.

Definition 13.22 (Generalized Subsequences) T is a subsequence of S if and only if there exists a map $N : \mathbb{N} \to \mathbb{N}$ such that $T = S \circ N$, and for each integer m there exists an integer n such that $i \geq n$ implies $N_i \geq m$.

The previous discussion introduces the following problem: we want to construct *generalized* sequences on any topological space in such a way that the convergence of such generalized sequences characterize the topology of the space.

Definition 13.23 A binary relation $\geq$ directs a set D if and only if $D \neq \varnothing$ and

(a) if m, n and p are elements of D such that $m \geq n$ and $n \geq p$, then $m \geq p$;
(b) if $m \in D$ then $m \geq m$;
(c) if m and n are elements of D, then there exists $p \in D$ such that $p \geq m$ and $p \geq n$.

Example 13.18 The real numbers and the natural numbers are directed by their usual order $\geq$.

Example 13.19 The family of all neighborhoods of a point in a topological space is directed by reverse inclusion: $U \geq V$ if and only if $U \subset V$. Condition (c) is a consequence of the fact that the intersection of two neighborhoods is a neighborhood. This is a particularly important example, since it will join the usual definition of convergence is a topological space with the definition of convergence of a net.

Example 13.20 The family of all finite partitions of the closed interval $[a, b]$ is a directed set, when ordered by the relation $P_2 \geq P_1$ if and only if P_2 is a refinement of P_1.

Exercise 13.21 Prove that the family of all finite subsets of a set is directed by direct inclusion: $A \geq B$ if and only if $A \supset B$.

Definition 13.24 (Nets) A directed set is a pair $(D, \geq)$ such that $\geq$ directs D. A net is a pair $(S, \geq)$ such that S is a function and $\geq$ directs the domain of S. If S is a function whose domain contains D and D is directed by $\geq$, then $\{S_n, n \in D, \geq\}$ is the net $(S_{|D}, \geq)$. A net $\{S_n, n \in D, \geq\}$ is in a set A if and only if $S_n \in A$ for each $n \in D$; it is eventually in A if and only if there exists $m \in D$ such that $S_n \in A$ for each $n \geq m$. The net is frequently in A if and only if for each $m \in D$ there exists $n \in D$ such that $n \geq m$ and $S_n \in A$.

Important: Notation

The full notation $\{S_n, n \in D, \geq\}$ is too cumbersome to be currently used. We will write $\{S_n, n \in D\}$ when the domain of S plays an explicit role and no confusion can arise about the direction $\geq$ in D.

We have formally recovered most set-theoretic definitions of sequences in a general topological setting. We are ready for the most important one.

Definition 13.25 (Convergent Nets) A net $(S, \geq)$ in a topological space (X, τ) converges to a point $s \in X$ if and only if for each neighborhood U of s, S is eventually in U.

Important: Nets Are not Sequences!

When working with sequences, convergence is often stated in the following way: a sequence $\{S_n \mid n \in \mathbb{N}\}$ converges to x if and only if for each neighborhood V of x, the sequence $\{S_n \mid n \in \mathbb{N}\}$ is in V for all but finitely many indices n. This equivalent definition is based on the fact that the complement of each *tail* $\{n \in \mathbb{N} \mid n \geq n_0\}$ is a finite set. This is in general false, if $\mathbb{N}$ is replaced by a directed set D.

Example 13.21 Let $\{a_n\}_n$ be a sequence in $\mathbb{R}$. The sequence converges to a (finite) limit L if and only if the associated net $(a, \geq)$ converges to L in the sense of Definition 13.25. Indeed, both statement mean: for each neighborhood U of L there exists an integer N such that $n \geq N$ implies $a_n \in U$.

Example 13.22 As another interesting construction based on nets, we consider *summability*. Let A be a set, and $f : A \to \mathbb{R}$ be a real-valued function. We direct finite subsets of A by $\supset$, and for each finite set $F \subset A$ we define

$$S(F, f) = \sum \{f(a) \mid a \in F\}.$$

In this way we construct a net $(S(F, f), F, \supset)$. We say that f is summable on A if this net converges to a real number I. This appears to be an *unordered* sum, since the elements of finite sets need not be arranged in any increasing order. But let us think of the familiar case $A = \mathbb{N}$, so that f is actually a sequence of real numbers. Finite subsets of $\mathbb{N}$ need not be finite *intervals*, i.e. sets of the form $\{1, 2, 3, \ldots, N\}$ for some $N \in \mathbb{N}$. When we introduced the idea of numerical series, we considered the convergence of partial sums, which are constructed by summing the values $f(n)$ as n ranges from 1 to some N. Here we are summing the values $f(n)$ as n ranges over finite subsets of the natural numbers. Is there any difference?

The two approaches lead to the same idea of convergence. Indeed, we know that if two numerical sequences are identical from a certain index onwards, then they have the same character. The same holds for numerical series, which are nothing but special numerical sequences. Using this simple remark, saying that for each $\varepsilon > 0$ there exists a finite subset F_ε of $\mathbb{N}$ with the property that $|S(f, f) - I| < \varepsilon$ holds for each finite subset F of $\mathbb{N}$ such that $F \supset F_\varepsilon$ is equivalent to saying that for each $\varepsilon > 0$ there exists a finite interval $F_\varepsilon = \{1, 2, \ldots, N_\varepsilon\}$ such that $|S(F, f) - I| < \varepsilon$ holds for each finite interval F of $\mathbb{N}$ such that $F = \{1, 2, \ldots, N_\varepsilon, \ldots, n\}$ with $n > N_\varepsilon$. This in turn means that

$$\lim_{n \to +\infty} \sum_{k=1}^{n} f(k) = I.$$

Example 13.23 We already know that the set $\mathcal{P}$ of all finite partitions of an interval $[a, b]$ is a directed set. If $f: [a, b] \to \mathbb{R}$ is a given function, we can define a net $\{L(f, P), P \in \mathcal{P}\}$ by letting $L(f, P)$ denote the lower Riemann sum of f with respect to the partition P. The same can be done for the upper Riemann sum, of course. Convergence of both nets to a common limit $c \in \mathbb{R}$ is then equivalent to the integrability of f on $[a, b]$, and $\int_a^b f \, dx = c$.

Exercise 13.22 Let $f: [0, 1] \to \mathbb{R}$ be a bounded function. We define the set D of all ordered pairs (P, ξ) such that $P = \{a_0, \ldots, a_n\}$ is a partition of $[0, 1]$ and $\xi = \{\xi_1, \ldots, \xi_n\}$ is a finite set of points such that $a_{k-1} \leq \xi_k \leq a_k$ for $k = 1, \ldots, n$. If (P, ξ) and (Q, η) are elements of D, we define $(P, \xi) \geq (Q, \eta)$ if and only if P is a refinement of Q. It is easy to prove that $\geq$ directs D. We define the Riemann net $S: D \to \mathbb{R}$ such that

$$S(P, \xi) = \sum_{k=1}^{n} f(\xi_k)(a_k - a_{k-1}).$$

Finally, let

$$U(P) = \sum_{k=1}^{n} \sup \{f(x) \mid a_{k-1} \leq x \leq a_k\}$$

$$L(P) = \sum_{k=1}^{n} \inf \{f(x) \mid a_{k-1} \leq x \leq a_k\}$$

$$\mu(P) = \sup\{a_k - a_{k-1} \mid k = 1, \ldots, n\}$$
$$\delta(P) = \inf\{a_k - a_{k-1} \mid k = 1, \ldots, n\}.$$

1. For every $(P, \xi) \in D$, prove that $L(P) \leq S(P, \xi) \leq U(P)$. Prove also that, for every $\varepsilon > 0$ and for every partition P, there exist ξ and η such that $(P, \xi) \in D$, $(P, \eta) \in D$, $S(P\xi) > U(P) - \varepsilon$ and $S(P, \xi) < L(P) + \varepsilon$.
2. If P, Q are partitions of $[0, 1]$ and if P is a refinement of Q, prove that $\mu(P) \leq \mu(Q)$. Deduce that if the Riemann integral of f exists and equals I, then the Riemann net S converges to I.
3. Prove that the Riemann integral of f exists if and only if for every $\varepsilon > 0$ there exists $\delta > 0$ such that whenever P is a partition of $[0, 1]$ with $\mu(P) < \delta$, then $U(P) - L(P) < \varepsilon$.
4. If P, Q are partitions of $[0, 1]$ such that $\mu(P) < (1/2)\delta(Q)$, prove that $U(P) - L(P) < 2(U(Q) - L(Q))$. *Hint:* observe that every interval of P is contained in at most two intervals of Q.
5. Prove that if the Riemann net S converges to I, then the integral of f exists and equals I. *Hint:* it follows from 3 and 4 that it is sufficient to construct a partition Q such that $U(Q) - L(Q)$ is small.

Exercise 13.23 Let X be a discrete space; prove that a net S converges to x if and only if it is eventually equal to x. If X is an indiscrete space, prove that any net converges to any point of X.

Theorem 13.14 *Let X be a topological space.*

(a) *A point x is an accumulation point of a subset A of X if and only if there exists a net in $A \setminus \{x\}$ which converges to s.*
(b) *A point x belongs to the closure of a subset A of X if and only if there is a net in A converging to x.*

Proof Suppose x is an accumulation point of A. For every neighborhood U of x there exists a point S_U of A which belongs to $U \setminus \{x\}$. Recall that the family $\mathcal{U}$ of all neighborhoods of x is directed by $\subset$. Now, if U and V are neighborhoods of x such that $V \subset U$, then $S_V \in V \subset U$. The net $\{S_U, U \in \mathcal{U}, \subset\}$ therefore converges to x. Conversely, if a net in $A \setminus \{x\}$ converges to x, then this net has values in every neighborhood of x, and $A \setminus \{x\}$ intersects each neighborhood of x.

To prove (b), we remember that the closure of a subset consists of A together with the set of all accumulation points of A. If x is an accumulation point of A, by the preceding discussion there exists a net in A converging to x. Furthermore, if x is a point of A, the net which is constantly equal to x converges to x. In any case, each point of the closure of A has a net in A which converges to it. On the other hand, if a net in A converges to x, then every neighborhood of x intersects A, and x thus belongs to $\overline{A}$.   □

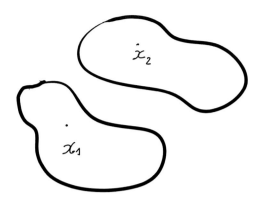

Fig. 13.5 Two points x_1 and x_2 separated by disjoint neighborhoods

Remark 13.7 The previous proof is a typical example of the use of the Axiom of Choice. Although this is usually taken as granted, when we define the net $U \mapsto S_U$ we actually *choose* a point S_U from the neighborhood U, and this choice can only be motivated by the Axiom of Choice.

We have seen that in an indiscrete space, any net converges to any point. This is in striking contrast to the familiar case of $\mathbb{R}^N$, in which a sequence can have at most one limit. Honestly, it is hard to imagine a sequence converging to any point, since our mind immediately draws picture in the physical space $\mathbb{R}^3$.

Definition 13.26 A topological space is a Hausdorff space, or a T_2-space, if and only if whenever x and y are distinct points of X, there exist disjoint neighborhoods of x and y. See Fig. 13.5.

In our intuition, this separation property is usually taken for granted. As we said before, we always think in $\mathbb{R}^3$, which is a metric space. This makes the difference.

Example 13.24 For any integer N, the space $\mathbb{R}^N$ (with the usual topology) is a Hausdorff space. Indeed, let x and y be two distinct point of $\mathbb{R}^N$. We call $\delta = \|x - y\| > 0$ their distance. The sets $B(x, \delta/3)$ and $B(y, \delta/3)$ are disjoint neighborhoods of x and y.

Theorem 13.15 *A topological space is a Hausdorff space if and only if each net in the space converges to at most one point.*

Proof One half of the proof is standard. Indeed, assume that X is a Hausdorff space and x, y are distinct points of X. By definition, we can pick disjoint neighborhoods U and V of x and y. Since a net cannot be eventually in each of the disjoint sets U and V, it is clear that a converging net in X cannot have distinct limits. Conversely, we proceed by contradiction, assuming that X is not a Hausdorff space and x and y are distinct points such that every neighborhood of x intersects every neighborhood of y. We now construct a net in X that converges to both x and y. Let $\mathcal{U}_x$ be the family of all neighborhoods of x, and $\mathcal{U}_y$ be the corresponding family of y. Both families are directed by inclusion. The cartesian product $\mathcal{U}_x \times \mathcal{U}_y$ is ordered by agreeing that $(T, U) \geq (V, W)$ if and only if $T \subset V$ and $U \subset W$. The cartesian product is directed by $\geq$, as is easy to check. By assumption, for each (T, U) of $\mathcal{U}_x \times \mathcal{U}_y$, the intersection $T \cap U$ is non-empty; hence we select a point $S_{(T,U)}$ in this intersection. If $(V, W) \geq (T, U)$, then $S_{(V,W)} \in V \cap W \subset T \cap U$, and consequently the net $\{S_{T,U)}, (T, U) \in \mathcal{U}_x \times \mathcal{U}_y, \geq\}$ converges to both x and y. $\quad\square$

Definition 13.27 Let (X, τ) be a Hausdorff space. If a net $\{S_n, n \in D, \geq\}$ converges to a point x, we write

$$\tau - \lim\{S_n, n \in D, \geq\} = x.$$

When no confusion may arise, we simply write $\lim_n S_n = x$.

Remark 13.8 Following [5], we reserve the symbol of limit to net in a Hausdorff space. It is tempting to denote by $\tau - \lim\{S_n, n \in D, \geq\}$ the *set* of all limits of the net. This is possible, but we must be coherent and systematically write $x \in \tau - \lim\{S_n, n \in D, \geq\}$. This piece of notation would be too different than the familiar $x = \lim_n s_n$ that we learn in Calculus.

Exercise 13.24 Prove that the trick introduced in the previous proof is a general one. Precisely, if $(D, \geq_D)$ and $(E, \geq_E)$ are directed sets, prove that $D \times E$ is directed by $(d, e) \geq_{D \times E} (f, g)$ if and only if $d \geq_E f$ and $e \geq_E g$. This is the *product directed set*.

Example 13.25 Cartesian products can be directed even in a much more general situation. The cartesian product $\prod\{D_a \mid a \in A\}$ of a family of sets is the set of all functions d on A such that $d(a) \in D_a$ for each $a \in A$. Suppose now each D_a is directed by $\geq_a$. The product $\prod\{D_a \mid a \in A\}$ is directed by $d \geq e$ if and only if $d(a) \geq e(a)$ for each $a \in A$. This is a very natural order relation on the product set. We now verify that we have constructed a directed set. Let d and e be two members of the product set. For each $a \in A$ there exists a member $f(a)$ of D_a such that $f(a) \geq_a d(a)$ and $f(a) \geq_a e(a)$. Consequently the function f whose value at a is $f(a)$ follows both d and e in the product order.

Exercise 13.25 Consider the product $\prod\{D \mid a \in A\}$ of all functions defined on A whose values are in D. Prove that the product order coincides with the familiar order on functions: for $f, g : A \to D$, $f \geq g$ if and only if $f(a) \geq g(a)$ for every $a \in A$.

Let us go back to the problem of subsequences.

Definition 13.28 A net $\{T_m, m \in E\}$ is a subnet of a net $\{S_n, n \in D\}$ if and only if there exists a function $N : E \to D$ such that

1. $T = S \circ N$, or equivalently $T_i = S_{N_i}$, for each $i \in E$;
2. for each $m \in D$ there exists $n \in E$ with the property that, if $p \geq n$, then $N(p) \geq m$.

Remark 13.9 The word "subnet"should not be taken too seriously. While a subsequence is always indexed by a selection of natural numbers, a subnet may be indexed by a set which has nothing to do with the set of indices of the original net. This is an important fact to remember, if we want to avoid silly mistakes.

We have already seen that the usual approach to subsequences is to require that the function N be strictly increasing. This of course guarantees that the second condition is satisfied.

We now try to relate cluster points and convergent subnets in general topological spaces.

Lemma 13.1 *Let S be a net and $\mathcal{A}$ be a family of sets such that S is frequently in each member of $\mathcal{A}$, and such that the intersection of two members of $\mathcal{A}$ contains a member of $\mathcal{A}$. Then there exists a subnet of S that is eventually in each member of $\mathcal{A}$.*

Proof By assumption, $\mathcal{A}$ is directed by $\subset$. Let $\{S_n, n \in D\}$ be a net which is frequently in each member of $\mathcal{A}$, and let E be the set of all pairs (m, A) such that $m \in D$, $A \in \mathcal{A}$, and $S_m \in A$. Using the assumption that the intersection of any two members of $\mathcal{A}$ contains a member of A, it is easy to check that E is directed by the product order of $D \times \mathcal{A}$. For every (m, A) of E, let $N(m, A) = m$. The map $N: E \to D$ is increasing; since the net S is frequently in each member of $\mathcal{A}$, $S \circ N$ is a subnet of S. To complete the proof, suppose $A \in \mathcal{A}$ and $m \in D$ is such that $S_m \in A$. If $(n, B) \geq (m, A)$, then $S \circ N(n, B) = S_n \in B \subset A$, and the net $S \circ N$ is eventually in A.   □

Definition 13.29 A point x of the space is a cluster point of a net S if and only if S is frequently in every neighborhood of x.

Theorem 13.16 *A point x is a topological space is a cluster point of a net S if and only if some subnet of S converges to x.*

Proof Let x be a cluster point of a net S, and let $\mathcal{U}$ be the family of all neighborhoods of x. The intersection of two members of $\mathcal{U}$ is again a member of $\mathcal{U}$. The preceding Lemma applies, and there is a subset of S which is eventually in each neighborhood of x. This means that this subnet converges to x. On the contrary, if x is not a cluster point of S, then there exists a neighborhood U of x such that S is not frequently in U, and therefore S is eventually in the complement of U. Then each subnet of S is eventually in the complement of U and hence cannot converge to x.   □

We conclude this section with some remarks about the limit of a function in general spaces. We have seen that nets are a genuine generalization of sequences, in the sense that any sequence is a net since $\mathbb{N}$ is directed by the usual order. But what about limits of functions?

Example 13.26 Suppose that Y is a topological space, (X, d) is a metric space, A is a subset of X, x_0 an accumulation point of A, and $f : A \to Y$ a function. The set $A \setminus \{x_0\}$ is directed as follows: $x \geq y$ if and only if $d(x, x_0) \leq d(y, x_0)$. Since x_0

is an accumulation point of A, it is easy to check (exercise!) that $\geq$ directs $A \setminus \{x_0\}$. How do we interpret the convergence of the net $\{f(x), x \in A \setminus \{x_0\}, \geq\}$? This net converges to a point $y_0 \in Y$ if and only if for each neighborhood V of y_0 there exists $x_V \in A \setminus \{x_0\}$ such that $x \in A \setminus \{x_0\}$ and $x \geq x_V$ implies $f(x) \in V$. This in turn means that $x \in A$, $0 < d(x, x_0) \leq d(x_V, x_0)$ imply $f(x) \in V$. Setting $\delta = d(x_V, x_0) > 0$ we recover the usual definition of $\lim_{x \to x_0} f(x) = y_0$.

The last example shows that we can use nets for defining limits of functions if the domain of the function is a metric space. In particular we recover the theory of limits for functions of a real variable. However, the situation is less clear if the domain of the function is merely a topological space. The following is the *ad hoc* definition that we can find in many textbooks.

Definition 13.30 Suppose that Y is a topological space, X is a topological space, A is a subset of X, x_0 an accumulation point of A, and $f : A \to Y$ a function. We say that $y_0 \in Y$ is a limit of f as x tends to x_0 if and only if for every neighborhood V of y_0 there exists a neighborhood U of x_0 such that $f(A \cap U \setminus \{x_0\}) \subset V$.

If Y is a Hausdorff space, it is easy to prove that the limit, if it exists, must be unique. Hence we can write $y_0 = \lim_{x \to x_0} f(x)$, and we are happy.

The question now is: can we interpret this definition of limit in the setting of nets? Well, in order that $\{f, A \setminus \{x_0\}\}$ be a net, the domain $A \setminus \{x_0\}$ must be directed in such a way that $x \geq y$ means "x is closer to x_0 than y." In general there is no canonical way to ensure that such an order relation exists: topologies do not always allow to *measure* the distance between points. Should we give up? No, although the answer is probably not so elegant as we may hope.

Definition 13.31 (Limits Through Nets) Suppose that Y is a topological space, X is a topological space, A is a subset of X, x_0 an accumulation point of A, and $f : A \to Y$ a function. We say that $y_0 \in Y$ is a limit of f as x tends to x_0 if and only if for every net $\{x_\alpha, \alpha \in D\}$ in $A \setminus \{x_0\}$ converging to x_0, the net $\{f(x_\alpha), \alpha \in D\}$ converges to y_0 in Y.

Notice that this is the straightforward generalization of Definition 7.1.

Exercise 13.26 Prove that Definitions 13.30 and 13.31 are indeed equivalent. *Hint:* in one direction you will need to construct a suitable net in $A \setminus \{x_0\}$ by using the family of neighborhoods of x_0 directed by reverse inclusion. Compare with Theorem 7.1.

Remark 13.10 Let A, B be disjoint subsets of a topological space X, and let $Y = A \cap B$. The set A (resp. B) is closed in Y if and only if for each net $\{S_n, n \in D\}$ in A (resp. in B) converging to a limit x in the relative topology of Y there results $x \in A$ (resp. $x \in B$). Now, $S_n \to x$ if and only if S_n eventually lies in every open (with respect to the topology of Y) neighborhood of x, i.e. if and only if for every open (with respect to the topology of X) neighborhood U of x there results $S_n \in U \cap Y$

eventually. Since $U \cap Y = (U \cap A) \cup (U \cap B)$ and $A \cap B = \emptyset$, this means that x is an accumulation point of either A or B in the topology of X. We conclude that $x \in A$ (resp. $x \in B$) if and only if $A \cap \overline{B} = \emptyset$ (resp. $\overline{A} \cap B = \emptyset$). This is al alternative proof of Theorem 13.9 with the language of nets.

13.7 Continuous Maps and Homeomorphisms

Topology is often described as the branch of mathematics which studies invariant properties under the action of continuous functions. The first task, therefore, is to define continuity.

Definition 13.32 A function f of a topological space (X, τ) to a topological space (Y, σ) is continuous if and only if $f^{-1}(V) \in \tau$ for every $V \in \sigma$. In words, the preimage of every open set of Y must be an open set of X.

Topologists often prefer the global approach to the local one. Analysts often do the opposite: in Calculus we all learn that continuity is a definition that applies to a single point.

Definition 13.33 A function f of a topological space X to a topological space Y is continuous at the point $x \in X$ if and only if the preimage of any neighborhood of $f(x)$ is a neighborhood of x.

Exercise 13.27 Prove that a function is continuous if and only if it is continuous at every point of the domain.

Theorem 13.17

(a) If f is continuous from X to Y, and if g is continuous from Y to Z, then $g \circ f$ is continuous from X to Z.
(b) If f is continuous from X to Y, and if A is a subset of X, then $f_{|A}$ is continuous from A (with the relative topology) to Y.

Proof Let V be an open set in Z; since $(g \circ f)^{-1}(V) = f^{-1}(g^{-1}(V))$, using first the continuity of g and then the continuity of f we see that $(g \circ f)^{-1}(V)$ is open in X. This proves (a).

If V is open in Y, then $f_{|A}^{-1}(V) = A \cap f^{-1}(V)$, and therefore $f_{|A}^{-1}(V)$ is an open set in the relative topology of A. □

We summarize eight equivalent definitions of continuity. Which one to choose is clearly a matter of taste

Theorem 13.18 *If X, Y are topological spaces and $f : X \to Y$ is a function, the following statements are equivalent:*

1. *f is continuous;*
2. *the preimage of each closed set is closed;*
3. *the preimage of each member of a subbase for the topology of Y is open;*

4. *for each* $x \in X$ *the preimage of every neighborhood of* $f(x)$ *is a neighborhood of* x;
5. *for each* $x \in X$ *and each neighborhood* V *of* $f(x)$, *there exists a neighborhood* U *of* x *such that* $f(U) \subset V$;
6. *for each net* $\{S_n, n \in D\}$ *in* X *which converges to a point* x, *the composition* $\{f \circ S_n, n \in D\}$ *converges to* $f(x)$;
7. *for each subset* A *of* X, *there results* $f(\overline{A}) \subset \overline{f(A)}$;
8. *for each subset* B *of* Y, *there results* $\overline{f^{-1}(B)} \subset f^{-1}(\overline{B})$.

Proof 1 is equivalent to 2: a trivial consequence of the identity $f^{-1}(Y \setminus B) = X \setminus f^{-1}(B)$ for every $B \subset Y$.

1 is equivalent to 3: if f is continuous, condition 3 is satisfied because any member of a subbase is an open set. Conversely, since every open set V in Y is the union of finite intersections of subbase members, $f^{-1}(Y)$ is the union of finite intersections of the preimages of subbase members; if these are open, then $f^{-1}(V)$ is open.

1 implies 4: for each $x \in X$ and each neighborhood V of $f(x)$, V contains an open neighborhood W of $f(x)$ and $f^{-1}(W)$ is an open neighborhood of x which is a subset of $f^{-1}(V)$. Therefore $f^{-1}(V)$ is a neighborhood of x.

4 implies 5: if V is a neighborhood of $f(x)$, then $f^{-1}(V)$ is a neighborhood of x such that $f(f^{-1}(V)) \subset V$.

5 implies 6: consider a net $\{S_n, n \in D\}$ in X that converges to x. If V is a neighborhood of $f(x)$, there exists a neighborhood U of x such that $f(U) \subset V$, and since S is eventually in U, $f \circ S$ is eventually in V.

6 implies 7: Let A be a subset of X and x a point of $\overline{A}$. There exists a net S in A which converges to x, and $f \circ S$ converges to $f(x)$. Hence $f(x) \in \overline{f(A)}$.

7 implies 8: if $A = f^{-1}(B)$, then $f(\overline{A}) \subset \overline{f(A)} \subset \overline{B}$. Hence $\overline{A} \subset f^{-1}(\overline{B})$. Equivalently, $\overline{f^{-1}(B)} \subset f^{-1}(\overline{B})$.

8 implies 2: If B is any closed set in Y, then $\overline{f^{-1}(B)} \subset f^{-1}(\overline{B}) = f^{-1}(B)$, and $f^{-1}(B)$ is thus a closed set.   □

Definition 13.34 A homeomorphism is a continuous bijective function f from a topological space X onto a topological space Y such that f^{-1} is continuous. If this is the case, X and Y are called homeomorphic, or topologically equivalent.

Example 13.27 Two discrete spaces X and Y are homeomorphic if and only if there exists a bijective function from X to Y. Therefore two discrete spaces are homeomorphic if and only if they have the same cardinality. The same statement holds for two indiscrete spaces.

Example 13.28 The space $\mathbb{R}$ with the usual topology is homeomorphic to the interval $(0, 1)$ with the relative topology. Indeed the map $x \mapsto (2x - 1)/(x(x - 1))$ from $(0, 1)$ to $\mathbb{R}$ is clearly a homeomorphism.

Exercise 13.28 Prove that any two open intervals of $\mathbb{R}$ are homeomorphic. *Hint:* by modifying the previous example, prove first that any open interval is homeomorphic to $\mathbb{R}$.

How can we prove that two spaces are, or are not, homeomorphic? It doesn't come as a surprise that this is a difficult task. As we said before, topology is exactly the study of properties that are left invariant under homeomorphisms.

Definition 13.35 A topological invariant is a property which when possessed by a space is also possessed by each homeomorphic space.

This definition offers the following consequence: if we can exhibit a topological invariant which is possessed by one space but not by another space, then the two spaces are hot homeomorphic.

Theorem 13.19 *If X is a connected space and $f : X \to Y$ is a continuous function, then $f(X)$ is also connected.*

Proof First of all, we may assume that f is surjective, since the map $f : X \to f(X)$ is continuous and surjective. Assume that B is both open and closed in Y. Hence $f^{-1}(B)$ is both open and closed in X by Theorem 13.18. Since X is connected, either $f^{-1}(B) = \emptyset$, or $f^{-1}(B) = X$. Recalling that f is surjective, we conclude that either $B = \emptyset$ or $B = Y$.   □

As an immediate corollary, if X and Y are homeomorphic and X is connected, then Y is also connected. We can use this fact to disprove that two spaces are homeomorphic.

Example 13.29 The interval $[0, 1]$ is not homeomorphic to $[0, 1/3] \cup [1/2, 1]$. Indeed $[0, 1]$ is connected, while $[0, 1/3] \cup [1/2, 1]$ has two connected components.

A characterization of connected spaces in terms of continuous function is also possible, as the next result shows.

Theorem 13.20 *A topological space X is connected if and only if every continuous function $f : X \to \mathbb{R}$ assumes for each pair of values $f(a) < f(b)$ also every η satisfying $f(a) \leq \eta \leq f(b)$.*

Proof First we suppose the existence of a continuous function f from X to $\mathbb{R}$ which leaves out some value η lying between $f(a)$ and $f(b)$. Hence $f(X)$ is not an interval, thus $f(X)$ is not connected as a subset of $\mathbb{R}$. It follows that X is not connected.

Conversely, we suppose that X is not connected. Hence there exists two non-empty open sets U_1 and U_2 of X such that $X = U_1 \cup U_2$ and $U_1 \cap U_2 = \emptyset$. We define $f(x) = 1$ for every $x \in U_1$ and $f(x) = 2$ for every $x \in U_2$. If $x \in X$, then x belongs to either U_1 or U_2, and f is identically equal to either 1 or 2, respectively. In any case there exists a neighborhood U of x such that $f = f(x)$ on U. This shows that f is continuous at x. It is evident that f cannot assume those values $\eta \in (1, 2)$, and the proof is complete.   □

13.8 Product Spaces, Quotient Spaces, and Inadequacy of Sequences

Once we have a good topological definition of continuity, we can use it to construct new spaces from old ones. There are essentially two abstract constructions: the first one is based on cartesian products, in such a way that all the projections onto the factors of the product should be continuous. The second one is based on a single surjective function that generates the largest topology on its codomain for which the function is continuous.

Definition 13.36 Let $X_0, X_1, \ldots, X_n$ be topological spaces. Let $\mathcal{B}$ the collection of all sets of the form $U_0 \times U_1 \times \cdots \times U_n$ such that each U_i is open in X_i, $i = 0, \ldots, n$. The product topology on $\prod_{i=0}^{n} X_i$ is the topology for which $\mathcal{B}$ is a base: a subset of $\prod_{i=0}^{n} X_i$ is open if and only if it is the union of members of $\mathcal{B}$.

Exercise 13.29 Prove that a set W is open in $\prod_{i=0}^{n} X_i$ if and only if for each element $(x_0, x_1, \ldots, x_n) \in U$ there exist open subsets $U_i \subset X_i$, $i = 0, \ldots, n$, such that $(x_0, x_1, \ldots, x_n) \in U_0 \times U_1 \times \cdots \times U_n \subset W$. See Fig. 13.6.

Definition 13.37 The canonical projections on $\prod_{i=0}^{n} X_i$ are the functions

$$P_i : \prod_{i=0}^{n} X_i \to X_i$$

defined by $P_i(x_0, \ldots, x_n) = x_i$ for $i = 0, \ldots, n$.

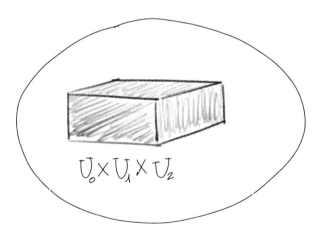

Fig. 13.6 An open subset of $X_1 \times X_2 \times X_3$

Theorem 13.21 *If $\prod_{i=0}^{n} X_i$ is endowed with the product topology, then each P_i is a continuous function.*

Proof Indeed, if U_i is an open subset of X_i, then

$$P_i^{-1}(U_i) = X_0 \times \cdots \times X_{i-1} \times U_i \times X_{i+1} \times \cdots \times X_n,$$

which is open in the product topology. □

The product topology is actually the smallest topology on the cartesian product such that each projection is continuous. We prove this in the particular case of two factors, but there is no difficulty in generalizing the proof to any finite number of factors.

Theorem 13.22 *Let X and Y be two topological spaces, The product topology of $X \times Y$ is the smallest topology such that the two projections $P_X : X \times Y \to X$ and $P_Y : X \times Y \to Y$ are continuous.*

Proof Let τ be any topology on $X \times Y$ such that P_X and P_Y are both continuous. If U is open in X and V is open in Y, then $U \times V$ is open in τ, since $U \times V = P_X^{-1}(U) \cap P_Y^{-1}(V)$, and this intersection is open as the intersection of two open subsets. Hence the product topology is smaller than τ, and we conclude by the arbitrariness of τ. □

Example 13.30 The space $\mathbb{R}^N$ is the product space of N copies of $\mathbb{R}^1 = \mathbb{R}$, each factor being endowed with the usual topology. The N-cells of the form $\prod_{i=0}^{N}(a_i, b_i)$, where a_i and b_i are real numbers, are a base of the product topology of $\mathbb{R}^N$.

Exercise 13.30 Using some geometric intuition, prove that the product topology of $\mathbb{R}^N$ is actually equivalent to the usual topology generated by the distance

$$d(x, y) = \sqrt{\sum_{i=0}^{N} |x_i - y_i|^2}.$$

As a hint, you should prove that each open ball contains a N-cell and is contained in another N-cell. This is fairly obvious, but you should make an effort to prove it.

In some applications to Functional Analysis, finite products of spaces are not enough. A typical situation is that of function spaces and weak topologies. As we have seen in Set Theory, the cartesian product of any collection of sets is already a set of functions.

Definition 13.38 (Product Set) Suppose we are given a set X_a for each a of some
index set A. The cartesian product $\prod_{a \in A} X_a = \prod \{X_a \mid a \in A\}$ is the set of all
functions x on A such that $x(a) \in X_a$ for each $a \in A$, i.e.

$$\prod_{a \in A} X_a = \left\{ x : A \to \bigcup \{X_a \mid a \in A\} \,\middle|\, \forall a \in A \; x(a) \in X_a \right\}$$

We will often write x_a instead of $x(a)$. The projections P_a are defined by $P_a(x) = x_a$ for each $a \in A$ and $x \in \prod_{a \in A} X_a$.

Definition 13.39 Suppose that each set X_a is endowed with a topology τ_a, $a \in A$.
The family of sets of the form $P_a^{-1}(U)$, where U is an open subset of X_a, is a
subbase of a topology, called the product topology of $\prod_{a \in A} X_a$.

A consequence of this definition is that a base for the product topology is the
family of all *finite* intersections of members of the subbase. Hence a typical member
of this base is a set of the form

$$U = \bigcap \left\{ P_a^{-1}(U_a) \,\middle|\, a \in F \right\} = \{x \mid x_a \in U_a \text{ for each } a \in F\},$$

where F is a *finite* subset of A, and U_a is open in X_a for each $a \in F$. A strange fact
arises from this discussion: a subset of the form $\prod_{a \in A} U_a$, where each U_a is open in
X_a, need not be open in the product topology! This however is true if only finitely
many sets U_a are different than X_a.

Exercise 13.31 Prove that the projections $P_a : \prod_{a \in A} X_a \to X_a$ are continuous.

Exercise 13.32 Compare the previous construction with Theorem 13.7.

Theorem 13.23 *A function f from a topological space to a product space $\prod_{a \in A} X_a$*
is continuous if and only if the composition $P_a \circ f$ is continuous for each projection
P_a.

Proof If f is continuous, the $P_a \circ f$ is the composition of two continuous functions.
If $P_a \circ f$ is continuous for each $a \in A$, for each open subset U of X_a the set
$(P_a \circ f)^{-1}(U_a) = f^{-1}(P_a^{-1}(U))$ is open. Since finite intersections of $P_a^{-1}(U)$
form a base of the product topology, it follows that the inverse image under f of
any member of a base is an open subset. Hence f is continuous.   □

We now consider nets in a product space.

Theorem 13.24 *A net S in a product space converges to a point x if and only if its*
projections onto each factor converges to the projection of x.

Proof If $\{S_n, n \in D\}$ is a net in the product space $\prod_{a \in A} x_a$ which converges to a
point x, since each projection is continuous it follows that the net $\{P_a \circ S_N, n \in D\}$
converges to $P_a(x) = x_a$. Conversely, suppose that $\{S_n, n \in D\}$ is a net in the
product space such that $P_a(S_n)$ converges to x_a for each $a \in A$. Given any open
neighborhood U_a of x_a, $\{P_a(S_n), n \in D\}$ is eventually in U_a, and therefore $\{S_n, n \in$

$D\}$ is eventually in $P_a^{-1}(U_a)$. We deduce that $\{S_n, n \in D\}$ must be eventually in any finite intersection of sets of the form $P_a^{-1}(U_a)$. Since these sets form a base of the product topology, the net $\{S_n, n \in D\}$ converges to x. □

Important: The Box Topology Is not Good

The previous proof explains why the so-called *box topology* on a product $\prod_{a \in A} X_a$ is not good enough for most purposes. It is defined by stating that a set is open if and only if it has the form

$$\prod_{a \in A} U_a,$$

for some open sets $U_a \in X_a$, $a \in A$. This looks rather natural, since it is the straightforward generalization of the product topology in $X \times Y$ (or in any product space with *finitely many* factors). Nevertheless, there is no reason why Theorem 13.24 should hold for such a topology. Indeed, looking at the proof, it is true that $\{S_n, n \in D\}$ is eventually in $P_a^{-1}(U_a)$, say $S_n \in P_a^{-1}(U_a)$ as soon as $n \geq n_a$. If however A contains infinitely many elements, it may very well happen that n_a "escapes to infinity", and it is therefore impossible to select a *single* element $\bar{n} \in D$ such that $S_n \in P_a^{-1}(U_a)$ when $n \geq \bar{n}$ for *every* $a \in A$. It was possible for a finite number of indices a, of course.

Remark 13.11 The previous result somehow motivates the terminology of *pointwise convergence* for convergence in the product topology. This is even clearer when we are dealing with a product of identical factors, $\prod_{a \in A} X = X^A$, which is simply the collection of all functions from A to X. A net $\{F_n, n \in D\}$ converges to f in X^A if and only if $\{F_n(a), n \in D\}$ converges to $f(a)$ for each $a \in A$.

Theorem 13.25 *The product of Hausdorff spaces is a Hausdorff space.*

Proof Let x and y be distinct members of $\prod_{a \in A} X_a$. Then $x_a \neq y_a$ for some a. By assumption, there exist disjoint open neighborhoods U and V of x_a and y_a respectively. Then $P_a^{-1}(U)$ and $P_a^{-1}(V)$ are disjoint open neighborhoods of x and y respectively, so that the product space is Hausdorff. □

We now take a break, and go back to the comparison of sequences and nets. We have insisted that nets completely describe a topology, while sequences in general do not. But why? The following is an example that leans on the product topology.

Example 13.31 We consider the space $X = \mathbb{R}^{\mathbb{R}}$ with the product topology. Let

$$E = \{f \in X \mid f(x) \in \{0, 1\} \text{ and } f(x) = 0 \text{ only finitely often}\}$$

and $g \in X$ be the function which is 0 everywhere. A generic neighborhood of g is therefore of the form

$$U = \{h \in X \mid |h(y) - g(y)| < \varepsilon \text{ if } y \in F\}$$

for some finite set $F \subset \mathbb{R}$ and some $\varepsilon > 0$. Such a neighborhood intersects E in the function h which is 0 on elements of F and 1 elsewhere. Hence $g \in \overline{E}$. We now claim that no sequence in E can converge to g. Indeed, if $\{f_n\}_n$ is a sequence in E with f_n being equal to zero on a set A_n, then any function which is a limit of $\{f_n\}_n$ can be zero at most on the countable set $\bigcup_{n=1}^{\infty} A_n$. Clearly g does not meet this condition, and the claim is proved.

Sequences do not suffice to assign a topology on a set. A natural question is whether we can add any additional requirement in order to restore the full power of converging sequences.

Definition 13.40 A topological space satisfies the first countability axiom, or briefly is first countable, if the neighborhood system of each point of the space has a countable base.

Theorem 13.26 *Let X be a first countable space.*

(a) *A point x is an accumulation point of a subset A if and only if there exists a sequence in $A \setminus \{x\}$ which converges to x.*
(b) *A subset A is open if and only if each sequence which converges to a point of A is eventually in A.*
(c) *If x is an accumulation point of a sequence S, there is a subsequence of S converging to x.*

Proof Suppose first that x is an accumulation point of a subset A, and that U_0, $U_1, \ldots, U_n, \ldots$ is a countable base of the neighborhood system at x. Let $V_n = \bigcap_{i=0}^{n} U_i$. Then the sequence $V_0, V_1, \ldots, V_n, \ldots$ is also a base of the neighborhood system at x with the additional property that $V_{n+1} \subset V_n$ for each n. For each n we select a point S_n in $V_n \cap (A_n \setminus \{x\})$, thus obtaining a sequence $\{S_n\}_n$ which clearly convergens to x. The converse of (a) is trivial.

If a subset A is not open in X, then there is a sequence in $X \setminus A$ which converges to a point of A. Such a sequence fails to be eventually in A, and (b) is proved.

Finally, suppose that x is an accumulation point of a sequence S and that $V_0, V_1, \ldots$, is a countable base for the neighborhood system at x such that $V_{n+1} \subset V_n$ for each n. For any integer i we choose N_i such that $N_i \geq i$ and $S_{N_i} \in V_i$. Thus $\{S_{N_i}\}_i$ is a subsequence of S which converges to x.   □

Remark 13.12 Any metric space (X, d) satisfies the first axiom of countability. Indeed the sequence

$$U_n = \left\{ x \;\middle|\; d(x, x_0) < \frac{1}{n} \right\}$$

is a countable base of the neighborhood system at a given point $x_0 \in X$. This is the reason why mathematical analysis in $\mathbb{R}^N$ can be completely explained in terms of sequences.

A nice consequence of the density of the rational numbers in $\mathbb{R}$ is contained in the following result.

Theorem 13.27 (Lindelöf) *Let C be a collection of open sets of $\mathbb{R}$. Then there exists a countable sub-collection $\{O_n\}_n$ of C such that*

$$\bigcup \{O \mid O \in C\} = \bigcup_{n=1}^{\infty} O_n.$$

Proof We call $U = \bigcup\{O \mid O \in C\}$ and we pick any $x \in U$. As such, there exists $O \in C$ such that $x \in O$. But O is an open set, hence there exists an open interval I_x such that $x \in I_x \subset O$. By the density of $\mathbb{Q}$ in $\mathbb{R}$, we can construct an open interval J_x whose end-points are rational numbers and which satisfies $x \in J_x \subset I_x$. The collection of all open intervals with rational end-points is countable, we see that $\{J_x \mid x \in U\}$ is countable and $U = \bigcup\{J_x \mid x \in U\}$. Now, for each interval in $\{J_x \mid x \in U\}$ we select a set O from C which contains it. In this way we have constructed a countable sub-collection $\{O_n\}_n$ of C such that $U = \bigcup_{n=1}^{\infty} O_n$, and the proof is complete.   □

This suggests a general definition connecting open covers and countability properties of the local neighborhoods.

Definition 13.41 A topological space X is a Lindelöf space if and only if every open cover of X has a countable sub-cover.

After this discussion on the importance of nets, we return to the second construction of a new space from an old one. Suppose that we are given a function f from a topological space X onto a set Y. Can we topologize Y so that f is continuous?

Definition 13.42 Let X be a topological space, and f be a function defined on X with range Y. The family $\mathcal{U}$ of all sets $U \subset Y$ such that $f^{-1}(U)$ is open in X is the quotient topology of Y induced by f.

Exercise 13.33 By using the elementary properties of preimages, prove that the quotient topology is indeed a topology, i.e. that $\mathcal{U}$ satisfies the axioms of a topology.

Suppose that Y has a topology τ such that f is a continuous function. For each $U \in \tau$, we then have that $f^{-1}(U)$ is open in X, which proves that U is also open for the quotient topology. We have thus proved that the quotient topology is the largest topology on Y such that f is a continuous function.

Exercise 13.34 Prove that a set B is closed in the quotient topology if and only if $f^{-1}(B)$ is a closed subset of X. *Hint:* $f^{-1}(Y \setminus B) = X \setminus f^{-1}(B)$.

We now establish a counterpart of Theorem 13.23 for the quotient topology.

Theorem 13.28 *Let f be a continuous map of a space X onto a space Y and let Y have the quotient topology. A function g from Y to a topological space Z is continuous if and only if the composition $g \circ f$ is continuous.*

Proof The composition of two continuous functions is always a continuous function. To prove the converse, let U be open in Z and $g \circ f$ be continuous. Then $(g \circ f)^{-1}(U) = f^{-1}(g^{-1}(U))$ is open in X, and therefore $g^{-1}(U)$ is open in the quotient topology of Y by definition. $\square$

13.9 Initial and Final Topologies

Cartesian product and quotient spaces are actually special cases of more general constructions of topologies which preserve the continuity of given families of functions.

Problem Suppose that X_α is a topological space for each index $\alpha \in A$, and suppose that Y is a set. Functions $f_\alpha : X_\alpha \to Y$ are given. We wish to find a topology on Y such that each f_α is a continuous function.

There is a trivial answer, of course, so we add a requirement: we wish to find the *largest* topology on Y such that each f_α is continuous.

Definition 13.43 The largest topology on Y such that each f_α is continuous is called the final topology on Y with respect to $\{f_\alpha \mid \alpha \in A\}$. This topology can be explicitly described: a subset U of Y is open if and only if $f_\alpha^{-1}(U)$ is open in X_α for each $\alpha \in A$.

The following result characterizes the final topology.

Theorem 13.29 (Universal Property of the Final Topology) *For each $\alpha \in A$, suppose that X_α is a topological space. The topology of a space Y is the final topology with respect to the functions $f_\alpha : X_\alpha \to Y$ if and only if the following condition is satisfied: for any topological space Z, a function $g : Y \to Z$ is continuous if and only if $g \circ f_\alpha : X_\alpha \to Z$ is continuous for every $\alpha \in A$.*

Proof Suppose first that Y is endowed with the final topology with respect to $\{f_\alpha \mid \alpha \in A\}$. If g is continuous, then $g \circ f_\alpha$ is continuous for every α as the composition of two continuous functions. Conversely, if $g \circ f_\alpha$ is continuous for any α, then

$$f_\alpha^{-1}\left(g^{-1}(U)\right) = (g \circ f_\alpha)^{-1}(U)$$

is open in X_α for any α and for any open $U \subset Z$. Hence $g^{-1}(U)$ is open in Y and g is continuous.

Suppose now that the topology of Y satisfies the condition in the Theorem. Then each $f_\alpha : X_\alpha \to Y$ is continuous, since the identity map on Y is continuous. Calling $\bar{Y}$ the set Y endowed with the final topology with respect to $\{f_\alpha \mid \alpha \in A\}$, the functions $f_\alpha : X_\alpha \to \bar{Y}$ are continuous. If $\iota : Y \to \bar{Y}$ is the identity function, then the composition $\iota \circ f_\alpha : X_\alpha \to \bar{Y}$ is continuous for every α. By our assumption ι is continuous. Also, each composition $\iota^{-1} \circ f_\alpha : X_\alpha \to Y$ is continuous. Since the topology of $\bar{Y}$ is the final topology of $\bar{Y}$ with respect to $\{f_\alpha \mid \alpha \in A\}$, ι^{-1} is continuous. Thus $\iota : Y \to \bar{Y}$ is a homeomorphism, and $Y = \bar{Y}$. $\square$

If any X_α is a subspace of Y, and if Y already has a topology, we expect some coherence of the initial topology with respect to the inclusions $\iota_\alpha : X_\alpha \to Y$. This suggests the following definition.

Definition 13.44 Let Y be a topological space, and $\{X_\alpha \mid \alpha \in A\}$ be a collection of subspaces of Y. The topology of Y is coherent with $\{X_\alpha \mid \alpha \in A\}$ if and only if it coincides with the initial topology with respect to the inclusion maps $\iota_\alpha : X_\alpha \to Y$.

The proof of the next characterization is left as a simple exercise.

Theorem 13.30 *A necessary and sufficient condition that a space Y has a topology coherent with a collection of its subspaces $\{X_\alpha \mid \alpha \in A\}$ is that $U \subset Y$ is open (resp. closed) if and only if $U \cap X_\alpha$ is open (resp. closed) in X_α for every index $\alpha \in A$.*

Let us now consider a *dual* situation.

Definition 13.45 Suppose that X is a set and Y_α is a topological space for every index $\alpha \in A$. For each $\alpha \in A$, a function $f_\alpha : X \to Y_\alpha$ is assigned. The initial topology induced by $\{f_\alpha \mid \alpha \in A\}$ is the smallest topology on X such that each f_α is a continuous function. Equivalently, the family

$$\left\{ f_\alpha^{-1}(U_\alpha) \,\middle|\, \alpha \in A, \ U_\alpha \text{ is open in } Y_\alpha \right\}$$

is a subbasis of this topology.

This definition should also be compared to Theorem 13.7.

Example 13.32 If $X = \prod_{\alpha \in A} Y_\alpha$, the initial topology induced by the canonical projections $P_\alpha : X \to Y_\alpha$ is just the product topology.

Theorem 13.31 (Universal Property of the Initial Topology) *Assume that X is endowed with the initial topology induced by the functions $f_\alpha : X \to Y_\alpha$, $\alpha \in A$. A function $g : Z \to X$, Z a topological space, is continuous if and only if $f_\alpha \circ g : Z \to Y_\alpha$ is continuous for each $\alpha \in A$. Moreover, this property characterizes the initial topology of X.*

Proof Suppose first that g is continuous. Each $f_\alpha : g$ is then continuous. On the other hand, suppose that each $f_\alpha : g$ is continuous. Then, for any open $U_\alpha \subset Y_\alpha$,

$$g^{-1}\left(f_\alpha^{-1}(U_\alpha)\right) = (f_\alpha \circ g)^{-1}(U_\alpha)$$

is open in Z. By the definition of the initial topology, g is then continuous.

To establish the last statement, suppose that X has the initial topology with respect to $\{f_\alpha \mid \alpha \in A\}$, and let $\bar{X}$ be the set X endowed with a topology such that a function $g : Z \to X$ is continuous if and only if $f_\alpha \circ g : Z \to Y_\alpha$ is continuous for each $\alpha \in A$. The functions $f_\alpha : \bar{X} \to Y_\alpha$ are then continuous, since the identity map on $\bar{X}$ is continuous. If $\iota : X \to \bar{X}$ is the identity map, then the continuity of the compositions $f_\alpha \circ \iota = f_\alpha : X \to Y_\alpha$ implies that ι is continuous. Similarly, the continuity of the compositions $f_\alpha \circ \iota^{-1} : \bar{X} \to Y_\alpha$ implies that $\iota^{-1} : \bar{X} \to X$ is continuous. Hence X and $\bar{X}$ are homeomorphic, and the proof is complete. □

A word of warning: the initial and the final topologies are often known under different names. The term *weak topology* is sometimes used for both of them, while we will reserve this name for a very special topology on Banach spaces. The names *projective topology*, *induced topology* are also used in the literature.

13.10 Compact Spaces

Mathematical analysts tend to believe that compactness is the most important topological property at all. In some sense this is correct: a good compactness property is of fundamental importance in many fields, from Functional Analysis to Calculus of Variations. We have already encountered sequential compactness in $\mathbb{R}$, although the characterization in terms of closedness and boundedness somehow hides the role of compactness. In a topological space it is generically impossible to provide a necessary and sufficient condition for a set to be compact.

Definition 13.46 (Open Cover) Let X be a topological space, and let E be a subset. An open cover of A is a family[4] $\{U_\alpha \mid \alpha \in A\}$ of open subsets such that

$$E \subset \bigcup_{\alpha \in A} U_\alpha.$$

Definition 13.47 (Compact Space) A topological space X is compact if and only if every open cover of X has a finite subcover. A subset E is compact if and only if it is compact as a space with the relative topology induced by X.

---

[4] The indexed notation is preferable. Otherwise we should say that a collection $\mathcal{U}$ of open subsets is an open cover of E if and only if $E \subset \bigcup \mathcal{U}$.

More explicitly, a space X is compact if and only if for every open cover $\{U_\alpha \mid \alpha \in A\}$ of X it is possible to find a finite set $F = \{\alpha_1, \dots, \alpha_n\} \subset A$ such that

$$X \subset U_{\alpha_1} \cup \cdots \cup U_{\alpha_n}.$$

Remark 13.13 The difficulty of the definition is that there is no restriction on the *cardinality* of the open cover. Compactness is a highly demanding property.

As a first step, we characterize compactness in terms of an intersection property.

Definition 13.48 A family $\mathcal{A}$ of sets has the finite intersection property if and only if the intersection of the members of each finite subfamily of $\mathcal{A}$ is non-empty.

Theorem 13.32 *A topological space is compact if and only if each family of closed subsets which has the finite intersection property has a non-empty intersection.*

Proof If $\mathcal{A}$ is a family of subsets of a topological space X, then

$$X \setminus \bigcup_{A \in \mathcal{A}} A = \bigcap_{A \in \mathcal{A}} (X \setminus A).$$

As a consequence, the family $\mathcal{A}$ is a cover of X if and only if the intersection of the complements of the members of $\mathcal{A}$ is empty. The space X is compact if and only if each family of open sets such that no finite subfamily covers X fails to be a cover. This is true if and only if each family of closed sets which possesses the finite intersection property has a non-empty intersection.   □

In the framework of $\mathbb{R}$, we have defined compactness via the existence of convergent subsequences. The next result shows, in greater generality, that compactness is indeed a property of converging subnets.

Theorem 13.33 *A topological space X is compact if and only if each net in X has a subnet which converges to some point of X.*

Proof It suffices to show that each net in X has an accumulation point. Indeed, a point is an accumulation point of a net if and only if some subnet converges to it. Let $\{S_n, n \in D\}$ be a net in the compact space X, and for each $n \in D$ let $A_n = \{S_m \mid m \geq n\}$.[5] The family $\{A_n\}_n$ has the finite intersection property, since $\geq$ directs D. Consequently the family $\{\overline{A}_n\}_n$ has the finite intersection property. Recalling that X is compact, there exists a point x which belongs to every $\overline{A}_n$, and such a point is an accumulation point of $\{S_n, n \in D\}$.

Conversely, we suppose that X is a topological space in which every net has an accumulation point. Let $\mathcal{A}$ be a family of closed subsets of X that has the finite intersection property. We define $\mathcal{B}$ as the family of all finite intersections of members

---

[5] The set A_n can be called the n-tail of the net $\{S_n, n \in D\}$.

of $\mathcal{A}$. Clearly $\mathcal{B}$ has the finite intersection property, and $\mathcal{A} \subset \mathcal{B}$. It is therefore sufficient to prove that $\{\overline{B} \mid B \in \mathcal{B}\} \neq \emptyset$. Observe that the intersection of two members of $\mathcal{B}$ is again a member of $\mathcal{B}$, so that $\mathcal{B}$ is directed by $\subset$. We choose[6] an element S_B from each $B \in \mathcal{B}$, obtaining a net $\{S_B, B \in \mathcal{B}, \subset\}$ in X. By assumption this net has an accumulation point x. Let B and C be any two members of $\mathcal{B}$ such that $C \subset B$; then $S_C \in B$, and thus the net $\{S_b, B \in \mathcal{B}\}$ is eventually in the closed subset $\overline{B}$. This implies that the accumulation point x lies in $\overline{B}$. Since B was arbitrary, the point x is a member of each member of $\overline{\mathcal{B}}$, and the intersection of all members of $\overline{\mathcal{B}}$ is thus non-empty.   □

The following a simple but important corollary.

Theorem 13.34 *A closed subset of a compact space is compact.*

Proof Let X be a compact space, and A be a closed subset of X. We suppose that a net $\{S_n, n \in D\}$ satisfies $S_n \in A$ for each $n \in D$. Since X is compact, there exists a subnet $S \circ N$ which converges to a point $x \in X$. But A is closed, thus $x \in A$, and we conclude.   □

Exercise 13.35 Prove the last theorem by means of open covers. *Hint:* let A be closed in the compact space X. If $\{U_\alpha \mid \alpha \in A\}$ is an open cover of A, we can add $X \setminus A$ to it and get an open cover of X. Now use the compactness of X.

The converse statement is less trivial, and generally false. It becomes true under a separation assumption.

Theorem 13.35 *If A is a compact subset of a Hausdorff space X and x is a point of $X \setminus A$, then there are disjoint neighborhoods of x and A. In particular, each compact subset of a Hausdorff space is closed.*

Proof Since X is a Hausdorff space, there is a neighborhood U of each point y of A such that x does not belong to the closure $\overline{U}$. Because A is compact, there exists a finite family $U_0, U_1, \ldots, U_n$ of open sets covering A and such that $x \notin \overline{U_i}$ for $i \in \{0, 1, \ldots, n\}$. Letting $V = \bigcup_{i=0}^n U_i$, then $A \subset V$ and $x \notin \overline{V}$. Consequently $X \setminus \overline{V}$ and V are disjoint neighborhoods of x and A.   □

Theorem 13.36 *Let X be a compact space, Y be a topological space, and $f : X \to Y$ a continuous function. The Y is compact. Furthermore, if Y is Hausdorff and f is bijective, then f is a homeomorphism.*

Proof If $\mathcal{A}$ is an open cover of Y, then $\{f^{-1}(A) \mid A \in \mathcal{A}\}$ is an open cover of X which must have a finite sub-cover. The collection of the images (under f) of the members of this sub-cover is a finite sub-collection of $\mathcal{A}$ which covers Y, and consequently Y is compact. Suppose that Y is Hausdorff and f is bijective. Consider any closed $A \subset X$; then A is compact and $f(A)$ is compact as well. Hence $f(A)$

---

[6] This is an evident application of the Axiom of Choice, since we need a *function* $B \mapsto S_B$.

is closed in Y. This proves the $(f^{-1})^{-1}(A)$ is closed, and the continuity of f^{-1} follows. □

Exercise 13.36 Provide an alternative proof of the previous result by using nets and Theorem 13.33.

Theorem 13.37 *If A and B are disjoint compact subsets of a Hausdorff space X, then there exist disjoint neighborhoods of A and B.*

Proof By Theorem 13.35, to each $x \in A$ there corresponds a neighborhood of x and a neighborhood of B which are disjoint. As a consequence there exists a neighborhood U of x such that $\overline{U} \cap B = \emptyset$, and since B is compact there exists a finite family U_i, $i = 0, \ldots, n$ such that $\overline{U_i} \cap B = \emptyset$ for $i = 0, \ldots, n$ and $A \subset V = \bigcup\{U_i \mid i = 0, \ldots, n\}$. Then V is a neighborhood of A and $X \setminus \overline{V}$ is a neighborhood of B which is disjoint from V. □

The fundamental example of compact sets for a mathematical analyst is clearly the N-cell in $\mathbb{R}^N$. We develop here a proof which does not make use of Tychonoff's theorem on the product of compact sets.

Definition 13.49 An N-cell in $\mathbb{R}^N$ is a cartesian product of N closed and bounded intervals, i.e. a set of the form

$$\left\{ x \in \mathbb{R}^N \ \middle| \ \text{for each } i \in \{1, \ldots, N\}, \ a_i \leq x_i \leq b_i \right\},$$

where a_i and b_i are real numbers. Hence an N-cell is the cartesian product

$$[a_1, b_1] \times [a_2, b_2] \times \cdots \times [a_N, b_N].$$

Theorem 13.38 *Let N be a positive integer. If $\{I_n\}_n$ is a sequence of N-cells such that $I_{n+1} \subset I_n$ for each n, then $\bigcap_{n=1}^{\infty} I_n \neq \emptyset$.*

Proof We first prove the statement for $N = 1$. Suppose that $I_n = [a_n, b_n]$, and let E be the set of all a_n. Clearly $a_n \leq b_1$, so that E is bounded from above and non-empty. We can therefore set $\alpha = \sup E$. If now m and n are positive integers, then $a_n \leq a_{m+n} \leq b_{m+n} \leq b_m$, and we conclude that $\alpha \leq b_m$ for each m. Therefore $a_m \leq \alpha \leq b_m$ for each m, and therefore $\alpha \in I_m$ for each m.

We now consider the general case $N > 1$. Suppose that I_n consists of all points $x = (x_1, \ldots, x_N)$ such that $a_{n,j} \leq x_j \leq b_{n,j}$ for each n and $j \in \{1, \ldots, N\}$. We define $I_{n,j} = [a_{n,j}, b_{n,j}]$. For fixed j, the sequence $I_{n,j}$ of 1-cells has non-empty intersection. Hence there exist real numbers x_j^*, $j = 1, \ldots, N$, such that $a_{n,j} \leq x_n^* \leq b_{n,j}$ for each n and $j \in \{1, \ldots, N\}$. The point $x^* = (x_1^*, \ldots, x_N^*)$ lies in each I_n, and the proof is complete. □

Unlike most topological spaces, the Euclidean space $\mathbb{R}^N$ possesses a complete characterization of all compact subsets. This is an important result for analysis, and we provide a statement that anticipates a more general result about sequences in compact spaces.

Theorem 13.39 *For a subset K of $\mathbb{R}^N$ the following statements are equivalent to each other:*

(a) K is closed and bounded;
(b) K is compact;
(c) every sequence in K has an accumulation point in K.

Proof Since any bounded subset is contained in a suitable N-cell, we use Theorem 13.34 to show that (a) implies (b). Suppose now that K is compact and E is an infinite subset of K. If no point of K is an accumulation point of E, then each point of K has an open neighborhood which contains at most one point of E. The union of all these neighborhoods is an open cover of K which has no finite subcover, since E is an infinite set. Thus (b) implies (c). We prove that (c) implies (a). If K is unbounded, we can construct a sequence of points $x_n \in K$ such that $|x_n| > n$ for each n. The set of these points is infinite and has no accumulation point in K, in contradiction with (c). Hence K is bounded. Suppose that K is not closed, i.e. there exists a point x_0 of $\mathbb{R}^N$ which is an accumulation point of K but does not belong to K. For each positive integer n, there exist points $x_n \in K$ such that $|x_n - x_0| < 1/n$. Let $S = \{x_n \mid n \in \mathbb{N}\}$. The set S is infinite, otherwise the positive number $|x_n - x_0|$ would be constant for infinitely many values of n. Furthermore x_0 is an accumulation point of S, and we claim that no other point of $\mathbb{R}^N$ is an accumulation point of K. Indeed, if $y \in \mathbb{R}^N$, $y \neq x_0$, then

$$|x_n - y| \geq |x_0 - y| - |x_n - x_0|$$

$$\geq |x_0 - y| - \frac{1}{n} \geq \frac{1}{2}|x_0 - y|.$$

for all but finitely many values of n. This shows that y cannot be an accumulation point of S. We have reached a contradiction with (c), so that K must be closed.   □

What about sequences? Recall that we called sequentially compact any subset K of $\mathbb{R}$ with the following property: any sequence in K has a converging subsequence in K. It is evident that the last property can be generalized to any setting.

Definition 13.50 A topological space is sequentially compact if and only if every sequence in the space has a subsequence which converges to some point of the space. A subset is sequentially compact if and only if it is sequentially compact in the relative topology.

We already know that sequences are not sufficient to describe the topology in the general case. As a matter of facts, sequential compactness is not equivalent to compactness in a generic topological space.

Definition 13.51 A topological space X satisfies the second axiom of countability if and only if the topology of X has a countable base.

Theorem 13.40 *If X satisfies the second axiom of countability, then the following statements are equivalent:*

(a) Every sequence in X has an accumulation point;
(b) for each sequence in X there exists a subsequence converging to a point of X;
(c) X is compact.

Proof If the topology of X has a countable base, then every open cover of X has a countable subcover. It is sufficient to prove that (a) implies (c), since (a) and (b) are equivalent by previous results. So, we must show that every open cover of X has a finite subcover. By assumption we may assume that the open cover is a sequence of open sets

$$A_0, A_1, A_2, \ldots, A_n, \ldots$$

By induction we set $B_0 = A_0$, and for each $p \in \mathbb{N}$ we define B_p as the first member of the cover which is not covered by

$$B_0 \cup B_1 \cup \cdots \cup B_{p-1}.$$

If such a choice is impossible at any stage, we conclude that $\{B_0, \ldots, B_{p-1}\}$ is a finite subcover. Otherwise it is possible to select a point $b_p \in B_p$ for each $p \in \mathbb{N}$ such that $b_p \notin B_i$ for $i < p$. Let x be an accumulation point of this sequence, so that there exists p with $x \in B_p$. Since x is an accumulation point, we must have $b_q \in B_p$ for some $q > p$, a contradiction. $\square$

Using Theorem 13.32 we can generalize Weierstrass' theorem on the existence of minima and maxima.

Definition 13.52 Let X be a topological space, and let $f : X \to \mathbb{R}$ be a function. We say that f is lower semicontinuous if and only if the set $f^{\leq \alpha} = \{x \in X \mid f(x) \leq \alpha\}$ is closed for each $\alpha \in \mathbb{R}$. We say that f is upper semicontinuous if and only if the set $f^{\geq \alpha} = \{x \in X \mid f(x) \geq \alpha\}$ is closed for each $\alpha \in \mathbb{R}$.

Theorem 13.41 (Generalized Weierstrass) *Suppose that X is a compact topological space, $f : X \to \mathbb{R}$ is a function, and $m = \inf_X f$. Then $m \in \mathbb{R}$ and $m \in f(X)$.*

Proof In principle $m \in [-\infty, +\infty)$. For each $\alpha > m$ the set $f^{\leq \alpha}$ is closed. The family $\{f^{\leq \alpha} \mid \alpha > m\}$ has the finite intersection property, since its members are clearly ordered by inclusion. The compactness of X implies the existence of a point x which belongs to each $f^{\leq \alpha}$, $\alpha > m$. Hence $m \leq f(x) \leq \alpha$ for each $\alpha > m$, which means $f(x) = m$. In particular $m > -\infty$. $\square$

We conclude this section with one of the most important results of General Topology. Roughly speaking we want to prove that the cartesian product of compact spaces is compact. The proof is reasonably elementary in the case of two compact spaces. The proof in the case of a generic product, finite or infinite, countable or uncountable, requires much more care. We begin with a more refined criterion for compactness via nets.

Definition 13.53 A net $\{S_n, n \in D\}$ in a topological space X is a universal net if and only if for each $E \subset X$, it is either eventually in E or eventually in $X \setminus E$.

Remark 13.14 An universal net in a topological space converges to any of its accumulation points. Indeed, if the net is frequently in a set, then it is eventually in this set.

Theorem 13.42 (Kelley) *Let X be a non-empty set. Every net in X has a universal subnet.*

Proof Let $S = \{S_n, n \in D\}$ be a net in X and let

$$\Phi = \{F \subset X \mid S \text{ is eventually in } F\}.$$

It is clear that $A \in \Phi$ and $B \in \Phi$ imply $A \cap B \in \Phi$. As a consequence the following properties hold for Φ:

(1) S is frequently in every element of Φ;
(2) Φ has the finite intersection property.

Let $\mathcal{S}$ be the set of all families of sets in X which contain Φ and have the properties (1) and (2). We order $\mathcal{S}$ by inclusion $\subset$, and remark that every totally ordered collection in $\mathcal{S}$ has an upper bound given by the union of all of its elements. Hence Zorn's Lemma applies and provides us with a maximal collection Ω which contains Φ and has properties (1) and (2).

Let $A \subset X$ and suppose that A does not belong to Ω. We claim that $X \setminus A \in \Omega$. Indeed, either S is eventually in $X \setminus A$ or there exists $B \in \Omega$ such that $A \cap B = \emptyset$. If S is eventually in $X \setminus A$ then $X \setminus A \in \Phi \subset \Omega$. Therefore we assume that there exists $B \in \Omega$ such that $A \cap B = \emptyset$. Then S must be frequently in $X \setminus A$ or else it could not be frequently in B and $B \subset X \setminus A$. As a consequence we can add $X \setminus A$ to get a larger collection Ω' which also has the properties (1) and (2). But Ω is maximal with respect to these properties, hence $\Omega' = \Omega$ and $X \setminus A \in \Omega$.

Since S is frequently in every element of Ω and Ω has the finite intersection property, there exists a subset of $S \circ N$ which is eventually in every element of Ω. Let $A \subset X$. If $A \in \Omega$, then $S \circ N$ is eventually in A. If, on the contrary, A is not in Ω, then we have seen before that $X \setminus A \in \Omega$ and therefore $S \circ N$ is eventually in $X \setminus A$. This means that $S \circ N$ is a universal subnet of S, and the proof is complete. □

Remark 13.15 Kelley's result can be proved in a more straightforward way.[7] We describe here the main steps. As a rule, if D is a directed set, we write

$$D_{\geq n} = \{p \in D \mid p \geq n\}.$$

---

[7] This proof appears in [1].

1. Let S be the collection of all subnets of the net S. Since $S \in \mathcal{S}$, we see that $\mathcal{S} \neq \emptyset$.

2. If $T_1 \in \mathcal{S}$ and $T_2 \in \mathcal{S}$, we define $T_1 \geq T_2$ if and only if T_1 is a subnet of T_2. It can be proved that $\geq$ is a partial order on $\mathcal{S}$.

3. Let $\{T_i \mid i \in I\}$ be[8] a totally ordered subset of $\mathcal{S}$, where each T_i is defined on a directed set E^i. We set

$$E = \left\{ E^i_{\geq m_i} \,\middle|\, m_i \in E^i, \ i \in I \right\},$$

and we order E by defining $E^j_{\geq m_j} \geq E^i_{\geq m_i}$ if and only if $T_j \geq T_i$ and $T_j(E^j_{\geq m_j}) \subset T_i(E^i_{\geq m_i})$.

4. If $E^i_{\geq m_i}$ and $E^j_{\geq m_j}$ are given, and if $T_j \geq T_i$, we can find $m'_j \in E^j$ such that $m'_j \geq m_j$ and $T_j(E^j_{\geq m'_j}) \subset T_i(E^i_{\geq m_i})$. Hence $E^j_{\geq m'_j} \geq E^i_{\geq m_i}$ and $E^j_{\geq m'_j} \geq E^j_{\geq m_j}$, and therefore E is a directed set.

5. We define $T^* \colon E \to X$ by

$$T^*(E^i_{\geq m_i}) = T_i(m_i).$$

If $E^j_{\geq m_j} \geq E^i_{\geq m_i}$ we have $T^*(E^j_{\geq m_j})$. Thus $T^* \geq T_i$ for every $i \in I$, or T^* is an upper bound for the collection $\{T_i \mid i \in I\}$, since T^* is evidently a subnet of S.

6. Zorn's Lemma applies, and there exists a maximal element $T \in \mathcal{S}$. It is easy to check that T is a universal net and a subnet of S.

Theorem 13.43 *A topological space X is compact if and only if each universal net in X converges.*

Proof We already know that X is compact if and only if each net has an accumulation point. By the previous result, this happens if and only if each universal net converges in X. □

Theorem 13.44 (Tychonoff) *A nonempty product space is compact if and only if each factor space is compact.*

Proof Let $\prod_{\alpha \in A} X_\alpha$ be a nonempty product space. If it is compact, and since all the projection maps are continuous, then each X_α, $\alpha \in A$, is a compact space.

Conversely, let $\{S_n, n \in D\}$ be a universal net in $\prod_{\alpha \in A} X_\alpha$. For each $\alpha \in A$, the net $\{P_\alpha \circ S_n, n \in D\}$ is a universal net in the compact space X_α, and therefore it converges to some point. We have proved that each component of the original

---

[8] Here we are using a bound-variable notation of a collection of nets. The variable i is *not* the dummy variable which runs over a directed set, but a dummy variable which *labels* the elements of the collection. Unfortunately it would be quite difficult to switch to an intrinsic notation.

universal net converges, and therefore the universal net itself must converge by
Theorem 13.24.   □

We present a slightly different proof of Tychonoff's Theorem, due to Paul R.
Chernoff [2].

Proof Let $\{X_\alpha \mid \alpha \in A\}$ be an indexed family of non-empty topological spaces,
each of which is compact. A basic neighborhood N of an element $f \in X =$
$\prod_{\alpha \in A} X_\alpha$ is determined by a finite subset $F \subset A$, together with neighborhoods
U_α of $f(\alpha)$ in X_α for every $\alpha \in F$. Hence N consists of all $h \in X$ such that for
all $\alpha \in F$, $h(\alpha) \in U_\alpha$. We will say that N is supported on F, and we will write
$N = N\{U_\alpha \mid \alpha \in F\}$. A partially defined element g of X is a function g with
domain $J \subset A$ such that, for every $\alpha \in J$, $g(\alpha) \in X_\alpha$.

   To prove the theorem, let $\{S_n, n \in D\}$, be a net in X. Suppose that g, with domain
$J \subset A$, is a partially defined element of X. We say that g is a partial accumulation
point of our net if and only if for every $n \in D$, for every finite subset $F \subset J$ and for
every basic neighborhood $N\{U_\alpha \mid \alpha \in F\}$ of g in $\prod_{\alpha \in J} X_\alpha$, there exists $m \in N$,
$m \geq n$, such that $S_m(\alpha) \in U_\alpha$ for every $\alpha \in J$. Of course, if the domain of g
coincides with A, then g is an accumulation point in X of the net $\{S_n, n \in D\}$. We
claim that such a g exists.

   Let $\mathcal{P}$ be the set of all partial accumulation points of the given net $\{S_n, n \in D\}$.
Since $\emptyset \in \mathcal{P}$, we see that $\mathcal{P} \neq \emptyset$. We introduce a partial order on $\mathcal{P}$ as follows:
$g_1 \leq g_2$ if and only if the domain of g_1 is contained in the domain of g_2, and
$g_1 = g_2$ on their common domain.[9] Let $\mathcal{L} = \{g_\alpha \mid \lambda \in \Lambda\}$ be a totally ordered
subset of $\mathcal{P}$, and define

$$g_0 = \bigcup \{g_\lambda \mid \lambda \in \Lambda\}.$$

Since any two elements of $\mathcal{L}$ must agree on their common domain, g_0 is a partially
defined element of X. Furthermore, $g_0 \in \mathcal{P}$, since every basic neighborhood of g_0
has finite support F, and thus F is contained in the domain of g_λ for some $\lambda \in \Lambda$.
To summarize, $g_0 \in \mathcal{P}$ and g_0 is an upper bound of $\mathcal{L}$.

   We can now use Zorn's Lemma, which yields a maximal element g in $\mathcal{P}$. We
want to show that the domain J of g coincides with A. Otherwise, we may choose
$k \in A \setminus J$. Now g is an accumulation point in $\prod_{\alpha \in J} X_\alpha$ of the net $\{(S_n)|J \mid n \in D\}$,
and thus g is the limit of some subnet $\{(S_{\varphi(\beta)})|J \mid \beta \in B\}$.

   Now, every X_k is a compact space, the net $\{S_{\varphi(\beta)} \mid \beta \in B\}$ has an accumulation
point $p \in X_k$. We define a function h with domain $J \cup \{k\}$ by setting $h = g$ on J and
$h(k) = p$. It is clear that h is a partial accumulation point of the net $\{S_n, n \in D\}$,

---

[9] Such a definition will return in the proof of the Hahn-Banach Theorem.

hence $h \in \mathcal{P}$ and h is strictly larger than g. This contradicts the maximality of g in $\mathcal{P}$, hence the domain of g is A, g is an accumulation point of $\{S_n, n \in D\}$, and the proof is complete. $\qquad \square$

Yet another proof of Tychonoff's Theorem is based on the following result, of independent interest.

Theorem 13.45 (Alexander Sub-Base Theorem) *Let X be a topological space with a sub-base $\mathcal{B}$. Then the following are equivalent:*

 (i) Every open cover has a finite subcover (i.e. X is compact);
(ii) Every sub-basic open cover has a finite subcover.

With a clear choice of words, a sub-basic open cover is merely a cover which consists of elements taken from the sub-base $\mathcal{B}$.

Proof We propose a proof by T. Tao. Call an open cover *bad* if it has no finite subcover, and *good* otherwise. It suffices to show that if every sub-basic open cover is good, then every basic open cover is also good, where *basic* refers to the basis

$$\mathcal{B}^* = \{B_1 \cap \cdots \cap B_k \mid B_1, \ldots, B_k \in \mathcal{B}, \ k \in \mathbb{N}\}$$

is the standard basis associated to the sub-basis $\mathcal{B}$. Suppose for contradiction that every sub-basic open cover was good, but at least one basic open cover was bad. If we order the bad basic open covers by set inclusion, observe that every chain of bad basic open covers has an upper bound that is also a bad basic open cover, namely the union of all the covers in the chain. Thus, by Zorn's lemma, there exists a maximal bad basic open cover

$$C = \{U_\alpha \mid \alpha \in A\}.$$

Thus this cover has no finite subcover, but if one adds any new basic open set to this cover, then there must now be a finite subcover.

Pick a basic open set U_α from this cover C. Then we can write

$$U_\alpha = B_1 \cap \cdots \cap B_k$$

for some choice of the sub-basic open sets $B_1, \ldots, B_k$. We claim that at least one of the $B_1, \ldots, B_k$ also lies in C. Suppose not, and observe that adding any of the B_i to C enlarges the basic open cover and thus creates a finite subcover; thus B_i together with finitely many sets from C cover X, or equivalently one can cover $X \setminus B_i$ with finitely many sets from C. Thus one can also cover

$$X \setminus U_\alpha = \bigcup_{i+1}^{k} (X \setminus B_i)$$

with finitely many sets from C and thus X itself can be covered by finitely many sets from C, a contradiction.

From the above discussion and the axiom of choice, we see that for each basic set U_α in C there exists a sub-basic set B_α containing U_α that also lies in C. (Two different basic sets U_α, U_β could lead to the same sub-basic set $B_\alpha = B_\beta$, but this will not concern us.) Since the U_α cover X, the B_α do also. By hypothesis, a finite number of B_α can cover X, and so C is good, which gives the desired a contradiction. □

Proof of Tychonoff's Theorem via Sub-Bases Let $X = \prod \{X_\alpha \mid \alpha \in A\}$ a product of compact spaces. In virtue of the Alexander sub-base Theorem, it suffices to show that any open cover of X by sub-basic open sets $\left\{\pi_{\alpha_\beta}^{-1}(U_\beta)) \mid \beta \in B\right\}$ has a finite sub-cover, where B is some index set, and for each $\beta \in B$, $\alpha_\beta \in A$ and U_β is open in X_{α_β}.

For each $\alpha \in A$, consider the sub-basic open sets $\pi_\alpha^{-1}(U_\beta)$ that are associated to those $\beta \in B$ with $\alpha_\beta = \alpha$. If the open sets U_β here cover X_α, then by compactness of X_α, a finite number of the U_β already suffice to cover X_α, and so a finite number of the $\pi_\alpha^{-1}(U_\beta)$ cover X, and we are done. So we may assume that the U_β do not cover X_α, thus there exists $x_\alpha \in X_\alpha$ that avoids all the U_β with $\alpha_\beta = \alpha$. One then sees that the point $(x_\alpha)_{\alpha \in A}$ in X avoids all of the $\pi_\alpha^{-1}(U_\beta)$, a contradiction. The claim follows. □

H. Lebesgue proved an interesting result: if $\mathcal{U}$ is an open cover of a closed interval of $\mathbb{R}$, then there exists a radius $r > 0$ such that, if $|x - y| < r$, then x and y are both contained in some member of the cover $\mathcal{U}$. It is not so easy to provide an intuitive proof without mentioning compactness: if it is evident that each open set of $\mathbb{R}$ contains an open interval of some length, in general this length depends on the member of the open cover.

We prove a generalization of Lebesgue's result valid in any metric space.

Theorem 13.46 (Lebesgue Covering Lemma) *If $\mathcal{U}$ is an open cover of a compact subset A of a metric space (X, d), then there exists a positive number r such that the open sphere of radius r about each point of A is contained in some member of $\mathcal{U}$.*

Proof By compactness, we may assume that $\mathcal{U} = \{U_1, \ldots, U_n\}$. We set

$$f_i(x) = d(x, X \setminus U_i) = \inf\{d(x, y) \mid y \in X \setminus U_i\},$$
$$f(x) = \max\{f_1(x), \ldots, f_n(x)\}.$$

Each f_i is a continuous function, and consequently f is a continuous function. Each point x of A belongs to some member U_i of $\mathcal{U}$, hence $f(x) \geq f_i(x) > 0$. The set $f(A)$ is a compact subset of $(0, +\infty)$, so there exists $r > 0$ such that $f(A) \subset (r, +\infty)$. Therefore, for each $x \in A$ there is an index i such that $f_i(x) > r$, and it follows that the open sphere of radius r about x is contained in U_i. □

Compactness can imply cardinality properties. We say that a point x of a topological space X is *isolated*, if the set $\{x\}$ is open in X.

Theorem 13.47 *Suppose that X is a compact Hausdorff space. If no isolated points exist in X, then X is uncountable.*

Proof First of all, we show that to any non-empty open $U \subset X$ and any point $x \in X$ there corresponds an open $V \subset U$ such that $x \notin \overline{V}$. Indeed, let $y \in U$ be a point with $y \neq x$. This choice is possible if $x \in U$ because x is not an isolated point, and it is also possible if $x \notin U$ because U is non-empty. By assumption we can choose open neighborhoods W_x and W_y of x and y respectively, such that $W_x \cap W_y = \emptyset$. Then $V = U \cap W_y$ is the desired open set.

To complete the proof, we pick any function $f : \mathbb{N} \to X$ and we prove that f cannot be surjective. This clearly implies that X is uncountable. For every $n \in \mathbb{N}$, write $x_n = f(n)$. We apply the previous claim to $U = X$ and choose a non-empty open $V_1 \subset X$ such that $\overline{V_1}$ does not contain x_1. By induction, if V_{n-1} has been selected, we choose a non-empty open $V_n \subset V_{n-1}$ such that $\overline{V_n}$ does not contain x_n. This construction produces a nested sequence of closed sets $\overline{V_1} \supset \overline{V_2} \supset \cdots$. Recalling that X is a compact space, there exists a point $x \in \bigcap_{n=1}^{\infty} \overline{V_n}$. This point x is different from every x_n, since $x_n \notin \overline{V_n}$. The proof is complete. $\square$

Corollary 13.1 *Every closed interval of $\mathbb{R}$ is uncountable. In particular, $\mathbb{R}$ is uncountable.*

Proof Immediate, since singletons are not open in the standard topology of $\mathbb{R}$. $\square$

13.10.1 The Fundamental Theorem of Algebra

Every Calculus student knows that a polynomial equation like

$$P(x) = 0$$

may not have a solution $x \in \mathbb{R}$. For instance, the equation $x^2 + 1 + 0$ does not have any real solution, by the obvious fact that $x^2 + 1 \geq 0 + 1 = 1 > 0$. The *Fundamental Theorem of Algebra* states that every polynomial with complex coefficients possesses at least a *complex* solution. The proof of this important result is often postponed to a course in Complex Analysis.

In this section we present an elementary proof due to Charles Fefferman, see [3].

Theorem 13.48 *Let $n \in \mathbb{N}$, $a_0, \ldots, a_n \in \mathbb{C}$ and*

$$P(z) = a_0 + a_1 z + \cdots + a_n z^n$$

be a polynomial in the complex indeterminate z. Then P has a zero.

Proof We first prove that the function $z \in \mathbb{C} \mapsto |P(z)|$ attains a minimum. To prove this claim, we notice that

$$|P(z)| = |z|^n \left| a_n + \frac{a_{n-1}}{z} + \cdots + \frac{a_0}{z^n} \right|$$

for every $z \in \mathbb{C} \setminus \{0\}$. Hence there exists a number $M > 0$ such that

$$|z| > M \implies |P(z)| \geq |a_0|. \tag{13.1}$$

Since the set $B[0, M] = \{z \in \mathbb{C} \mid |z| \leq M\}$ is compact in $\mathbb{C}$, the continuous function $z \mapsto |P(z)|$ attains a global minimum at some $z_0 \in B[0, M]$. Hence

$$|P(z)| \geq |P(z_0)| \quad \text{for every } z \in B[0, M]. \tag{13.2}$$

Since $0 \in B[0, M]$, we see that $|P(z_0)| \leq |P(0)| = |a_0|$, and (13.1) implies that $|P(z_0)| \leq |P(z)|$ as soon as $|z| > M$. A comparison with (13.2) shows that

$$|P(z_0)| \leq |P(z)| \quad \text{for every } z \in \mathbb{C}. \tag{13.3}$$

The claim is then proved. As a second and last step, we will show that $P(z_0) = 0$.

Indeed, we exploit the identity $P(z) = P(z_0 + (z - z_0))$ to write $P(z)$ as a sum of powers of $z - z_0$. More formally, there exists a polynomial Q such that

$$P(z) = Q(z - z_0).$$

Hence (13.3) becomes

$$|Q(0)| \leq |Q(z)| \quad \text{for every } z \in \mathbb{C}.$$

We need to prove that $Q(0) = 0$. Let j the smallest positive integer such that z^j has a non-zero coefficient in the expansion of the polynomial Q. Then we can write

$$Q(z) = c_0 + c_j z^j + \cdots + c_n z^n,$$

where $c_j \neq 0$. Factoring z^{j+1} in the last terms, we may write

$$Q(z) = c_0 + c_j z^j + z^{j+1} R(z),$$

for some complex polynomial R. Writing

$$-\frac{c_0}{c_j} = r e^{i\theta},$$

the number $z_1 = r^{1/j} e^{i\theta/j}$ satisfies

$$c_j z_1^j = -c_0. \tag{13.4}$$

Let $\varepsilon > 0$, so that

$$Q(\varepsilon z_1) = c_0 + c_j \varepsilon^j z_1^j + \varepsilon^{j+1} z_1^{j+1} R(\varepsilon z_1).$$

Pick $N > 0$ so large that $|R(\varepsilon z_1)| \le N$ for every $\varepsilon \in (0, 1)$. Recalling (13.4) we see that

$$
\begin{aligned}
|Q(\varepsilon z_1) &\le \left| c_0 + c_j \varepsilon^j z_1^j \right| + \varepsilon^{j+1} |z_1|^{j+1} |R(\varepsilon z_1)| \\
&\le \left| c_0 + \varepsilon^j (c_j z_1^j) \right| + \varepsilon^{j+1} |z_1|^{j+1} N \\
&= \left| c_0 + \varepsilon^j (-c_0) \right| + \varepsilon^{j+1} |z_1|^{j+1} N \\
&= (1 - \varepsilon^j) |c_0| + \varepsilon^{j+1} |z_1|^{j+1} N \\
&= |c_0| - \varepsilon^j |c_0| + \varepsilon^{j+1} |z_1|^{j+1} N.
\end{aligned}
$$

Suppose now that $c_0 \ne 0$. Since ε is arbitrary, we can pick it so small that

$$|Q(\varepsilon z_1)| \le |c_0| - \varepsilon^j |c_0| + \varepsilon^{j+1} |z_1|^{j+1} N < |c_0| = |Q(0)|.$$

This contradicts the fact that Q attains a global minimum at 0. Hence $c_0 = 0$, and the proof is complete.   □

13.10.2 Local Compactness

Compactness is a very strong property of a topological space. If we think back of the real line $\mathbb{R}$ with its usual topology, we may observe that any open neighborhood contains a *compact* neighborhood. Indeed, if U is a neighborhood of a point x, then there exists an open ball $B(x, r)$ contained in U. Then the closed ball $\overline{B(x, r/2)}$ is a compact subset contained in U, and clearly it contains the open ball $B(x, r/4)$. This is a very specific example of the following definition.

Definition 13.54 A topological space is locally compact if and only if each point has at least a compact neighborhood.

Exercise 13.37 Prove that a compact space is locally compact.

Exercise 13.38 Prove that every discrete space is locally compact.

Exercise 13.39 Prove that each closed subspace of a locally compact space is locally compact. *Hint:* the intersection of a closed set and a compact set is a closed subset of the latter, and therefore a compact subset.

Example 13.33 It is false that the continuous image of a locally compact space must be locally compact. This follows from the interesting fact that any topological space is the continuous one-to-one image of a discrete space. Indeed, let (X, τ) be a topological space, and let Y be the topological space consisting of the set X with the discrete topology. The identity map from Y to X is continuous and bijective. But every discrete space is locally compact, while X is an arbitrary topological space.

Definition 13.55 (Nowhere Dense) A set in a topological space is nowhere dense if and only if its closure has an empty interior.

Theorem 13.49 *Let* $X = \prod \{X_\alpha \mid \alpha \in A\}$ *be a topological product space. If an infinite number of the coordinate spaces* X_α *are non-compact, then each compact subset of* X *is nowhere dense.*

Proof Suppose B is a compact subset of X with an interior point x. Then B contains a neighborhood U of x which is of the form $U = \bigcap \{P_\alpha^{-1}(V_\alpha) \mid \alpha \in F\}$, for some finite subset F of A and some open sets V_α in X_α. If $\beta \in A \setminus F$, then $P_\beta(B) = X_\beta$ and X_β is compact as the continuous image of a compact space. As a consequence, all but finitely many of the coordinate spaces X_α are compact. The proof is complete. $\qquad \square$

Theorem 13.50 (Local Compactness of Product Spaces) *If a product space is locally compact, then each coordinate space is locally compact and all but a finite number of coordinate spaces are compact.*

Proof Suppose that a product space is locally compact. Since the projection into a coordinate space is an open map, each coordinate space is locally compact. Indeed, if a function is both continuous and open, the image of a compact neighborhood of a point is a compact neighborhood of the image point.

If infinitely many coordinate spaces are non-compact, then each compact subset of the product space is nowhere dense by Theorem 13.49. Hence no point can have a compact neighborhood. The proof is complete. $\qquad \square$

13.11 Compactification of a Space

As a matter of facts, non-compact spaces exist. So the question is: can we somehow *embed* a non-compact space into a compact one? To be more precise: can we construct a compact space which contains the non-compact space as a subspace?

This is a typical problem in mathematical analysis. For instance, it is common to attach two "points" $-\infty$ and $+\infty$ to the real line, so that the resulting set is a compact space. In Complex Analysis, the complex unit sphere is constructed by adjoining a single point ∞ to the bi-dimensional space $\mathbb{C}$ and specifying that the neighborhoods of ∞ are the complements of bounded subsets of $\mathbb{C}$. In this section we present an abstract construction along the same lines.

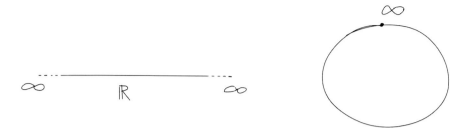

Fig. 13.7 The compactification of $\mathbb{R}$ is homeomorphic to the unit circle

Definition 13.56 Let X be a topological space. The one-point compactification of X is the set $X^* = X \cup \{\infty\}$ with the topology whose members are the open subsets of X and all subsets U of X^* such that $X^* \setminus U$ is a closed compact subset of X.

Remark 13.16 We have used the undefined symbol ∞ to denote any element which is not a member of X. For instance $\infty = X$ can be a good choice, although very few analysts think of ∞ in this way.

Figure 13.7 provides an intuition of the one-point compactification of the real line.

Theorem 13.51 (Alexandroff) *The one-point compactification X^* of a topological space X is compact and X is a subspace. The space X^* is Hausdorff if and only if X is locally compact and Hausdorff.*

Proof We follow [5]. A set U is open in X^* if and only if (a) $U \cap X$ is open in X and (b) whenever $\infty \in U$, then $X \setminus U$ is compact. As a consequence, finite intersections and arbitrary unions of sets open in X^* intersect X in open sets. If ∞ is a member of the intersection of two open subsets of X^*, then the complement of the intersection is the union of two closed compact subsets of X and is therefore closed and compact. If ∞ belongs to the union of the members of a family of open subsets of X^*, then ∞ belongs to some member U of the family, and the complement of the union is a closed subset of the compact set $X \setminus U$ and is therefore closed and compact. Consequently X^* is a topological space and X is a subspace.

Let $\mathcal{U}$ be an open cover of X^*. Then ∞ is a member of some U in $\mathcal{U}$ and $X \setminus U$ is compact, and hence there is a finite subcover of $\mathcal{U}$. This proves that X^* is compact.

If X^* is a Hausdorff space, then its open subspace X is a locally compact Hausdorff space. To conclude, we must show that X^* is a Hausdorff space if X is a locally compact Hausdorff space. It is sufficient to show that, if $x \in X$, then there exist disjoint neighborhoods of x and ∞. Since X is locally compact and Hausdorff, there is a closed compact neighborhood U of x in X, and $X^* \setminus U$ is the required neighborhood of ∞. $\square$

Exercise 13.40 Prove that X is a compact space if and only if ∞ is an isolated point of the one-point compactification X^* of X. In this sense, the one-point compactification is "useless" (since X is already compact) if and only if we are adjoining an isolated point to X.

13.12   Filters and Convergence

As we have seen, nets are a powerful generalization of sequences, and a topology can be recovered in terms of convergent nets. Furthermore, nets depict topologies in a rather *dynamic* way.

Another tool of convergence exists besides nets: it was introduced by H. Cartan under the name of *filters*. We provide now a quick introduction to this topic, following the book [8].

Definition 13.57 (Filters) Let X be any set. We say that a family $\mathcal{F}$ of subsets of E is a filter if and only if

1. $\emptyset \notin \mathcal{F}$;
2. if $A \in \mathcal{F}$ and $B \in \mathcal{F}$, then $A \cap B \in \mathcal{F}$;
3. if $A \in \mathcal{F}$ and $A \subset B$, then $B \in \mathcal{F}$;

Example 13.34 The following are example of filters.

(a) The family of all subsets of X which contain a fixed subset X_0 of X.
(b) The family of all neighborhoods of a point in a topological space X: this is the *neighborhood filter* of that point.
(c) The family of all the complements of finite subsets of an infinite set X. In the special case $X = \mathbb{N}$ this is the *Fréchet filter*.

Definition 13.58 A family $\mathcal{B}$ of subsets of X is a filter basis if and only if

1. if $A \in \mathcal{B}$ and $B \in \mathcal{B}$, there exists $C \in \mathcal{B}$ such that $C \subset A \cap B$.
2. $\emptyset \notin \mathcal{B}$.

Definition 13.59 A family $\mathcal{B}$ of subsets of X is a filter basis of the filter $\mathcal{F}$ if and only if for each $V \in \mathcal{F}$, there exists $A \in \mathcal{B}$ such that $A \subset V$. We will say that the filter $\mathcal{F}$ is generated by $\mathcal{B}$, or that $\mathcal{B}$ generates $\mathcal{F}$.

Concisely, the filter generated by a filter basis $\mathcal{B}$ consists of all supersets of the elements of $\mathcal{B}$.

Exactly as topologies, filters can be compared.

Definition 13.60 A filter $\mathcal{F}_1$ is finer than a filter $\mathcal{F}_2$ if $\mathcal{F}_2 \subset \mathcal{F}_1$. In other words, every element of $\mathcal{F}_1$ is also an element of $\mathcal{F}_2$. Similarly, $\mathcal{F}_1$ is strictly finer than $\mathcal{F}_2$ if $\mathcal{F}_2 \subset \mathcal{F}_1$ and $\mathcal{F}_2 \neq \mathcal{F}_1$.

Remark 13.17 Again, it would be preferable to say that $\mathcal{F}_1$ is larger than $\mathcal{F}_2$. We adhere to the French tradition which uses $\mathcal{F}_1$ *est plus fin que* $\mathcal{F}_2$.

Definition 13.61 A collection $\mathcal{U}$ of subsets of X is a ultrafilter if and only if $\mathcal{U}$ is a filter and there is no filter strictly finer than $\mathcal{U}$.

Example 13.35 Let X be a set, and a be a point of X. The family $\mathcal{U} = \{A \mid (A \subset X) \wedge (a \in A)\}$ is a ultrafilter. This is a rather useless ultrafilter, and its name is actually the trivial ultrafilter defined by the point a.

Exercise 13.41 Prove that the trivial ultrafilter defined by a is indeed a ultrafilter. *Hint:* suppose $\mathcal{U}'$ is a strictly finer filter. Then $\mathcal{U}'$ must contain a set $A \subset X$ which does not contain a. Since $\{a\} \in \mathcal{U}$, the ultrafilter $\mathcal{U}'$ must contain $\emptyset$, which is impossible.

The existence of non-trivial ultrafilters leans on the axiom of choice, or better on Zorn's Lemma.

Theorem 13.52 *For each filter, there exists a ultrafilter which contains it.*

Proof Let $\mathcal{F}$ be a filter, and let $\mathcal{S}$ be the set of all filters strictly finer than $\mathcal{F}$. We claim that $\mathcal{S}$ is inductive, in the sense that any totally ordered subset of $\mathcal{S}$ has an upper bound. Indeed, a family $\{\mathcal{F}_i \mid i \in I\}$ of filters has an upper bound if for each sub-family $\{\mathcal{F}_i \mid i \in J\}$, $J \subset I$ finite, and for each system $A_i \in \mathcal{F}_i$, the intersection $\bigcap\{A_i \mid i \in J\} \neq \emptyset$. Now, if the family $\{\mathcal{F}_i \mid i \in I\}$ is totally ordered, then the previous condition is clearly true, and therefore $\mathcal{S}$ is inductive.

By Zorn's Lemma, $\mathcal{S}$ has a maximal element which is a ultrafilter.   □

Example 13.36 Let X be a Hausdorff topological space, and let a be a point of X which is not an isolated point. This amounts to saying that $\{a\}$ is not an open set. The family

$$\{V \setminus \{a\} \mid V \text{ is a neighborhood of } a\}$$

is a filter base which defines a filter $\mathcal{F}$. As we have seen, there exists a ultrafilter $\mathcal{U}$ finer that $\mathcal{F}$. The elements of $\mathcal{F}$ are mutually disjoint, and the same must be true for the elements of $\mathcal{U}$. Actually, if $b \neq a$, there exists a neighborhood V of a which does not contain b, so that $V \setminus \{a\}$ belongs to $\mathcal{F}$ and does not contain b. Clearly, $V \setminus \{a\}$ does not contain a.

Theorem 13.53 *Let X be a set. A filter $\mathcal{U}$ is a ultrafilter if and only if for each $A \subset X$, either $A \in \mathcal{U}$ or $X \setminus A \in \mathcal{U}$.*

Proof Let $\mathcal{U}$ be a filter which contains either A or $X \setminus A$ for each $A \subset X$. If $\mathcal{U}$ is not a ultrafilter, there exists a filter $\mathcal{F}$ strictly finer than $\mathcal{U}$. Hence there exists a subset A of X such that $A \in \mathcal{F}$ and $A \notin \mathcal{U}$. Then $X \setminus A \in \mathcal{U}$ and therefore $X \setminus A \in \mathcal{F}$ since $\mathcal{F}$ is finer than $\mathcal{U}$. As a consequence, $\emptyset = A \cap (X \setminus A) \in \mathcal{F}$, a contradiction. Hence $\mathcal{U}$ is a ultrafilter.

Conversely, let $\mathcal{U}$ be a ultrafilter, and suppose that $A \notin \mathcal{U}$. We set $B = X \setminus A$. For each $V \in \mathcal{U}$, V is not a subset of A, so that $V \cap B \neq \emptyset$. The set of all $V \cap B$ as V ranges over $\mathcal{U}$ is a filter base which generates a filter $\mathcal{W}$. Now, we deduce from $V \cap B \subset V$ that $\mathcal{W}$ is finer than $\mathcal{U}$, and therefore $\mathcal{W} = \mathcal{U}$ since $\mathcal{U}$ is a ultrafilter. Finally, $B = X \cap B$ is an element of $\mathcal{W}$ since it is an element of $\mathcal{U}$.   □

We now see how filters behave under the action of functions.

Definition 13.62 Let $f : X \to Y$ and let $\mathcal{B}$ be a filter base on Y. If $f^{-1}(B) \neq \emptyset$ for each $B \in \mathcal{B}$, then $f^{-1}(\mathcal{B}) = \{f^{-1}(B) \mid B \in \mathcal{B}\}$ is a filter base on X which we call the counter-image of $\mathcal{B}$ under f. The filter generated by $f^{-1}(\mathcal{B})$ is called the counter-image of the filter generated by $\mathcal{B}$.

Example 13.37 When $X \subset Y$, we can consider f as the inclusion map. Every filter $\mathcal{F}$ on Y has a counter-image under f if and only if for each $B \in \mathcal{F}$, there results $X \cap B \neq \emptyset$. This is the filter on X induced by $\mathcal{F}$. Compare it with the induced topology on a subset.

Definition 13.63 Let $\mathcal{B}$ be a filter base on X, and $f : X \to Y$ be a map. The family $f(\mathcal{B}) = \{f(A) \mid A \in \mathcal{B}\}$ is a filter base on Y called the direct image of $\mathcal{B}$ under f. The filter generated by $f(\mathcal{B})$ is called the direct image of the filter generated by $\mathcal{B}$.

It should be remarked that $f(\mathcal{B})$ need not be a filter, even in the favorable case in which $\mathcal{B}$ is a filter.

Example 13.38 Let $\{x_n\}_n$ be a sequence in X, i.e. a function from $\mathbb{N}$ into X. The Fréchet filter on $\mathbb{N}$ has a direct image $\mathcal{F}$ in X, which we call the Fréchet filter of the sequence, or the elementary filter associated to the given sequence: this is the set of all subsets of X which contain all but finitely many terms x_n.

Theorem 13.54 *If $\mathcal{B}$ is a ultrafilter base of X, its direct image under a map $f : X \to Y$ is a ultrafilter base on Y.*

Proof Let $B \subset Y$; the sets $f^{-1}(B)$ and $f^{-1}(Y \setminus B)$ are complementary, hence at least one of them belongs to the filter generated by $\mathcal{B}$, since $\mathcal{B}$ is a ultrafilter base. Hence either B of $Y \setminus B$ belongs to the filter generated by $f(\mathcal{B})$, which is a ultrafilter base. $\qquad\square$

Filters were introduced for the same reason as nets: to describe *convergence* in a general setting.

Definition 13.64 Let X be a topological space. We say that a filter $\mathcal{F}$ converges to a point $x \in X$ if and only if $\mathcal{F}$ is finer than the neighborhood filter of x. Concretely, this means that every neighborhood of x belongs to $\mathcal{F}$. In this case we write $x \in \lim \mathcal{F}$ or $\mathcal{F} \to x$.[10]

Remark 13.18 The convergence of a filter extends the definition of convergent sequence. Indeed, the filter associated to a sequence $\{x_n\}_n$ converges to a point x if and only if each neighborhood of x contains all but finitely many terms of the sequence $\{x_n\}_n$. In other words, this filter converges to x if and only if $x_n \to x$ in the topology of the space.

Theorem 13.55 *In a Hausdorff space, a filter can converge to at most one point.*

---

[10] Here we are pedantic: without any further assumption on the topology, a filter can converge to different points, and this is why we write $x \in \lim \mathcal{F}$. Nevertheless, the notation $x = \lim \mathcal{F}$ is often used in the literature.

Proof Indeed, if a filter $\mathcal{F}$ converges to both x and y, there exists disjoint neighborhoods of x and y whose intersection belongs to $\mathcal{F}$. But the intersection is empty, a contradiction.   □

Filters characterize accumulation points.

Theorem 13.56 *Let X be a topological space, and A be a subset of X. A point $a \in X$ is an accumulation point of A if and only if there exists a filter $\mathcal{F}$ which has a base consisting of subsets of A, or such that $A \in \mathcal{F}$, and which converges to a.*

Proof Let us suppose that such a filter $\mathcal{F}$ exists. Now $A \in \mathcal{F}$, and since $\mathcal{F}$ is finer that the neighborhood filter of the point a, we have that $A \cap V \in \mathcal{F}$ for each neighborhood V of a. Hence $a \in \overline{A}$.

Conversely, if $a \in \overline{A}$, we have $A \cap V \neq \emptyset$ for each neighborhood V of a. The neighborhood filter of a induces a filter base $\mathcal{B}$ on A, consisting of subsets of A. On the other hand, if V is a neighborhood of a, V belongs to the filter $\mathcal{F}$ generated by $\mathcal{B}$, because V contains $A \cap V$. In particular $\mathcal{F} \to a$.   □

We conclude with the characterization of continuity in terms of converging filters.

Theorem 13.57 *Let $f : X \to Y$ be a map between two topological spaces. The map f is continuous if and only if for each filter $\mathcal{F}$ converging to x in X, the filter $f(\mathcal{F})$ converges to $f(x)$ in Y.*

Proof Let us suppose that the last condition is satisfied. In particular we can select the neighborhood filter $\mathcal{F}$ of x in X. Its image $f(\mathcal{F})$ converges to $f(x)$, and thus f is continuous at x.

Conversely, let us suppose that f is continuous at x. For each neighborhood W of $f(x)$ in Y we can pick a neighborhood V of x in X such that $f(V) \subset W$. This shows that W belongs to the image of the neighborhood filter of x, and therefore this image filter is finer than the neighborhood filter of $f(x)$ in Y. Now, if a filter $\mathcal{F}$ is finer than the neighborhood filter of x, its image $f(\mathcal{F})$ is finer than the neighborhood filter of $f(x)$, and thus it converges to $f(x)$.   □

An interesting remark we make is that filters and nets are somehow homomorphic structures.

Theorem 13.58 *Consider a set X.*

(a) If $\mathcal{F}$ is a filter on X then the set $I_{\mathcal{F}}$ of pairs (A, p) such that $A \in \mathcal{F}$ and $p \in A$ is directed by $(A, p) \leq (B, q)$ if and only if $A \supset B$. Furthermore the function

$$(A, p) \in I_{\mathcal{F}} \mapsto p \in X$$

is a net associated to the filter $\mathcal{F}$.

(b) If $\{S_n, n \in D\}$ is a net in X, then[11]

$$\mathcal{F} = \{A \subset X \mid \exists N \in D \,\forall n \in D(n \geq N \implies S_n \in A)\}$$

is a filter on X, associated to $\{S_n, n \in D\}$. The subsets

$$A_n = \{S_j \mid j \geq n\}$$

are a basis for the filter $\mathcal{F}$.

Exercise 13.42 Prove Theorem 13.58. Establish a comparison between the convergence of a net and the convergence of the associated filter.

We will not pursue further the study of filters, since it is by now apparent that they produce the same results as nets. Choosing either nets or filters for a description of the topology is essentially a matter of taste.

13.13 Epilogue: The Limit of a Function

A serious objection to any reasonable definition of the limit of a function between topological spaces is that it not *elegant*. Although this is a matter of taste, let us compare several possible definitions. In each of them, X and Y are topological spaces, $f : X \to Y$ is a function, and p is an accumulation point of X. The value q of the limit is a point of Y.

Definition 13.65 (Traditional Definition) We say that $\lim_{x \to p} f(x) = q$ if and only if for each neighborhood V of q there exists a neighborhood U of p such that $f(V \cap X \setminus \{p\}) \subset V$.

Definition 13.66 (Limit à la Cartan) Let $\mathcal{F}$ be a filter in X. We say that $\lim_{\mathcal{F}} f = q$ if and only if the filter $f(\mathcal{F})$ converges to q. In particular, if $\mathcal{F}$ is the filter of neighborhoods of the point p, $\lim_{x \to p} f(x) = q$ if and only if for each neighborhood V of q there exists a neighborhood U of p such that $f(U) \subset V$.

Let us compare the two definitions. If p does not belong to X (but is an accumulation point of X), these definitions agree. On the other hand, if $p \in X$, these definitions are not equivalent. In particular, the limit à la Cartan exists if and only if f is continuous at p, and therefore $q = f(p)$. The question is: can we tolerate this fact?

---

[11] $A \in \mathcal{F}$ if and only if the net $\{S_n, n \in D\}$ is eventually in A.

While most textbooks in the USA propose the traditional definition, in France the school of Bourbaki imposed the second definition from high schools to universities. The reason is fairly philosophical: Bourbaki believes that (Real) Analysis is a daughter of General Topology, and in General Topology the true concept is continuity. What many people call *removable discontinuities* is a strange animal that we can discard from the theory of limits.

If we are in love with nets, we may propose the following definition.

Definition 13.67 (Limits with Nets) We say that $\lim_{x \to p} f(x) = q$ if and only if for each net $\{x_n, n \in D\}$ such that $x_n \neq p$ for each $n \in D$ and which converges to p, the net $\{f(x_n), n \in D\}$ converges to q.

This turns out to be equivalent to the traditional definition, and this does not come as a surprise: it is a mere generalization of the characterization of limits in metric spaces in terms of sequences. But the price to pay is that the condition

$$x_n \neq p \text{ for each } n \in D$$

does not belong to the general theory of convergent nets. If the domain X is a subspace of a metric space, there is a nice way out: we can say that $x > y$ if and only if $|x - p| < |y - p|$. In this way we exclude p in an elegant fashion, and convergence with respect to the direction $>$ reduces to the traditional definition of limit. What about filters? Well, the first idea is to use the "filter" of punctured neighborhoods of p, but this is *not* a filter: the point p may get back through the window if we pass to a superset of a punctured neighborhood.

So, the only way out is to "screw up" the whole topological space X to which p belongs by removing p once and for all: now punctured neighborhoods of p form a filter. But this amounts to considering the new function $f_{|X\setminus\{p\}}$ which agrees with f away from the point p. But then $q = \lim_{x \to p} f(x)$ is equivalent to the requirement that

$$g(x) = \begin{cases} f(x) & \text{if } x \neq p \\ q & \text{if } x = p \end{cases}$$

be continuous at p.

Let us summarize:

In a general theory, the traditional definition of limit is artificial both in the language of filters and in the language of nets. Excluding the limit point p must be an additional requirement, and we must choose if we can accept it.

13.14 Separation and Existence of Continuous Extensions

A common problem in Analysis is related to the existence of continuous functions
that *extend* a given (continuous) function on a subset. Furthermore, the extension
should also preserve important properties of the extended function. In this Section
we will prove some general results which relate the solvability of such a problem to
the separation properties of the underlying topological space.

Definition 13.68 Let X be a topological space.

(a) We say that X is regular if and only if for each point x and for each
    neighborhood U of x there exists a closed neighborhood V of x such that
    $x \in V \subset U$. Equivalently, X is regular if and only if for each point x and
    for each closed $A \subset X$, there exist disjoint open sets U and V such that $x \in U$,
    $A \subset V$.
(b) We say that X is normal if the following condition is satisfied: if A and B are
    closed subsets of X, there exist open sets U, V such that $A \subset U$, $B \subset V$, and
    $U \cap V = \emptyset$. See Fig. 13.8.

For reference we summarize the most useful separation properties in General
Topology.

Definition 13.69 A topological space X is

 (i) a T_0-space if and only if for every couple x, y of distinct points in X there
     exists an open set which contains only one of them;
(ii) a T_1-space if and only if for every couple x, y of distinct points in X there exist
     open sets U_1 and U_2 such that $x \in U_1$, $y \in U_2$, but $y \notin U_1$ and $x \notin U_2$;
(iii) a T_2-space, or a Hausdorff space, if and only if for every couple x, y of distinct
      points in X there exists open sets U_1, U_2 such that $x \in U_1$, $y \in U_2$, and
      $U_1 \cap U_2 = \emptyset$;
(iv) a T_3-space if and only if X is T_1 and regular;
 (v) a T_4 space if and only if X is T_1 and normal.

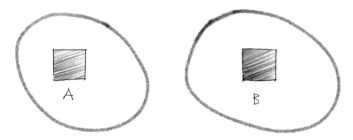

Fig. 13.8 Separation in normal spaces

It should be noted that *intermediate* separation properties have been introduced, but they are of little importance in our setting.

Exercise 13.43 Prove that a space X is T_1 if and only if every singleton $\{x\}$ is a closed set.

In the basic topological space $\mathbb{R}$ (endowed with its natural distance), and even in any metric space, a point y is an accumulation point for a subset A if and only if each neighborhood of y contains *infinitely many* elements of A. The proof is standard: just consider open balls $B(y, 1/n)$ which *shrink* to y as $n \to +\infty$. However this argument depends on the metrizability of the topology. The following result shows that a rather weak separation assumption ensures the validity of the previous characterization of accumulation points.

Theorem 13.59 *Suppose that y is an accumulation point of a subset A of a T_1-space X. Then every neighborhood of y contains infinitely many points of A.*

Proof Let U be a neighborhood of y, and let $F = A \cap (U \setminus \{y\})$. We claim that F is an infinite set. Otherwise, $X \setminus F$ would be an open set (since singletons are closed in a T_1-space), and $y \in X \setminus F$. But the neighborhood $X \setminus F$ of y cannot contain points of A other than y itself, and this contradicts the assumption that y is an accumulation point of A. The proof is complete.   □

The following technical Lemmas will be useful in the sequel.

Lemma 13.2 *Suppose that for each element t of a dense subset D of $(0, +\infty)$, F_t is a subset of a set X such that*

(a) $t < s$ implies $F_t \subset F_s$,
(b) $X = \bigcup \{F_t \mid t \in D\}$.

For $x \in X$ let $f(x) = \inf \{t \in D \mid x \in F_t\}$. Then

$$\{x \in X \mid f(x) < s\} = \bigcup \{F_t \mid t \in D, \ t < s\}$$

$$\{x \in X \mid f(x) \le s\} = \bigcap \{F_t \mid t \in D, \ t > s\}$$

for each real number s.

Proof By definition,

$$\{x \in X \mid f(x) < s\} = \{x \in X \mid \inf \{t \in D \mid x \in F_t\} < s\},$$

and the properties of the infimum imply that $\{x \in X \mid f(x) < s\}$ is the set of all $x \in X$ such that for some $t \in D$, $t < s$ and $x \in F_t$. This proves the first identity of the conclusion.

To prove the second one, we remark that $\inf \{t \in D \mid x \in F_t\} \le s$ if and only if for each $u > s$ there exists $t < u$ such that $x \in F_t$. Conversely, if for each $t \in D$ such that $t > s$ it is true that $x \in F_t$, then $\inf \{t \in D \mid x \in F_t\} \le s$

because D is dense in $(0, +\infty)$. We conclude that the set of all x such that $f(x) = \inf\{t \in D \mid x \in F_t\} \leq s$ coincides with

$$\{x \in X \mid \text{if } t \in D, \ t > s, \text{ then } x \in F_t\},$$

and the proof is complete.   □

Lemma 13.3 *Suppose that for each element t of a dense subset D of $(0, +\infty)$, F_t is an open subset of a topological space X such that*

(a) $t < s$ implies $\overline{F_t} \subset F_s$,
(b) $X = \bigcup\{F_t \mid t \in D\}$.
Then the function f such that $f(x) = \inf\{t \in D \mid x \in F_t\}$ is continuous.

Proof It is sufficient to prove that the set $\{x \in X \mid f(x) < s\}$ is open and the set $\{x \in D \mid f(x) \leq s\}$ is closed for each $s \in R$. By Lemma 13.2, the set $\{x \in X \mid f(x) < s\}$ is the union of open sets. Moreover,

$$\{x \in D \mid f(x) \leq s\} = \bigcap\{F_t \mid t \in D, \ t > s\},$$

and the proof will be complete once we show that this set is identical with

$$\bigcap\{\overline{F_t} \mid t \in D, \ t > s\}.$$

For each $t \in D$, $F_t \subset \overline{F_t}$, so that

$$\bigcap\{F_t \mid t \in D, \ t > s\} \subset \bigcap\{\overline{F_t} \mid t \in D, \ t > s\}.$$

On the other hand, for each $t \in D$ with $t > s$ there exists $r \in D$ such that $s < r < t$, and thus such that $\overline{F_r} \subset F_s$. he reverse inclusion follows, and the proof is complete.
                                                                              □

Theorem 13.60 (Urysohn's Lemma) *Let Y be a Hausdorff space. The following statements are equivalent:*

1. *Y is normal;*
2. *for each pair A, B of disjoint closed sets in Y, there exists a continuous function $f: Y \to \mathbb{R}$, which we call a Urysohn function, such that (a) $0 \leq f \leq 1$ on Y, (b) $f = 0$ on A, (c) $f = 1$ on B.*

Proof We show that 2. implies 1. Fix two closed sets A and B such that $A \cap B = \emptyset$. Denoting by f a Urysohn function for this pair, we set $U = \{y \in Y \mid f(y) < 1/2\}$ and $V = \{y \in Y \mid f(y) > 1/2\}$. Then U and V are disjoint open sets such that $A \subset U$, $B \subset V$.

The converse implication is proved as follows. Let D be the set of positive dyadic rational numbers, i.e. the set of all numbers of the form $p2^{-q}$ and p and q range over all positive integers. For every $t \in D$ with $t > 1$, we let $F(t) = X$, $F(1) = X \setminus B$, and $F(0)$ be an open set which contains A and such that $\overline{F(0)} \cap B = \emptyset$. For every $t \in D$ with $0 < t < 1$, we write $t = (2m+1)2^{-n}$ and choose, inductively on n, $F(t)$ to be an open set which contains $\overline{F(2m2^{-n})}$ and such that $\overline{F(t)} \subset F((2m+2)2^{-n})$. Of course this construction if possible because X is a normal space. Finally, let $f(x) = \inf\{t \in D \mid x \in F(t)\}$. Lemma 13.3 shows that f is a continuous function. The function f is zero on A because $A \subset F(t)$ for every $t \in D$, and f is one on B because $F(t) \subset X \setminus B$ for every $t \leq 1$ and $F(t) = X$ for $t > 1$. Hence f is a Urysohn function, and the proof is complete.   □

Although normality is a complete characterization of the existence of Urysohn's functions, sometimes a sufficient condition is needed.

Theorem 13.61 *If X is a locally compact Hausdorff space, then the family of closed compact neighborhoods of each point is a base for its neighborhood system.*

Proof Fix any point x of X, and let C be a compact neighborhood of x, U be an arbitrary neighborhood of x. Since X is Hausdorff and W is the interior of $U \cap C$, then $\overline{W}$ is a compact Hausdorff space, and W contains a closed compact set V which is a neighborhood of x in $\overline{W}$. But V is also a neighborhood of x in W and is therefore a neighborhood of x in X.   □

Theorem 13.62 *If X is a regular topological space, A is a compact subset, and U is an open set containing A, then there exists a closed neighborhood V of A such that $V \subset U$. In particular, a compact regular space is normal.*

Proof Since X is regular, to each point $x \in A$ there corresponds an open neighborhood W of x such that $\overline{W} \subset U$, and by compactness we may assume that there exist finitely many such neighborhoods $W_0, W_1, \ldots, W_n$ such that $\overline{W_i} \subset U$ for each i. Then $V = \bigcup\{\overline{W_i} \mid i = 0, \ldots, n\}$ is the required closed neighborhood of A.   □

Theorem 13.63 *If U is a neighborhood of a closed compact subset A of a regular locally compact topological space X, then there exists a closed compact neighborhood V of A such that $A \subset V \subset U$.*

Moreover, there exists a continuous function $f : X \to [0, 1]$ such that $f = 0$ on A and $f = 1$ on $X \setminus V$.

Proof To each $x \in A$ there corresponds a neighborhood W which is a closed compact subset of U. Since A is compact, a finite union of such neighborhoods covers A, and their union is a closed compact neighborhood V of A. Then V with the relative topology is a regular compact spaces, which is normal by Theorem 13.62. Hence there exists a continuous function $g : V \to [0, 1]$ such that $g = 0$ on A and $g = 1$ on $V \setminus V^\circ$, where V° denotes the interior of V. Let f be equal to g on V and equal to 1 on $X \setminus V$. Since V° and $X \setminus V$ are separated sets, it follows easily that f is continuous on X.   □

We conclude this Section with another strong characterization of normal spaces. We state a useful lemma.

Lemma 13.4 *Let A be a closed subset of a Hausdorff normal space X, let* $g : A \to$ *$\mathbb{R}$ be a continuous function such that* $|g(x)| \leq c$ *for every* $x \in A$. *Then there exists a continuous function* $h : X \to \mathbb{R}$ *such that*

(a) $|h(x)| \leq c/3$ *for every* $x \in X$;
(b) $|g(x) - h(x)| \leq (2/3)c$ *for every* $x \in A$.

Proof Let

$$
A_+ = \left\{ x \in a \;\middle|\; g(x) \geq \frac{c}{3} \right\}, \quad A_- = \left\{ x \in A \;\middle|\; g(a) \leq -\frac{c}{3} \right\}.
$$

These two sets are disjoint and closed in the closed $A \subset X$, so that both A_- and A_+ are closed in X. Since X is normal, a Urysohn function $h : X \to \mathbb{R}$ exists having value $c/3$ on $A+$ and $-c/3$ on A_-. Furthermore $-c/3 \leq h(x) \leq c/3$ for every $x \in X$.   □

Theorem 13.64 (Tietze Extension Theorem) *Let X be a Hausdorff topological space. The following statements are equivalent:*

1. *X is normal;*
2. *for every closed set* $A \subset X$ *and every continuous function* $f : A \to \mathbb{R}$, *there exists a function* $F : X \to \mathbb{R}$ *such that* F *coincides with* f *on* A. *Furthermore, if* $|f(x)| < c$ *for every* $x \in A$, *then* $|F(x)| < c$ *for every* $x \in X$.

The function F is an *extension* of f, since coincides with f on the common domain.

Proof Suppose that 2. holds, and let A, B be disjoint closed subsets of X. The map $f : A \cup B \to \mathbb{R}$ which sends A to a value y_0 and B to a value $y_1 \neq y_0$ has an extension to a continuous $F : X \to \mathbb{R}$. If U and V are open neighborhoods of y_0 and y_1 respectively, then $F^{-1}(U)$, $F^{-1}(V)$ are disjoint open neighborhoods of A and B, respectively.

Let us prove that 1. implies 2.

Step 1. $|f(x)| \leq c$ for every $x \in A$. We apply Lemma 13.4 with f in place of g, and call $h_0 : X \to \mathbb{R}$ the corresponding function. On A we thus have $|f - h_0| \leq (2/3)c$. We apply the same Lemma once more to the function $f - h_0$ on A, to get $h_1 : X \to \mathbb{R}$ such that

$$
|h_1(x)| \leq \frac{1}{3} \cdot \frac{2}{3} c \quad x \in X
$$

$$
|f(x) - h_0(x) - h_1(x)| \leq \frac{2}{3} \cdot \frac{2}{3} c \quad x \in A.
$$

Now we proceed by induction, and assume that $h_0, h_1, \ldots, h_n$ have been defined. Lemma 13.4 applied to $g = f - h_0 - \cdots - h_n$ on A yields $h_{n+1} : X \to \mathbb{R}$ such that

$$|h_{n+1}(x)| \leq \frac{1}{3} \cdot \left(\frac{2}{3}\right)^n c \quad x \in X$$

$$|f(x) - h_0(x) - \cdots - h_{n+1}(x)| \leq \frac{2}{3} \cdot \left(\frac{2}{3}\right)^n c \quad x \in A.$$

We thus have a function $h_n : X \to \mathbb{R}$ for each $n \in \mathbb{N}$. The function F such that $F(x) = \sum_{n=0}^{\infty} h_n(x)$ is continuous on X, $F(x) = f(x)$ for every $x \in A$, and

$$|F(x)| \leq \frac{1}{3} c \sum_{n=0}^{\infty} \left(\frac{2}{3}\right)^n = c.$$

Step 2. $|f(x)| < c$ for every $x \in A$. Indeed, the extension F constructed in Step 1 satisfies $|F(x)| \leq c$ for every $x \in X$. We set $A_0 = \{x \in X \mid |F(x)| = c\}$. This set is closed in X and disjoint from A. Therefore there exists a Urysohn function $\varphi : X \to \mathbb{R}$ having value 1 on A and value 0 on A_0, with $0 \leq \varphi \leq 1$ everywhere. We define $G(x) = \varphi(x)F(x)$, a continuous function such that $G(x) = F(x) = f(x)$ for $x \in A$. Thus G extends f; furthermore $|G(x)| < c$ for every $x \in X$. Indeed $G(x) = 0$ if $x \in A_0$, while $|\varphi(x)| \leq 1$ if $X \in X \setminus A_0$, and $|F(x)| < c$.

Step 3. f is not necessarily bounded. In this case we introduce the function $h : \mathbb{R} \to (-1, 1)$ such that

$$h(x) = \frac{x}{1 + |x|}.$$

By Step 2, the map $h \circ f : A \to (-1, 1)$ possesses an extension $F : X \to (-1, 1)$ and then $h^{-1} \circ F$ is an extension of f, since $h^{-1} \circ F : x \in A \mapsto h^{-1} \circ h \circ f(x) = f(x)$. The proof is now complete.

$\square$

The previous results require some refined separation property like normality or regularity. For the purposes of Measure Theory, it will be useful to prove Urysohn's Lemma in a very particular environment. We propose here a proof based on the on-point compactification. Later on we will propose a more concrete proof.

Let X be a locally compact Hausdorff (LCH) space. We denote by X^* its one-point compactification, see Theorem 13.51.

Proposition 13.1   *If K is a compact set and if U is an open set such that $K \subset U \subset X$, then there exists an open set V with compact closure such that $K \subset V \subset \overline{V} \subset U$.*

Proof   As a compact subset of X^*, K is closed in X^*, while U is open in X^*. Since X^* is a compact Hausdorff space, it is normal, and there exists an open subset V of

X^* such that $K \subset V \subset \overline{V} \subset U$. The closure of V in X^* coincides with the closure of V in X: indeed the former is a subset of X and the latter is equal to the former intersected with X. Since $\overline{V}$ is closed in the compact space X^*, it is also compact, and since $V \subset X$ is open in X^*, it is also open in X. Thus V is the desired open set.   □

Theorem 13.65 (Urysohn's Lemma for LCH Spaces) *If X is a locally compact Hausdorff space, $K \subset U \subset X$, K is compact and U is open, then there exists a continuous function f on X such that $0 \le f \le 1$, $f \equiv 1$ on K, and* supp $f \subset V$.

Proof By the previous Proposition, we may choose an open set V with compact closure such that $K \subset V \subset \overline{V} \subset U$, since K and $X^* \setminus V$ are disjoint closed subspaces of the normal space X. By Theorem 13.60, there is a continuous function g on X such that $0 \le g \le 1$, $g \equiv 1$ on K, and $g \equiv 0$ on $X^* \setminus V$. We define f as the restriction of g to X. Clearly f is continuous on X, $0 \le f \le 1$, and $f \equiv 1$ on K. Since g vanishes outside $\overline{V}$, so does f, and this implies

$$\text{supp } f = \overline{\{x \in X \mid f(x) \ne 0\}} \subset \overline{V} \subset U,$$

since $\overline{V}$ is closed. The proof is complete.   □

13.15   Partitions of Unity and Paracompact Spaces

One of the most important tools of Mathematical Analysis is the possibility of *gluing together* functions with compact support. As a rough idea, several constructions like Analysis on Manifolds or Partial Differential Equations proceed from local to global: around every point we are able to construct something, and the we would like to extend such a construction to the whole space.

If we work in a compact setting, it is intuitive that a compact space behaves much like a finite space, in the sense that every open cover may be assumed to be finite from the beginning. However this is only the most favorable case, and quite often compactness may not be assumed. In this section we introduce the definition of paracompactness, originally due to J. Dieudonné. We will show that it generalizes the definition of compactness and that it allows us to define partitions of unity: a collection of continuous, compactly supported functions whose sum equals one at any point.

Definition 13.70 A collection $\mathcal{A}$ of subsets of a topological space X is locally finite in X if and only if every point of X possesses a neighborhood which intersects only finitely many elements of $\mathcal{A}$.

Exercise 13.44 In $\mathbb{R}$ with the Euclidean topology, prove that $\mathcal{A} = \{(n, n+2) \mid n \in \mathbb{Z}\}$ is locally finite.

Exercise 13.45 Prove that $\mathcal{A} = \{(0, 1/n) \mid n = 1, 2, 3, \ldots\}$ is locally finite in $(0, 1)$ (with the induced topology), but not in $\mathbb{R}$ with the Euclidean topology.

Definition 13.71 Let $\mathcal{A}$ be a collection of subsets of a topological space X. A collection $\mathcal{B}$ of subsets of X is a refinement of $\mathcal{A}$ if and only if for every element B of $\mathcal{B}$ there exists an element A of $\mathcal{A}$ such that $B \subset A$. If all the elements of $\mathcal{B}$ are open sets, we say that $\mathcal{B}$ is an open refinement of $\mathcal{A}$.

Definition 13.72 A topological space X is paracompact if and only if every open cover $\mathcal{A}$ of X possesses a locally finite open refinement $\mathcal{B}$ which covers X.

Theorem 13.66 *The space $\mathbb{R}^n$ with the Euclidean topology is paracompact.*

Proof Consider an open cover $\mathcal{A}$ of $\mathbb{R}^n$. We begin with $B_0 = \emptyset$ and for every positive integer m we call B_m the open ball of radius m centered at the origin. Since $\overline{B_m}$ is compact, we choose finitely many elements of $\mathcal{A}$ which cover $\overline{B_m}$ and we intersect each of them with $\mathbb{R}^n \setminus \overline{B_{m-1}}$. We call such a finite collection C_m. Clearly $C = \bigcup_{m=1}^\infty C_m$ is an open refinement of $\mathcal{A}$. It is evidently locally finite, since the open ball B_m intersects only finitely many elements of C, i.e. those elements which belong to $C_1 \cup \cdots \cup C_m$. Finally, for every $x \in \mathbb{R}^n$ we select the smallest positive integer m such that $x \in B_m$. It follows that x belongs to an element of C_m, and therefore C covers $\mathbb{R}^n$. The proof is complete.   □

Theorem 13.67 *Every closed subspace of a paracompact space is paracompact.*

Proof Let Y be a closed subspace of a paracompact space X, and let $\mathcal{A}$ be an open (relative to the induced topology) cover of Y. To each $A \in \mathcal{A}$ we associate an open set A_1 of X such that $A_1 \cap Y = A$. We now cover X by the open sets A_1 and by $X \setminus Y$. By assumption there exists a locally finite open refinement $\mathcal{B}$ which covers X. Hence the collection $C = \{B \cap Y \mid B \in \mathcal{B}\}$ is the required locally finite open refinement of $\mathcal{A}$.   □

Theorem 13.68 *Every paracompact Hausdorff space X is normal.*

Proof Consider a point a of X and a closed set B which does not contain a. For every $b \in B$ we choose an open neighborhood U_b of b such that $\overline{U_b}$ does not contain a. Putting together these sets U_b and $X \setminus B$ we cover X, so that we may take a locally finite refinement C which covers X. We consider the collection $\mathcal{D}$ of those elements of C which intersect B. Then $\mathcal{D}$ covers X, and the closure $\overline{D}$ of every $D \in \mathcal{D}$ is disjoint from a. Since D intersects B, D must lie in some set U_b whose closure is disjoint from a. Let

$$V = \bigcup \{D \mid D \in \mathcal{D}\};$$

then V is open and contains B. Since $\mathcal{D}$ is locally finite,

$$\overline{V} = \bigcup \{\overline{D} \mid D \in \mathcal{D}\},$$

and V is disjoint from a. We have proved that X is a regular space. To prove normality, we repeat the previous argument, replacing a by a closed set A and using regularity instead of the Hausdorff property.   □

Although we have proved that $\mathbb{R}^n$ is paracompact by hands, it is a particular case of a more general result. The proof is rather difficult.

Theorem 13.69 (A. H. Stone) *Every metric space is paracompact.*

Proof Let X be a metric space. We say that a collection $\mathcal{A}$ of subsets of X is countably locally finite if and only if $\mathcal{A}$ is a countable union of collections $\mathcal{A}_n$, $n \in \mathbb{N}$, each of which is locally finite.

We claim that if $\mathcal{A}$ is any open cover of X, there exists an open cover $\mathcal{E}$ of X which is a refinement of $\mathcal{A}$ and which is countably locally finite.

Indeed, by the Well-ordering principle we may choose a well ordering $<$ for the collection $\mathcal{A}$. Consider a positive integer n. If U belongs to $\mathcal{A}$, we define

$$S_n(U) = \{x \in X \mid B(x, 1/n) \subset U\}.$$

In order to exploit the well-ordering property of $<$, we introduce

$$T_n(U) = S_n(U) \setminus \bigcup \{V \mid V < U\}.$$

If V and W are distinct elements of $\mathcal{A}$, then $d(x, y) \geq 1/n$ for every $x \in T_n(V)$ and every $y \in T_n(W)$. Indeed, since $<$ is a well-ordering for $\mathcal{A}$, we may assume that $V < W$. Since $x \in T_n(V)$, we have $x \in S_n(V)$, so the open ball centered at x with radius $1/n$ lies in V. On the other hand, since $y \in T_n(W)$ and $V < W$, we see that $y \notin V$. Hence y does not lie in the open ball of center x and radius $1/n$.

Since there is no reason why the sets $T_n(U)$ should be open, we enlarge them a little bit, and we set

$$E_n(U) = \left\{ x \in \mathbb{R} \ \middle| \ d(x, T_n(U)) < \frac{1}{3n} \right\} = \bigcup \left\{ B\left(x, \frac{1}{3n}\right) \ \middle| \ x \subset T_n(U) \right\},$$

where $d(x, T_n(U)) = \inf\{d(x, z) \mid z \in T_n(U)\}$. It is a simple exercise to show that $E_n(U)$ and $E_n(V)$ are disjoint provided that U and V are distinct elements of $\mathcal{A}$. Finally, we set

$$\mathcal{E}_n = \{E_n(U) \mid U \in \mathcal{A}\}.$$

The collection $\mathcal{E}_n$ is a locally finite collection of open sets which refines $\mathcal{A}$. Indeed, $E_n(V) \subset V$ for every $V \in \mathcal{A}$. Furthermore, for every $x \in \mathbb{X}$, the open ball of center x and radius $1/(6n)$ intersects at most one element of $\mathcal{E}_n$. To obtained the desired countably locally finite refinement of $\mathcal{A}$, we define

$$\mathcal{E} = \bigcup_{n=1}^{\infty} \mathcal{E}_n.$$

We only need to check that $\mathcal{E}$ covers X. If x is any point of X, let U be the smallest element of $\mathcal{A}$ which contains x, with respect to the well-order $<$. Since U is an open set, there exists a positive integer n such that $B(x, 1/n) \subset U$. By definition $x \in S_n(U)$. It follows that $x \in T_n(U)$, since U is the smallest element of $\mathcal{A}$ which contains x, and thus $x \in E_n(U) \in \mathcal{E}_n$, and the claim is proved.

The conclusion follows now from Theorem 13.70.   □

Theorem 13.70 *Suppose X is a regular topological space. The following conditions are equivalent: Every open cover of X has a refinement which is*

(a) *an open cover of X and countably locally finite;*
(b) *a cover of X and locally finite;*
(c) *a closed cover of X and locally finite;*
(d) *an open cover of X and locally finite.*

Proof Since (d) implies (a) in a trivial way, we will prove that (a) implies (b) implies (c) implies (d).

(a) implies (b)   Consider any open cover $\mathcal{A}$ of X, and let $\mathcal{B}$ be an open refinement which covers X and is countably locally finite. We may suppose that $\mathcal{B} = \bigcup_{n=1}^{\infty} \mathcal{B}_n$, where each $\mathcal{B}_n$ is locally finite. For every index i we define $V_i = \bigcup \{U \mid U \in \mathcal{B}_i\}$. For every positive integer n and every $U \in \mathcal{B})n$, define

$$ S_n(U) = U \setminus \bigcup_{i<n} V_i. $$

The collection $C_n = \{S_n(U) \mid U \in \mathcal{B}_n\}$ is a refinement of $\mathcal{B}_n$, since $U \in \mathcal{B}_n$ implies $S_n(U) \subset U$. If $C = \bigcup_{n=1}^{\infty} C_n$, we claim that C is the desired locally finite refinement of $\mathcal{A}$. To prove that C covers X, we pick $x \in X$ and we select the smallest positive integer N such that x belongs to some element U of $\mathcal{B}_N$. Since N is the smallest positive integer with such a property, $x \in S_N(U) \in C$. Furthermore, for every $n = 1, \dots, N$ we can select a neighborhood W_n of x which intersects only finitely many elements of $\mathcal{B}_n$. If $W_n \cap S_n(V) \neq \emptyset$, then W_n must intersect any element V of $\mathcal{B}_n$. As a consequence W_n intersects only finitely many elements of C_n. But U is in $\mathcal{B}_N$, hence U does not intersect any element of C_n for $n > N$. To summarize, the neighborhood

$$ W_1 \cap W_2 \cap \cdots \cap W_N \cap U $$

of the point x intersects only finitely many elements of C_n.

(b) implies (c)   Let $\mathcal{A}$ be an open cover of X, and let $\mathcal{B}$ be the family of all open sets $U \subset X$ such that $\overline{U}$ is contained in some element of $\mathcal{A}$. Since X is a regular space, $\mathcal{B}$ is an open cover of X. By assumption (b), there exists a refinement C of $\mathcal{B}$ which covers X and is also locally finite. Setting $\mathcal{D} = \{\overline{C} \mid C \in C\}$, we see that $\mathcal{D}$ is a cover of X which refines $\mathcal{A}$ and is locally finite.

(c) implies (d)   This proof is slightly longer. We begin with an open cover $\mathcal{A}$ of X and we use (c) to construct a refinement $\mathcal{B}$ which covers X and is locally finite. The trick here is to "enlarge" the elements of $\mathcal{B}$ in order to get open sets. The geometric intuition is somewhat misleading, since we do not have a distance function to play with. The proof is subtle.

So, for every $x \in X$ there exists an open neighborhood of X which intersects only finitely many elements of $\mathcal{B}$. The union of these neighborhoods is therefore an open cover of X. By assumption (c), we may define a closed refinement C of such an open cover which is still a cover of X and is also locally finite. Clearly, every element of C can intersect only finitely many elements of $\mathcal{B}$.

Now, for every $B \in \mathcal{B}$, we define

$$C(B) = \{C \mid C \in C \land C \subset X \setminus B\},$$

and

$$E(B) = X \setminus \bigcup \{C \mid C \in C(B)\}.$$

The set $E(B)$ is an open subset of X.[12] By definition, $B \subset E(B)$. The collection of all sets $E(B)$ need not be a refinement of $\mathcal{A}$, but we can always choose some $F(B) \in \mathcal{A}$ which contains B. Then $\mathcal{D} = \{E(B) \cap F(B) \mid B \in \mathcal{B}\}$ is a refinements of $\mathcal{A}$ which covers X, since $B \subset E(B) \cap F(B)$.

To conclude, we still have to prove that $\mathcal{D}$ is locally finite. Pick a point $x \in X$, and choose an open neighborhood W of x which intersects only finitely many elements $C_1, \ldots, C_k$ of C. Since C is a cover of X, $W \subset \bigcup_{i=1}^{k} C_i$. To prove that W intersects only finitely many elements of $\mathcal{D}$, it is now sufficient to prove that every $C \in C$ intersects only finitely many elements of $\mathcal{D}$. Suppose that C intersects $E(B) \cap F(B)$. Then C is not contained in $X \setminus B$, which means that $C \cap B \neq \emptyset$. But C intersects only finitely many elements of $\mathcal{B}$, hence it can intersect no more than the same number of elements of $\mathcal{D}$. The proof is complete.

$\square$

Proof of Theorem 13.69, Different Flavor   Consider an open cover $\{U_i \mid i \in I\}$ of the metric space X. As we saw above, we may suppose that the index set I is well-ordered by $\leq$. In particular, for every $x \in X$ there exists a unique index $i \in I$ such that $x \in U_i \setminus \bigcup \{U_j \mid j < i\}$. Explicitly,

$$i = \min \{j \in I \mid x \in U_j\}.$$

For every $i \in I$ and every positive integer n, we define

---

[12] Here the fact that C is a locally finite collection of closed sets is essential to ensure that $\bigcup \{C \mid C \in C(B)\}$ is a closed set. Recall that arbitrary unions of closed sets are not a closed set, in general.

$$V_{i,n} = \bigcup \{B(x, 2^{-n}) \mid x \in X_{i,n}\}$$

$$X_{i,n} = \left\{ x \in X \;\middle|\; \begin{array}{l} B(x, 3 \cdot 2^{-n}) \subset U_i, \; x \notin \bigcup \{U_j \mid j < i\}, \\ x \notin \bigcup \{V_{j,k} \mid j \in I, \; k < n\} \end{array} \right\}.$$

Our claim is that $\{V_{i,n} \mid i \in I, n \in \mathbb{N}\}$ is a locally finite open refinement of $\{U_i \mid i \in I\}$. Of course each element of $\{V_{i,n} \mid i \in I, n \in \mathbb{N}\}$ is open. By definition, $B(x, 2^{-n}) \subset B(x, 3 \cdot 2^{-n}) \subset U_i$ for every ball contributing to $V_{i,n}$, hence $V_{i,n} \subset U_i$: this proves that $\{V_{i,n} \mid i \in I, n \in \mathbb{N}\}$ is an open refinement of $\{U_i \mid i \in I\}$. To complete the proof, we must show that this is a locally finite refinement.

Pick any $x \in X$, fix $i = \min \{j \in I \mid x \in U_j\}$, and select $n \in \mathbb{N}$ such that $V_{i,n} = \bigcup \{B(x, 2^{-n}) \mid x \in X_{i,n}\}$. Two cases are possible: either $x \in V_{j,k}$ for some $j \in I$ and $k < n$, or $x \in X_{i,n} \subset V_{i,n}$. In any case, $\{V_{i,n} \mid i \in I, n \in \mathbb{N}\}$ is a cover of X.

For every $x \in X$, define $i = \min \{j \in I \mid x \in \bigcup V_{j,n} \mid n \in \mathbb{N}\}$. We can choose positive integers k and n such that $B(x, 2^{-k}) \subset V_{i,n}$. We are going to prove the following statements:

(i) if $\ell \geq n + k$, then $B(x, 2^{-n-k})$ does not intersect any $V_{j,\ell}$;
(ii) if $\ell < n + k$, then $B(x, 2^{-n-k})$ intersects $V_{j,\ell}$ for at most one index $j \in I$.

These claims imply that the open neighborhood $B(x, 2^{-n-k})$ can meet at most $n + k - 1$ elements of $\{V_{i,n} \mid i \in I, n \in \mathbb{N}\}$, and the latter is then a locally finite refinement.

To prove (i), we pick $y \in X_{j,\ell}$. Since $\ell > n$, $y \notin V_{i,n}$. But $B(x, 2^{-k}) \subset V_{i,n}$, hence $d(x, y) \geq 2^{-k}$. Since $\ell \geq k + 1$ and $n + k \geq k + 1$, the condition

$$z \in B(x, 2^{-n-k}) \cap B(y, 2^{-\ell})$$

would imply

$$d(x, y) \leq d(x, z) + d(z, y) \leq 2^{-n-k} + 2^{-\ell} \leq 2^{-k-1} + 2^{-k-1} = 2^{-k},$$

which is a contradiction. Thus $B(x, 2^{-n-k})$ is disjoint from each ball $B(y, 2^{-\ell})$, $y \in X_{j,\ell}$. Since these balls cover $V_{j,\ell}$, (i) is proved.

Let us now turn to (ii). For $i < j$ we pick $x \in V_{i,\ell}$, $y \in V_{j,\ell}$. So there exist points x' and y' in X such that

$$x \in B(x', 2^{-\ell}) \subset V_{i,\ell}, \quad y \in B(y', 2^{-\ell}) \subset V_{j,\ell}.$$

Hence $B(x', 3 \cdot 2^{-\ell}) \subset U_i$, but $y' \notin U_i$. Therefore $d(x', y') \geq 3 \cdot 2^{-\ell}$, and the triangle inequality yields

$$3 \cdot 2^{-\ell} \leq d(x', y') \leq d(x', x) + d(x, y) + d(y, y') \leq d(x, y) + 2^{-\ell} + 2^{-\ell},$$

or $d(x, y) > 2^{-\ell}$. As a trivial consequence, $V_{i,\ell} \cap V_{j,\ell} = \emptyset$ whenever $i \neq j$, and every ball of radius $2^{-\ell-1}$ intersects $V_{i,\ell}$ for at most one index $i \in I$. Since $\ell < n+k$, i.e. $n+k \geq \ell+1$, the same conclusion holds for every ball $B(x, 2^{-n-k})$. The proof is complete. $\square$

Important: About A. H. Stone's Theorem

In its astonishing simplicity, Theorem 13.69 remains one of the most important result of Topology and also of Abstract Analysis, and was obtained in [9]. The first proof we have presented follows Munkres [6], and is based on the Well-ordering Principle. The second proof is essentially due to [7], with the only difference that M. E. Rudin began her proof by indexing the open cover $\{U_i\}$ over the ordinals. Since the ordinals are well-ordered by definition, Rudin actually uses the Axiom of Choice in place of the Well-ordering Principle.

The main application of paracompactness in Analysis is the construction of general *partitions of unity*. In the chapter on Measure Theory we will need a convenient version in locally compact spaces, which will be easier since local compactness essentially reduces the proof to a finite case.

Definition 13.73 Let $\{U_\alpha \mid \alpha \in J\}$ be an open cover of a topological space X.[13] A collection of functions $\phi_\alpha : X \to [0, 1]$, $\alpha \in J$, is a partition of unity dominated by $\{U_\alpha \mid \alpha \in J\}$ if

(1) $\operatorname{supp} \phi_\alpha \subset U_\alpha$ for every $\alpha \in J$;
(2) the collection $\{\operatorname{supp} \phi_\alpha \mid \alpha \in J\}$ is locally finite;
(3) $\sum_{\alpha \in J} \phi_\alpha(x) = 1$ for every $x \in X$.[14]

To prove the existence of partitions of unity in paracompact spaces, we begin with a sort of "shrinking" result.

Lemma 13.5 *Suppose that X is a paracompact Hausdorff space and $\{U_\alpha \mid \alpha \in J\}$ is an open cover of X. Then there exists a locally finite open cover $\{V_\alpha \mid \alpha \in J\}$ of X such that $\overline{V_\alpha} \subset U_\alpha$ for every $\alpha \in J$.*[15]

Proof Let us define

$$\mathcal{A} = \left\{ A \; \middle| \; \begin{array}{l} A \text{ is open and } \overline{A} \text{ is contained in} \\ \text{some element of } \{U_\alpha \mid \alpha \in J\} \end{array} \right\}.$$

---

[13] We prefer here the use of indices to label the elements of the cover, since an intrinsic notation would hide the correspondence between the index of the cover and the index of the function in the partition of unity.

[14] The sum is indeed a finite sum by condition (2), in the sense that only finitely many terms are different than zero.

[15] Sometimes $\{V_\alpha \mid \alpha \in J\}$ is called a precise refinement of $\{U_\alpha \mid \alpha \in J\}$.

Since X is regular, see Theorem 13.68, $\mathcal{A}$ is a cover of X. By paracompactness we can find a locally finite refinement $\mathcal{B}$ of open sets which cover X. We write $\mathcal{B} = \{B_\beta \mid \beta \in K\}$. Recalling that $\mathcal{B}$ is a refinement of $\mathcal{A}$, we define a function $f : K \to J$ by choosing, for every $\beta \in K$, and element $f(\beta) \in J$ such that $\overline{B_\beta} \subset U_{f(\beta)}$.[16] For every $\alpha \in J$, let V_α be the union of all the elements of

$$\mathcal{B}_\alpha = \{B_\beta \mid f(\beta) = \alpha\}.$$

Of course we set $V_\alpha = \emptyset$ if the condition $f(\beta) = \alpha$ cannot be satisfied.

By definition, for every B_β in $\mathcal{B}_\alpha$ we have $\overline{B_\beta} \subset U_\alpha$. Since $\mathcal{B}_\alpha$ is locally finite, $\overline{V_\alpha}$ coincides with the closure of the union of the elements of $\mathcal{B}_\alpha$, hence $\overline{V_\alpha} \subset U_\alpha$.

To conclude, for every point $x \in X$ we choose a neighborhood W of x which intersects only finitely many elements $B_{\beta_1}, \ldots, B_{\beta_k}$. Then W intersects V_α only if α belongs to the finite set $\{f(\beta_1), \ldots, f(\beta_k)\}$, and therefore $\{V_\alpha \mid \alpha \in J\}$ is locally finite. □

Theorem 13.71 *Suppose that X is a paracompact Hausdorff space. If $\{U_\alpha \mid \alpha \in J\}$ is an open cover of X, then there exists a partition of unity dominated by $\{U_\alpha \mid \alpha \in J\}$.*

Proof Invoking the previous Lemma twice, we construct locally finite open covers $\{W_\alpha \mid \alpha \in J\}$ and $\{V_\alpha \mid \alpha \in J\}$ such that

$$\overline{W_\alpha} \subset V_\alpha, \qquad \overline{V_\alpha} \subset U_\alpha$$

for every $\alpha \in J$. By normality, to every $\alpha \in J$ we may attach a continuous function $\psi_\alpha : X \to [0, 1]$ such that $\psi_\alpha(\overline{W_\alpha}) = \{1\}$ and $\psi_\alpha(X \setminus V_\alpha) = \{0\}$. Then the support of ψ_α is contained in $\overline{V_\alpha}$, and hence in U_α. The collection $\{\overline{V_\alpha} \mid \alpha \in J\}$ is locally finite, the collection $\{\text{supp}\,\psi_\alpha \mid \alpha \in J\}$ is also locally finite. Furthermore, for any $x \in X$ at least one of the functions ψ_α must be positive at x, since $\{W_\alpha \mid \alpha \in J\}$ covers X.

Hence, for every $x \in X$ we can form the sum

$$\Psi(x) = \sum_{\alpha \in J} \psi_\alpha(x),$$

since x has a neighborhood which intersects only finitely many of the sets $\text{supp}\,\psi_\alpha$. Being a finite sum of continuous functions, Ψ is a continuous function at any point of X. If we set

$$\phi_\alpha(x) = \frac{\psi_\alpha(x)}{\Psi(x)},$$

---

[16] Here we are using the Axiom of Choice.

it is immediate to check that $\{\phi_\alpha \mid \alpha \in J\}$ is the desired partition of unity dominated by $\{U_\alpha \mid \alpha \in J\}$. The proof is complete.   □

Partitions of unity are typically used to construct continuous[17] functions which enjoy additional requirements. Here is a nice example.

Theorem 13.72 *Let X be a paracompact space, and let $g: X \to \mathbb{R}$ be a lower semicontinuous function, $h: X \to \mathbb{R}$ be an upper semicontinuous function. If $h(x) < g(x)$ for every $x \in X$, there exists a continuous function $p: X \to \mathbb{R}$ such that $h(x) < p(x) < g(x)$ for every $x \in X$.*

Proof Let

$$\mathcal{U} = \left\{ U_\alpha \mid U_\alpha \text{ is open and } \sup_{U_\alpha} h < \inf_{U_\alpha} g \right\}.$$

We already know that there exists a partition of unity $\{\phi_\alpha \mid \alpha \in J\}$ dominated by $\mathcal{U}$. For each $\alpha \in J$ and each $x \in X$ such that $\varphi_\alpha(x) \neq 0$, we choose a number k_α with $h(x) < k_\alpha < g(x)$. The function $p: X \to \mathbb{R}$ defined by

$$p(x) = \sum_{\alpha \in J} k_\alpha \varphi_\alpha(x)$$

is continuous and clearly satisfies $h(x) < p(x) < g(x)$ at any $x \in X$.   □

Some constructions in mathematical analysis and also in differential geometry are based on a common idea: one can "fill up" the Euclidean space $\mathbb{R}^n$ by a sequence $\{K_n\}_n$ of compact sets such that $K_n \subset K_{n+1}^\circ$, i.e. each K_n is continued in the interior of K_{n+1}. This is a trivial fact, since one can consider $K_n = \{x \in \mathbb{R}^n \mid \|x\| \leq n\}$. We now show that the same idea also works in more general situations.

Definition 13.74 (Exhaustion by Compact Sets) An exhaustion by compact sets in a locally compact Hausdorff space X consists of a sequence $\{K_n\}_n$ of compact sets of X such that

$$K_n \subset K_{n+1}^\circ \quad \text{and} \quad X = \bigcup_{n=1}^{\infty} K_n.$$

Theorem 13.73 *Suppose that X is a locally compact Hausdorff space.*

1. *If the topology of X has a countable base, then X has an exhaustion by compact sets.*
2. *If X is paracompact and connected, then X has an exhaustion by compact sets.*
3. *If X has an exhaustion by compact sets, then X is paracompact.*

---

[17] In many situations, continuity in not sufficient, and smooth partitions of unity must be used. This happens in Differential Geometry, but also in the theory of Function Spaces. Smoothness is ensured by convolution with suitable kernels.

Proof We choose a countable base $\{B_n\}_n$ of X, such that each $\overline{B_n}$ is compact. We define $K_1 = \overline{B_1}$ and we assume that $K_2, \ldots, K_n$ have been chosen. Let $m > n$ be the smallest integer such that $K_n \subset B_1 \cup \cdots \cup B_m$, and let

$$K_{n+1} = \overline{B_1} \cup \cdots \cup \overline{B_m}.$$

It is easy to check that $\{K_n\}_n$ is an exhaustion of X by compact sets. This proves 1.

To prove 2, we choose a locally finite open cover $\{U_i \mid i \in I\}$ of X, such that each $\overline{U_i}$ is compact. Then every compact subset of X intersects only finitely many of the U_i. Choose $i(1) \in I$ such that $U_{i(1)} \neq \emptyset$ and define $K_1 = \overline{U_{i(1)}}$. Then K_1 is compact and therefore intersects only finitely many of the U_i. We now enumerate these, i.e.

$$K_1 \cap U_{i(j)} \neq \emptyset \quad \text{for } i(1), \ldots, i(j_2) \in I.$$

We define $K_2 = \overline{U_{i(1)}} \cup \cdots \cup \overline{U_{i(j_2)}}$. As before, K_2 is compact and intersects finitely many of the U_i, i.e.

$$K_2 \cap U_{i(j)} \neq \emptyset \quad \text{for } i(1), \ldots, i(j_3) \in I.$$

We then define $K_3 = \overline{U_{i(1)}} \cup \cdots \cup \overline{U_{i(j_3)}}$. Recursively, we obtain a sequence of compact sets K_n such that

$$K_n \subset U_{i(1)} \cup \cdots \cup U_{i(n+1)} = K_{n+1}^\circ.$$

In particular, $X = \bigcup_{n=1}^\infty K_n$ is open in X. Let X be an accumulation point of $\bigcup_{n=1}^\infty K_n$. A compact neighborhood of x in X intersects only finitely many of the $U_{i(j)}$, so x lies in the finite union of the closures of these $U_{i(j)}$. This union is contained in one of the K_n, and so is x. We have proved that $\bigcup_{n=1}^\infty K_n$ contains its accumulation points, hence it is closed in X. Since X is connected and $\bigcup_{n=1}^\infty K_n \neq \emptyset$, it follows that $X = \bigcup_{n=1}^\infty K_n$.

To prove 3, we fix an exhaustion $\{K_n\}_n$ of X by compact sets. Let $\mathcal{U} = \{U_i \mid i \in I\}$ be an open cover of X. Each $K_n \setminus K_{n+1}^\circ$ is compact, hence there exist finite open covers $\{V_{nj} \mid 1 \leq j \leq k_n\}$ of $K_n \setminus K_{n+1}^\circ$ with the properties that all V_{nj} are contained in[18] $K_{n+1}^\circ \setminus K_{n+2}$ and for every V_{nj} there exists a U_j with $V_{nj} \subset U_j$. The collection of all the V_{nj} is then an open cover of X, it is locally finite, and it is a refinement of the cover $\mathcal{U}$. Hence X is paracompact, and the proof is complete. □

---

[18] We define here $K_{-1} = K_0 = \emptyset$.

13.16   Function Spaces

One of the most important applications of General Topology to Mathematical Analysis concerns *spaces of maps*, i.e. topological spaces whose elements are functions.

Definition 13.75 Let X, Y be sets. The set

$$\Omega = \text{Map}(X, Y)$$

is the set of all *continuous* functions $f : X \to Y$.

We want to topologize Ω, and this can be done in several ways. First of all, let us recall that the set Y^X of all functions $f : X \to Y$ (continuous or not) has a natural topology as a product space as soon as Y is a topological space.

Definition 13.76 Suppose that Y is a topological space. The topology induced on Ω by the product topology of Y^X is called the topology of pointwise convergence in Ω.

Exercise 13.46 For every $x \in X$ and every open set $U \subset Y$, define

$$E(x, U) = \left\{ f \in Y^X \mid f(x) \in U \right\}$$
$$M(x, U) = \{ f \in \text{Map}(X, Y) \mid f(x) \in U \}.$$

1. Prove that the sets $E(x, U)$, as $x \in X$ and $U \subset Y$ is open, form a subbasis of the topology of Y^X.
2. Since $M(x, U) = \Omega \cap E(x, U)$, deduce that $\{M(x, U) \mid x \in X, \ U \subset Y \text{ is open}\}$ is a subbasis of the pointwise topology of Ω.

Proposition 13.2 *For each finite set $F \subset X$ and every open set $U \subset Y$, let*

$$M(F, U) = \{ f \in \text{Map}(X, Y) \mid f(F) \subset U \} = \bigcap \{M(x, U) \mid x \in F\}.$$

The set $M(F, U)$ is open in the topology of pointwise convergence in Ω. In particular, the collection of the sets $M(F, U)$ as F ranges of the finite subsets of X and U ranges over the open sets in Y is a subbasis of the topology of pointwise convergence in Ω.

Proof The proof is immediate: indeed, $M(F, U)$ is open in the topology of pointwise convergence as a finite intersection of open sets of the form $M(x, U)$, for $x \in F$. □

The topology of pointwise convergence is often referred to as the *finite-open* topology, as the previous result suggests.

The basic weakness of the finite-open topology on Ω is clearly the fact that the construction is independent of any topological structure of the set X. This topology typically contains very few open sets, and few open sets correspond to few continuous maps.

Definition 13.77 Let X and Y be topological spaces. For any compact $K \subset X$ and any open $U \subset Y$, we define

$$M(K, U) = \{f \in \Omega \mid f(K) \subset U\}.$$

The compact-open topology of Ω is the smallest topology on Ω which contains all sets of the form $M(K, U)$. In particular, a set is open in the compact-open topology if and only if it is the union of a collection of finite intersections of sets $M(K, U)$.

We now want to investigate some topologies τ on $\Omega = \mathrm{Map}(X, Y)$. We begin with the natural *evaluation map*

$$\omega \colon \ (\Omega, \tau) \times X \to Y$$

$$\omega(f, x) = f(x).$$

Definition 13.78 A topology τ on Ω is admissible if and only if the map ω is continuous.

We cannot expect the compact-open topology to be always admissible. The next results show, however, that this is correct under some regularity assumption on the space.

Definition 13.79 A Hausdorff space X is regular if each $x \in X$ each closed set A not containing x have disjoint neighborhoods. More precisely, if $A \subset X$ is closed and if $x \notin A$, there exist open sets U containing x and $V \supset A$ such that $U \cap V = \emptyset$.

In a metric space, we can always *shrink* neighborhoods by playing with their radii. The category of regular spaces is precisely the category of topological spaces in which the same trick is possible.

Proposition 13.3 *The following statements are equivalent:*

1. *the topological space X is regular;*
2. *for each $x \in X$ and for each neighborhood U of x, there exists a neighborhood V of x such that $x \in V \subset \overline{V} \subset U$;*
3. *for each $x \in X$ and for each closed A not containing x, there exists a neighborhood V of x such that $\overline{V} \cap A = \emptyset$.*

Proof Given U, by definition x and $X \setminus U$ have disjoint neighborhoods V and W. Thus $V \subset X \setminus W$, so that $\overline{V} \subset X \setminus W$. From $\overline{V} \cap W \subset \overline{V} \cap W = \emptyset$ we deduce $\overline{V} \subset U$. Hence 1. implies 2.

Using now x and its neighborhood $X \setminus A$, we find an open set V such that $x \in V \subset \overline{V} \subset X \setminus A$. Hence $\overline{V} \cap A = \emptyset$, and 2. implies 3.

Let A be closed, $x \notin A$. Pick a neighborhood V of x such that $\overline{V} \cap A = \emptyset$. Then $A \subset X \setminus \overline{V}$, and $V \cap (X \setminus \overline{V}) = \emptyset$. This proves that 3. implies 1.   □

Theorem 13.74 (Arens) *If X is a locally compact regular space, then the compact-open topology on Ω is admissible. More precisely, it is the smallest of all admissible topologies on Ω.*

Proof Pick $f \in \Omega$, $x \in X$, and an open set W of Y containing $f(x)$. Since f is continuous, $f^{-1}(W)$ is an open set in X which contains x. Since X is a regular locally compact space, there exists an open neighborhood V of x such that $\overline{V}$ is compact in X and is contained in $f^{-1}(W)$. Then $U = M(\overline{V}, W)$ is an element of the subbasis of the compact-open topology of Ω which contains f. It follows that the evaluation ω sends $U \times V$ into W, so that the compact-open topology of Ω is admissible.

To complete the proof, we will show that the compact-open topology of Ω is smaller than any other admissible topology τ on Ω without using our assumptions on the space X. Pick any set of the form $M(K, W)$, where K is compact in X and W is open in Y, and pick an element $f \in M(K, W)$. Since the topology τ is admissible, the evaluation map $\omega : (\Omega, \tau) \times X \to Y$ is continuous. Hence, for every $x \in K$, there exist an open neighborhood V_x of x in X and an open neighborhood U_x of f in (Ω, τ) such that $\omega(U_x \times V_x) \subset W$. Since K is compact, there are finitely many points $x_1, \ldots, x_n$ such that

$$K \subset V_{x_1} \cup \cdots \cup V_{x_n}.$$

Let

$$U = U_{x_1} \cap \cdots \cap U_{x_n}.$$

Then U is an open neighborhood of f in (Ω, τ). We claim that $U \subset M(K, W)$. Indeed, let $g \in U$. We have $g(K) = \omega(g \times K) \subset W$. This implies that $g \in M(K, W)$ and hence that $U \subset M(K, W)$. We have proved that $M(K, W)$ is open in (Ω, τ). It follows that the compact-open topology is smaller than τ, and the proof is complete.   □

Corollary 13.2 *The topology of pointwise convergence is not admissible if it is different than the compact-open topology of Ω.*

Proof This follows at once from the fact that the properties of X are not used in the proof of the second statement of Theorem 13.74.   □

We have seen in a previous chapter of this book that the pointwise convergence of a sequence of continuous functions (of a real variable) is typically insufficient to ensure good properties of the limit function. On the contrary, continuity, integrability and (to some extent) differentiability are stable under *uniform* convergence. We want to describe now a topology on a space of maps which describes uniform convergence of sequences.

Although a natural setting for this purpose would be the theory of uniform structures, we believe that a simplified setting is more than enough to the Analyst's eye. So we will consider a metric space Y. The following result brings in another useful property.

Proposition 13.4 *Every metric space is homeomorphic to a bounded metric space.*

Proof Let (Y, d) be a metric space. We define $d^* : Y \times Y \to [0, +\infty)$ such that

$$d^*(x, y) = \frac{d(x, y)}{1 + d(x, y)}, \quad (x, y) \in Y \times Y.$$

It is easy to check that d^* is a metric on Y. Indeed, $d^*(x, y) = 0$ means $d(x, y) = 0$. Moreover,

$$\begin{aligned}
d^*(x, y) + d^*(y - z) &\geq \frac{d(x, y)}{1 + d(x, y)} + \frac{d(x, z)}{1 + d(x, y) + d(x, z)} \\
&= \frac{d(x, y) + d(x, z)}{1 + d(x, y) + d(x, z)} = \frac{1}{1 + \frac{1}{d(x,y)+d(x,z)}} \\
&\geq \frac{1}{1 + \frac{1}{d(y,z)}} = d^*(y, z).
\end{aligned}$$

Since $t \mapsto t/(1 + t)$ is a homeomorphism of $[0, +\infty)$ onto itself, the spaces (Y, d) and (Y, d^*) are homeomorphic. Finally, the fact that $d^*(x, y) \leq 1$ for every $x \in Y$, $y \in Y$ shows that (Y, d^*) is a bounded metric space.   □

In virtue of the previous Proposition *we may assume that (Y, d) is a bounded metric space.*

Definition 13.80 Consider $\Omega = \mathrm{Map}(X, Y)$. The function $d^* : \Omega \times \Omega \to [0, +\infty)$ such that

$$d^*(f, g) = \sup \{d(f(x), f(y)) \mid x \in X\}$$

is a metric on Ω which induces a topology called the topology u of uniform convergence.

Exercise 13.47 Prove that d^* is a metric on Ω.

Theorem 13.75 *The topology u is admissible.*

Proof We need to prove that the evaluation $\omega\colon (\Omega, \mathfrak{u}) \times X \to Y$ is continuous. Let $f_0 \in \Omega$ and $x_0 \in X$ be given, and pick any $\delta > 0$. We denote $y_0 = f(x_0)$. Since f_0 is continuous, there exists a neighborhood V of x_0 in X such that $d(y_0, f_0(x)) < \delta/2$ for every $x \in V$. Let

$$U = \left\{ f \in \Omega \,\middle|\, d^*(f, f_0) < \frac{\delta}{2} \right\}.$$

There results

$$
\begin{aligned}
d(\omega(f_0, x_0), \omega(f, x)) &= d(f_0(x), f(x)) \\
&\le d(f_0(x_0), f_0(x)) + d(f_0(x), f(x)) \\
&\le d(y_0, f_0(x)) + d^*(f_0, f) < \delta
\end{aligned}
$$

for every $f \in U$ and every $x \in V$. This proves the continuity of ω at (f_0, x_0). □

A natural question is whether the topology of uniform convergence coincides with the compact-open topology.

Theorem 13.76 *If X is a compact space, the compact-open topology and the topology of uniform convergence coincide on Ω.*

Proof By Theorem 13.75, every open set for the compact-open topology is also open for the topology of uniform convergence. Conversely, we need to prove that, for any $f \in \Omega$ and any $\delta > 0$, there exists an open set V for the compact-open topology such that

$$f \in V \subset U = \left\{ g \in \Omega \,\middle|\, d^*(f, g) < \delta \right\}.$$

For $x \in X$, we denote

$$W_x = \left\{ y \in Y \,\middle|\, d(f(x), y) < \frac{\delta}{2} \right\}.$$

Since f is continuous and $f(x) \in W_x$, there exists an open neighborhood G_x of x in X such that $f(K_x) \subset W_x$, where $K_x = \overline{G_x}$. Now the compactness of X comes into play: there exist finitely many points $x_1, \ldots, x_n$ in X such that

$$X = G_{x_1} \cup \cdots \cup G_{x_n}.$$

Each set $K_{x_i}, i = 1, \ldots, n$, is compact as a closed subset of a compact space. Hence

$$V = M(K_{x_1}, W_{x_1}) \cap \cdots \cap M(K_{x_n}, W_{x_n})$$

is an open set of the compact-open topology of Ω, and $f \in V$. It remains to prove that $V \subset U$. Let $g \in V$ and $x \in X$. For some $i = 1, \ldots, n$ there results $x \in G_{x_i} \subset K_{x_1}$. This implies $f(x_i) \in W_{x_i}$ and $g(x) \in W_{x_i}$. Then

$$d(f(x_i), f(x)) < \frac{\delta}{3}, \quad d(f(x_i), g(x)) < \frac{\delta}{3}.$$

We obtain

$$d(f(x), g(x)) \leq d(f(x), f(x_i)) + d(f(x_i), g(x)) < \frac{2}{3}\delta.$$

Since $x \in X$ is arbitrary, $d^*(f, g) \leq (2/3)\delta < \delta$, and thus $g \in U$. The proof is complete. $\qquad\square$

Remark 13.19 We point out that the boundedness of (Y, d) is irrelevant in Theorem 13.76. Indeed the boundedness of d^* is ensured by the compactness of X.

As we have already seen, compactness is probably the most useful property of a topological space for applications to Analysis. We want to introduce now a celebrated compactness condition in the space of continuous functions.

We begin with a general property of compact products, and with some definitions which are typical of metric spaces.

Theorem 13.77 (Wallace) *Let X, Y be topological spaces, $A \subset X$, $B \subset Y$ be compact subspaces and $W \subset X \times Y$ be an open set such that $A \times B \subset W$. Then there exist open sets $U \subset X$, $V \subset Y$ such that $A \subset U$, $B \subset V$, and $U \times V \subset W$.*

Proof We first prove the theorem in the particular case $A = \{a\}$. For any point $b \in B$ there exist two open sets $U_b \subset X$, $V_b \subset Y$, such that $(a, b) \in (U_b, V_b) \subset W$. The collection of open sets $\{V_b \mid b \in B\}$ covers B, and by compactness there are points $b_1, \ldots, b_n \in B$ such that

$$B \subset V_{b_1} \cup \cdots \cup V_{b_n}.$$

Then

$$\{a\} \times B \subset U \times V \subset \bigcup_{i=1}^{n} U_{b_i} \times V_{b_i} \subset W.$$

We now prove the statement in the general case. We have just proved that for every $a \in A$ there exist open sets $U_a \subset X$, $V_a \subset Y$, such that $\{a\} \times B \subset U_a \times V_a \subset W$. Again, the collection $\{U_a \mid a \in A\}$ of open sets covers A, and by compactness $A \subset U_{a_1} \cup \cdots U_{a_n}$ for some points $a_1, \ldots, a_n \in A$.

To conclude, the open sets $U = U_{a_1} \cup \cdots \cup U_{a_n}$ and $V = V_{a_1} \cap \cdots \cap V_{a_n}$ satisfy

$$A \times B \subset U \times V \subset W.$$

$\square$

Definition 13.81 A metric space is totally bounded if it can be covered by a finite number of open balls of radius r, for every $r > 0$.

Part of the following result was proved in a more general setting. We present a version that contains a new characterization of compactness in metric spaces.

Theorem 13.78 *If K is a closed subset of a complete metric space (X, d), the following statements are equivalent:*

(a) K is compact;
(b) every infinite subset of K has an accumulation point;
(c) K is totally bounded.

Proof Suppose (a). If E is an infinite subset of K and no point of K is an accumulation point of E, there is an open cover $\{V_\alpha \mid \alpha \in I\}$ of K such that each V_α contains at most one point of E. Therefore $\{V_\alpha \mid \alpha \in I\}$ has no finite subcover, contradicting the compactness of K.

Suppose (b). Let $\varepsilon > 0$. Pick $x_1 \in K$ at random. Suppose $x_1, \ldots, x_n$ have been chosen in K with the condition that $d(x_i, x_j) \geq \varepsilon$ if $i \neq j$. If possible, choose $x_{n+1} \in K$ so that $d(x_i, x_{n+1}) \geq \varepsilon$ for $1 \leq i \leq n$. This process must stop after a finite number of steps, because of assumption (b). Thus the open balls with radius ε centered at $x_1, \ldots, x_n$ cover K, and K is totally bounded.

Suppose now (c). Consider an open cover Γ of K which has no finite subcover. By assumption (c), K is a union of finitely many closed sets of diameters less than or equal to one. One of these, say K_1, cannot be covered by finitely many members of Γ. Now we repeat this scheme with K_1 instead of K, and continue. The result is a sequence of closed set K_i such that

(i) $K \supset K_1 \supset K_2 \supset \ldots$
(ii) $\operatorname{diam}(K_n) = \sup \{d(x, y) \mid x \in K_n, \ y \in K_n\} \leq 1/n$ for $n = 1, 2, \ldots$
(iii) no K_n can be covered by finitely many members of Γ.

Let $x_n \in K_n$. By (i) and (ii), the sequence $\{x_n\}_n$ is a Cauchy sequence in the complete metric space X, so it converges to a limit $x \in \bigcap_{n=1}^\infty K_n$. Hence $x \in V$ for some $V \in \Gamma$. By (ii), $K_n \subset V$ provided that n is sufficiently large. This contradicts (iii), and K is therefore compact. $\square$

Definition 13.82 Let X be a topological space, and (Y, d) be a metric space. A family $\mathcal{F}$ of elements of $\operatorname{Map}(X, Y)$ is equicontinuous if for every $x_0 \in X$ and every $\varepsilon > 0$ there exists a neighborhood U of x_0 in X such that $d(f(x), f(x_0)) < \varepsilon$ for every $f \in \mathcal{F}$ and every $x \in U$. The family $\mathcal{F}$ is pointwise totally bounded if the set $\{f(x) \mid f \in \mathcal{F}\}$ is totally bounded in Y for every $x \in X$.

Theorem 13.79 (Ascoli-Arzelà) *Let X be a compact topological space and (Y, d) be a complete metric space. A family $\mathcal{F} \subset \mathrm{Map}(X, Y)$ is relatively compact in the topology of uniform convergence if and only if:*

(i) $\mathcal{F}$ is equicontinuous, and
(ii) $\mathcal{F}$ is pointwise totally bounded.

Proof We prove that (i) and (ii) are necessary conditions, and we assume that $\mathcal{F}$ is contained in a compact set $K \subset \mathrm{Map}(X, Y)$. Since the evaluation map $\omega(\cdot, x)\colon \mathrm{Map}(X, Y) \to Y$ defined by $f \mapsto f(x)$ is continuous for every $x \in X$, the set $\{ f(x) \mid f \in \mathcal{F} \}$ is contained in $e_x(K)$, a compact set.

Let $x_0 \in X$ be fixed, and consider the continuous function $\alpha\colon \mathrm{Map}(X, Y) \times X \to [0, +\infty)$ such that

$$\alpha(f, x) = d(f(x_0), f(x)).$$

For every $\varepsilon > 0$ the compact set $K \times \{x_0\}$ is a subset of the open set $\alpha^{-1}([0, \varepsilon))$, and Wallace's Theorem 13.77 yields an open set $U \subset X$ such that $x_0 \in U$ and $K \times U \subset \alpha^{-1}([0, \varepsilon))$. In particular $d(f(x_0), f(x)) < \varepsilon$ for every $f \in \mathcal{F}$ and any $x \in U$.

We now prove the converse implication, assuming that (i) and (ii) hold. It suffices to prove that $\mathcal{F}$ is totally bounded in $\mathrm{Map}(X, Y)$. Pick any $\varepsilon > 0$; the equicontinuity of the family $\mathcal{F}$ implies that for any $x \in X$ there exists an open neighborhood U_x of x such that $d(f(x), f(y)) < \varepsilon$ for every $f \in \mathcal{F}$ and every $y \in U_x$. Since the space X is compact, there exist finitely many points $x_1, \ldots, x_n \in X$ such that

$$X = U_{x_1} \cup \cdots \cup U_{x_n}.$$

The image of the map $\mathcal{F} \to Y \times \cdots \times Y$ such that $f \mapsto (f(x_1), \ldots, f(x_n))$ is therefore contained in the product $\prod \omega(\mathcal{F}, x_i)$ of totally bounded sets, so it is totally bounded. We can find a finite set $F \subset \mathcal{F}$ such that , for every $f \in \mathcal{F}$, there exists $g \in F$ with the property that $d(f(x_i), g(x_i)) < \varepsilon$ for every $i = 1, \ldots, n$.

We claim that $\mathcal{F}$ is the union of open balls centered at $g \in F$ with radius 3ε. Indeed, pick $f \in \mathcal{F}$ and $g \in F$ such that $d(f(x_i), g(x_i)) < \varepsilon$ for every i. Then for any $x \in X$ there exists an index i such that $x \in U_{x_i}$, whence

$$d(f(x), g(x)) \leq d(f(x), f(x_i)) + d(f(x_i), g(x_i)) + d(g(x_i), g(x)) < 3\varepsilon,$$

and the claim follows.   □

Corollary 13.3 (Classical Version of Ascoli-Arzelà's Theorem) *Let X be a compact space. A subspace $\mathcal{F}$ of $\mathrm{Map}(X, \mathbb{R}^n)$ is relatively compact in the topology of uniform convergence if and only if $\mathcal{F}$ is equicontinuous and pointwise totally bounded (in the standard metric topology of $\mathbb{R}^n$).*

Corollary 13.4 (Arzelà) *Let X be a compact space, and let $\{f_k\}_k$ be a sequence of continuous functions from X to $\mathbb{R}^n$. If the sequence $\{f_k\}_k$ is pointwise bounded and equicontinuous, then there exists a uniformly convergent subsequence.*

13.17   Cubes and Metrizability

Metric spaces are for sure the most ubiquitous topological structure in Mathematical Analysis. We dare say that our minds always think in terms of a distance, although this might be conceptually wrong.

Anyway, metric topologies are particularly easy to work with, and in this section we try to analyze them and discover when a given topology is induced by a metric.

Definition 13.83 A pseudo-metric (or *ècart*) for a set X is a function $d : X \times X \to \mathbb{R}$ satisfying the following conditions: for every $x \in X$, $y \in X$ and $z \in X$,

(a) $d(x, y) = d(y, x)$;
(b) $d(x, y) \leq d(x, z) + d(z, y)$ (triangle inequality);
(c) $d(x, y) = 0$ if $x = y$.

The function d is a metric if, in addition to the previous properties, it also satisfies

(d) if $d(x, y) = 0$, then $x = y$.

A pseudo-metric space is a couple (X, d) in which X is a set and d is a pseudo-metric on X.

Remark 13.20 Although the *non-degeneracy* condition (d) is usually assumed in any Analysis textbook, it turns out that its relevance from a topological viewpoint is really small. Of course the results of this section hold for any metric space.

Definition 13.84 (Balls) The open ball of center x and radius $r > 0$ in a pseudo-metric space X is the set

$$B(x, r) = \{y \in X \mid d(x, y) < r\}.$$

The closed ball of center x and radius $r > 0$ is the set

$$\overline{B}(x, r) = \{y \in X \mid d(x, y) \leq r\}.$$

Definition 13.85 (Topology Defined by a Pseudo-Metric) A subset U of a pseudo-metric space X is open if and only if for every point $x \in U$ there exists an open ball $B(x, r)$ such that $B(x, r) \subset U$. This property gives rise to a topology for X, called the pseudo-metric topology induced by d.

Definition 13.86 (Distance from a Set) Let A be a subset of a pseudo-metric space X. The distance from a point $x \in X$ to A is

$$d(x, A) = \inf \{d(x, a) \mid a \in A\}.$$

Theorem 13.80 *Let $A \subset X$. The function $x \mapsto d(x, A)$ is continuous with respect to the pseudo-metric topology of X.*

Proof This follows easily from the triangle inequality. Indeed, since $d(x, a) \leq d(x, y) + d(y, a)$, we can minimize over $a \in A$ and get

$$d(x, A) \leq d(x, y) + d(y, A).$$

Swapping x and y we see that $d(y, A) \leq d(y, x) + d(x, A)$, and therefore

$$|d(x, A) - d(y, A)| \leq d(x, y).$$

If $y \in B(x, r)$, then $|d(x, A) - d(y, A)| \leq r$, and the proof is complete. □

Theorem 13.81 *In a pseudo-metric space X, the closure of A is the set of point whose distance to A is equal to zero.*

Proof We have to prove that

$$\overline{A} = \{x \in X \mid d(x, A) = 0\}.$$

By continuity, the set $\{x \in X \mid d(x, A) = 0\}$ is closed in X and trivially contains A. Hence $\overline{A} \subset \{x \in X \mid d(x, A) = 0\}$. Conversely, if $y \notin \overline{A}$, there exists a neighborhood of y, which we may assume to be an open sphere $B(y, r)$ for some $r > 0$, such that $B(y, r) \cap A = \emptyset$. Thus $d(y, A) \geq r > 0$, and $\{x \in X \mid d(x, A) = 0\} \subset \overline{A}$. The proof is complete. □

Theorem 13.82 *Every pseudo-metric space is normal.*

Proof Consider two disjoint closed subsets A and B of a pseudo-metric space X. We define

$$U = \{x \in X \mid d(x, A) - d(x, B) < 0\}$$
$$V = \{x \in X \mid d(x, A) - d(x, B) > 0\}.$$

By continuity, U and V are open sets, and they are clearly disjoint. Since A and B are closed, it follows from Theorem 13.81 that $A \subset U$ and $B \subset V$. This concludes the proof. □

Theorem 13.83 *A net $\{S_n, n \in D\}$ is a pseudo-metric space X converges to a point x if and only if the net $\{d(x, S_n), n \in D\}$ converges to zero.*

Proof Indeed, a net $\{S_n, n \in D\}$ converges to x if and only if it is eventually in every open ball $B(x, r)$, and this happens if and only if $\{d(x, S_n), n \in D\}$ is eventually in the open interval $(-r, r)$ centered at $0 \in \mathbb{R}$.   □

Definition 13.87 (Diameter) The diameter of a subset A of a pseudo-metric space X is

$$\text{diam } A = \sup \{d(x, y) \mid x \in A, \ y \in A\}.$$

If the supremum on the right-hand side is infinite, we way that A has infinite diameter.

We have already proved the following result in Proposition 13.4. We propose here a different proof.

Theorem 13.84 *Let (X, d) be a pseudo-metric space, and let*

$$e(x, y) = \min \{1, d(x, y)\}.$$

Then (X, e) is a pseudo-metric space whose topology coincides with the topology induced by d.

Proof Let $a \geq 0, b \geq 0$ and $c \geq 0$ be such that $a + b \geq c$. We claim that

$$\min\{1, a\} + \min\{1, b\} \geq \min\{1, c\}. \tag{13.5}$$

If either $\min\{1, a\}$ of $\min\{1, b\}$ is equal to 1, then the claim reduces to $\min\{1, c\} \leq 1$, which is surely true. If neither of these is equal to 1, the validity of (13.5) follows from $a + b \geq c \geq \min\{1, c\}$.

If we set $a = d(x, y)$, $b = d(y, z)$ and $c = d(z, y)$, we see that e is a pseudo-metric on X. Now, the collection of all open balls of radius $r \leq 1$ is a base for the topology induced by any pseudo-metric. Since this collection is the same for d and for e, the topologies induced by d and e must coincide. This concludes the proof.   □

Theorem 13.85 (Countable Products of Pseudo-Metric Spaces) *Consider a sequence $\{(X_n, d_n) \mid n \in \mathbb{N}\}$ of pseudo-metric spaces, each of diameter at most one, and define d by*

$$d(x, y) = \sum_{n=1}^{\infty} \frac{d_n(x_n, y_n)}{2^n}$$

Then d is a pseudo-metric for $X = \prod_{n=1}^{\infty} X_n$, and the topology induced by d is the product topology.

Proof The fact that d is indeed a pseudo-metric on the cartesian product is an easy exercise. We only show that the product topology coincides with the pseudo-metric topology. To this aim, we remark that if $V = B(x, 2^{-p})$ is an open ball centered at $x \in X$ and if

$$U = \left\{ y \in X \mid d_n(x_n, y_n) < 2^{-n-p-2} \text{ for every } n \le p + 2 \right\},$$

there results $U \subset V$. Indeed, if $y \in U$, then

$$d(x, y) < \sum_{n=1}^{p+2} \frac{1}{2^{n+p+2}} + \sum_{n=p+3}^{\infty} \frac{1}{2^n} < \frac{1}{2^p}.$$

It follows that each set which is open in the pseudo-metric topology is also open in the product topology of X. Conversely, we consider any element U of the sub-base which defines the product topology of X. Hence $U = \{x \in X \mid x_n \in W\}$ for some open set $W \subset X_n$. Any $x \in U$ has an open ball of radius r, centered at x_n and contained in W. Since

$$d(x, y) \ge \frac{d_n(x_n, y_n)}{2^n},$$

the open ball centered at x with radius $r/2^n$ is a subset of U. This proves that any element of the sub-base of the product topology is open relative to the pseudo-metric topology, and the proof is complete.   □

Definition 13.88 (Isometries) A function $f \colon X \to Y$ between the pseudo-metric spaces (X, d) and (Y, e) is an isometry if and only if

$$e(f(x), f(y)) = d(x, y)$$

holds for every $x \in X$, $y \in X$.

Exercise 13.48 Prove that any isometry is continuous. Prove that the composition of two isometries is again an isometry, and that the inverse of an isometry (if it exists) is an isometry.

So far we have seen that all the basic properties of *metric* spaces are properties of *pseudo-metric* spaces. There is however a point at which condition (d) of Definition 13.83 is needed to exclude undesired limitations.

Definition 13.89 (Distance Between Sets) The distance of two non-empty subsets A and B of a pseudo-metric space X is

$$d(A, B) = \inf \{d(a, b) \mid a \in A, b \in B\}.$$

Unfortunately, we immediately realize that $d(X, A) = 0$ for every $A \subset X$, and the triangle inequality fails. To overcome this issue, we introduce an equivalence relation on X.

Definition 13.90 If (X, d) is a pseudo-metric space, we define the relation

$$R = \{(x, y) \mid d(x, y) = 0\}$$

on $X \times X$. The quotient space X/R is endowed with the quotient topology associated to R.

Theorem 13.86 *Let (X, d) be a pseudo-metric space, and let*

$$\mathcal{D} = \left\{ \overline{\{x\}} \mid x \in X \right\}.$$

For every $A \in \mathcal{D}$ and $b \in \mathcal{D}$ we define

$$d(A, B) = \inf \{d(a, b) \mid a \in A, \ b \in B\}.$$

Then $(\mathcal{D}, d)$ is a metric space whose topology is the quotient topology of X/R. Furthermore, the projection of X onto $\mathcal{D}$ is an isometry.

Proof We begin with a remark: $u \in \overline{\{x\}}$ if and only if $d(u, x) = 0$, which is true if and only if $x \in \overline{\{u\}}$. If $u \in \overline{\{x\}}$ and $v \in \overline{\{y\}}$, then

$$d(u, v) \le d(u, x) + d(x, y) + d(y, v) = d(x, y).$$

But $x \in \overline{\{u\}}$ and $y \in \overline{\{v\}}$, hence $d(u, v) = d(x, y)$. As a consequence, for every $A \in \mathcal{D}$, $B \in \mathcal{D}$, the value of $d(A, B)$ coincides with the value of $d(x, y)$ for every $x \in A$ and $y \in B$. This proves that $\mathcal{D}, d)$ is a metric space, and the projection of X onto $\mathcal{D}$ is an isometry.

Let U be an open set in X and $x \in U$. There exists $r > 0$ such that $x \in B(x, r) \subset U$, hence $\overline{\{x\}} \subset U$. The projection of X onto $\mathcal{D}$ is thus an open map with respect to the quotient topology, but the projection is also an open map with respect to the metric topology induced by the distance d between sets. Hence these topologies must coincide. $\qquad\square$

A basic question arises at this point: when is a given topology on a set X the topology associated to a (pseudo)metric?

Definition 13.91 (Metrizable Spaces) A topological space X is metrizable if and only if its topology coincides with the topology induced by a metric on X. Similarly, X is pseudo-metrizable if and only if f its topology coincides with the topology induced by a pseudo-metric on X.

Exercise 13.49 Prove that a pseudo-metric is a metric if and only if the associated topology is T_1. Deduce that a topological space is metrizable if and only if it is pseudo-metrizable and its topology is T_1.

A typical approach to metrizability results is via good embeddings of a given topological space.

Definition 13.92 (Cubes) Any cartesian product of the unit interval $[0, 1]$, endowed with the product topology, is called a cube.

Explicitly, a cube is a topological space of the form $[0, 1]^A$, where A is a set, and its topology is the topology of pointwise convergence. As a product space, each element of a cube is a function whose domain is a specified set. In view of this generality, cubes may fail to have good properties, and this is the reason why we need to add suitable assumptions to the functions of our cubes.

Definition 13.93 Let X be a topological space, and let F be a collection of functions $f_j : X \to Y_j, i \in J$, such that Y_j is a topological space. The evaluation map $e : X \to \prod \{Y_j \mid j \in J\}$ is defined as follows: for every $x \in X$, $e(x)$ is the function $j \in J \mapsto f_j(x)$.

Thus, roughly speaking the j-th coordinate of $e(x)$ is $f_j(x)$, or $e(x)_j = f_j(x)$.

Definition 13.94 The collection $F = \{f_j : X \to Y_j \mid j \in J\}$ distinguishes points if and only if for every $x \in X$, $y \in X$ such that $x \neq y$ there exists $j \in J$ such that $f_j(x) \neq f_j(y)$.

Definition 13.95 The collection $F = \{f_j : X \to Y_j \mid j \in J\}$ distinguishes points and closed sets if and only if for every $x \in X$ and every closed set A such that $x \in X \setminus A$ there exists $j \in J$ such that $f_j(x) \notin \overline{f(A)}$.

We summarize in the next result the main topological features of the evaluation map.

Theorem 13.87 *Let $F = \{f_j : X \to Y_j \mid j \in J\}$ be a collection of continuous functions. Then*

(a) the evaluation map $e : X \to \prod \{Y_j \mid j \in J\}$ is continuous;
(b) The evaluation e is an open map of X onto $e(X)$ if and only if F distinguishes points and closed sets.
(c) The evaluation map e is injective if and only if F distinguishes points.

Proof Since $P_j \circ e(x) = f_j(x)$ for every $j \in J$, by definition of the product topology it follows that e is continuous, and (a) is proved. To prove (b), we show that the image under e of an open neighborhood U of a point x contains the intersection of $e(X)$ and a neighborhood of $e(x)$ in the product topology. Let $j \in J$ such that $f_j(x)$ does not belong to the closure of $f_j(X \setminus U)$. Now, the set of all $y \in \prod \{Y_j \mid j \in J\}$ such that $P_j(y) \notin \overline{f_j(X \setminus U)}$ is open, and evidently the intersection with $e(X)$ is a subset of $e(U)$. This proves that e is a open map of X onto $e(X)$.

Finally, $e(x) = e(y)$ if and only if $f_j(x) = f_j(y)$ for every $j \in J$. Hence e is injective if and only if F distinguishes points. $\qquad \square$

In virtue of the last result, a topological space can be embedded into a cube (i.e. it is homeomorphic to a subset of a cube) provided that it is possible to construct a sufficiently rich collection of continuous functions defined on the space. By the very definition, the existence of such collections seems to be related to some separation properties of the topology.

Definition 13.96 (Completely Regular Spaces) A topological space X is completely regular if and only if for every $x \in X$ and every neighborhood U of x, there exists a continuous function $f : X \to [0, 1]$ such that $f(x) = 0$ and $f = 1$ on $X \setminus U$. A completely regular T_1 space is a Tychonoff space.

Exercise 13.50 Prove that the collection of all continuous functions $f : X \to [0, 1]$ defined on a completely regular space X distinguishes points and closed sets.

Exercise 13.51 Suppose X is a Tychonoff space, and let F be the collection of all continuous functions from X to $[0, 1]$. Prove that the evaluation map e is a homeomorphism of X with a subspace of $[0, 1]^F$. *Hint:* use Theorem 13.87.

Our task now is to prove the converse of the last exercise. We need a preliminary result.

Theorem 13.88 *Any product of Tychonoff spaces is a Tychonoff space.*

Proof We say that a continuous function $f : X \to [0, 1]$ defined on a topological space X is *for* a pair (x, U) if and only if x is a point of X, U is a neighborhood of x, $f(x) = 0$ and $f = 1$ on $X \setminus U$. Now, if $f_1, \ldots, f_n$ are for $(x, U_1), \ldots, (x, U_n)$, then we can set

$$g(x) = \sup \{ f_i(x) \mid i = 1, \ldots, n \}$$

and conclude that g is for $\left(x, \bigcap_{i=1}^{n} U_i\right)$. This shows that X is completely regular if for every x and every neighborhood U of x belonging to a sub-base of the topology of X, there exists a function for the pair (x, U).

Consider now the case in which X is a product $\prod \{X_\alpha \mid \alpha \in A\}$ of Tychonoff spaces. Let $x \in X$ and U_a be a neighborhood of $x_a = P_a(x)$ in X_a. If f is a function for (x_a, U_a), then $f \circ P_a$ is a function for $(x, P_a^{-1}(U_a))$. The collection of all sets $P_a^{-1}(U_a)$ forms a sub-base of the product topology of X, hence X is completely regular. The fact that any product of T_1 spaces is a T_1 space completes the proof. $\square$

Theorem 13.89 (Embedding into Cubes) *For a topological space X, the following are equivalent:*

(a) X is a Tychonoff space;
(b) X is homeomorphic to a subspace of a cube.

Proof Clearly (b) implies (a). Conversely, we remark that the space $[0, 1]$ is a Tychonoff space, thus any cube is a Tychonoff space by Theorem 13.88. Therefore any subspace of a cube is a Tychonoff space. Exercise 13.51 and Theorem 13.87 show that the evaluation map e is a homeomorphism of X into a cube, and the proof is complete. $\square$

We are ready to prove a sufficient condition for a topology to be metrizable.

Theorem 13.90 (Urysohn) *A regular T_1 topological space whose topology has a countable base is homeomorphic to a subspace of the cube $[0, 1]^{\mathbb{R}}$. In particular, it is metrizable.*

Proof Let us explain the strategy of the proof. A product of countably many pseudo-metrizable spaces is pseudo-metrizable, see Theorem 13.85. By Theorem 13.87, if F is a collection of continuous function on a T_1 space X, where an element f of F maps X to Y_f, then the evaluation map is continuous from X to $\prod \{Y_f \mid f \in F\}$, and it is a homeomorphism as soon as F distinguishes points and closed sets. To show that X is metrizable, it suffices to construct a countable family (i.e. a sequence) F of continuous functions, each from X into $[0, 1]$, such that F distinguishes points and closed sets.

Let $\mathcal{B}$ be a countable base for the topology of X, and let

$$\mathcal{A} = \left\{ (U, V) \mid U \in \mathcal{B},\ V \in \mathcal{B},\ \overline{U} \subset V \right\}.$$

It is clear that $\mathcal{A}$ is a countable set. To each $(U, V) \in \mathcal{A}$ we associate a continuous function $f \colon X \to [0, 1]$ such that $f = 0$ on U and $f = 1$ on $X \setminus V$. We call F the collection of all such functions. Since F is indexed over a countable set, F is countable. We claim that F distinguishes points and closed sets. So, let B be a closed set and $x \in X \setminus B$. We choose a neighborhood $V \in \mathcal{B}$ of x such that $x \in V \subset X \setminus B$. Furthermore, we choose $U \in \mathcal{B}$ such that $\overline{U} \subset V$ (this is possible since X is regular). But then $(U, V) \in \mathcal{A}$, and if f is the element of F associated to (U, V), then $f(x) = 0 \notin \{1\} = \overline{f(B)}$. The proof is complete. □

A more general metrizability result can be proved, but we need to introduce a new definition and a couple of preliminary results.

Definition 13.97 A collection $\mathcal{G}$ of subsets of a topological space is σ-locally finite if and only if $\mathcal{G}$ is the union of a countable collection of locally finite collections $\mathcal{G}_k$, $k \in \mathbb{N}$.

We recall that a collection $\mathcal{G}$ of subsets of a topological space X is locally finite if and only if each point $x \in X$ has a neighborhood which intersects only finitely many elements of $\mathcal{G}$.

Theorem 13.91 (Stone) *Every open cover $\mathcal{U}$ of a metric space (X, d) has a σ-locally finite refinement $\mathcal{V} = \bigcup \{\mathcal{V}_k \mid k \in \mathbb{N}\}$.*

Proof For every $k \in \mathbb{N}$ and every $U \in \mathcal{U}$ we define

$$U_k = \left\{ x \in X \mid d(x, X \setminus U) \geq \frac{1}{2^k} \right\} \subset U.$$

An immediate consequence of the triangle inequality is that

$$d(U_k, X \setminus U_{k+1}) \geq \frac{1}{2^k} - \frac{1}{2^{k+1}} = \frac{1}{2^{k+1}}.$$

Here comes the hard part of the proof: we select a well-order $\ll$ on $\mathcal{U}$. For every $k \in \mathbb{N}$ and every $U \in \mathcal{U}$ we introduce

$$U'_k = U_k \setminus \bigcup \{V_{k+1} \mid V \in \mathcal{U}, \ V \ll U\} \subset U_k.$$

In the well-order $\ll$, any pair (U, V) of distinct sets in $\mathcal{U}$ satisfies either $U \ll V$ or $V \ll U$, but not both. Hence either $U'_k \subset X \setminus V_{k+1}$ or $V'_k \subset X \setminus U_{k+1}$. In other words, either

$$d(U'_k, V'_k) \geq d(X \setminus V_{k+1}, V_k) \geq \frac{1}{2^{k+1}}$$

or

$$d(U'_k, V'_k) \geq d(U_k, X \setminus U_{k+1}) \geq \frac{1}{2^{k+1}}.$$

Then

$$U''_k = \left\{ x \in X \ \middle| \ d(x, U'_k) \leq \frac{1}{2^{k+3}} \right\}$$

is contained in U. Moreover, for $U \neq V$,

$$d(U''_k, V''_k) \geq \frac{1}{2^{k+1}} - 2 \cdot \frac{1}{2^{k+3}} = \frac{1}{2^{k+2}}.$$

Thus each collection $\mathcal{V}_k = \{U''_k \mid U \in \mathcal{U}\}$ is locally finite. Indeed, for every point $x \in X$, the open ball $B(x, 2^{-k-3})$ intersects at most one element of $\mathcal{V}_k$.

   The collection $\mathcal{V} = \{\mathcal{V}_k \mid k \in \mathbb{N}\}$ is a cover of X. Every $x \in X$ belongs to some $U \in \mathcal{U}$, hence it belongs to U_k for some $k \geq 1$. As a consequence, $x \in U'_k \subset U''_k$, when U is the smallest element of $\mathcal{U}$ which contains x. Since U''_k is open and is a subset of U, the collection $\mathcal{V}$ is a σ-locally finite open refinement of $\mathcal{U}$. The proof is complete.   □

Theorem 13.92 *Every regular topological space X whose topology has a σ-locally finite base $\mathcal{B} = \bigcup \{\mathcal{B}_k \mid k \in \mathbb{N}\}$ is a normal space.*

Proof Fix two disjoint closed subsets A and B of X. By regularity, a cover $\mathcal{U}$ of A exists made by open sets that have closure disjoint from B, and a cover $\mathcal{V}$ of B exists made by open sets that have closure disjoint from A. By assumption we may express

$$\mathcal{U} = \bigcup \{\mathcal{U}_k \mid k \in \mathbb{N}\}, \quad \mathcal{V} = \bigcup \{\mathcal{V}_k \mid k \in \mathbb{N}\},$$

where $\mathcal{U}_k \subset \mathcal{B}_k$ and $\mathcal{V}_k \subset \mathcal{B}_k$ for every k. For every $k \in \mathbb{N}$, we set

$$U_k = \bigcup \{U \mid U \in \mathcal{U}_k\}, \qquad V_k = \bigcup \{V \mid V \in \mathcal{V}_k\}.$$

Since $\mathcal{B}_k$ is locally finite,

$$\overline{U}_k = \bigcup \{\overline{U} \mid U \in \mathcal{U}_k\}, \qquad \overline{V}_k = \bigcup \{\overline{V} \mid V \in \mathcal{V}_k\}.$$

We thus see that A has the countable cover $\{U_k \mid k \in \mathbb{N}\}$ by open sets with closure $\overline{U}_k \subset X \setminus B$, and B has the countable cover $\{V_k \mid k \in \mathbb{N}\}$ by open sets with closure $\overline{V}_k \subset X \setminus A$. By replacing U_k and V_k with the union of their predecessors in the indexing, we obtain two covers of A and B by nested open sets:

$$U_1 \subset U_2 \subset \ldots \subset U_k \subset \ldots$$
$$V_1 \subset V_2 \subset \ldots \subset V_k \subset \ldots$$

These sets still have the properties that $\overline{U}_k \subset X \setminus B$ and $\overline{V}_k \subset X \setminus A$. Then the sets

$$U = \bigcup \{U_k \setminus \overline{V}_k \mid k \in \mathbb{N}\}, \qquad V = \bigcup \{V_k \setminus \overline{U}_k \mid k \in \mathbb{N}\}$$

are open neighborhoods of A and B, respectively. But they are also disjoint, since

$$U_k \cap (X \setminus \overline{U}_n) \supset (U_k \setminus \overline{V})k) \cap (V_n \setminus \overline{U}_n) \subset (X \setminus \overline{V}_k) \cap V_n,$$

where $U_k \cap (X \setminus \overline{U}_n) = \emptyset$ when $k \leq n$, and $(X \setminus \overline{V}_k) \cap V_n = \emptyset$ when $n \leq k$. □

Theorem 13.93 (Nagata-Smirnov) *For a topological space X the following properties are equivalent:*

(a) X is metrizable;
(b) X is regular and has a σ-locally finite base $\mathcal{B} = \bigcup \{\mathcal{B}_k \mid k \geq 1\}$.

Proof (a) implies (b). Indeed, for every integer $n \geq 1$ the open cover $\mathcal{U}(b)$ by open balls of radius $1/n$ has a σ-locally finite refinement $\mathcal{B}(n) = \bigcup \{\mathcal{B}_k(n) \mid k \geq 1\}$ by Stone's Theorem. Then $\mathcal{B} = \bigcup \{\mathcal{B}(n) \mid n \geq 1\}$ is a σ-locally finite base for the metric topology of X.

(b) implies (a). Indeed, we begin with a σ-locally finite base $\mathcal{B} = \bigcup \{\mathcal{B}_k \mid k \geq 1\}$ for the topology of X. For every couple $(m, n) \in \mathbb{N} \times \mathbb{N}$ and every $U \in \mathcal{B}_m$, we set

$$G = \bigcup \{V \in \mathcal{B}_n \mid \overline{V} \subset U\}.$$

Since $\mathcal{B}_n$ is locally finite,

$$\overline{G} = \bigcup \{\overline{V} \in \mathcal{B}_n \mid \overline{V} \subset U\},$$

so that $G \subset U$. By Theorem 13.92, X is normal. We select a Urysohn function $u_U : X \to [0, 1]$ such that $u_U(\overline{G}) = \{1\}$ and $u_U(X \setminus U) = \{0\}$. We consider the function $d_{m,n} : X \times X \to \mathbb{R}$ such that

$$d_{m,n}(x, y) = \sum \{|u_U(x) - u_U(y)| \mid U \in \mathcal{B}_m\}.$$

Since $\mathcal{B}_m$ is locally finite, each element $(x, y) \in X \times X$ has a neighborhood on which the summation in the definition of $d_{m,n}$ is a finite sum. As a consequence, $d_{m,n}$ is a continuous function. It is easy to check that $d_{m,n}$ is a pseudo-metric on X. We call $Y_{m,n}$ the set X endowed with the topology induced by the pseudo-metric $d_{m,n}$, and let $g_{m,n} : X \to Y_{m,n}$ be the inclusion map. It is clear that $g_{m,n}$ is a continuous function.

We claim that the countable collection $\mathcal{F} = \{g_{m,n} \mid (m, n) \in \mathbb{N} \times \mathbb{N}\}$ separates points and closed sets. Indeed, pick any $x \in X$ and any closed set A which does not contain x. There exists a member $U \in \mathcal{B}_m \subset \mathbb{B}$ of the base such that $x \in U \subset X \setminus A$. By regularity, there exists a second member $V \in \mathcal{B}_m$ of the base such that $x \in V \subset \overline{V} \subset U \subset X \setminus A$. But then $d_{m,n}(x, A) \geq 1$, since $u_U(\overline{V}) = \{1\}$ and u_U vanishes on A. As a consequence, $g_{m,n}(x) = x$ does not belong to the closure (with respect to $d_{m,n}$) of $g_{m,n}(A) = A$. The claim is proved.

The evaluation map

$$e : X \to \prod \{Y_{m,n} \mid (m, n) \in \mathbb{N} \times \mathbb{N}\}$$

is an embedding of X onto a subspace of the countable product of pseudo-metrizable spaces. By Theorem 13.85 the product $\prod \{Y_{m,n} \mid (m, n) \in \mathbb{N} \times \mathbb{N}\}$ is pseudo-metrizable by a metric d. But this pseudo-metric is actually a metric on the image $e(X)$, since $e(X)$ consists of constant sequences. Indeed, there exists a pair $(m, n) \in \mathbb{N} \times \mathbb{N}$ such that $d_{m,n}(x, y) \geq 1$ whenever $x \neq y$, by taking $A = \{y\}$ in the separation property of the collection $\mathcal{F}$. The proof is complete.   □

13.18 Problems

13.1 Let S be a subset of a topological space X. Show that a sequence $\{s_n\}_n$ in S converges to $s \in S$ in the relative topology if and only if, considered as a sequence in X, the sequence $\{s_n\}_n$ converges to s.

13.2 Prove that the space

$$\mathbb{R}^2 \setminus \{(x, y) \mid x \in \mathbb{N} \wedge y \in \mathbb{N}\}$$

is homeomorphic to the space

$$\mathbb{R}^2 \setminus \left\{ (x, y) \,\middle|\, (\exists n)(\exists m)(n \in \mathbb{N} \wedge m \in \mathbb{N}) \implies (x - n)^2 + (y - m)^2 < \frac{1}{10} \right\}.$$

13.3 Let (X, d) be a compact metric space, and let $f : X \to X$ be an isometry, i.e. $d(x, y) = d(f(x), f(y))$ for every $x \in X$, $y \in X$. Prove that f is surjective. *Hint:* set $A = f(X)$, and suppose there exists $x_0 \in X \setminus A$. Define inductively $x_n = f(x_{n-1})$. Now elaborate on the sequence $\{x_n\}_n$.

13.4 Let (X, d) be a metric space. In analogy with the field of real numbers, we say that a sequence $\{x_n\}_n$ in X is a Cauchy sequence if and only if for each $\varepsilon > 0$ there exists $N \in \mathbb{N}$ such that $N \geq N$, $m \geq N$ imply $d(x_n, x_n) < \varepsilon$. Furthermore, we say that (X, d) is a complete metric space if and only if each Cauchy sequence in X converges to a point of X. Prove that X is complete if it is a compact metric space.

13.5 Let (X, d) be a metric space. The diameter of any non-empty subset E of X is defined to be the supremum of $d(x, y)$ as x and y range over E. Prove that a sequence $\{x_n\}_n$ in X is a Cauchy sequence if and only if

$$\lim_{n \to +\infty} \operatorname{diam} \{x_n, x_{n+1}, x_{n+2}, \ldots\} = 0.$$

13.6 Let (X, d) be a metric space, and let E be a subset of X. Prove that $\operatorname{diam} \overline{E} = E$.

13.7 Let (X, d) be a metric space.

1. Call two Cauchy sequences $\{x_n\}_n$, $\{y_n\}_n$ in X equivalent if and only if

$$\lim_{n \to +\infty} d(x_n, y_n) = 0.$$

Prove that this is indeed an equivalence relation.
2. Let X^* be the set of all equivalence classes so obtained. If $P \in X^*$ and $Q \in X^*$, define

$$\Delta(P, Q) = \lim_{n \to +\infty} d(x_n, y_n),$$

where $P = [\{x_n\}_n]$ and $Q = [\{y_n\}_n]$. Prove that this limit exists, and that $\Delta(P, Q)$ depends only on P and Q, but not on the representatives $\{x_n\}_n$ and $\{y_n\}_n$ of P and Q.
3. Prove that (X^*, Δ) is a complete metric space.
4. For each point $x \in X$, there is a Cauhcy sequence all of whose terms are equal to x: let P_x be the element of X^* which contains this sequence. Prove that

$$\Delta(P_x, P_y) = d(x, y)$$

for all x, $y \in X$. In other words, the mapping φ defined by $\varphi(x) = P_x$ is an isometry of X into X^*.

5. Prove that $\varphi(X)$ is dense in X^*, and that $\varphi(X) = X^*$ if X is complete. The metric space (X^*, Δ) is called the *completion* of (X, d).

13.8 A proper interval in $\mathbb{R}$ is a half-open, open, or closed interval which contains more than one point. If $\mathcal{A}$ is an arbitrary family of proper intervals, prove that there exists a countable subfamily $\mathcal{B}$ of $\mathcal{A}$ such that

$$\bigcup \{B \mid B \in \mathcal{B}\} = \bigcup \{A \mid A \in \mathcal{A}\} \, .$$

Hint: prove that a disjoint family of proper intervals is countable, and show that all but a countable number of points of $\bigcup \{A \mid A \in \mathcal{A}\}$ are interior points of members of $\mathcal{A}$.

13.9 Let X be a topological space. If A is dense and U is open, prove that $U \subset \overline{A \cup U}$.

13.10 For each $a \in \mathbb{Z}$, $b \in \mathbb{Z}$, let

$$N_{a,b} = \{a + kb \mid k \in \mathbb{Z}\} \, .$$

1. Prove that the collection $\mathcal{B} = \{N_{a,b} \mid a \in \mathbb{Z}, \ b \in \mathbb{Z}, \ b > 0\}$ is a base for a topology τ on $\mathbb{Z}$.
2. Prove that each $N_{a,b}$ is both open and closed in $(\mathbb{Z}, \tau)$.
3. Let $P = \{2, 3, \ldots\}$ the set of prime numbers. Prove that

$$\mathbb{Z} \setminus \{-1, 1\} = \bigcup \{N_{0,p} \mid p \in P\} \, .$$

Deduce that if P were a finite set, then $\{-1, 1\}$ would be an open set in $(\mathbb{Z}, \tau)$. Hence P is an infinite set.

13.11 Let Y be a dense subset of a Hausdorff topological space X. If Y is locally compact, prove that Y is open in X.

13.12 Let (X, d) be a metric space. If $A \subset X$ and $\varepsilon > 0$, we set

$$U(A, \varepsilon = \{x \in X \mid d(x, a) < \varepsilon \text{ for every } a \in A\} \, .$$

Let $\mathcal{H}$ be the collection of all non-empty closed, bounded subsets of X. For A, $B \in \mathcal{H}$, we set

$$D(A, B) = \inf \{\varepsilon > 0 \mid A \subset U(B, \varepsilon) \text{ and } B \subset U(A, \varepsilon)\} \, .$$

(a) Prove that D is a metric on $\mathcal{H}$, which we call the Hausdorff metric.
(b) Prove that if (X, d) is complete, so is $(\mathcal{H}, D)$. *Hint:* Let $\{A_n\}_n$ be a Cauchy sequence in $\mathcal{H}$; by passing to a subsequence, we may assume that $D(A_n, A_{n+1}) < 2^{-n}$. Define A to be the set of all points x which are the limits of sequences $\{x_n\}_n$ such that $x_n \in A_n$ for every n and $d(x_n, x_{n+1}) < 2^{-n}$. Prove that $A_n \to A$ with respect to D.
(c) Prove that if (X, d) is totally bounded, so is $(\mathcal{H}, D)$. *Hint:* Given $\varepsilon > 0$, pick $\delta < \varepsilon$ and let S be a finite subset of X such that X is covered by $\{B(x, \delta) \mid x \in S\}$. Let $\mathcal{A}$ be the collection of all non-empty subsets of S. Prove that $\{B(A, \varepsilon) \mid A \in \mathcal{A}\}$ covers $\mathcal{H}$.
(d) If X is compact in the metric d, prove that $\mathcal{H}$ is compact in the Hausdorff metric D.

13.13 Let X be a topological space, and (Y, d) be a metric space. If the subset $\mathcal{F}$ of $\mathrm{Map}(X, Y)$ is totally bounded with respect to the metric of uniform convergence, prove that $\mathcal{F}$ is equicontinuous.

13.14 A real number $p > 0$ is called an almost-period relative to $\varepsilon > 0$ for the function $f : \mathbb{R} \to \mathbb{R}$ if and only if $|f(x + p) - f(x)| < \varepsilon$ for every $x \in \mathbb{R}$. The function f is almost-periodic if and only if for every $\varepsilon > 0$ there exists a length $\ell > 0$ such that in every interval of length ℓ it is possible to find an almost-period p relative to ε. Prove that an almost-periodic function must be bounded.

13.19 Comments

I believe that the chapter on General Topology has a fundamental role in a book like this one. This is why this chapter is particularly long and full of ideas and results. The reader may have noticed that I did not insist on a strictly *economic* exposition: a few definitions appear twice, and the order of appearance of the main characters is not always coherent with the tradition. As an example, separation axioms just come into play when they are needed from the viewpoint of a Mathematical Analyst. As long as it looks possible, the Hausdorff separation axiom is used alone, because this is exactly the basic condition ensuring the most important fact of Analysis: a function converges to at most one point. When the construction of continuous functions which separate sets becomes necessary, we introduce normality and regularity. The locally compact Hausdorff case is dealt with separately, although one might optimize several proofs by reducing to the normal case. I have decided to do so because locally compact Hausdorff space are the natural setting of Measure Theory, a topic that will be discussed in the next chapters.

While writing this chapter, I had in mind the classical reference [5]. Kelley's book remains a great source for the young Analyst who wishes to learn "what every Analyst should know about topology", but its main feature is that the whole book should be read like a romance, from cover to cover. This is nowadays uncommon, since textbooks are written for the time-lacking reader.

Another standard reference is [6], whose style is opposed to Kelley's. There are plenty of pictures, diagrams, sketches, although the book is somehow redundant for our purposes.

The definitive *bible* of General Topology is surely [4], whose bibliography is an encyclopedia of references.

References

1. J.F. Aarnes, P.R. Andenaes, On nets and filters. Math. Scand. **31**, 285–292 (1972)
2. P.R. Chernoff, A simple proof of Tychonoff's theorem via nets. Am. Math. Mon. **99**(10), 932–934 (1992)
3. C. Fefferman, An easy proof of the fundamental theorem of algebra. Am. Math. Mon. **74**, 854–855 (1967)
4. R. Engelking, *General Topology*. Sigma Series in Pure Mathematics Series Profile, vol. 6 (Heldermann Verlag, Berlin, 1989)
5. J.L. Kelley, *General Topology*. Graduate Texts in Mathematics, vol. 27 (Springer-Verlag, New York-Berlin, 1975). Reprint of the 1955 edition [Van Nostrand, Toronto, Ont.]
6. J.R. Munkres, *Topology* (Prentice Hall, Upper Saddle River, 2000), xvi, 537 p.
7. M.E. Rudin, A new proof that metric spaces are paracompact. Proc. Am. Math. Soc. **20**, 603 (1969)
8. L. Schwartz, Analyse, in *Topologie Générale et Analyse Fonctionnelle* (Hermann, Paris, 1993)
9. A.H. Stone, Paracompactness and product spaces. Bull. Am. Math. Soc. **54**, 977–982 (1948)

Chapter 14
Differentiating Again: Linearization in Normed Spaces

Abstract This chapter is devoted to an overview of basic linear analysis in normed spaces.

14.1 Normed Vector Spaces

Let us start with some old friends.

Definition 14.1 A vector space V over $\mathbb{R}$ is a commutative group under a binary operation $+$, together with an operation of scalar multiplication $\mathbb{R} \times V \to V$, denoted by juxtaposition, such that

1. $a(x + y) = ax + ay$
2. $(a + b)x = ax + bx$
3. $a(bx) = (ab)x$
4. $1x = x$

for every $x \in V$, $y \in V$, $a \in \mathbb{R}$, $b \in \mathbb{R}$. The symbol 0 will be used both for the real number zero and for the additive identity in V.

Remark 14.1 The systematic use of bold-face fonts to denote vectors, like in **x** or **v**, is no longer popular among mathematicians.

Definition 14.2 Let V be a vector space. Given $E \subset \mathbb{R}$, $a \in \mathbb{R}$, $A \subset V$, $B \subset V$, $x_0 \in V$, we set

$$A + B = \{x + y \mid x \in A, \ y \in B\}$$

$$x_0 + B = = \{x_0 + y \mid y \in B\}$$

$$EA = \{ax \mid a \in E, \ x \in A\}$$

$$aA = \{ax \mid x \in A\}.$$

Definition 14.3 Let V be a vector space, and let $W \subset V$. We say that W is a vector subspace of V if $W + W \subset W$ and $\mathbb{R}W \subset W$. Furthermore, we say that W is convex if $ax + (1 - a)y \in W$ whenever $a \in \mathbb{R}$, $x \in W$, $y \in W$ and $0 \le a \le 1$. We say that W is symmetric if $-1W = -W = W$, and that W is balanced if $aW \subset W$ for every $\mathbb{R}$ such that $|a| \le 1$.

Exercise 14.1 Prove that W is a vector subspace of V if and only if for every $a \in \mathbb{R}$, $b \in \mathbb{R}$, $u \in W$ and $v \in W$ there results $au + bv \in W$.

Definition 14.4 Let V be a vector space, and suppose that $p \colon V \to \mathbb{R}$. We say that p is a seminorm on V if

1. $p(x + y) \le p(x) + p(y)$
2. $p(ax) = |a| p(x)$

for every $x \in V$, $y \in V$, $a \in \mathbb{R}$. Of course the first property is the triangle inequality for p, while the second property is a homogeneity property.

Exercise 14.2 Let p be a seminorm on the vector space V. Prove that $p(0) = 0$, $p(x) \ge 0$ for every $x \in V$, and $p(x - y) \ge |p(x) - p(y)|$ for every $x \in V$, $y \in V$.

Definition 14.5 Let V be a vector space, and let p be a seminorm on V. If

$$\forall x \, (x \in V \wedge p(x) = 0) \implies (x = 0),$$

i.e. $x = 0$ is the unique zero of the seminorm p, we say that p is a norm on V.

If p is a norm on V, we will usually employ a less generic notation. It is customary to write $\| \cdot \|$ instead of $p(\cdot)$. Normed vector spaces will be our primary object of interest in the rest of this chapter.

Remark 14.2 If $(V, \| \cdot \|)$ is a normed vector space, it is a trivial exercise to check that V inherits a topology as a metric space with the metric $d \colon V \times V \to \mathbb{R}$ such that

$$d(x, y) = \|x - y\|.$$

This topology will be called the *norm topology*.

Definition 14.6 A normed vector space is a Banach space if it is complete (as a metric space) in the norm topology.

Let us present a few examples that play an important role in Analysis.

Example 14.1 Let X be a Hausdorff topological space. By $C(X)$, $C_0(X)$ and $C_c(X)$ we denote, respectively, the vector spaces of all continuous real-valued functions on X that are bounded, vanish at infinity, or have compact support. The algebraic operations are, of course, defined pointwise on X.

The set $C_0(X)$ requires some further explanation, since X does not carry a metric structure. We say that a function $f \colon X \to \mathbb{R}$ vanishes at infinity if, for every $\varepsilon > 0$, there exists a compact subset K_ε of X such that $|f(x)| < \varepsilon$ for every $x \in X \setminus K_\varepsilon$. If X is a normed vector space (with norm $\| \cdot \|$), this means that for every $\varepsilon > 0$ there exists $M \in \mathbb{R}$ such that $\|x\| > M$ implies $|f(x)| < \varepsilon$, or equivalently that $\lim_{\|x\| \to +\infty} f(x) = 0$.

If f is an element of any of these spaces, we set

$$\|f\| = \|f\|_\infty = \sup \{ f(t) \mid t \in X \}.$$

Exercise 14.3 Prove that $C(X)$ and $C_0(X)$ are Banach spaces. If X is a compact Hausdorff space, prove that the normed spaces $C(X)$, $C_0(X)$ and $C_c(X)$ coincide.

Example 14.2 Let $a < b$ and let $n = 0, 1, 2, \ldots$ We denote by $C^n([a, b])$ the vector space of n-times continuously differentiable functions from $[a, b]$ to $\mathbb{R}$. We define

$$\|f\|_n = \sum_{k=0}^{n} \left\| f^{(k)} \right\|_\infty$$

for $f \in C^n([a, b])$, where $f^{(k)}$ stands for the k-th derivative of f, $0 \leq k \leq n$. Finally, we define

$$C^\infty([a, b]) = \bigcap_{n=0}^{\infty} C^n([a, b]).$$

Exercise 14.4 Prove that $C^n([a, b])$ is a Banach space.

Remark 14.3 The space $C^\infty([a, b])$ has a more complicated topological structure which we are not going to describe in this book.

Example 14.3 Let $0 < p < \infty$ be a fixed number. We define $\ell^p = \ell^p(\mathbb{N})$ as the vector space of all sequences $\{x_n\}_n$ of real numbers such that $\sum_{n=1}^{\infty} |x_n|^p \in \mathbb{R}$. In this case, for $x = \{x_n\}_n$ we define the norm

$$\|x\|_p = \left(\sum_{n=1}^{\infty} |x_n|^p \right)^{1/p}.$$

If $p = \infty$, the space $\ell^\infty = \ell^\infty(\mathbb{N})$ is the vector space of all bounded sequences of real numbers. The norm of $x = \{x_n\}_n \in \ell^\infty$ is

$$\|x\|_\infty = \sup_{n \in \mathbb{N}} |x_n|.$$

Exercise 14.5 Prove that ℓ^p and ℓ^∞ are Banach spaces.

Definition 14.7 Let V be a (real) vector space. An inner (or a scalar) product on V is a function which assigns to every $(x, y) \in V \times V$ a real number $\langle x \mid y \rangle$ such that

1. (linearity) $\langle ax + by \mid z \rangle = a \langle x \mid z \rangle + b \langle y \mid z \rangle$
2. (symmetry) $\langle x, y \rangle = \langle y \mid x \rangle$
3. (positivity) $\langle x \mid x \rangle \geq 0$
4. (non-degeneracy) $\langle x \mid x \rangle = 0$ implies $x = 0$

for every $x \in V$, $y \in V$, $z \in V$, $a \in \mathbb{R}$, $b \in \mathbb{R}$. The norm induced by the inner product is defined by

$$\|x\| = \sqrt{\langle x \mid x \rangle} \qquad x \in V.$$

A Hilbert space is a vector space which is a complete metric space with respect to the norm induced by an inner product.

Remark 14.4 If V is a complex vector space, in the sense that vectors are multiplied by complex numbers, condition 2 in the definition of an inner product must be changed to

2. $\langle x, y \rangle = \overline{\langle y \mid x \rangle}$;

in this case the inner product is usually called *hermitian* instead of *symmetric*.

14.2   Bounded Linear Operators

Important: Norms on Different Spaces

With an abuse of notation, more often than not we use the same symbol for norms of different spaces. The context usually permits to avoid any confusion.

Definition 14.8 Let X, Y be Banach spaces. A function $T : X \to Y$ is a bounded linear operator if

1. T is linear, i.e. $T(ax + by) = aTx + bTy$ for every $x \in X$, $y \in X$, $a \in \mathbb{R}$, $b \in \mathbb{R}$;
2. there results

$$\sup_{x \in X \setminus \{0\}} \frac{\|Tx\|}{\|x\|} \in \mathbb{R}. \tag{14.1}$$

If this is the case, the real number in (14.1) is called the *operator norm* of T, and is denoted by $\|T\|$ (as usual!). The set of all bounded linear operators from X to Y is denoted by $L(X, Y)$.

Exercise 14.6 Prove that $L(X, Y)$ is a Banach space under the operator norm. It should be remarked that only the completeness of Y is necessary here.

Definition 14.9 Let X be a Banach space. The (topological) dual space of X is $X^* = L(X, \mathbb{R})$, i.e. the Banach space of all bounded linear operators from X to $\mathbb{R}$. Any such operator will be called a *bounded linear functional* on X. The operator norm of $f \in X^*$ will be also denoted by $\|f\|_{X^*}$.

Remark 14.5 Sometimes X' is used instead of X^*, but we will always stick to our symbol.

Theorem 14.1 *Let X, Y be Banach spaces, and let $T : X \to Y$ be a linear operator. The following are equivalent:*

(a) T is continuous at some point $x_0 \in X$;
(b) T is continuous at $0 \in X$;
(c) T is a bounded operator;
(d) T is uniformly continuous.

Proof Obviously (d) implies (c) implies (b) implies (a). Suppose now that (a) holds. If V is an open neighborhood of the origin of Y, then $V_1 = V + Tx_0$ is an open neighborhood of Tx_0. By assumption there exists an open neighborhood U_1 of x_0 such that $T(U_1) \subset V_1$. But then $U = U_1 - x_0$ is an open neighborhood of the origin of X, and if y and z are elements of X such that $y - z \in U$, then

$$Ty - Tz = T(y - z) \in V_1 - Tx_0 = V.$$

This shows that T is uniformly continuous, and thus (a) implies (d). The proof is complete.   □

14.3   The Hahn-Banach Theorem

Definition 14.10 Let V be a vector space. A gauge on X is a function $p : V \to \mathbb{R}$ such that

1. $p(x + y) \le p(x) + p(y)$
2. $p(\lambda x) = \lambda p(x)$

for every $x \in V$, $y \in V$, $\lambda > 0$.

Theorem 14.2 (Hahn-Banach, Analytic Form) *Suppose p is a gauge on a vector space V, and suppose that W is a vector subspace of V. Let $g\colon W \to \mathbb{R}$ be such that*

$$g(x) \le p(x) \quad \text{for all } x \in W.$$

Then there exists a linear map[1] *$f\colon V \to \mathbb{R}$ such that $g(x) = f(x)$ for every $x \in W$, and*

$$f(x) \le p(x) \quad \text{for all } x \in V.$$

Proof We consider the set

$$P = \left\{ h \,\middle|\, \begin{array}{l} h\colon D(h) \subset V \to \mathbb{R} \text{ where } D(h) \text{ is a vector} \\ \text{subspace of } V,\ h \text{ is linear, } G \subset D(h),\ h = g \\ \text{on } W \text{ and } h(x) \le p(x) \text{ for every } x \in D(h) \end{array} \right\}.$$

An order relation is defined on P as follows: $h_1 \le h_2$ if and only if $D(h_1) \subset D(h_2)$ and $h_1 = h_2$ on $D(h_1)$. Then $P \ne \emptyset$, since $g \in P$. Let $Q \subset P$ be a totally ordered subset. We set $Q = \{h_i \mid i \in I\}$ and

$$D(h) = \bigcup_{i \in I} D(h_i)$$
$$h(x) = h_i(x) \quad \text{if } x \in D(h_i).$$

It is easy to check that $h \in P$, and that h is an upper bound of Q. We can apply Zorn's Lemma to produce a maximal element f. We claim that $D(f) = V$, so that the proof will be complete.

We assume on the contrary that $D(f) \ne V$, and let $x_0 \in V \setminus D(f)$. We define $D(h) = D(f) + \mathbb{R}x_0$ and $h(x + tx_0) = f(x) + t\alpha$ for every $x \in D(f)$, $\in \mathbb{R}$. Here α is a real number that will be fixed conveniently in a moment. Hence $h \in P$, and we need to prove that

$$f(x) + t\alpha \le p(x + tx_0)$$

for every $x \in D(f)$ and $t \in \mathbb{R}$. Since p is a gauge, it is sufficient to prove that

$$f(x) + \alpha \le p(x + x_0)$$
$$f(x) - \alpha \le p(x - x_0).$$

---

[1] Here we use linearity in a pure algebraic sense. No reference to any norm is understood.

This is certainly true provided that

$$\sup\{f(y) - p(y - x_0) \mid y \in D(f)\} \leq \alpha \leq \inf\{p(x + x_0) - f(x) \mid x \in D(f)\}.$$

Such a choice of α is possible because

$$f(y) - p(y - x_0) \leq p(x + x_0) - f(x)$$

for every $x \in D(f)$, $y \in D(f)$, which follows from

$$f(x) + f(y) \leq p(x + y) \leq p(x + x_0) + p(y - x_0).$$

We have just proved that $f \leq h$, against the maximality of f. The proof is now complete. □

Corollary 14.1 *Let G be a vector subspace of X and let $g : G \rightarrow \mathbb{R}$ be a continuous linear map whose norm is*

$$\|g\|_{G^*} = \sup\left\{ \frac{g(x)}{\|x\|} \;\middle|\; x \in G, \; x \neq 0 \right\}.$$

Then there exists $f \in X^*$ such that $f = g$ on G and $\|f\|_{X^*} = \|g\|_{G^*}$.

Proof We just apply Theorem 14.2 with the gauge $p(x) = \|g\|_{G^*}\|x\|$. □

Corollary 14.2 *For every $x_0 \in X$ there exists $f_0 \in X^*$ such that*

$$\|f_0\| = \|x_0\|, \qquad f_0(x_0) = \|x_0\|^2.$$

Proof Let $G = \mathbb{R}x_0$ and $g(tx_0) = t\|x_0\|^2$. We have $|g|_{G^*} = \|x_0\|^2$, so that Corollary 14.1 applies. □

Corollary 14.3 *For every $x \in X$ there results*

$$\|x\| = \sup\left\{ \frac{|f(x)|}{\|f\|_{X^*}} \;\middle|\; f \in X^* \right\} = \max\left\{ \frac{|f(x)|}{|f\|_{X^*}} \;\middle|\; f \in X^* \right\}.$$

Proof The conclusion is trivial if $x = 0$, since $f(0) = 0$ whenever $f \in X^*$. Assume $x \neq 0$. Clearly

$$\sup\left\{ \frac{|f(x)|}{\|f\|_{X^*}} \;\middle|\; f \in X^* \right\} \leq \|x\|.$$

On the other hand, there exists $f_0 \in X^*$ such that $\|f_0\| = \|x\|$ and $f_0(x) = \|x\|^2$. We define $f_1 = \|x\|^{-1} f_0$ in such a way that $\|f_1| = 1$ and $f_1(x) = \|x\|$. □

Theorem 14.2 can be also presented in a more geometric way.

Definition 14.11   A hyperplane is a set of the form

$$H = \{x \in X \mid f(x) = \alpha\},$$

where $f : X \to \mathbb{R}$ is a linear map[2] and $\alpha \in \mathbb{R}$. We will always assume that f is not identically zero.

For the sake of brevity, we will also write

$$[f = \alpha] = \{x \in X \mid f(x) = \alpha\}.$$

Theorem 14.3   *The hyperplane $[f = \alpha]$ is closed if and only if $f \in X^*$.*

Proof   If $f \in X^*$, then $[f = \alpha]$ is closed as the preimage of the closed subset $\{\alpha\}$ in $\mathbb{R}$. Conversely, suppose that $H = [f = \alpha]$ is closed in X. Fix $x_0 \in X \setminus H$, and suppose for definiteness that $f(x_0) < \alpha$. Pick $r > 0$ such that $B(x_0, r) \subset X \setminus H$, where

$$B(x_0, r) = \{x \in X \mid \|x - x_0\| < r\}.$$

We claim that $f(x) < \alpha$ for all $x \in B(x_0, r)$. Indeed, suppose that $f(x_1) > \alpha$ for some $x_1 \in B(x_0, r)$. For every $t \in [0, 1]$, the point $x_t = (1 - t)x_0 + tx_1$ lies in $B(x_0, r)$, so that $f(x_t) \neq \alpha$ for every $t \in [0, 1]$. But $f(x_t) = \alpha$ for

$$t = \frac{f(x_1) - \alpha}{f(x_1) - f(x_0)},$$

which is a contradiction. The claim is proved, and it follows that $f(x_0 + rz) < \alpha$ for every $z \in B(0, 1)$. As a consequence $f \in X^*$ with

$$\|f\|_{X^*} < \frac{\alpha - f(x_0)}{r}.$$

$\square$

Definition 14.12   Let $A \subset X$, $B \subset X$. The hyperplane $H = [f = \alpha]$ separates A and B if and only if

$$x \in A \implies f(x) \leq \alpha$$
$$x \in B \implies f(x) \geq \alpha.$$

---

[2] Continuity is not required.

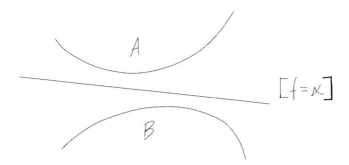

Fig. 14.1 Separation of closed convex subsets

We say that H strictly separates A and B if there exists $\varepsilon > 0$ such that

$$x \in A \implies f(x) \leq \alpha - \varepsilon$$
$$x \in B \implies f(x) \geq \alpha + \varepsilon.$$

Figure 14.1 describes the definition of separating hyperplane.

Lemma 14.1 *Let $C \subset X$ be an open convex subset such that $0 \in C$. We define the gauge of C as the function $p_C : X \to \mathbb{R}$ such that*

$$p_C(x) = \inf \left\{ \alpha > 0 \ \middle|\ \frac{1}{\alpha}x \in C \right\}$$

for every $x \in X$. Then p_C is a gauge. Furthermore there exists $M \in \mathbb{R}$ such that $0 \leq p_C(x) \leq M\|x\|$ for every $x \in X$, and

$$C = \{x \in X \mid p_C(x) < 1\}.$$

Proof Since C is open, there exists $r > 0$ such that $B(0, r) \subset C$. By definition,

$$p_C(x) \leq \frac{1}{r}|x| \quad \text{for every } x \in X.$$

Hence $M = 1/r$ works. The first property of a gauge is trivial.

Now, assume that $x \in C$. since C is open, we have $(1 + \varepsilon)x \in C$ for $\varepsilon > 0$ sufficiently small. Hence $p_C(x) \leq 1/(1 + \varepsilon) < 1$. Conversely, if $p_C(x) < 1$ then there exists $0 < \alpha < 1$ such that $*1/\alpha)x \in C$ and thus

$$x = \alpha \cdot \frac{1}{\alpha}x + (1 - \alpha) \cdot 0 \in C.$$

Finally, let $x \in X$, $y \in X$, $\varepsilon > 0$. It follows from the previous considerations that

$$\frac{x}{p_C(x) + \varepsilon} \in C, \qquad \frac{y}{p_C(y) + \varepsilon} \in C.$$

Therefore

$$\frac{tx}{p_C(x) + \varepsilon} + \frac{(1 - t)y}{p_C(y) + \varepsilon} \in C$$

for every $t \in [0, 1]$, since C is convex. Choosing

$$t = \frac{p_C(x) + \varepsilon}{p_C(x) + p_C(y) + 2\varepsilon}$$

we obtain

$$\frac{x + y}{p_C(x) + p_C(y) + 2\varepsilon} \in C.$$

Hence

$$p_C\left(\frac{x + y}{p_C(x) + p_C(y) + 2\varepsilon}\right) < 1,$$

or $p_C(x + y) < p_C(x) + p_C(y) + 2\varepsilon$, for every $\varepsilon > 0$. We deduce that $p_C(x + y) \leq p_C(x) + p_C(y)$, and the proof is complete. $\qquad\square$

Lemma 14.2 *Let $C \subset X$ be a non-empty, open and convex set, and let $x_0 \in X \setminus C$. There exists $f \in X^*$ such that $f(x) < f(x_0)$ for every $x \in C$. In particular, the hyperplane $[f = f(x_0)]$ separates $\{x_0\}$ and C.*

Proof Without loss of generality, we assume $0 \in C$ (otherwise we replace C by $C - x_0$). We set $G = \mathbb{R}x_0$ and $g : G \to \mathbb{R}$ such that

$$g(tx_0) = tp_C(x_0),$$

where p_C is the gauge of C. If $t > 0$, then $g(tx_0) = tp_C(x_0) = p_C(tx_0)$. Since $0 = p_C(0) = p_C(x_0 - x_0) \leq p_C(x_0) + p_C(-x_0)$ implies $p_C(-x_0) \leq p_C(x_0)$, it follows that $g(tx_0) \leq p_C(tx_0)$ for every $t \leq 0$. In any case, $g(x) \leq p_C(x)$ for every $x \in G$.

By the Hahn-Banach Theorem, there exists $f \in X^*$ such that $f = g$ on G and $f(x) \leq p_C(x)$ for every $x \in X$. In particular $f(x_0) = 1$. Lemma 14.1 yields $f(x) < 1$ for every $x \in C$. $\qquad\square$

Theorem 14.4 (Hahn-Banach, Geometric Form) *Let* $A \subset X$ *and* $B \subset X$ *be non-empty, convex and disjoint sets.*

(a) If A is open, then there exists a closed hyperplane which separates A and B.
(b) If A is closed and B is compact, then there exists a closed hyperplane which strictly separates A and B.

Proof

(a) We set $C = A - B$, so that C is convex. Since $A \cap B = \emptyset$ and

$$C = \bigcup \{(A - y) \mid y \in B\},$$

we see that $0 \notin C$ and C is open. Lemma 13.2 yields $f \in X^*$ such that $f(z) < 0$ for every $z \in C$, i.e. $f(x) < f(y)$ for every $x \in A$, $y \in B$. Pick a number

$$\sup \{f(x) \mid x \in A\} \le \alpha \le \sup \{f(y) \mid y \in B\},$$

and conclude that $[f = \alpha]$ separates A and B.

(b) Let $\varepsilon > 0$ be given. We define

$$A_\varepsilon = A + B(0, \varepsilon)$$
$$B_\varepsilon = B + B(0, \varepsilon),$$

so that A_ε and B_ε are convex, open and non-empty sets. We claim that if $\varepsilon > 0$ is sufficiently small, then $A_\varepsilon \cap B_\varepsilon = \emptyset$. This is the only place where the topological assumptions on A and B are used. Indeed, if the claim were false, there would exist sequences $\varepsilon_n \to 0$, $x_n \in A$ and $y_n \in B$ such that $\|x_n - y_n\| \le 2\varepsilon_0$. By compactness, there would exist a subsequence $y_{n_k} \to x \in B$, and therefore $x_{n_k} \to x$ as well. But A is closed, hence $x \in A \cap B = \emptyset$, a contradiction. We can now use part (a) to produce $f \in X^*$ such that $[f = \alpha]$ separates A_ε and B_ε. As a consequence,

$$f(x + \varepsilon z) \le \alpha \le f(y + \varepsilon z)$$

for every $x \in A$, $y \in B$, $z \in B(0, 1)$. It follows that

$$f(x) + \varepsilon \|f\| \le \alpha \le f(y) - \varepsilon \|f\|$$

for every $x \in A$, $y \in B$. Since $\|f\| \ne 0$, we see that A and B are strictly separated by the hyperplane $[f = \alpha]$, and the proof is complete.

$\square$

A very useful consequence of the Hahn-Banach separation theorem is a sufficient condition for the (topological) *density* of a vector subspace.

Corollary 14.4 *Let Y be a vector subspace such that $\overline{Y} \neq X$. Then there exists $f \in X^*$, $f \neq 0$, such that*

$$f(y) = 0 \quad \text{for every } y \in Y.$$

Proof Indeed, pick $x_0 \in X \setminus \overline{Y}$. We apply the Hahn-Banach Theorem in geometric form with $A = \overline{Y}$ and $B = \{x_0\}$. There exists $f \in X^*$, $f \neq 0$, such that the hyperplane $[f = \alpha]$ strictly separates A and B. Hence

$$f(y) < \alpha < f(x_0) \quad \text{for all } y \in Y,$$

and this implies $f(y) = 0$ for every $y \in Y$, since $\lambda f(y) < \alpha$ for every $\lambda \in \mathbb{R}$. □

14.4   Baire's Theorem and Uniform Boundedness

We begin with a general result in metric topology.

Theorem 14.5 (Baire's Theorem) *Let X be a complete metric space, and let $\{U_n \mid n \in \mathbb{N}\}$ be a sequence of open subsets of X. If each U_n is dense in X, then $G = \bigcap_{n \in \mathbb{N}} U_n$ is dense in X.*

Proof Let ω be a non-empty open set in X. We will show that $\omega \cap G \neq \emptyset$. Pick any point $x_0 \in \omega$ and a number $r_0 > 0$ such that

$$\overline{B(x_0, r_0)} \subset \omega.$$

Since U_1 is dense, it must intersect the open set $B(x_0, r_0)$, so we may choose a point $x_1 \in B(x_0, r_0) \cap U_1$ and a number $r_1 > 0$ such that $0 < r_1 < r_0/2$ and

$$\overline{B(x_1, r_1)} \subset B(x_0, r_0) \cap U_1.$$

Iterating, we construct sequences $\{r_n\}_n$ and $\{x_n\}_n$ such that

$$0 < r_{n+1} < \frac{r_n}{2}, \quad \overline{B(x_{n+1}, r_{n+1})} \subset B(x_n, r_n) \cap U_{n+1}.$$

It follows immediately that $\{x_n\}_n$ is a Cauchy sequence, and therefore $x_n \to \ell$ as $n \to +\infty$. Since $x_{n+p} \in B(x_n, r_n)$ for every $n \in \mathbb{R}$ and every $p \geq 0$, we obtain in the limit as $p \to +\infty$

$$\ell \in \overline{B(x_n, r_n)} \quad \text{for all } n \in \mathbb{N}.$$

In particular $\ell \in \omega \cap G$, and the proof is complete. □

Corollary 14.5 *Let X be a non-empty complete metric space. If $\{X_n \mid n \in \mathbb{N}\}$ is a sequence of closed subsets such that*

$$X = \bigcup_{n=1}^{\infty} X_n,$$

then there exists $n_0 \in \mathbb{N}$ such that X_{n_0} has non-empty interior.

Proof Consider $U_n = X \setminus X_n$. By assumption $\bigcap_{n \in \mathbb{N}} U_n$ is not dense, so there exists n_0 such that U_{n_0} is not dense, and this means that the interior of X_{n_0} is non-empty. ∎

We introduce the principle of uniform boundedness in a non-linear setting. Although this is not the common statement in Functional Analysis, we believe that it fits into our topological approach to mathematical analysis.

Definition 14.13 Let X be a topological space. A function $f \colon X \to \mathbb{R}$ is lower semicontinuous if the set $\{x \in X \mid f(x) \leq t\}$ is closed in X for evert $t \in \mathbb{R}$.

Theorem 14.6 (Osgood) *Let X be a topological space, and suppose that $\{f_\alpha \mid \alpha \in A\}$ is any family of real-valued lower semicontinuous functions defined on X. If for every $x \in X$ there exists $M_x > 0$ such that*

$$\sup\{f_\alpha(x) \mid \alpha \in A\} \leq M_x,$$

then there exist a non-empty open set $U \subset X$ and a number $M > 0$ such that

$$\sup\{f_\alpha(x) \mid \alpha \in A, \ x \in U\} \leq M.$$

Proof For every $n \in \mathbb{N}$, let

$$X_n = \{x \in X \mid f_\alpha(x) \leq n \text{ for every } \alpha \in A\}.$$

Each X_n is closed because f_α is lower semicontinuous. Moreover, $X = \bigcup_{n \in \mathbb{N}} X_n$. Thus there exists $n_0 \in \mathbb{N}$ such that the interior of X_{n_0} is non-empty. Denoting by U the interior of X_{n_0}, there results

$$\sup\{f_\alpha(x) \mid \alpha \in A, \ x \in U\} \leq n_0 = M.$$

∎

Remark 14.6 The meaning of Osgood's Theorem is that a *uniform* bound on a suitable open set U can be deduced from a *pointwise* bound (recall that M_x depends on x).

We now apply Osgood's Theorem in a *linear* setting. The role of the open set U becomes irrelevant as a consequence of linearity.

Theorem 14.7 (Uniform Boundedness Theorem) *Suppose that X and Y are Banach space, and suppose that $\{T_\alpha \mid \alpha \in A\}$ is any family of bounded linear operators. If for every $x \in X$ there exists $M_x > 0$ such that*

$$\sup\{\|T_\alpha x\| \mid \alpha \in A\} \leq M_x,$$

then there exists $M > 0$ such that

$$\sup\left\{\|T_\alpha\|_{L(X,Y)} \mid \alpha \in A\right\} \leq M.$$

Proof For every $\alpha \in A$, the function $f_\alpha(x) = \|Tx\|$ is lower semicontinuous (and indeed continuous) on X. For every $x \in X$, $\sup\{f_\alpha(x) \mid \alpha \in A\} \leq M_x$. By Osgood's Theorem there exist a number $M' > 0$ and a non-empty open set $U \subset X$ such that

$$\sup\{\|T_\alpha x\| \mid \alpha \in A, \ x \in U\} \leq M'.$$

In particular, there exist some $x_0 \in U$ and some $\delta > 0$ such that $B(x_0, \delta) \subset U$ and

$$\sup\{\|Tx\| \mid \alpha \in A, \ x \in B(x_0, \delta)\} \leq M'.$$

If $y \in X$, $\|y\| \leq 1$, then $y + x_0 \in B(x_0, \delta)$, and so for every $\alpha \in A$,

$$\|T_\alpha y\| \leq \|T_\alpha(y + x_0)\| + \|T_\alpha x_0\| \leq 2M'.$$

Finally, if $z \in X$, $z \neq 0$, we set

$$y = \frac{\delta}{2\|z\|}z.$$

Clearly $\|y\| \leq \delta$, and the previous equation shows that

$$\|Tz\| = \frac{2\|z\|}{\delta}\|T_\alpha y\| \leq \frac{4M'}{\delta}\|z\|$$

for every $\alpha \in A$. This implies $\|T_\alpha\|_{L(X,Y)} \leq M = 4M'/\delta$ for every $\alpha \in A$, and the proof is complete. □

Corollary 14.6 *Let X, Y be two Banach spaces, and let $\{T_n\}_n$ be a sequence of bounded linear operators from X to Y such that for every $x \in X$, $T_n x$ converges as $n \to +\infty$ to a limit Tx. Then there results*

(a) $\sup\{\|T_n\| \mid n \in \mathbb{N}\} \in \mathbb{R}$,
(b) $T \in L(X, Y)$,
(c) $\|T\| \leq \liminf_{n \to +\infty} \|T_n\|_{L(X,Y)}$.

Proof Statement (a) follows directly from the Uniform Boundedness Theorem, so there exists $M > 0$ such that

$$(n \in \mathbb{N}) \wedge (x \in X) \implies \|T_n x\| \leq \|x\|.$$

Taking the limit as $n \to +\infty$ we find $\|Tx\| \leq M\|x\|$ for every $x \in X$. Since the linearity of T follows from the basic algebra of limits, the proof of (b) is complete. Finally we have

$$\|T_n x\| \leq \|T_n\|_{L(X,Y)} \cdot \|x\|$$

for every $x \in X$, and (c) follows at once. The proof is complete.   □

14.5   The Open Mapping Theorem

Theorem 14.8 *Let X and Y be Banach spaces. If $T : X \to Y$ is a bounded linear operator and if T is surjective, then there exists $c > 0$ such that*

$$B(0, c) \subset T(B(0, 1)).$$

Proof We claim that there exists $c > 0$ such that

$$B(0, 2c) \subset \overline{T(B(0, 1))}.$$

Indeed, for every positive integer n we set $X_n = n\overline{T(B(0, 1))}$. The surjectivity of T implies that $Y = \bigcup_{n=1}^{\infty} X_n$, and Baire's Lemma provides us with an integer n_0 such that X_{n_0} has non-empty interior. Trivially, $\overline{T(B(0, 1))}$ has non-empty interior, and therefore there exist $c > 0$ and $y_0 \in Y$ such that

$$B(y_0, 4c) \subset \overline{T(B(0, 1))}.$$

Since $y_0 \in \overline{T(B(0, 1))}$ we have by symmetry that $-y_0 \in \overline{T(B(0, 1))}$. It follows that

$$B(0, 4c) \subset \overline{T(B(0, 1))} + \overline{T(B(0, 1))} = 2\overline{T(B(0, 1))},$$

since $\overline{T(B(0, 1))}$ is a convex set. The claim is proved.

We now complete the proof via an iterative scheme. Let $y \in Y$ such that $\|y\| < c$. We are looking for $x \in X$ such that $Tx = y$. The previous claim ensures that for every $\varepsilon > 0$ there exists $z \in X$ such that $\|z\| < 1/2$ and $\|y - Tz\| < \varepsilon$. We choose $\varepsilon = c/2$ and we get $z_1 \in X$ such that $\|z_1\| < 1/2$ and $\|y - Tz_1\| < c/2$.

Applying the same reasoning to $y - Tz_1$ instead of y and with $\varepsilon = c/2$, we get $z_2 \in X$ such that $\|z_2\| < 1/4$ and $\|y - Tz_1 - Tz_2\| < c/4$. In this way we construct a sequence $\{z_n\}_n$ of points of X such that $\|z_n\| < 2^{-n}$ and

$$\|y - T(z_1 + \cdots + z_n)\| < \frac{c}{2^n}$$

for every n. Hence the sequence $x_n = z_1 + \cdots + z_n$ is a Cauchy sequence which must converge to some $x \in X$. In particular $\|x\| < 1$ and $y = Tx$ since T is continuous. The proof is complete.   □

Exercise 14.7 Justify the name of the Open Mapping Theorem: let U an open subset of X, and pick $y_0 \in T(U)$. Prove that there exists a ball $B(y_0, \delta) \subset T(U)$. *Hint:* let $x_0 \in X$ such that $Tx_0 = y_0$, and fix $r > 0$ such that $B(x_0, r) \subset U$. Then $y_0 + B(0, r) \subset T(U)$, and the Open Mapping Theorem yields a number $c > 0$ such that $B(0, rc) \subset T(B(0, r))$. Hence $B(y_0, rc) \subset T(U)$.

Theorem 14.9 (Isomorphism Theorem) *Let X and Y be Banach spaces. If $T: X \to Y$ is a bijective bounded linear operator, then T^{-1} is a bounded linear operator.*

Proof The statement of the Open Mapping Theorem can be formulated as follows: for every $x \in X$ such that $\|Tx\| < c$, there results $\|x\| < 1$. Hence

$$\|x\| < \frac{1}{c}\|Tx\| \quad \text{for every } x \in X.$$

This clearly means that T^{-1} is a bounded linear operator. The proof is complete.   □

Example 14.4 Let X be a Banach space, and let $\|\cdot\|_1, \|\cdot\|_2$ be two norms on X. Suppose that X is complete under both norms, and suppose also that there exists $C \geq 0$ such that

$$\|x\|_2 \leq C\|x\|_1 \quad \text{for every } x \subset X.$$

Then there exists $c > 0$ such that

$$\|x\|_1 \leq C\|x\|_2 \quad \text{for every } x \in X.$$

In other words, the two norms are equivalent. Indeed, the identity operator between $(X, \|\cdot\|_1)$ and $(X, \|\cdot\|_2)$ is bounded and bijective. By the Isomorphism Theorem, its inverse is also bounded, and the existence of the constant c follows.

Theorem 14.10 (Closed Graph Theorem) *Let X, Y be Banach spaces, and $T: X \to Y$ be a linear operator. If*

$$G(T) = \{(x, Tx) \mid x \in X\}$$

is a closed subset of $X \times Y$, then T is continuous.

Proof We endow X with two norms

$$\|x\|_1 = \|x\| + \|Tx\|, \quad \|x\|_2 = \|x\|.$$

Since $G(T)$ is closed $(X, \|\cdot\|_1)$ is complete, and trivially $\|\cdot\|_2 \le \|\cdot\|_1$. Hence there exists $c > 0$ such that $\|x\|_1 \le c\|x\|_2$ for every $x \in X$. This implies that $\|Tx\| \le c\|x\|$ for every $x \in X$, and the proof is complete. $\square$

14.6 Weak and Weak* Topologies

A fundamental application of the constructions proposed in Sect. 13.9 yields a new topology on a Banach space X and a new topology on the dual space X^*.

Definition 14.14 Let X be a Banach space, and let $x^* \in X^*$ be an element of the dual space of X. We consider the function $\varphi_{x^*}: X \to \mathbb{R}$ such that

$$\varphi_{x^*}(x) = f(x) \quad \text{for every } x \in X.$$

The weak topology $\sigma(X, X^*)$ is the smallest topology on X such that every function $\varphi_{x^*}, x^* \in X^*$, is continuous.

To be more explicit, we are considering here the setting of Definition 13.45 with $Y_\alpha = \mathbb{R}$ for every α, and $A = X^*$. Hence the weak topology on X is the initial topology induced by the dual space X^*.

Theorem 14.11 *Let $x_0 \in X$. A neighborhood base at x_0 for the weak topology $\sigma(X, X^*)$ consists of the sets of the form*

$$V = \left\{ x \in X \mid |x_i^*(x - x_0)| < \varepsilon \text{ for every } i \in I \right\},$$

where I is a finite set, $x_i^* \in X^*$ for every $i \in I$, and $\varepsilon > 0$.

Proof Clearly enough, a set of the form

$$V = \bigcap \left\{ \varphi_{x_i^*}^{-1} \left((a_i - \varepsilon, a_i + \varepsilon) \right) \mid i \in I \right\}$$

with $a_i = x_i^*(x_0)$ is an open set for $\sigma(X, X^*)$. Conversely, let U be a neighborhood of x_0 for $\sigma(X, X^*)$. By definition of the initial topology, there exists a neighborhood W of x_0 such that $W \subset U$ and

$$W = \bigcap \left\{ \varphi_{x_i^*}^{-1} (\omega_i) \mid i \in I \right\},$$

where I is a finite set and ω_i is a neighborhood of the number $a_i = x_i^*(x_0)$ in $\mathbb{R}$. Therefore there exists $\varepsilon > 0$ such that $(a_i - \varepsilon, a_i + \varepsilon) \subset \omega_i$ for every $i \in I$, and this implies that $x_0 \in V \subset W \subset U$. The proof is complete.   □

Corollary 14.7 *A net* $\{S_n, n \in D\}$ *in* X *converges weakly to a point* x *if and only if* $x^*(S_n) \to x^*(x)$ *for every* $x^* \in X^*$.

Theorem 14.12 *The weak topology* $\sigma(X, X^*)$ *is Hausdorff.*

Proof Let $x_1 \in X$, $x_2 \in X$ be two distinct points. According to the Hahn-Banach Theorem, there exists a closed hyperplane which separates $\{x_1\}$ and $\{x_2\}$ strictly. Hence there exist $x^* \in X^*$ and $\alpha \in \mathbb{R}$ such that

$$x^*(x_1) < \alpha < x^*(x_2).$$

If we define

$$U_1 = \{x \in X \mid x^*(x) < \alpha\}$$
$$U_2 = \{x \in X \mid x^*(x) > \alpha\},$$

then U_1 and U_2 are open sets for $\sigma(X, X^*)$, and clearly $x_1 \in U_1, x_2 \in U_2, U_1 \cap U_2 = \emptyset$.   □

In a similar way we introduce a new topology on the dual space X^*. Let us recall once more that X^* is always topologized by the operator norm

$$\|x^*\| = \sup\{|x^*(x)| \mid \|x\| \le 1\} \quad \text{for every } x^* \in X^*.$$

We can then consider the *bidual* space X^{**} of X, i.e. the dual of X^*, and this space is endowed with the norm

$$\|x\| = \sup\{|x^*(x)| \mid x^* \in X^*, \|x^*\| \le 1\} \quad \text{for every } x \in X.$$

Theorem 14.13 *There exists a canonical injection* $J : X \to X^{**}$ *such that*

$$\|J(x)\| = \|x\| \quad \text{for every } x \in X.$$

Proof Indeed, for every $x \in X$ we define the linear bounded map

$$x^* \in X^* \mapsto x^*(x).$$

In this way we associate to every $x \in X$ a unique element $J(x) \in X^{**}$. Concisely,

$$J(x)(x^*) = x^*(x) \quad \text{for every } x \in X \text{ and } x^* \in X^*.$$

By direct computation,

$$\|J(x)\| = \sup\left\{|J(x)(x^*)| \mid \|x^*\| \le 1\right\} = \sup\left\{|x^*(x)| \mid \|x^*\| \le 1\right\} = \|x\|$$

as a consequence of Corollary 14.3. The proof is complete. □

Important: Warning

In general, $J : X \to X^{**}$ is not surjective. If it is, the space X is called *reflexive*. Be careful that a space is reflexive if and only if J is a bijective map; different bijective maps may exist between X and X^{**}, but they do not enter into the definition of reflexivity.

Definition 14.15 For every $x \in X$ we consider the application $\varphi_x : X^* \to \mathbb{R}$ such that $\varphi_x(x^*) = x^*(x)$ for every $x^* \in X^*$. The weak* topology of X^*, also denoted by $\sigma(X^*, X)$, is the smallest topology such that all the applications φ_x, $x \in X$, are continuous.

Once again, the weak* topology is an initial topology with $A = X$ and $Y_\alpha = \mathbb{R}$ for every $\alpha \in A$. As before, we may provide a useful characterization of open sets in $\sigma(X^*, X)$.

Theorem 14.14 *Let $x_0^* \in X^*$. A neighborhood base at x_0^* for $\sigma(X^*, X)$ consists of all sets of the form*

$$V = \left\{x^* \in X^* \mid |(x^* - x_0^*)(x_i)| < \varepsilon \text{ for every } i \in I\right\}$$

where I is a finite set, $x_i \in X$ for every $i \in I$, and $\varepsilon > 0$.

Theorem 14.15 *The topology $\sigma(X^*, X)$ is Hausdorff.*

Proof Let $x_1^* \in X^*$ and $x_2^* \in X^*$ two distinct points of the dual space X^*. By definition of equality of two functions, there exists at least one point $x \in X$ such that $x_1^*(x) \ne x_2^*(x)$. Without loss of generality, we may assume that $x_1^*(x) < x_2^*(x)$. Fix a number α such that

$$x_1^*(x) < \alpha < x_2^*(x).$$

If

$$U_1^* = \left\{x^* \in X^* \mid x^*(x) < \alpha\right\}$$
$$U_2^* = \left\{x^* \in X^* \mid x^*(x) > \alpha\right\},$$

then U_1^* and U_2^* are disjoint open sets such that $x_1^* \in U_1^*$ and $x_2^* \in U_2^*$. The proof is complete. □

As an application of Tychonoff's Theorem, we prove a fundamental result in the theory of Banach spaces.

Theorem 14.16 (Alaoglu) *If X is a Banach space, then the unit ball B^* of X^* is compact with respect to $\sigma(X^*, X)$.*

Proof In this proof we denote with B and B^* the unit balls of X and X^*, respectively. By Tychonoff's Theorem, the topological space $[-1, 1]^B$ consisting of functions from B to $[-1, 1]$ with the product topology is a compact space. We define the restriction application $R: B^* \to [-1, 1]^B$ by $R(\psi) = \psi|B$ for every $\psi \in B^*$. Suppose that

(i) $R(B^*)$ is a closed subset of $[-1, 1]^B$, and
(ii) R is a topological homeomorphism from B^* with the weak* topology onto $R(B^*)$ with the product topology.

Once (i) and (ii) have been established, we deduce that $R(B^*)$ is compact. Hence B^* must be compact as a homeomorphic copy of $R(B^*)$. It remains to prove that (i) and (ii) hold.

First of all, R is injective: if $R(\psi) = R(\eta)$ for some $\psi \in B^*$, $\eta \in B^*$ such that $\psi \neq \eta$, then there exists a point $x \in B$ such that $\psi(x) \neq \eta(x)$. Hence $R(\psi) \neq R(\eta)$. The fact that R is a homeomorphism of B^* onto $R(B^*)$ follows from a direct comparison of the basic neighborhoods in the weak* topology and in the product topology. To conclude, we need to prove that $R(B^*)$ is a closed subset of $[-1, 1]^B$ in the product topology.

Let $f: B \to [-1, 1]$ be a point of the closure of $R(B^*)$ in the product topology. By definition of the application R, we only need to check that for every $u \in B$, $v \in B$, $\lambda \in \mathbb{R}$ such that $u + v \in B$ and $\lambda u \in B$, there results

$$f(u + v) = f(u) + f(v), \qquad f(\lambda u) = \lambda f(u).$$

Now, for every $\varepsilon > 0$, the weak* neighborhood consisting of those $g \in [-1, 1]^B$ such that

$$|g(u) - f(u)| < \varepsilon$$
$$|g(v) - f(v)| < \varepsilon$$
$$|g(u + v) - f(u + v)| < \varepsilon$$

contains some element $R(\psi_\varepsilon)$, and since ψ_ε is linear, we must have

$$|f(u + v) - f(u) - f(v)| < 3\varepsilon.$$

The proof that $|f(\lambda u) - \lambda f(u)| < 2\varepsilon$ is similar, and thus $f \in R(B^*)$. The proof is complete $\square$

14.7 Isomorphisms

Theorem 14.17 *Let X be a Banach space, and let $T \in L(X, X)$ be such that $\|T\| < 1$. Then the operator $I - T$ is invertible, and its inverse operator belongs to $L(X, X)$.*

Proof Consider the series

$$\sum_{n=0}^{\infty} T^n = I + T + T^2 + \cdots + T^n + \cdots ,$$

where $T^n = T \circ T^{n-1}$ for $n = 2, 3, \ldots$ Since $\|T^n\| \leq \|T\|^n$, this series converges in X by comparison with the geometric series $\sum_{n=0}^{\infty} \|T\|^n$.[3] We call S the sum of this series, and we remark that $S \in L(X, X)$. Then $S \circ T = T \circ S$ is the sum of the series $\sum_{n=1}^{\infty} T^n$, and

$$S \circ (I - T) = (I - T) \circ S = I,$$

and this implies that S is the inverse operator of $I - T$. The proof is complete. $\quad\square$

Definition 14.16 If X and Y are Banach spaces, we denote by $\mathrm{Iso}(X, Y)$ the subset of $L(X, Y)$ consisting of all T such that T is invertible and the inverse of T belongs to $L(Y, X)$.

Remark 14.7 Members of $\mathrm{Iso}(X, Y)$ are usually called isomorphisms between X and Y. In abstract algebra the word isomorphism refers to invertible functions which preserves some prescribed *algebraic* structure in X and in Y. In our definition we add a *topological* condition, i.e. the continuity of T and of its inverse, and we should look for a less generic word. It must be said that continuity is somehow the smallest property that we want to preserve besides linearity in Functional Analysis, and for this reason we force the word isomorphism to include the continuity preservation.

Theorem 14.18 *Let X and Y be Banach spaces.*

(a) $\mathrm{Iso}(X, Y)$ *is an open subset of* $L(X, Y)$.
(b) *The function* $T \mapsto T^{-1}$ *of* $\mathrm{Iso}(X, Y)$ *to* $L(Y, X)$ *is continuous.*

Proof Since the empty set is always open, we will assume that $\mathrm{Iso}(X, Y) \neq \emptyset$. Pick $T_0 \in \mathrm{Iso}(X, Y)$. For $T : X \to Y$ to be an isomorphism it is necessary and sufficient that $T_0^{-1} \circ T : X \to Y$ be an isomorphism. We set $T_0^{-1} \circ T = I - \Lambda$. If we can

---

[3] More precisely, $\left\| \sum_{k=m}^{n} T^k \right\| \leq \sum_{k=m}^{n} \left\| T^k \right\| \leq \sum_{k=m}^{n} \|T\|^k$, and the conclusion follows from the completeness of $L(X, Y)$.

ensure that $\|\Lambda\| < 1$, then Theorem 14.17 implies that Λ is an isomorphism. Since

$$\Lambda = I - T_0^{-1} \circ T = T_0^{-1} \circ (T_0 - T),$$

we have $\|\Lambda\| \leq \|T_0^{-1}\| \|T - T_0\|$. Therefore, if

$$\|T - T_0\| < \frac{1}{\|T_0^{-1}\|},$$

then $\|\Lambda\| < 1$ and also T is an isomorphism. This proves (a).

To prove (b), we remark that

$$T^{-1} = (T_0 \circ (I - T))^{-1} = (I - \Lambda)^{-1} \circ T_0^{-1},$$

hence

$$T^{-1} - T_0^{-1} = [(I - \Lambda)^{-1} - I] \circ T_0^{-1}.$$

But $(I - \Lambda)^{-1} = \sum_{n=0}^{\infty} \Lambda^n$, hence $(I - \Lambda)^{-1} - I = \sum_{n=1}^{\infty} \Lambda^n$, and

$$\left\|(I - \Lambda)^{-1} - I\right\| \leq \sum_{n=1}^{\infty} \|\Lambda\|^n = \frac{|\Lambda|}{1 - \|\Lambda\|}.$$

This implies that

$$\left\|T^{-1} - T_0^{-1}\right\| \leq \|T_0^{-1}\| \frac{|\Lambda|}{1 - \|\Lambda\|}.$$

As $T \to T_0$, $\|\Lambda\| \to 0$, and thus $T^{-1} \to T_0^{-1}$. This proves that T^{-1} is a continuous function of T if T remains in $\mathrm{Iso}(X, Y)$. The proof is complete.   □

14.8   Continuous Multilinear Applications

Let $X_1, \ldots, X_n$ and Y be vector spaces. A function $f : X_1 \times \cdots \times X_n \to Y$ is multilinear if and only if for every $k \in [1, n] \cap \mathbb{N}$ and every system $a_i \in X_i$, $i \neq k$, of vectors, the application

$$x \mapsto f(a_1, \ldots, a_{k-1}, x, a_{k+1}, \ldots, a_n)$$

is linear from X_k to Y. Roughly speaking, multilinear means linear in each variable separately.

Exercise 14.8 If f is multilinear, prove that

$$f(\lambda_1 x_1, \ldots, \lambda_n x_n) = \lambda_1 \lambda_2 \cdots \lambda_n f(x_1, \ldots, x_n)$$

for every $\lambda_i \in \mathbb{R}, i = 1, \ldots, n$.

Suppose now that all vector spaces are Banach spaces. How can we describe the continuity of a multilinear application? The answer is similar to the case of a linear operator.

Theorem 14.19 *Let $X_1, \ldots, X_n$ and Y be Banach spaces, and let $f : X_1 \times \cdots \times X_n \to Y$ be multilinear. The following statements are equivalent:*

(a) f is continuous at any point;
(b) f is continuous at $(0, \ldots, 0)$;
(c) the set

$$\{\|f(x_1, \ldots, x_n)\| \mid \|x_i\| \le 1 \text{ for } i = 1, \ldots, n\}$$

is bounded.

Proof Clearly (a) implies (b). Suppose that (b) holds. Since f is continuous at the origin, the pre-image of the unit ball of Y is an open neighborhood of the origin in $X_1 \times \cdots \times X_n$. Hence there exists $r > 0$ such that $|x_i| \le r$ for every i implies $\|f(x_1, \ldots, x_n)\| \le 1$. By homogeneity, $\|x_i\| \le 1$ for every i implies $\|f(x_1, \ldots, x_n)\| \le r^{-n}$. Thus (c) holds.

Assume now that (c) holds, and let $M > 0$ be such that $\|f(x_1, \ldots, x_n)\| \le M$ whenever $\|x_i\| \le 1$ for every i. By homogeneity, for every $x_1, \ldots, x_n$,

$$\|f(x_1, \ldots, x_n)\| \le M \|x_1\| \cdots \|x_n\|.$$

Let $(a_1, \ldots, a_n)$ be a given point. We write

$$f(x_1, \ldots, x_n) - f(a_1, \ldots, a_n)$$

$$= f(x_1 - a_1, x_2, \ldots, x_n) + f(a_1, x_2 - a_2, x_3, \ldots, x_n)$$

$$+ \cdots + f(a_1, \ldots, a_{n-1}, x_n - a_n).$$

Hence

$$\|f(x_1, \ldots, x_n) - f(a_1, \ldots, a_n)\|$$

$$\le M \|x_1 - a_1\| \|x_2\| \cdots \|x_n\| + M \|x_2 - a_2\| \|a_2\| \|x_3\| \cdots \|x_n\|$$

$$+ \cdots + M \|x_n - a_n\| \|a_1\| \cdots \|a_{n-1}\|.$$

Suppose that $\|x_i - a_i\| \leq \varepsilon$ for every i. It follows that $\|x_i\| \leq \|a_i\| + \varepsilon$, and there exists $A > 0$ such that $\|x_i - a_i\| \leq \varepsilon$ for every i implies $\|x_i\| \leq A$ for every i. We deduce that

$$\|f(x_1, \ldots, x_n) - f(a_1, \ldots, a_n)\| \leq MA^{n-1}\left(\sum_{i=1}^{n} \|x_i - a_i\|\right) \leq nMA^{n-1}\varepsilon.$$

whenever $\|x_i - a_i\| \leq \varepsilon$ for every i. Since we may choose $A > 0$ which does not depend on ε provided that $\varepsilon > 0$ is sufficiently small, the continuity of f at $(a_1, \ldots, a_n)$ follows. $\square$

Definition 14.17 If $X_1, \ldots, X_n$ and Y are Banach spaces, we denote by

$$L(X_1, \ldots, X_n; Y)$$

the set of all continuous multilinear applications from $X_1 \times \cdots \times X_n$ to Y. For any $f \in L(X_1, \ldots, X_n; Y)$ we define the norm

$$\|f\| = \sup\left\{|f(x_1, \ldots, x_n)| \mid \|x_i\|_{X_i} \leq 1 \text{ for } i = 1, \ldots, n\right\}.$$

Exercise 14.9 Prove that $L(X_1, \ldots, X_n; Y)$ is a Banach space with respect to the norm just defined.

Example 14.5 Consider three Banach spaces X, Y and Z. We define $\varphi \colon L(Y, Z) \times L(X, Y) \to L(X, Z)$ such that $\varphi(f, g) = g \circ f$. It is (almost) obvious that φ is bilinear. For every $x \in X$ we notice that

$$\|g \circ f(x)\| = \|g(f(x))\| \leq \|g\| \cdot \|f(x)\|$$
$$\leq \|g\| \cdot \|f\| \cdot \|x\|.$$

Hence φ is continuous, and $\|\varphi\| \leq 1$.

We now introduce a canonical isomorphism which we will exploit in order to interpret in different ways the definition of derivatives of order higher than one.

Let X, Y and Z be three Banach spaces. We define

$$\varphi \colon L(X, Y; Z) \to L(X, L(Y, Z))$$

in the following way: any $f \in L(X, Y; Z)$ is a bilinear application $f(x, y)$ of the variables $x \in X$ and $y \in Y$. If we keep x fixed, the application $f_x \colon y \mapsto f(x, y)$ is linear from Y to Z. Furthermore

$$\|f_x\| = \|f(x, y)\| \leq \|f\| \cdot \|x\| \cdot \|y\|,$$

so that $\|f_x\| \leq \|f\| \cdot \|x\|$. This inequality proves the continuity of f_x, and allows us to define the application $g: x \in X \mapsto f_x \in L(Y, Z)$. The previous inequality becomes now

$$\|g(x)\| \leq \|f\| \cdot \|x\|.$$

Therefore $\|g\| \leq \|f\|$.

To summarize, to each $f \in L(X, Y; Z)$ we have associated an application $g: X \mapsto L(Y, Z)$, which we may denote by $\varphi(f)$. It is immediate to check that φ is linear and that $\|\varphi\| \leq 1$. Our next step consists in constructing an inverse of φ, namely a continuous application

$$\psi: L(X, L(Y, Z)) \to L(X, Y; Z)$$

which inverts φ. We start with $g: X \to L(Y, Z)$, and we notice that g associates to each $x \in X$ a bounded linear application $g(x)$ from Y to Z. Hence

$$f: (x, y) \in X \times Y \mapsto g(x)(y)$$

is bilinear from $X \times Y$ to Z. Moreover

$$\|g(x)\| \leq \|g\| \cdot \|x\|,$$

which implies

$$\|f(x, y)\| = \|g(x)(y)\| \leq \|g(x)\| \cdot \|y\| \leq \|g\| \cdot \|x\| \cdot \|y\|.$$

This shows that $f \in L(X, Y; Z)$ and $\|f\| \leq \|g\|$. We summarize as follows: the application ψ associates to each $g \in L(X, L(Y, Z))$ the application $f \in L(X, Y; Z)$ in such a way that $\|\psi\| \leq 1$.

Theorem 14.20 *There exists an isometry[4] between $L(X, Y; Z)$ and $L(X, L(Y, Z))$.*

Proof It is clear that $\varphi \circ \psi$ and $\psi \circ \varphi$ are the identities in the corresponding spaces. In particular the operator norm of $\psi \circ \varphi$ must be equal to one. Hence $1 = \|\psi \circ \varphi\| \leq \|\psi\| \cdot \|\varphi\|$, and the fact that $\|\varphi\| \leq 1$ and $\|\psi\| \leq 1$ implies $\|\psi\| = 1$, $\|\varphi\| = 1$. This proves that φ is a bounded linear operator which preserves norms, and the proof is complete.   □

---

[4] A bounded linear application with bounded inverse, which preserves norms.

14.9 Inner Product Spaces

Theorem 14.21 (Cauchy-Schwarz Inequality) *Let H be an inner product space. For every $u \in H$ and $v \in H$, the following inequality holds:*

$$|\langle u, v \rangle| \le \sqrt{\langle u, u \rangle}\sqrt{\langle v, v \rangle}.$$

Proof Let us define $p\colon \mathbb{R} \to \mathbb{R}$ such that $p(\lambda) = \langle \lambda u + v, \lambda u + v \rangle$. From the definition of the inner product it follows that $p(\lambda) \ge 0$ for every λ. But

$$p(\lambda) = \langle u, u \rangle \lambda^2 + 2\langle u, v \rangle \lambda + \langle v, v \rangle.$$

We thus see that $p(\lambda)$ is a non-negative polynomial of degree two in λ, hence it cannot have real zeroes. As a consequence

$$4\left(\langle u, v \rangle^2 - \langle u, u \rangle \langle v, v \rangle\right) < 0,$$

which immediately gives the conclusion.   □

Exercise 14.10 Deduce from the previous proof that $|\langle u, v \rangle| = \sqrt{\langle u, u \rangle}\sqrt{\langle v, v \rangle}$ holds if and only if the vectors u and v are linearly dependent. *Hint:* the discriminant Δ of the polynomial p must be zero in this case.

Theorem 14.22 (Triangle Inequality) *Let H be an inner product space. For every $u \in H$ and $v \in H$, the following inequality holds:*

$$\sqrt{\langle u + v, u + v \rangle} \le \sqrt{\langle u, u \rangle} + \sqrt{\langle v, v \rangle}.$$

Proof Indeed,

$$
\begin{aligned}
0 \le \langle u + v, u + v \rangle &= \langle u, u \rangle + 2\langle u, v \rangle + \langle v, v \rangle \\
&\le \langle u, u \rangle + 2\sqrt{\langle u, u \rangle}\sqrt{\langle v, v \rangle} + \langle v, v \rangle \\
&= \left(\sqrt{\langle u, u \rangle} + \sqrt{\langle v, v \rangle}\right)^2.
\end{aligned}
$$

□

Definition 14.18 (Norms in Inner Product Spaces) The norm induced by the inner product of a space H is defined by

$$\|u\| = \sqrt{\langle u, u \rangle} \tag{14.2}$$

for every $u \in H$.

The fact that this is actually a norm on H follows at once from the triangle inequality and the algebraic properties of the inner product. Unless otherwise stated, the norm of an inner product space will refer to (14.2).

Theorem 14.23 (Parallelogram Identity) *If H is an inner product space, and if $u \in H$, $v \in H$, then*

$$\left\| \frac{u+v}{2} \right\|^2 + \left\| \frac{u-v}{2} \right\|^2 = \frac{\|u\|^2 + \|v\|^2}{2}.$$

Proof The identity follows from an expansion of the left-hand side according to the bilinearity properties of the inner product. The details are left as an exercise. □

Definition 14.19 (Hilbert Spaces) An inner product H is a Hilbert space if and only if it is a complete metric space with respect to the distance $d(u, v) = \|u - v\|$ associated to the norm (14.2).

Theorem 14.24 (Projection on Closed Convex Subsets) *Let H be a Hilbert space, and let K be a closed convex subset of H. For every $f \in H$ there exists one and only one element $u \in K$ such that*

$$\|f - u\| = \min\{\|f - v\| \mid v \in K\}.$$

Furthermore, $u \in K$ is characterized by the properties

$$\begin{cases} u \in K, \\ \langle f - u, v - u \rangle \leq 0 \quad \text{for every } v \in K. \end{cases}$$

Proof Let us set $d = \min\{\|f - v\| \mid v \in K\}$, and let $\{v_n\}_n$ be a sequence in K such that $d_n = |f - v_n| \to d$ as $n \to +\infty$. The parallelogram identity shows that

$$\left\| f - \frac{v_n + v_m}{2} \right\|^2 + \left\| \frac{v_n - v_m}{2} \right\|^2 = \frac{d_n^2 + d_m^2}{2}$$

for every n, m. Since K is convex, $\frac{v_n+v_m}{2} \in K$, and therefore $\left\| f - \frac{v_n+v_m}{2} \right\|^2 \geq d^2$. Hence

$$\left\| \frac{v_n - v_m}{2} \right\|^2 \leq \frac{d_n^2 + d_m^2}{2} - d^2.$$

This shows that $\|v_n - v_m\|^2$ can be made as small as we please by choosing n and m sufficiently large. In other words, $\{v_n\}_n$ is a Cauchy sequence in H. Since H is a complete metric space, $v_n \to u$ as $n \to +\infty$, and $u \in K$ because K is closed. But then $d = \|f - u\|$, and we have proved the existence of the desired element of K.

Suppose now that $u \in K$ satisfies $\|f - u\| = \min\{\|f - v\| \mid v \in K\}$, and fix $w \in K$. For every $t \in (0, 1]$ we have $v = (1 - t)u + tw \in K$, hence

$$\|f - u\| \leq \|f - ((1 - t)u + tw)\| = \|(f - u) - t(w - u)\|.$$

It follows that

$$|f - u\|^2 \leq \|f - u\|^2 - 2t\langle f - u, w - u\rangle + t^2|w - u\|^2.$$

Simplifying we see that $2\langle f - u, w - u\rangle \leq t\|w - u\|^2$, and we conclude by letting $t \to 0$.

Conversely, suppose that $u \in K$ and $\langle f - u, v - u\rangle \leq 0$ for every $v \in K$. Then

$$\|u - f\|^2 - \|v - f\|^2 = 2\langle f - u, v - u\rangle - \|u - v\|^2 \leq 0$$

for every $v \in K$, and this shows that $\|f - u\| = \min\{\|f - v\| \mid v \in K\}$. To complete the proof, we must now show that the element $u \in K$ is unique. Assume that $u_1 \in K$ and $u_2 \in K$ satisfy $\langle f - u_1, v - u_1\rangle \leq 0$ and $\langle f - u_2, v - u_2\rangle \leq 0$ for every $v \in K$. In particular

$$\langle f - u_1, u_2 - u_1\rangle \leq 0$$
$$\langle f - u_2, u_1 - u_2\rangle \leq 0$$

and thus $\|u_1 - u_2\|^2 \leq 0$, Hence $u_1 = u_2$.   □

Definition 14.20 (Projection Operator) With the same notation as in Theorem 14.24, we call $u \in K$ the projection of f on K, and we denote it with the symbol $P_K f$.

The condition

$$\langle f - u, v - u\rangle \leq 0 \quad \text{for every } v \in K$$

is a condition about the *angle* between the vectors $f - u$ and $v - u$. If K is a vector subspace of H, it is not surprising that this condition becomes stronger.

Theorem 14.25 *Let H be a Hilbert space.*

(i) Suppose that K is a closed convex subset of H. If $f_1 \in H$ and $f_2 \in H$, then

$$\|P_K f_1 - P_k f_2\| \leq \|f_1 - f_2\|.$$

(ii) If K is a closed subspace of H and $f \in H$, then $u = P_K f$ is characterized by

$$\begin{cases} u \in H \\ \langle f - u, v \rangle = 0 \quad \text{for every } v \in K. \end{cases}$$

In this case, P_K is a linear operator.

Proof Writing $u_1 = P_K f_1$, $u_2 = P_K f_2$, we have by definition

$$\langle f_1 - u_1, v - u_1 \rangle \leq 0 \quad \text{for every } v \in K$$
$$\langle f_2 - u_2, v - u_2 \rangle \leq 0 \quad \text{for every } v \in K.$$

Choosing $v = u_2$ in the first inequality and $v = u_1$ in the second inequality, we see that

$$\|u_1 - u_2\|^2 \leq \langle f_1 - f_2, u_1 - u_2 \rangle.$$

The Cauchy-Schwarz inequality yields $\|u_1 - u_2\| \leq \|f_1 - f_2\|$. This proves (i).
    To prove (ii), we start with

$$\langle f - u, v - u \rangle \leq 0 \quad \text{for every } v \in K.$$

Since K is a vector space, $\lambda v \in K$ for every $\lambda \in \mathbb{R}$, so that

$$\langle f - u, \lambda v - u \rangle \leq 0 \quad \text{for every } v \in K \text{ and every } \lambda \in \mathbb{R}.$$

As a consequence, $\langle f - u, v \rangle = 0$ for every $v \in K$. Suppose conversely that $u \in K$ and $\langle f - u, v \rangle = 0$ for every $v \in K$. Then $\langle f - u, v - u \rangle = 0$ for every $v \in K$, and the proof is complete.   □

Theorem 14.26 (Riesz-Fréchet) *Let H be a Hilbert space. For every $\varphi \in H^*$ there exists a unique $f \in H$ such that $\langle f, v \rangle = \varphi(v)$ for every $v \in H$. In addition, $\|f\| = \|\varphi\|_{H^*}$.*

Proof Let us set $K = \ker \varphi = \varphi^{-1}(\{0\})$. If $K = H$, then $\varphi = 0$ identically, and clearly $f = 0$ suffices. We may now suppose that $H \setminus K \neq \emptyset$, so that there exists $g_0 \in H$ such that $g_0 \notin K$. We define $g_1 = P_K g_0$ and

$$g = \frac{g_0 - g_1}{\|g_0 - g_1\|}.$$

It follows immediately that $\|g\| = 1$, $g \notin K$, and $\langle g, w \rangle = 0$ for every $w \in K$.
    Now, every element $v \in H$ can be decomposed as $v = \lambda g + w$ for some $\lambda \in \mathbb{R}$ and some $w \in K$: indeed,

$$\lambda = \frac{\varphi(v)}{\varphi(g)}, \qquad w = v - \lambda g.$$

From the properties of g it follows that

$$0 = \langle g, w \rangle = \langle g, v - \lambda g \rangle,$$

or

$$\langle g, v \rangle = \lambda = \frac{\varphi(v)}{\varphi(g)}.$$

We conclude by defining $f = \varphi(g)g$.   □

The Riesz-Fischer Theorem is a representation theorem, since it provides a complete description, up to isometries, of the dual space of the Hilbert space H.

We conclude this section with an important result, due to the Italian mathematician Guido Stampacchia, which is of great importance in the analysis of Partial Differential Equations. Before stating and proving it, we introduce a celebrated result about fixed points in complete metric spaces.

Theorem 14.27 (Banach-Caccioppoli) *Let (X, d) be a complete metric space. If $T : X \to X$ is a contractive map, i.e. there exists a number $0 \le L < 1$ such that $d(T(x), T(y)) \le L d(x, y)$ for every $x \in X$, $y \in X$, then T has one and only one fixed point $z \in X$ such that $T(z) = z$.*

Proof The case $L = 0$ is trivial. Suppose now $0 < L < 1$, and pick any point $x_0 \in X$, and define recursively $x_{n+1} = T(x_n)$ for $n = 1, 2, \ldots$ An easy induction argument shows that

$$d(x_{n+1}, x_n) \le L^n d(x_1, x_0)$$

for every $n \in \mathbb{N}$, $n \ge 1$. Let $m > n$ be two integers. Then

$$
\begin{aligned}
d(x_m, x_n) &\le d(x_m, x_{m-1}) + d(x_{m-1}, x_{m-2}) + \cdots + d(x_{n+1}, x_n) \\
&\le L^{m-1} d(x_1, x_0) + L^{m-2} d(x_1, x_0) + \cdots + L^n d(x_1, x_0) \\
&= L^n d(x_1, x_0) \sum_{k=0}^{m-n-1} L^k \\
&\le L^n d(x_1, x_0) \sum_{k=0}^{\infty} L^k \\
&= d(x_1, x_0) \frac{L^n}{1 - L}.
\end{aligned}
$$

If $d(x_1, x_0) = 0$, then $x_1 = T(x_0) = x_0$ and x_0 is a fixed point of T. Otherwise, for every $\varepsilon > 0$ there exists $N \in \mathbb{N}$ such that

$$L^N < \frac{\varepsilon(1 - L)}{d(x_1, x_0)}.$$

Hence, if $m > N$ and $n > N$, we see that

$$d(x_m, x_n) \leq d(x_1, x_0)\frac{L^n}{1 - L} < \frac{\varepsilon(1 - L)}{d(x_1, x_0)}\frac{d(x_1, x_0)}{1 - L} = \varepsilon.$$

Therefore $\{x_n\}_n$ is a Cauchy sequence, and by completeness it must converge to some point $z \in X$. But

$$z = \lim_{n \to +\infty} x_n = \lim_{n \to +\infty} T(x_{n-1}) = T\left(\lim_{n \to +\infty} x_{n-1}\right) = T(z),$$

since T is continuous. This proves the existence of a fixed point for T. Uniqueness is easy, since $T(z_1) = z_1$ and $T(z_2) = z_2$ imply $d(z_1, z_2) = d(T(z_1), T(z_2)) < Ld(z_1, z_2)$, so that $d(z_1, z_2) = 0$. The proof is complete. $\qquad\square$

Theorem 14.28 (Stampacchia) *Let H be a Hilbert space, and let $a \colon H \times H \to \mathbb{R}$ a continuous bilinear form. We assume that there exists $\alpha > 0$ such that*

$$a(u, u) \geq \alpha\|u\|^2 \quad \text{for every } u \in H.$$

Let K be a non-empty closed convex subset of H. For every $\varphi \in H^*$ there exists a unique element $u \in K$ such that

$$a(u, v - u) \geq \varphi(v - u) \quad \text{for every } v \in K.$$

If $a(v, w) = a(w, v)$ for every v and w, then u is characterized by the variational property

$$\begin{cases} u \in K \\ \frac{1}{2}a(u, u) - \varphi(u) = \min\left\{\frac{1}{2}a(v, v) - \varphi(v) \,\middle|\, v \in K\right\}. \end{cases}$$

Proof We may represent $\varphi \in H^*$ by a unique element $f \in H$ in the sense that $\varphi(v) = \langle f, v \rangle$ for every $v \in H$. Similarly, for every $u \in H$, the function $v \mapsto a(u, v)$ is linear and continuous on H, hence there exists a unique element $Au \in H$ such that $a(u, v) = \langle Au, v \rangle$ for every $v \in H$. It is easy to check that $A \colon H \to H$ is linear, continuous, and

$$\langle Au, u \rangle \geq \alpha\|u\|u^2 \quad \text{for every } u \in H.$$

To complete the proof, it suffices to show that there exists a unique $u \in K$ such that

$$\langle Au, v - u \rangle \geq \langle f, v - u \rangle \quad \text{for every } v \in K.$$

Let $\varrho > 0$ be a number that will be chosen hereafter. We can rewrite our inequality as

$$\langle \varrho f - \varrho Au + u - u, v - u \rangle \leq 0 \quad \text{for every } v \in K.$$

But this just means that

$$u = P_K(\varrho f - \varrho Au + u).$$

This is a *fixed point* problem. More precisely, we are looking for a fixed point $u \in K$ of the function $S \colon v \in K \mapsto P_K(\varrho f - \varrho Av + v)$. We may now play with the free parameter $\varrho > 0$ to ensure that S is a strict contraction on the complete metric space K.

Indeed, we already know that

$$\|Sv_1 - Sv_2\| \leq \|(v_1 - v_2) - \varrho(Av_1 - Av_2)\|,$$

so that

$$\|Sv_1 - Sv_2\|^2 \leq \|v_1 - v_2\|^2 - 2\varrho\langle Av_1 - Av_2, v_1 - v_2 \rangle + \varrho^2\|Av_1 - Av_2\|^2$$

$$\leq \|v_1 - v_2\|^2 \left(1 - 2\varrho\alpha + \|A\|^2_{L(H,H)}\varrho^2\right).$$

We now choose

$$0 < \varrho < \frac{2\alpha}{\|A\|_{L(H,H)}},$$

so that $1 - 2\varrho\alpha + \|A\|^2_{L(H,H)}\varrho^2 < 1$. We may now apply the Banach-Caccioppoli Theorem 14.27 to conclude that S has a unique fixed point $u \in K$.

We suppose now that the bilinear form a is symmetric. In particular, a is an inner product on H which induces a norm equivalent to the original one. Hence H is a Hilbert space with respect to the inner product a. The Riesz-Fréchet Theorem yields $g \in H$ such that $\varphi(v) = a(g, v)$ for every $v \in H$. The relation $a(u, v - u) \geq \varphi(v - u)$ for every $v \in K$ reduces to

$$a(g - u, v - u) \leq 0 \quad \text{for every } v \in K,$$

or $u = P_K g$, where the projection is understood in the sense of the inner product a. As we know, the element u is a solution to the minimization problem

$$\min \left\{ \sqrt{a(g - v, g - v)} \ \middle| \ v \in K \right\},$$

which is equivalent to the minimization on K of $a(g - v, g - v)$, which is in turn equivalent to the minimization of

$$\frac{1}{2} a(v, v) - \varphi(v)$$

with respect to $v \in K$. The proof is complete.   □

Theorem 14.29 (Lax-Milgram) *Let H be a Hilbert space, $a: H \times H \to \mathbb{R}$ be a continuous bilinear form. Suppose that there exists $\alpha > 0$ such that*

$$a(u, u) \geq \alpha \|u\|^2$$

for every $u \in H$. For every $\varphi \in H^*$ there exists a unique element $u \in H$ such that

$$a(u, v) = \varphi(v) \quad \text{for every } v \in H.$$

Furthermore, if a is symmetric, i.e. $a(v, w) = a(w, v)$ for every v and w, then the element u is characterized by

$$\begin{cases} u \in H \\ \frac{1}{2} a(u, u) - \varphi(u) = \min \left\{ \frac{1}{2} a(v, v) - \varphi(v) \ \middle| \ v \in H \right\}. \end{cases}$$

Proof The conclusion follows from Stampacchia'a Theorem and Theorem 14.25.   □

14.10   Linearization in Normed Vector Spaces

Let X and Y be Banach spaces. In the very particular case $X = Y = \mathbb{R}^n$, the notion of derivative has been studied in the first part of the book. As we said there, it is customary to think of the derivative of a function f at a point a as a real number, defined as

$$f'(a) = \lim_{x \to a} \frac{f(x) - f(a)}{x - a},$$

provided that this limit exists in $\mathbb{R}$. In the general case we need to avoid the division by $x - a$, which is now a vector.

Definition 14.21 Let U be an open subset of X. Two functions $f_1 : U \to Y$ and $f_2 : U \to Y$ are tangent at the point $a \in U$ if and only if the function

$$m(r) = \sup \{\|f_1(x) - f_2(x)\| \mid \|x - a\| \le r\},$$

which is defined for any $r > 0$ sufficiently small, satisfies

$$\lim_{r \to 0+} \frac{m(r)}{r} = 0.$$

Exercise 14.11 We say that $f_1 \sim f_2$ if and only if f_1 and f_2 are tangent at a. Prove that this is an equivalence relation.

Exercise 14.12 Prove that if f_1 and f_2 are tangent at a, then $f_1 - f_2$ is continuous at a. If, in particular, f_1 is continuous at a, then f_2 is continuous as well, and $f_1(a) = f_2(a)$.

Example 14.6 Let g be a linear function $X \to Y$, and let $f(x) = g(x - a)$. It is clear that f is tangent to zero at a if and only if $\|g\| = 0$, i.e. if and only if g is identically zero. Indeed,

$$m(r) = \|g\|r.$$

The previous example proves an important result.

Theorem 14.30 *If $f : U \to Y$ and $a \in U$, then there exists at most one linear application $g : X \to Y$ such that*

$$x \mapsto f(x) - f(a)$$

and

$$x \mapsto g(x - a)$$

are tangent at a. If such a g exists, it is continuous at zero if and only if f is continuous at a.

Definition 14.22 Let U be open in X. A function $f : U \to Y$ is differentiable in the sense of Fréchet (or F-differentiable) at the point $a \in U$ if and only if

(i) f is continuous at a, and
(ii) there exists a linear application $g : X \to Y$ such that $x \mapsto f(x) - f(a)$ and $x \mapsto g(x - a)$ are tangent at a.

In this case, the application g is the derivative of f at a, and it is denoted by one of the symbols $f'(a)$, $Df(a)$, $df(a)$.

Finally, the function f is differentiable on U if and only if it is differentiable at any point of U.

If f is differentiable at the point a, the derivative $f'(a)$ is a continuous linear operator. The standard notation for the value of $f'(a)$ at $h \in X$ should be either $f'(a)(h)$ or $f'(a)h$. The first one is the ordinary symbol for functions, while the second one is the common symbol for linear operators. Some authors also use $f'(a)[h]$.

Exercise 14.13 Sometimes the definition of differentiability is stated as follows: f is differentiable at a if and only if there exists $g \in L(X, Y)$ such that

$$\| f(x) - f(a) - g(x - a)\| = o(\|x - a\|) \quad \text{as } x \to a.$$

Prove that these definitions are actually equivalent.

If f is differentiable on U, a new map is defined by

$$f' : U \to L(X, Y), \quad x \mapsto f'(x).$$

This is the *derivative* of f. Recalling that $L(X, Y)$ is a Banach space with respect to the operator norm, the following definition makes sense.

Definition 14.23 The function $f : U \to Y$ is of class $C^1(U)$ if and only if $f' : U \to L(X, Y)$ is a continuous function.[5]

A very natural question is whether the differentiability property depends on the norms used to topologize X and Y.

Definition 14.24 Two norms $\| \cdot \|$ and $\| \cdot \|_1$ on a vector space X are equivalent if and only if there exists a constant $M > 0$ such that

$$\frac{1}{M} \|x\| \leq \|x\|_1 \leq M \|x\|$$

for every $x \in X$.

Exercise 14.14 Prove that equivalent norms induce the same topology on X. *Hint:* any ball for the first norm contains and is contained in a ball for the second norm. Hence both norms produce the same open neighborhoods.

Theorem 14.31 *Suppose that $\| \cdot \|_1$ is an equivalent norm on X and that $\| \cdot \|_2$ is an equivalent norm on Y. A function $f : U \to Y$ is differentiable at $a \in U$ with respect to the norms $\| \cdot \|_1$ and $\| \cdot \|_2$ if and only if it is differentiable with respect to the original norms.*

---

[5] It is then understood that f is differentiable on U.

Proof Since there is a perfect symmetry between the old and the new norms, it is enough to prove that the differentiability with respect to the original norms implies the differentiability with respect to the equivalent norms. Assume therefore that there exists $g \in L(X, Y)$ such that

$$\lim_{x \to a} \frac{\| f(x) - f(a) - g(x - a) \|}{\| x - a \|} = 0.$$

By assumption

$$\frac{1}{\| x - a \|_1} \leq M \frac{1}{\| x - a \|}.$$

Since $\| \cdot \|_2$ is an equivalent norm on Y, there exists $M' > 0$ such that

$$\| f(x) - f(a) - g(x - a) \|_2 \leq M' \| f(x) - f(a) - g(x - a) \|.$$

Hence

$$\frac{\| f(x) - f(a) - g(x - a) \|_2}{\| x - a \|_1} \leq M \cdot M' \frac{\| f(x) - f(a) - g(x - a) \|}{\| x - a \|}.$$

The conclusion is now immediate.   □

Exercise 14.15 Suppose that f and g are differentiable at a. Prove that $f + g$ is differentiable at a, and fg is differentiable at a for any $k \in \mathbb{R}$. How do you express the derivatives of $f + g$ and fg in terms of $f'(a)$ and $g'(a)$?

The last exercise shows that differentiability passes through sums and products. Quotients are troublesome, since it is forbidden to divide a vector by a vector.

Theorem 14.32 (Chain Rule) *Suppose X, Y and Z are Banach spaces, that U is open in X, V is open in Y, and $a \in U$. Let $f : U \to Y$ and $g : V \to Z$ be continuous functions, and assume that $b = f(a) \in V$. Hence the composition $g \circ f$ is defined on an open neighborhood U' of a. If f is differentiable at a and if g is differentiable at b, then $h = g \circ f$ is differentiable at a, and there results*

$$h'(a) = g'(a) \circ f'(a).$$

Proof By assumption

$$f(x) = f(a) + f'(a)(x - a) + \varphi(x - a) \tag{14.3}$$

where $\| \varphi(x - a) \| = o(\| x - a \|)$ as $x \to a$. Similarly,

$$g(y) = g(b) + g'(b)(y - b) + \psi(y - b)$$

where $\|\psi(y - b)\| = o(\|y - b\|)$ as $y \to b$. Putting these identities together we have

$$h(x) - h(a) = g'(f(a))(f(x) - f(a)) + \psi(f(x) - f(a)).$$

Replacing $f(x) - f(a)$ by its value from (14.3) we get

$$h(x) - h(a) = (g'(f(a)) \circ f'(a))(x - a)$$
$$+ g'(f(a))(\varphi(x - a))$$
$$+ \psi(f(x) - f(a)).$$

We need to prove that the second and the third line satisfy

$$\|g'(f(a))(\varphi(x - a))\| = o(\|x - a\|)$$
$$\|\psi(f(x) - f(a))\| = o(\|x - a\|)$$

as $x \to a$. The first estimate follows at once from

$$\|g'(f(a))(\varphi(x - a))\| \le \|g'(f(a))\| \cdot \|\varphi(x - a)\|.$$

The second estimates follows from the fact that $\|\psi(f(x) - f(a))\| = o(\|f(x) - f(a)\|)$ and $\|f(x) - f(a)\| \le 2\|f'(a)\| \cdot \|-a\|$ holds as long as x is sufficiently close to a. The proof is complete. □

Example 14.7 If U is open in X and if $f : U \to Y$ is the restriction of a continuous linear application, then f is differentiable and $f'(x) = f$ for every $x \in U$. Indeed, $f(x) - f(a) = f(x - a) = f(f - a) + 0$ by linearity, and the conclusion follows.

Theorem 14.33 *Let $\varphi : \mathrm{Iso}(X, Y) \to \mathrm{Iso}(Y, X)$ be such that $\varphi(u) = u^{-1}$ for every u. Then $\varphi \in C^1(\mathrm{Iso}(X, Y))$, and*

$$\varphi'(u) : h \in L(X, Y) \mapsto -u^{-1} \circ h \circ u^{-1}.$$

Proof We already know that $\mathrm{Iso}(X, Y)$ is open in $L(X, Y)$. We can also consider φ as a map into $L(Y, X)$. Let us fix $u \in \mathrm{Iso}(X, Y)$ and $h \in L(X, Y)$. We have

$$\varphi(u + h) - \varphi(u) = (u + h)^{-1} - u^{-1}$$
$$= (u + h)^{-1} \circ (u - (u + h)) \circ u^{-1}$$
$$= -(u + h)^{-1} \circ h \circ u^{-1}.$$

It suffices to prove that the difference between $(u+h)^{-1} \circ h \circ u^{-1}$ and $u^{-1} \circ h \circ u^{-1}$ is $o(\|h\|)$. Now,

$$(u+h)^{-1} \circ h \circ u^{-1} - u^{-1} \circ h \circ u^{-1} = ((u+h)^{-1} - u^{-1}) \circ h \circ u^{-1},$$

whence

$$\|(u+h)^{-1} \circ h \circ u^{-1} - u^{-1} \circ h \circ u^{-1}\| \leq \|(u+h)^{-1} - u^{-1}\| \|h\| \|u^{-1}\|$$

We claim that $\|(u+h)^{-1} - u^{-1}\| \to 0$ as $\|h\| \to 0$. Indeed, this is a straightforward consequence of Theorem 14.18.

To prove that φ is of class C^1, we must prove that the function

$$\varphi' \colon \mathrm{Iso}(X, Y) \to L(L(X, Y), L(Y, Y))$$

is continuous. We will use a convenient notation: for every $v \in L(Y, X)$ and $w \in L(Y, X)$, we let

$$\psi(v, w) \colon h \in L(X, Y) \mapsto -v \circ h \circ w \in L(Y, X).$$

We have proved above that $\varphi'(u) = \psi(u^{-1}, u^{-1})$. The function $(v, w) \mapsto \psi(v, w)$ is bilinear from $L(Y, X) \times L(X, Y)$ to $L(L(X, Y), L(Y, X))$. Furthermore it is continuous, since

$$\|\psi(v, w)h\| = \|v \circ h \circ w\| \leq \|v\| \|h\| \|w\|.$$

Hence

$$\|\psi(v, w)\| = \sup_{h \in L(X,Y) \setminus \{0\}} \frac{\|\psi(v, w)h\|}{\|h\|} \leq \|v\| \|w\|.$$

These considerations show that $u \mapsto \varphi'(u) = \psi(u^{-1}, u^{-1})$ is the composition of the continuous function $u \mapsto (u^{-1}, u^{-1})$ of $\mathrm{Iso}(X, Y)$ to $L(Y, X) \times L(Y, X)$ and of the continuous function $(v, w) \mapsto \psi(v, w)$. Hence it is continuous, and the proof is complete.   □

Remark 14.8 If $X = Y = \mathbb{R}$, every linear application from $\mathbb{R}$ to $\mathbb{R}$ is identified with a real number. Hence $u \in \mathrm{Iso}(\mathbb{R}, \mathbb{R})$ if and only if $u \neq 0$, and $u^{-1} = 1/u$ via this identification. The previous theorem shows that $u \mapsto 1/u$ is differentiable at any $u \in \mathbb{R} \setminus \{0\}$, and its derivative is $-1/u^2$. A very complicated way to derive an elementary fact.

Example 14.8 (Derivative of Bilinear Applications) Let X, Y and Z be Banach spaces, and let $f \colon X \times Y \to Z$ be bilinear and continuous. We first introduce a norm on $X \times Y$ as follows: for $x \in X$, $y \in Y$, let

$$\|(x, y)\| = \|x\| + \|y\|.$$

It is easy to check that this is indeed a norm on $X \times Y$, and that $X \times Y$ becomes a Banach space with respect to this norm.

We now claim that f is differentiable at any point $(a, b) \in X \times Y$, and that

$$f'(a, b): (h, k) \in X \times Y \mapsto f(h, b) + f(a, k).$$

Indeed, we write

$$f(a + h, b + k) - f(a, b) = f(h, b) + f(a, k) + f(h, k).$$

It suffices to prove that $\|f(h, k)\| = o(\|(h, k)\|)$ as $\|(h, k)\| \to 0$. By definition, $\|(h, k)\| = \|h\| + \|k\|$, so that

$$\|f(h, k)\| \le \|f\| \|h\| \|k\| \le \|f\| (\|h\| + \|k\|)^2.$$

Since it is evident that $(\|h\| + \|k\|)^2 = o(\|h\| + \|k\|)$ as $\|(h, k)\| \to 0$, the proof is complete.

Exercise 14.16   Generalize the previous example to a bounded multilinear application $f: X_1 \times \cdots \times X_n \to Z$, where the norm in $X_1 \times \cdots \times X_n$ is defined by

$$\|(x_1, \ldots, x_n)\| = \|x_1\| + \cdots + \|x_n\|.$$

Our definition of Fréchet differentiability is the straightforward generalization of the basic idea of linear approximation. There are circumstances in which one might be satisfied with a weaker kind of approximation: this is typical for functions of several variables, and leads to the definition of the *directional derivative*.

Definition 14.25   Let U be an open subset of a Banach space, and let Y be a Banach space. A function $F: U \to Y$ is differentiable in the sense of Gâteaux (or G-differentiable) at a point $a \in U$ if and only if there exists $g \in L(X, Y)$ such that

$$\lim_{\varepsilon \to 0} \frac{F(a + \varepsilon h) - F(a)}{\varepsilon} = g(h) \qquad (14.4)$$

for every $h \in X$. If it exists, the application g is unique, and it is denoted by $D_G F(a)$ of by $d_G F(a)$.[6]

---

[6] The symbol $F'_G(a)$ is also used, but it may be confused with the Fréchet derivative of a function called F_G. We prefer to avoid such a symbol.

Remark 14.9 The limit which defines the Gâteaux derivative is a limit of a function of one real variable ε (which takes values in Y).

Important: Different Definitions Are Possible

A slightly more general definition of the Gâteaux derivative is often proposed in the literature by removing the *linearity* of g in (14.4). In other words, the limit in (14.4) must exist for every h, but its value need not depend linearly on h.

Exercise 14.17 Prove that F-differentiability implies G-differentiability (with the same derivative). Consider the function $F : \mathbb{R}^2 \to \mathbb{R}^2$ such that

$$F(x, y) = \begin{cases} \left(\frac{x^2 y}{x^4 + y^2}\right)^2 & \text{if } y \neq 0 \\ 0 & \text{otherwise,} \end{cases}$$

and conclude that a G-differentiable function need not be continuous.

Theorem 9.8 is a basic result of Calculus courses. Although such a strong result does not extend to our infinite-dimensional setting, the following is a fundamental replacement. Given two points u and v of a vector space, we denote by

$$[u, v] = \{tu + (1 - t)v \mid 0 \leq t \leq 1\}$$

the interval defined by u and v.

Theorem 14.34 (Mean-Value Inequality) *Let $F : U \to X$ be G-differentiable at any point of the open subset U of X. For every u and v in U, there results*

$$\|F(u) - F(v)\| \leq \sup\{\|D_G F(w)\| \mid w \in [u, v]\} \cdot \|u - v\|.$$

Proof We assume that $F(u) \neq F(v)$, otherwise the proof is trivial. By Corollary 14.3 there exists $\psi \in Y^*$ such that $\|\psi\| = 1$ and

$$\psi(F(u) - F(v)) = \|F(u) - F(v)\|.$$

We define $\gamma(t) = tu + (1 - t)v$ for any $t \in [0, 1]$ and

$$h(t) = \psi(F(\gamma(t))).$$

Since $\gamma(t + \tau) = \gamma(t) + \tau(u - v)$ it follows that

$$\frac{h(t + \tau) - h(t)}{\tau} = \psi\left(\frac{F(\gamma(t + \tau)) - F(\gamma(t))}{\tau}\right)$$

$$= \psi\left(\frac{F(\gamma(t) + \tau(u - v)) - F(\gamma(t))}{\tau}\right)$$

By assumption we may let $\tau \to 0$ and derive

$$h'(t) = \psi\left(D_G F(tu + (1 - t)v)(u - v)\right).$$

Since $h: [0, 1] \to \mathbb{R}$, Theorem 9.8 applies and there exists $\vartheta \in (0, 1)$ such that

$$h(1) - h(0) = h'(\vartheta).$$

To conclude we notice that

$$\|F(u) - F(v)\| = h(1) - h(0) = h'(\vartheta)$$

$$= \psi\left(D_G F(tu + (1 - t)v(u - v))\right)$$

$$\leq \|\psi\| \|D_G F(\vartheta u + (1 - \vartheta)v)\| \|u - v\|.$$

Since $\|\psi\| = 1$ and $\vartheta u + (1 - \vartheta)v = w \in [u, v]$, the proof is complete.   □

Remark 14.10 The previous Mean Value Inequality is actually a *one-dimensional* result, as the proof clearly shows. We have reduced the infinite-dimensional function F to the function h of one real variable. However, the most interesting technique of the proof consists in composing F with $\psi \in Y^*$. This is a standard trick to *project* the range of the function to $\mathbb{R}$, so that the basic Lagrange Theorem may be applied. The simple example

$$\vartheta \in [0, 1] \mapsto e^{2\pi i \vartheta} \in \mathbb{C}$$

shows that the Lagrange Theorem does not hold for vector-valued functions.

A basic use of Theorem 14.34 is explained in the next regularity result.

Theorem 14.35 *Suppose that $F: U \to Y$ is G-differentiable in U, and let*

$$D_G F: U \to L(X, Y), \quad u \mapsto D_G F(u)$$

be continuous (with the standard topologies of each space) at the point u^*. Then F is F-differentiable at u^*, and $F'(u^*) = D_G F(u^*)$.

Proof We only need to show that F is F-differentiable at u^*, since we already know that the F-derivative must then coincide with the G-derivative. For every $h \in X$, we define

$$R(h) = F(u^* + h) - F(u^*) - D_G F(u^*)h.$$

We need to prove that $R(h) = o(\|h\|)$ as $\|h\| \to 0$. For $\varepsilon > 0$ sufficiently small, the function R is G-differentiable in $B(0, \varepsilon)$, and the Chain Rule yields

$$D_G R(h): k \in X \mapsto D_G F(u^* + h)k - D_G F(u^*)k.$$

We apply Theorem 14.34 with $[u, v] = [0, h]$ to get

$$\|R(h)\| = \|R(h) - R(0)\| \leq \sup_{0 \leq t \leq 1} \|D_G R(th)\| \|h\|.$$

But

$$\|D_G R(th)\| = \|D_G F(u^* + th) - D_G F(u^*)\|,$$

hence

$$\|R(h)\| \leq \sup_{0 \leq t \leq 1} \|D_G F(u^* + th) - D_G F(u^*)\| \|h\|.$$

The continuity of $D_G F$ at u^* comes now into play for the first time, and yields

$$\lim_{\|h\| \to 0} \sup_{0 \leq t \leq 1} \|D_G F(u^* + th) - D_G F(u^*)\| = 0.$$

In other words, we have proved that $\|R(h)\| \leq o(1)\|h\|$ as $\|h\| \to 0$, and the proof is complete.   □

Remark 14.11 The previous result offers a convenient tool for checking the Fréchet-differentiability of a function. Since the Gâteaux derivative is just a limit in one real variable, it is usually easier to compute. Then one hopes that the G-derivative depends continuously on the point at which it is evaluated. Of course this is only a *sufficient* condition for Fréchet-differentiability, and in many situations the only possible approach to F-differentiability is via the basic definition.

14.11 Derivatives of Higher Order

For functions of a real variable there is no need to distinguish the nature of the first derivative and of the second derivative, since at each step we go back to some suitable function of a real variable. The situation changes drastically for real-valued functions of two or more variables: if the first derivative is usually defined as a suitable *vector*, the second derivative is a *matrix*. Something is lurking behind the very rich structure of $\mathbb{R}^n$, and in this section we want to investigate this situation in a general setting.

Suppose that $F \in C(U, Y)$ is differentiable (in the sense of Fréchet) at all points of the open set $U \subset X$. We consider $F' : U \to L(X, Y)$.

Definition 14.26 Let $u^* \in U$. We say that F is twice F-differentiable at u^* if and only if F' is differentiable at u^*. The second F-derivative of F at u^* is then

$$D^2 F(u^*) = DF'(u^*) = D(DF(u^*)).$$

The symbol $F''(u^*)$ is also used. If F is twice differentiable at any point of U, we say that F is twice differentiable in U. If F'' is continuous from U to $L(X, Y)$, we write $F \in C^2(X, Y)$.

This definition mimics the basic definition of the second derivative of a function of one real variable. But if we think twice (!) about it, we easily realize that we are considering the differentiability of the function F', which acts between U and $L(X, Y)$, a space which is usually much different than Y. So, the second derivative of F at u^* is a bounded linear operator between X and $L(X, Y)$, i.e.

$$F''(u^*) \in L(X, L(X, Y)).$$

Recalling Theorem 14.20, it is convenient to think of $F''(u^*)$ as a continuous bilinear application from $X \times X$ to Y. We will often write

$$F''(u^*)(h, k)$$

instead of[7]

$$F''(u^*)(h)(k).$$

The isomorphism between $L(X, L(X, Y))$ and $L(X, X; Y)$ is usually understood in notation, and $F''(u^*)$ is employed to represent both objects.

Exercise 14.18 Let $X = C([0, 1])$ be endowed with the usual norm of uniform convergence. For $n \in \mathbb{N}$, define $F : X \to X$ such that $u \mapsto u^n$ (here u^n means "u to the power n"). Prove that $F \in C^2(X)$ and that

$$F''(u) : (h, k) \in X \times X \to n(n - 1)u^{n-2} hk.$$

---

[7] This notation is formally correct but awful. It means that $F''(u^*)(h)$ is a bounded linear operator which associates to $k \in X$ the element $F''(u^*)(h)(k) \in Y$. Indeed, $F''(u^*)$ associates to each $h \in X$ an element of $F''(u^*)(h) \in L(X, Y)$.

When $n = 2$, $F''(u)$ is independent of u. Compare this with the basic fact that "the second derivative of $x \mapsto x^2$ is constant." Can you extend the results of this exercise to the case $n \in \mathbb{R}$, $n \geq 2$?

The actual computation of the second derivative can be performed by "fixing" the first increment h, and differentiating once more with respect to u. The precise statement is as follows.

Theorem 14.36 *Suppose that $F : U \to Y$ is twice differentiable at $u^* \in U$. For all $h \in X$, the function $F_h : X \to Y$ such that $F_h(u) = F'(u)h$ is differentiable at u^* and*

$$F'_h(u^*) \colon k \in X \mapsto F''(u^*)(h, k).$$

Proof We decompose F_h as $u \mapsto F'(u) \mapsto F'(u)h$ by means of $F' : U \to L(X, Y)$ and the evaluation map which associates to any $A \in L(X, Y)$ the value $Ah \in Y$. This evaluation map is linear, hence the result follows at once from the Chain Rule. □

No discussion of second derivatives is complete without the investigation of the *symmetry* of the bilinear application $F''(u)$. The next result extends a celebrated result about the symmetry of the Hessian matrix for functions of two real variables.

Theorem 14.37 *If $F : U \to Y$ is twice differentiable at $u \in U$, then $F''(u) \in L(X, X; Y)$ is symmetric:*

$$F''(u)(h, k) = F''(u)(k, h) \quad \textit{for every } h \in X, \ k \in X.$$

Proof Let $\varepsilon > 0$ be given. For any h and k in $B(0, \varepsilon)$, we define

$$\psi(h, k) = F(u + h + k) - F(u + h) - F(u + k) + F(u)$$

$$\gamma_h(\xi) = F(u + h + \xi) - F(u + \xi).$$

For every $h \in X$, the function $g_h : B(0, \varepsilon) \to Y$ such that

$$g_h(k) = \psi(h, k) - F''(u)(h, k) = \gamma_h(k) - \gamma_h(0) - F''(u)(h, k)$$

Applying Theorem 14.34 to $g_h(k) - g_h(0)$ we find

$$\|\psi(h, k) - F''(u)(h, k)\| \qquad \leq \qquad \sup_{0 \leq t \leq 1} \|\gamma'_h(tk) - F''(u)(h)\| \cdot \|k\|$$

$$= \qquad \sup_{0 \leq t \leq 1} \|F'(u + h + tk) - F'(u + tk$$

$$- F''(u)(h)\| \cdot \|k\|.$$

Now,

$$F'(u + h + tk) = F'(u) + F''(u)(h + tk) + \omega(h + tk),$$
$$F'(u + tk) = F'(u) + F''(u)(tk) + \omega(tk)$$

with $\omega(v) = o(\|v\|)$ as $\|v\| \to 0$. Therefore

$$F'(u + h + tk) - F'(u + tk) = F''(u)h + \omega(h + tk) - \omega(tk),$$

and

$$\|\psi(h, k) - F''(u)(h, k)\| \leq \sup_{0 \leq t \leq 1} \|\omega(h + tk) - \omega(tk)\| \cdot \|k\|$$
$$\leq \varepsilon(\|h\| + \|k\|)\|k\|$$

provided that $\|h\|$ and $\|k\|$ are sufficiently small.

By the same token, swapping h and k, we find

$$\|\psi(k, h) - F''(u)(k, h)\| \leq \sup_{0 \leq t \leq 1} \|\omega(k + th) - \omega(th)\| \cdot \|h\|$$
$$\leq \varepsilon(\|k\| + \|h\|)\|h\|.$$

The fact that $\psi(h, k) = \psi(k, h)$ yields

$$\|F''(u)(h, k) - F''(u)(k, h)\| \leq \varepsilon(2\|k\|^2 + 2|h\|^2 + 2|h\|\|k\|)$$
$$\leq 3\varepsilon(\|k\|^2 + \|h\|^2).$$

By homogeneity, the last inequality remains true for *every* h and k in X. Since $\varepsilon > 0$ was arbitrary, $F''(u)(h, k) = F''(u)(k, h)$, and the proof is complete. □

The definition of higher order derivatives is now clear. If $n \in \mathbb{N}$, the function $F : U \to Y$ is n-times differentiable at $u^* \in U$ if and only if $D^{n-1}F$ is differentiable at u^*. In particular, $D^n F(u^*)$ can be seen as a continuous multilinear application from $X \times \cdots \times X$ (n factors) to Y via a repeated application of Theorem 14.20.

In particular

Definition 14.27 Let X and Y be Banach spaces, and let U be an open subset of X. A function $F : U \to Y$ is of class $C^m(U, Y)$ ($m \geq 1$ being an integer) if and only if DF, D^2F, ..., $D^m F$ exist and are continuous functions on U. As in the real-variable case, we set

$$C^\infty(U, Y) = \bigcap_{m=0}^{\infty} C^m(U, Y).$$

14.12   Partial Derivatives

A very peculiar aspect of differential calculus in Banach spaces is that every function is actually considered as a function of a *single* vector variable. When we study Calculus in Several Variables, the fact that any point of $\mathbb{R}^n$ is an n-tuple of real numbers plays a fundamental rôle. This is one of the many places where the algebraic structure of $\mathbb{R}^n$ conflicts with its geometric structure, and things become unexpectedly obscure. Indeed, our definition of the Fréchet derivative applies very well to the case $X = \mathbb{R}^n$, and no need to distinguish variables appears.

There are however situations in which the domain of a function is given as a cartesian product of two Banach spaces, and then partial derivatives become a natural idea. This is the object of this section.

Suppose that X, Y and Z are Banach spaces, and let $(u^*, v^*) \in X \times Y$ be a point. Two applications can be naturally defined as follows:

$$\sigma_{v^*}(u) = (u, v^*)$$

$$\tau_{u^*}(v) = (u^*, v).$$

Hence $\sigma_{v^*} : X \to X \times Y$ and $\tau_{u^*} : Y \to X \times Y$. Both functions are linear, hence

$$D\sigma_{v^*}(u) : \ h \mapsto (h, 0)$$

$$D\tau_{u^*}(v) : \ k \mapsto (0, k).$$

Since these derivatives are independent of u^*, v^*, u and v, we will denote them by σ and τ, respectively.

We consider an open subset Q of $X \times Y$ and a function $F : Q \to Z$.

Definition 14.28 The function F is partially differentiable with respect to u at (u^*, v^*) if and only if $F \circ \sigma_{v^*}$ is differentiable at u^*. In this case the bounded linear operator $D(F \circ \sigma_{v^*})(u^*) \in L(Y, Z)$ is the partial derivative of F at (u^*, v^*) with respect to u, and is denoted by

$$\partial_1 F(u^*, v^*).$$

Similarly, F is partially differentiable with respect to v at (u^*, v^*) if and only if $F \circ \tau_{u^*}$ is differentiable at v^*. In this case the bounded linear operator $D(F \circ \tau_{u^*})(v^*) \in L(X, Z)$ is the partial derivative of F at (u^*, v^*) with respect to v, and is denoted by

$$\partial_2 F(u^*, v^*).$$

Remark 14.12 Of course u and v are *dummy* variables, and we prefer to avoid the popular notation $\partial_u F$, $\partial_v F$ to denote partial derivatives. What really matters here is just the *position* of the variable with respect to which we are differentiating, and our notation reflects this fact.

Exercise 14.19 Retain the setting of the previous definition. Prove that the definition of partial derivatives is equivalent to requiring that there exist $A_u \in L(X, Z)$ and $A_v \in L(Y, Z)$ such that

$$F(u^* + h, v^*) - F(u^*, v^*) = A_u h + o(\|h\|)$$
$$F(u^*, v^* + k) - F(u^*, v^*) = A_v k + o(\|k\|)$$

as $\|h\| \to 0$, $\|k\| \to 0$. In other words, partial derivatives are indeed what we expect.

Exercise 14.20 Suppose that F is differentiable at (u^*, v^*). Prove that F has partial derivatives at (u^*, v^*), and

$$\partial_1 F(u^*, v^*): \ h \mapsto F'(u^*, v^*)(h, 0)$$
$$\partial_2 F(u^*, v^*): \ k \mapsto F'(u^*, v^*)(0, k).$$

Hence partial derivatives are just *evaluations* of the Fré derivative, provided that F is F-differentiable.

14.13 The Taylor Formula

Theorem 14.38 *Let $f: U \to Y$ be a function of class $C^m(U, Y)$, where U is an open subset of a Banach space X and Y is a Banach space. If $p \in U$, $u \in U$ are such that $[p, u] = \{(1 - t)p + tu \mid 0 \leq t \leq 1\} \subset U$, then*

$$F(u) = F(p) + \frac{1}{1!}Df(p)(u - p) + \frac{1}{2!}D^2 F(p)(u - p, u - p) + \cdots$$

$$+ \frac{1}{(m-1)!}D^{m-1} F(p)(u - p, \ldots, u - p)+$$

$$+ \frac{1}{(m-1)!}\int_0^1 (1 - t)^{m-1} D^m F(p + t(u - p))(u - p, \ldots, u - p)\, dt.$$

If $Y = \mathbb{R}$, there exists a point $\xi \in [p, u]$ such that $\xi \neq p$, $\xi \neq u$ and

$$\frac{1}{(m-1)!} \int_0^1 (1-t)^{m-1} D^m F(p + t(u-p))(u-p, \ldots, u-p)\, dt$$

$$= \frac{1}{m!} D^m F(p)(u-p, \ldots, u-p).$$

Proof The idea is to reduce to a function of a single variable. Let $\sigma(t) = p + t(u - p)$ for every $t \in [0, 1]$. If $g = F \circ \sigma$, then $g \in C^m([0, 1], Y)$ and

$$g^{(\ell)}(t) = D^\ell F(\sigma(t))(u-p, \ldots, u-p)$$

for every positive integer $\ell \leq m$. Hence the usual Taylor formula with integral remainder yields

$$g(1) = \sum_{\ell=0}^{m-1} \frac{g^{(\ell)}(0)}{\ell!} 1^\ell + \frac{1}{(m-1)!} \int_0^1 (1-\tau)^{m-1} g^{(m)}(\tau)\, d\tau.$$

The proof is complete.   □

14.14   The Inverse and the Implicit Function Theorems

Many problems can be translated into an equation of the form

$$F(u) = v,$$

in which the unknown u must be found for a given datum v. If F happens to be invertible, then $u = F^{-1}(v)$ and the problem is solved. Unfortunately it is often difficult to ensure that the function F be invertible by direct inspection.

Furthermore, the equation $F(u) = v$ is a particular case of the more general equation $G(u, v) = 0$, in which the unknown u is (hopefully) defined *implicitly* in terms of v. Again, it might be very difficult to solve $G(u, v) = 0$ with respect to u in terms of elementary functions only.

These are the basic examples which introduce two[8] powerful tools of Nonlinear Analysis. We first introduce a classical version of these results, and then we investigate the possibility of inverting a nonlinear function *globally*.

---

[8] It is a matter of fact that these two results are indeed equivalent, so that they could be seen as different flavors of the same result.

14.14.1   Local Inversion

We begin with *local* inversion of a function around suitable points. We will always deal with continuous functions $F: X \to Y$ between Banach spaces. We recall that $\mathrm{Iso}(X, Y)$ denotes the Banach space of bounded linear operators from X to Y which possess a bounded linear inverse.

Definition 14.29 The function $F \in C(X, Y)$ is locally invertible at $u^* \in X$ if and only if there exist open neighborhoods U of u^* and V of $v^* = F(u^*)$ such that $F \in \mathrm{Iso}(U, V)$.

Theorem 14.39 *Suppose that $F \in C^1(X, Y)$ and $F'(u^*)$ is a bounded linear invertible operator. Then F is locally invertible at u^* with a local inverse of class C^1. More precisely, there exists open neighborhoods U of u^* and V of $v^* = F(u^*)$ such that*

(a) $F \in \mathrm{Iso}(U, V)$
(b) $F^{-1} \in C^1(V, X)$ and for every $v \in V$ there results $(F^{-1})'(v) = (F'(u))^{-1}$, where $u = F^{-1}(v)$.
(c) If $F \in C^k(X, Y)$ for some $k > 1$, then $F^{-1} \in C^k(V, X)$

Proof A few reductions may be convenient. Firstly, we assume that $u^* = 0$ and $v^* = 0$: the general case follows by composition with two translations. If we define the linear operator $A = (F'(0))^{-1}$ and we replace F with $A \circ F$, we see that we may always consider the particular case of a function of the form

$$F = I + \Psi,$$

where $I = I_X$ is the identity on X and $\Psi \in C^1(X, X)$ satisfies $\Psi'(0) = 0$. In the rest of the proof we will retain these assumptions.

Pick $r > 0$ so small that

$$\forall p (p \in X \wedge \|p\| < r) \implies \|\Psi'(p)\| < \frac{1}{2}.$$

Using the Mean Value Inequality, for every p and q in the ball $B(0, r)$ there results

$$\|\Psi(p) - \Psi(q)\| \leq \sup \{\|\Psi'(w)\| \mid w \in [p, q]\} \|p - q\|$$
$$\leq \frac{1}{2} \|p - q\|.$$

This shows that Ψ is a contraction and that $\|\Psi(p)\| \leq (1/2)\|p\|$ whenever $\|p\| < r$. Now, for every $v \in X$ we introduce the auxiliary function

$$\Phi_v(u) = v - \Psi(u).$$

The function Φ_v is also a contraction, and for every $u \in B(0, r)$ and $v \in B(0, r/2)$ there results

$$\|\Phi_v(u)\| \le \|v\| + \|\Psi(u)\| \le r.$$

As a consequence, whenever $\|v\| \le r/2$, Φ_v maps $B(0, r)$ into itself and is a contraction. It follows that Φ_v possesses a unique fixed point $u \in B(0, r)$ which satisfies $u = \Phi_v(u)$, i.e.

$$u = v - \Psi(u).$$

This is equivalent to $F(u) = v$, and therefore there exists an inverse $F^{-1} : B(0, r/2) \to B(0, r)$. To prove that F^{-1} is continuous, we call $u = F^{-1}(v)$ and $w = F^{-1}(z)$, i.e.

$$u + \Psi(u) = v, \qquad w + \Psi(w) = z.$$

These equalities yield

$$|u - w| \le |v - z| + |\Psi(u) - \Psi(w)| \le \|v - z\| + \frac{1}{2}\|u - w\|.$$

Hence $\|F^{-1}(v) - F^{-1}(z)\| \le 2\|v - z\|$, and F^{-1} is actually Lipschitz continuous. Letting

$$V = B\left(0, \frac{r}{2}\right), \qquad U = B(0, r) \cap F^{-1}(V),$$

we see that $F_{|U} \in \mathrm{Iso}(U, V)$. This proves (a).

To prove (b), we set again $u = F^{-1}(v)$ and from $v = u + \Psi(u)$ we derive $F^{-1}(v) = v - \Psi(F^{-1}(v))$. But $\Psi(u) = o(\|u\|)$ as $\|u\| \to 0$ and F is Lipschitz continuous, hence $\Psi(F^{-1}(v)) = o(\|v\|)$ and F^{-1} is differentiable at $v = 0$ with $(F^{-1})'(0) = I$. To treat the general case, we pick $v \in B(0, r/2)$ and $u = F^{-1}(v)$; we then translate both u and v to the origin of X, and we find that $(F^{-1})'(v) = (F'(u))^{-1}$. The application

$$(F')^{-1} : v \mapsto (F'(F^{-1}(v)))^{-1}$$

can be factored as

$$v \mapsto u = F^{-1}(v) \mapsto F'(u) \mapsto (F^{-1})^{-1},$$

which shows that $F^{-1} \in C^1$ as a composition of functions of class C^1. Notice that we are using Theorem 14.33. The proof of (b) is complete. The proof of (c) follows from (b) by an easy induction argument.   □

As a remarkable corollary of Theorem 14.39 we will prove a local Implicit function Theorem. Following [1] we employ an *asymmetric* notation: we consider two Banach spaces X and Y, an open subset Λ of some Banach space T, and a function

$$F: \Lambda \times X \to Y.$$

This suggests that Λ should be considered as a *space of parameters*, which is the typical situation in many applications to differential equations. Needless to say, this is just a suggestion, since names are just names.

Theorem 14.40 (Implicit Function Theorem) *Suppose that $F \in C^k(\Lambda \times U, Y)$ for some $k \geq 1$, where U is an open subset of X. Suppose that*

1. $F(\lambda^*, u^*) = 0$,
2. the partial derivative $\partial_2 F(\lambda^*, u^*)$ is a bounded linear invertible operator.

Then there exist open neighborhoods Θ of λ^* in T and U^* of u^* in X and a function $g \in C^k(\Theta, X)$ such that

(a) $F(\lambda, g(\lambda)) = 0$ for every $\lambda \in \Theta$;
(b) $F(\lambda, u) = 0$ and $(\lambda, u) \in \Theta \times U^*$ imply $u = g(\lambda)$;
(c) $g(\lambda) = -(\partial_2 F(\lambda, g(\lambda)))^{-1} \circ \partial_1 F(\lambda, g(\lambda))$ for every $\lambda \in \Theta$.

The strategy of the proof is to reduce the conclusion to the setting of Theorem 14.39. This is the technical content of the following Lemma.

Lemma 14.3 *Let $(\lambda^*, u^*) \in \Lambda \times U$. Suppose that*

1. F is continuous, $\partial_2 F$ exists in $\Lambda \times U$, and $\partial_2 F: \Lambda \times U \to L(X, Y)$ is continuous;
2. $\partial_2 F(\lambda^*, u^*)$ is a bounded linear invertible operator from X to Y.

Then the application $\Psi: \Lambda \times U \to T \times Y$ such that

$$\Psi(\lambda, u) = (\lambda, F(\lambda, u))$$

is locally invertible at (λ^*, u^*) with continuous inverse Φ. Furthermore, if $F \in C^1(\Lambda \times U, Y)$, then Φ is of class C^1.

Proof An argument similar to that in the proof of Theorem 14.39 shows that Ψ is locally invertible with a continuous inverse Φ. Now, assume that $F \in C^1(\Lambda \times U, Y)$ and define

$$A = \partial_1 F(\lambda^*, u^*), \qquad B = \partial_2 F(\lambda^*, u^*).$$

It is clear that $\Psi \in C^1(\Lambda \times U, T \times Y)$ and that

$$\Psi'(\lambda^*, u^*): (\xi, v) \mapsto (\xi, A\xi + Bv).$$

Hence the equation $\Psi'(\lambda^*, u^*)(\xi, v) = (\eta, \upsilon)$ yields $\xi = \eta$ and $A\eta + Bv = \upsilon$. Recalling that B is invertible, we derive $v = B^{-1}(\upsilon - A\eta)$. As a consequence $\Psi'(\lambda^*, u^*)$ is a bounded linear invertible operator from $T \times X$ to $T \times Y$ The conclusion follows from Theorem 14.39. $\qquad\qquad\qquad\qquad\qquad\qquad\qquad\square$

Proof (of Theorem 14.40) We consider again the function $\Psi(\lambda, u) = (\lambda, F(\lambda, u))$ introduced in Lemma 14.3. It follows that Ψ is locally invertible at (λ^*, u^*) and

$$\Psi(\lambda^*, u^*) = (\lambda^*, F * \lambda^*, u^*)) = (\lambda^*, 0).$$

The local inverse Φ os Ψ is of the form $\Phi(\lambda, v) = (\lambda, \varphi(v))$ for some function $\varphi: \Theta \times V \to X$ defined in an open neighborhood $\Theta \times V$ of $(\lambda^*, F(\lambda^*, u^*))$ and satisfying $F(\lambda, \varphi(\lambda, v)) = v$ for every $\lambda \in \Theta$. This follows at once from the fact that the first component of Ψ is the identity in Λ. An easy induction argument shows also that $\varphi \in C^k$ provided that $F \in C^k$.

We define $g(\lambda) = \varphi(\lambda, 0)$ for every $\lambda \in \Theta$. It follows that

$$F(\lambda, g(\lambda)) = F(\lambda, \varphi(\lambda, 0)) = 0$$

for every $\lambda \in \Theta$. Hence (a) follows, and (b) follows from the fact that Φ is injective. Differentiating the identity $F(\lambda, \varphi(\lambda, v)) = v$ yields

$$\partial_1 F + \partial_2 F \circ \partial_1 \varphi = 0$$

$$\partial_2 F \circ \partial_2 \varphi = I.$$

Hence $\partial_1 \varphi = -(\partial_2 F)^{-1} \circ \partial_1 F$, and (c) follows at once. $\qquad\qquad\qquad\square$

14.15   A Global Inverse Function Theorem

The nature of Theorem 14.39 in strongly *local*, since the assumption on the derivative at a single point does not allow us to prove that the given function is globally injective. It is then natural to investigate whether a *global* Inverse Function Theorem may exist at all. In this section we first present a very classical result in this direction, whose proof is inspiring in its own. As a further improvement, we prove a strong result which requires some definitions of Algebraic Topology.

Theorem 14.41 (Hadamard-Caccioppoli) *Let X and Y be Banach spaces, and let $F \in C^1(X, Y)$ such that $F'(x)^{-1} \in L(Y, X)$ for all $x \in X$. If there exist constants $A > 0$ and $B > 0$ such that*

$$\left\| F'(x)^{-1} \right\| \leq A \|x\| + B \quad \text{for all } x \in X,$$

then F is a diffeomorphism between X and Y.[9]

Proof We need to prove that f is surjective and injective. The regularity of F^{-1} follows as in the proof of Theorem 14.39.

Surjectivity We want to prove that for every $y \in Y$ there exists $x \in X$ such that $f(x) = y$. We consider the last equation as part of a *family* of equations as follows: given $x_0 \in X$, we define $\tilde{F} : [0, 1] \times X \to Y$ such that

$$\tilde{F}(t, x) = F(x) - [(1 - t)F(x_0) + ty].$$

We consider the set

$$S = \{t \in [0, 1] \mid F(t, \cdot) = 0 \text{ is solvable}\}.$$

It is clear that $0 \in S$, and Theorem 14.39 shows that S is an open set, since $(\partial_2 \tilde{F})^{-1}(t, x) = F'(x)^{-1} \in L(Y, X)$. We claim that S is closed, and this implies that $S = [0, 1]$ by connectedness. In any connected component (a, b) of S there exists a branch $t \mapsto x_t$ of solutions satisfying

$$\tilde{F}(t, x_t) = 0 \quad \text{for every } t \in (a, b).$$

This follows from the Implicit Function Theorem. Differentiating both sides we get

$$F'(x_t) \frac{dx_t}{dt} = y - F(x_0).$$

Therefore

$$\left\| \frac{dx_t}{dt} \right\| \leq \left\| F'(x_t)^{-1} \right\| \|y - F(x_0)\| \leq (A \|x_t\| + B) \|y - F(x_0)\|. \tag{14.5}$$

Let $c = (a + b)/2$, so that

$$\|x_t\| \leq \|x_c\| + \int_c^t \|y - F(x_0)\| (A \|x_s\| + B) \, ds$$

for every $t > c$. It follows easily[10] that there exists a constant $C > 0$ such that

$$\|x_t\| \leq C \quad \text{for every } t \in (a, b).$$

---

[9] Recall that this means that both F and F^{-1} are differentiable with continuous inverse.
[10] This is often called *Gronwall's inequality*.

Now (14.5) yields another constant $C_1 > 0$ such that

$$\left\| \frac{dx_t}{dt} \right\| \leq C \quad \text{for every } t \in (a, b),$$

and it follows that S is a closed subset of $[0, 1]$. Thus F is surjective.

Injectivity We proceed by contradiction, assuming the existence of $y \in Y$ and $x_0 \in X$, $x_1 \in X$ such that $F(x_0) = y = F(x_1)$. Let $\gamma : [0, 1] \to X$ be the segment

$$\gamma(s) = (1 - s)x_0 + sx_1, \quad 0 \leq s \leq 1.$$

In particular $f \circ \gamma$ is a closed loop passing through y. We want to construct a survey $x : [0, 1] \to X$ such that

$$x(0) = x_0$$
$$x(1) = x_1$$
$$f(x(s)) = y \quad \text{for every } s \in [0, 1].$$

This will immediately contradict the local invertibility of F. We set $I = [0, 1]$ and $T : I \times C_0(I, X) \to C_0(I, Y)$ such that

$$T : (t, u(\cdot)) \mapsto F(\gamma(\cdot) + u(\cdot)) - ty - (1 - t)F(\gamma(\cdot)),$$

where

$$C_0(I, X) = \{u \in C(I, X) \mid u(0) = u(1) = 0\}.$$

We need to solve the equation $T(t, u) = 0$. It is evident that $T(0, 0) = 0$; if, moreover, we have $u \in C_0(I, X)$ satisfying $T(1, u) = 0$, then $x = u + \gamma$ is what we need. Observe now that

1. the partial derivative

$$\partial_2 T(t, u) = F'(\gamma(\cdot) + u(\cdot)) \in L(C_0(I, X), C_0(I, Y)),$$

   which has a bounded inverse. Therefore the set

$$S = \{t \in I \mid T(t, u) = 0 \text{ is solvable}\}$$

   is open, as a consequence of the Implicit Function Theorem.
2. Let $s \mapsto u_t(s)$ be a solution at $t \in S$. Then

$$F'(\gamma(s) + u_t(s)) \frac{du_t}{dt} = y - F(\gamma(s)).$$

As before we obtain

$$\left\| \frac{du_t}{dt} \right\|_{C_0(I,X)} \le \left(A \, \|u_t\|_{C_0(I,X)} + B_1 \right) \|y - F \circ \gamma\|_{C_0(I,Y)},$$

and again there exists a constant $C > 0$ such that

$$\left\| \frac{du_t}{dt} \right\|_{C_0(I,X)} \le C \quad \text{for every } t \in S.$$

Hence $1 \in S$, and this is a contradiction. Since we have proved that F is injective, the proof is complete.

$\square$

A much stronger refinement of Theorem 14.41 is possible, but we need to introduce some terminology from Algebraic Topology.

Definition 14.30 A topological space X is path-connected if and only if for every $x \in X$, $y \in X$ there exists a continuous function $\gamma : [0, 1] \to X$ such that $\gamma(0) = x$ and $\gamma(1) = y$.

Definition 14.31 A topological space X is simply connected if and only if it is path-connected and for every continuous $p : [0, 1] \to X$, $q : [0, 1] \to X$ such that $p(0) = q(0)$, $p(1) = q(1)$, there exists a continuous $F : [0, 1] \times [0, 1] \to X$ such that $F(x, 0) = p(x)$ and $F(1, x) = q(x)$ for every $x \in X$.

Definition 14.32 Let X and Y be topological spaces. A function $f : X \to Y$ is a local homeomorphism if and only if for every $x \in X$ there exist open neighborhoods U of x, V of $f(x)$ such that $f : U \to V$ is a homeomorphism.

Theorem 14.42 (Global Inverse Function Theorem) *Let $f : X \to Y$ be a local homeomorphism between two path-connected Hausdorff spaces X and Y, and let Y be simply connected. The following statements are equivalent:*

(a) f is a homeomorphism;
(b) f is a proper function, i.e. for every compact $K \subset Y$, the pre-image $f^{-1}(K)$ is a compact subset of X.

The proof is rather long, and requires some preliminaries. In the rest of this section, X, Y and Z will denote Hausdorff spaces.

Definition 14.33 Let $f : X \to Y$ be a local homeomorphism, and let $p : Z \to Y$ be a continuous function. A continuous function $\tilde{p} : Z \to X$ is a lifting of p by f if and only if $f \circ \tilde{p} = p$.

Proposition 14.1 *Let $f : X \to Y$ be a local homeomorphism, and let $p : Z \to Y$ be a continuous function. If Z is connected and if $\tilde{p}_1, \tilde{p}_2 : Z \to X$ are both liftings of p by f, then either $\tilde{p}_1 = \tilde{p}_2$, or $\tilde{p}_1(z) \ne \tilde{p}_2(z)$ for every $z \in Z$.*

Proof Let $C = \{z \in Z \mid \tilde{p}_1(z) = \tilde{p}_2(z)\}$. We first claim that C is open in Z. If $C = \emptyset$, there is nothing to prove; otherwise we consider $z_0 \in C$ and let $x_0 = \tilde{p}_1(z_0) = \tilde{p}_2(z_0)$. Moreover, let U and V open neighborhoods of x_0 and $f(x_0)$ such that $f: U \to V$ has a continuous inverse $g: V \to U$. The set $W = \tilde{p}_1^{-1}(U) \cap \tilde{p}_2^{-1}(U)$ is an open neighborhood of z_0 and there results $\tilde{p}_1|W = \tilde{p}_2|W = g \circ p|W$. Thus $W \subset C$ and C is an open set.

By a very easy argument the complement $Z \setminus C$ is also open, hence $C = Z$ by connectedness. The proof is complete. $\qquad\qquad\square$

Definition 14.34 A local homeomorphism $f: X \to Y$ lifts the paths if and only if, for every continuous $\alpha: [0, 1] \to Y$ such that $\alpha(0) \in f(X)$ and for every $x_0 \in f^{-1}(\{0\})$, there exists a lifting $\tilde{\alpha}: [0, 1] \to X$ of α such that $\tilde{\alpha}(0) = x_0$.

Exercise 14.21 Prove that there exists at most one lifting $\tilde{\alpha}$ as described in the previous Definition. *Hint:* use Proposition 14.1.

Definition 14.35 Any continuous function $H: Z \times [0, 1] \to Y$ is a homotopy with base $H_0: Z \to Y$ such that $z \mapsto H(z, 0)$. A function $f: X \to Y$ lifts the homotopies if, for every homotopy H and any continuous function $\tilde{H}_0: Z \to X$ such that $\tilde{H}_0 = H_0$, there exists a continuous lifting $\tilde{H}$ with base $\tilde{H}_0$, i.e. $f \circ \tilde{H} = H$ and $\tilde{H}(z, 0) = \tilde{H}_0(z)$ for all $z \in Z$.

Proposition 14.2 *If a local homeomorphism $f: X \to Y$ between Hausdorff spaces lifts the paths, then it lifts the homotopies.*

Proof Let $t \mapsto \tilde{H}(z, t)$ be the unique lifting of the path $t \mapsto H(z, t)$ with origin $\tilde{H}_0(z)$, for any $z \in Z$. It is clear that $f \circ \tilde{H} = H$, and that $\tilde{H}(z, 0) = \tilde{H}_0(z)$. Since $z \in Z$ is arbitrary, we have thus defined $\tilde{H}: Z \times [0, 1] \to X$. We claim that $\tilde{H}$ is continuous.

Let $z_0 \in Z$ and let

$$D = \left\{ t \in [0, 1] \mid \tilde{H} \text{ is not continuous at } (z_0, t) \right\}.$$

If $D \neq \emptyset$, then we may introduce $\alpha = \inf D \geq 0$. By continuity of $t \mapsto \tilde{H}(z_0, t)$, for every open neighborhood U of $\tilde{H}(z_0, \alpha)$ there exists an open interval J_1 containing α such that $\tilde{H}(z_0, t) \in U$ for every $t \in J_1$. By shrinking U we may also assume that f induces a homeomorphism between U and some open neighborhood V of $H(z_0, \alpha)$. Since H is continuous, there exist an open neighborhood W_1 of z_0 in Z and an open interval J_2 containing α such that $H(W_1 \times J_2) \subset V$. Let $J = J_1 \cap J_2$, and let $b \in J$ be such that $b < \alpha$ if $\alpha > 0$; if $\alpha = 0$, we choose $b = 0$. In any case $z \mapsto \tilde{H}(z, b)$ is continuous at z_0, and since $\tilde{H}(z_0, b) \in U$, there exists an open neighborhood W_2 of z_0 in Z such that $\tilde{H}(W_2 \times \{b\}) \subset U$. Setting $W = W_1 \cap W_2$, we claim that

$$\tilde{H}|(W \times J) = (f|U)^{-1} \circ H|(W \times J).$$

These two functions agree on $W \times \{b\}$, hence for every $z \in W$ the functions

$$t \mapsto \tilde{H}(z, t)$$

$$t \mapsto (f|U)^{-1} \circ H(z, t)$$

are liftings of $t \mapsto H(z, t)$ which coincide at $b \in J$, and hence coincide on J. In conclusion $\tilde{H}$ is continuous at (z_0, t) for every $t \in J$ with $t \geq \alpha$, contradicting the minimality of α. The proof is complete.   □

The next result is the turning point of our investigation.

Proposition 14.3 *Suppose that $f: X \to Y$ is a local homeomorphism between Hausdorff spaces, and that f lifts the paths. If X and Y are path-connected and if Y is simply connected, then f is a homeomorphism.*

Proof We prove that f is surjective. Fix $y_0 \in f(X)$ and $x_0 \in f^{-1}(\{y_0\})$, and let $\alpha: [0, 1] \to Y$ be any continuous function such that $\alpha(0) = y_0$ and $\alpha(1) = y$. We know that there exists a unique lifting $\tilde{\alpha}$ of α such that $\tilde{\alpha}(0) = x_0$. Since $f \circ \tilde{\alpha} = \alpha$ yields $f(\tilde{\alpha}(1)) = y$. Since $y \in Y$ is arbitrary, f is surjective.

To prove that f is injective, let x_0 and x_1 be two distinct points of X such that $f(x_0) = f(x_1) = y_0$. Since X is path-connected, we can consider a continuous $\sigma: [0, 1] \to X$ such that $\sigma(0) = x_0$ and $\sigma(1) = x_1$. The formula $\alpha = f \circ \sigma$ defines a continuous function with $\alpha(0) = \alpha(1) = y_0$. Now we recall that Y is simply connected, so that there exists a continuous $h: [0, 1] \times [0, 1] \to Y$ such that $h(t, 0) = \alpha(t)$, $h(t, 1) = y_0$ for every $t \in [0, 1]$, and $h(0, s) = y_0 = h(1, s)$ for every $s \in [0, 1]$.

By assumption f lifts the paths, hence it lifts the homotopies by Proposition 14.2. Call $\tilde{h}: [0, 1] \times [0, 1] \to X$ be a homotopy which lifts h and which satisfies $\tilde{h}(t, 0) = \sigma(t)$ for every $t \in [0, 1]$.

It follows from the definitions that a constant path[11] is lifted to a constant path. In particular $\tilde{h}(0, s) = \sigma(0) = x_0$, $\tilde{h}(1, s) = \sigma(1) = x_1$ for every $s \in [0, 1]$. Exploiting the fact that $t \mapsto \tilde{h}(t, 1)$ is also constant, we have

$$x_0 = \tilde{h}(0, 1) = \tilde{h}(1, 1) = x_1.$$

This contradiction proves that f is injective, and therefore it is a homeomorphism.   □

Definition 14.36 Suppose that $f: X \to Y$ is a local homeomorphism. For a continuous path $\alpha: [0, 1] \to Y$ such that $\alpha(0) \in f(X)$ and $x_0 \in f^{-1}(\alpha(0))$, we define the maximal lifting $\phi: J \to X$ of α with $\phi(0) = x_0$ in the following way. There exists a continuous function $\phi_I: I \to X$, with $I = [0, b) \subset [0, 1]$, such that

---

[11] This refers to any continuous function $t \mapsto \alpha(t)$ such that $\alpha(t)$ does not depend on $t \in [0, 1]$.

$\phi_I(0) = x_0$ and $f \circ \phi_I = \alpha|I$. By Proposition 14.1 the formula $\phi|J = \phi_I$ defines the mapping $\phi: J \to X$ on the union J of all the intervals I.

Definition 14.37 Let $\phi: [0, b) \to X, 0 < b \leq +\infty$, be a continuous function. The ω-limit set of ϕ is

$$\omega_\phi = \bigcap \left\{ \overline{\phi([t, b))} \mid 0 \leq t < b \right\}.$$

Exercise 14.22 Prove that a point $x \in \omega_\phi$ if and only if x is an accumulation point of some sequence $\{\phi(t_n)\}_n$, where $t_n \to b$ and $0 \leq t_n < b$ for every n. If X is a metric space, $x \in \omega_\phi$ if and only if there exists a sequence $\{t_n\}_n$ such that $0 \leq t_n < b$ for every n, $t_n \to b$ and $\phi(t_n) \to x$.

Remark 14.13 The terminology is reminiscent of the terminology used in Dynamical Systems. We want to stress the fact that here no differential equation is involved, and the definition of the ω-limit set is purely topological.

Proposition 14.4 *Let $f: X \to Y$ be a local homeomorphism between Hausdorff spaces, and let $\phi: J \to X$ be the maximal lifting of $\alpha: [0, 1] \to Y$ with $\phi(0) = x_0 \in f^{-1}(\alpha(0))$. If $J \neq [0, 1]$, then J has the form $[0, b)$ for some $b \in (0, 1)$, and $\omega_\phi = \emptyset$.*

Proof We assume that $J = [0, a]$ for some $0 < a < 1$. At the point $f(\phi(a))$ the function f has a local inverse, and we can extend ϕ to a lifting defined on some larger interval. This contradicts the maximality of ϕ, and therefore the maximal lifting must be defined on an interval $[0, b)$ with $0 < b < 1$.

To prove the second statement, we assume that $x_0 \in \omega_\phi$. Then $f(x_0) = \alpha(b)$ since f is continuous, and thus $f(\overline{\phi([t, b))}) \subset \overline{f(\phi([t, b))}$ and

$$\bigcap \left\{ \overline{f(\phi([t, b))} \mid 0 \leq t < b \right\} = \bigcap \{\alpha([t, b]) \mid 0 \leq t < b\} = \{\alpha(b)\}.$$

Pick open neighborhoods U and V of x_0 and $f(x_0)$ respectively, such that $f|U: U \to V$ is a homeomorphism, and let g be the inverse of $f|U$. We select $a \in [0, b)$ such that $\alpha([a, b]) \subset V$ and such that $\phi(a) \in U$. We then define $\psi: [0, b] \to X$ as a lifting of $\alpha|[0, b]$ by $\psi|[0, a] = \phi|[0, a]$ and by $\psi|(a, b] = g \circ \alpha|(a, b]$. Again we reach a contradiction with the maximality of ϕ, and the proof is complete. □

Proof (of Theorem 14.42) Since (a) trivially implies (b), we only prove that (b) implies (a). Let f be a proper function. If we can prove that f lifts the paths, an application of Proposition 14.3 gives the desired result.

We argue by contradiction, assuming the existence of a continuous path

$$\alpha: [0, 1] \to Y$$

and of a point $x_0 \in f^{-1}(\alpha(0))$ such that the maximal lifting ϕ of α with $\phi(0) = x_0$ is defined on a proper subset $[0, b)$ of $[0, 1]$. Then we know that $\omega_\phi = \emptyset$. But

$\phi([0, b)) \subset f^{-1}(\alpha([0, 1]))$ and the latter set is compact since $\alpha([0, 1])$ is compact in Y and f is proper.

Every finite collection of closed sets $\{\phi([t_i, b))\}_i$ has non-empty intersection, then

$$\omega_\phi = \bigcap \left\{ \overline{\phi([t, b))} \mid 0 \le t < b \right\} \neq \emptyset,$$

by the finite-intersection property of compact spaces. This contradiction proves the statement.   □

14.16   Critical and Almost Critical Points

Every (or almost every) student should remember the definition of critical point for a function $f \colon \mathbb{R} \to \mathbb{R}$: we say that x_0 is a critical point of f if and only if $f'(x_0) = 0$ (which implies that f must be differentiable at x_0). What can we do for functions between normed vector spaces?

Definition 14.38 Let X, Y be Banach spaces, and let U be an open subset of X. A point $u \in U$ is a critical point of a function $F \colon U \to Y$ if and only if $DF(u) \in L(X, Y)$ is not surjective.

Example 14.9 If $Y = \mathbb{R}$, a critical point is just a point $u \in U$ such that $DF(u) = 0 \in X^*$, i.e.

$$DF(u)(v) = 0 \quad \text{for every } v \in X.$$

The quest of critical points is a formidable task, and it is an independent branch of mathematics called Critical Point Theory. In this chapter we want to present an important tool that can be used to *construct* critical points. We follow [6].

Theorem 14.43 (Ekeland Variational Principle) *Let (M, d) be a complete metric space and let $\Phi \colon M \to (-\infty, +\infty]$ be a lower semicontinuous function, bounded from below and not identical to $+\infty$. Let $\varepsilon > 0$ and $u \in M$ such that*

$$\Phi(u) \le \inf_M \Phi + \varepsilon.$$

Then there exists $v \in M$ such that

(a) $\Phi(v) \le \Phi(u)$;
(b) $d(u, v) \le 1$;
(c) for every $w \neq v$ in M, there results $\Phi(w) > \Phi(v) - \varepsilon d(w, v)$.

Proof The relation

$$w \le v \iff \Phi(w) + \varepsilon d(w, v) \le \Phi(v)$$

is a partial order on M. We call $u_0 = u$ and we suppose that u_n is know. Let

$$S_n = \{w \in M \mid w \leq u_n\}.$$

There exists $u_{n+1} \in S_n$ such that

$$\Phi(u_{n+1}) \leq \inf_{S_n} \Phi + \frac{1}{n+1}.$$

Since $u_{n+1} \leq u_n$ we see that $S_{n+1} \subset S_n$. Moreover S_n is a closed subset since Φ is lower semicontinuous. Now, for every $w \in S_{n+1}$ we have $w \leq u_{n+1} \leq u_n$, so that

$$\varepsilon d(w, u_{n+1}) \leq \Phi(u_{n+1}) - \Phi(w) \leq \inf_{S_n} \Phi + \frac{1}{n+1} - \inf_{S_n} \Phi = \frac{1}{n+1}.$$

As a consequence,

$$\operatorname{diam} S_{n+1} = \sup \{d(w_1, w_2) \mid w_1 \in S_{n+1}, \ w_2 \in S_{n+1}\} \leq \frac{2}{\varepsilon(n+1)}.$$

But M is complete, hence

$$\bigcap_{n=0}^{\infty} S_n = \{v\}$$

for some $v \in M$. In particular $v \in S_0$, i.e. $v \leq u_0 = u$. This implies

$$\Phi(v) \leq \Phi(u) - \varepsilon d(u, v) \leq \Phi(u)$$

and

$$d(u, v) \leq \frac{\Phi(u) - \Phi(v)}{\varepsilon} \leq \frac{\inf_M \Phi + \varepsilon \cdot \inf_M \Phi}{\varepsilon} = 1.$$

To conclude, it suffices to show that $w \leq v$ implies $w = v$.[12] By definition, $w \leq u_n$ for every n, hence $w \in S_n$ for every n. This implies $w = v$, as claimed. The proof is complete.   □

Remark 14.14 The same proof can be repeated with the equivalent metric λd, $\lambda > 0$, instead of d. Of course λ and ε may be related, as in the proof of the following corollary.

---

[12] In this sense, v is a minimal element.

Theorem 14.44 (Almost Critical Points) *Let X be a Banach space, $\varphi \colon X \to \mathbb{R}$ be a function bounded from below and differentiable on X. For every $\varepsilon > 0$ and for every $u \in X$ such that*

$$\varphi(u) \leq \inf_X \varphi + \varepsilon,$$

there exists $v \in X$ such that

$$\varphi(v) \leq \varphi(u)$$
$$\|u - v\| \leq \sqrt{\varepsilon}$$
$$\|D\varphi(v)\| \leq \sqrt{\varepsilon}.$$

Proof We apply the Ekeland Variational Principle in $M = X$ with $\Phi = \varphi$ and $\lambda = \varepsilon^{-1/2}$ as in the previous Remark. This produces an element v, and we only need to prove that $\|D\varphi(v)\| \leq \sqrt{\varepsilon}$.

Let $w = v + th$ for $t > 0$, $h \in X$ and $\|h\| = 1$. The properties of v give

$$\varphi(v + th) - \varphi(v) > -\sqrt{\varepsilon}t.$$

Dividing by t and letting $t \to 0$ we see that

$$-\sqrt{\varepsilon} \leq D\varphi(v)(h).$$

Replacing h with $-h$, we conclude that $-\sqrt{\varepsilon} \leq D\varphi(v)(h) \leq \sqrt{\varepsilon}$ for every $h \in X$, $\|h\| = 1$. The conclusion follows. $\square$

Corollary 14.8 *Let X be a Banach space, $\varphi \colon X \to \mathbb{R}$ be a function bounded from below and differentiable on X. For every minimizing sequence $\{u_n\}_n$ of φ there exists a minimizing sequence $\{v_n\}_n$ such that*

$$\varphi(v_n) \leq \varphi(u_n)$$
$$\|u_n - v_n\| \to 0$$
$$\|D\varphi(v_n)\| \to 0.$$

Proof We define

$$\varepsilon_n = \begin{cases} \varphi(u_n) - \inf_X \varphi & \text{if } \varphi(u_n) - \inf_X \varphi > 0 \\ 1/n & \text{if } \varphi(u_n) - \inf_X \varphi = 0 \end{cases}$$

and we choose u_n according to Theorem 14.44. $\square$

Exercise 14.23 Let (M, d) be a complete metric space, $\varphi: M \to \mathbb{R}$ a lower semicontinuous non-negative function, and $T: M \to M$ a function such that $d(u, T(u)) \leq \varphi(u) - \varphi(T(u))$ for every $u \in M$. Prove that T has a fixed point. *Hint:* Ekeland's Principle with $\varepsilon = 1/2$ gives $v \in M$ such that

$$\frac{1}{2}d(v, T(v)) \geq \varphi(v) - \varphi(T(v)).$$

Hence $d(v, T(v)) \leq 0$.

Exercise 14.24 Let (M, d) be a complete metric space, $T: M \to M$ such that there exists $L \in [0, 1)$ such that $d(T(u), T(v)) \leq Ld(u, v)$ for every $u \in M$ and $v \in M$. Use the previous exercise to show that T has a fixed point. *Hint:* try to choose a constant $c > 0$ such that $\varphi(u) = cd(u, T(u))$ satisfies the conditions of the previous exercise.

At this point, the natural question is: does the sequence $\{v_n\}_n$ of Corollary 14.8 converge?

The answer is negative, in general. This is the reason why the following definition was introduced.

Definition 14.39 Let X be a Banach space, $F: X \to \mathbb{R}$ be a differentiable function. The function F satisfies the Palais-Smale condition ((PS) for short) if and only if every sequence $\{u_n\}_n$ in X such that $\{F(u_n)\}_n$ is bounded and $\|DF(u_n)\| \to 0$ as $n \to +\infty$ has a convergent subsequence.

The (PS) condition—which can be considerably weakened—is the basic compactness condition of Critical Point Theory. We refer to [6] for a readable exposition of this theory.

14.17 Problems

14.1 Let $(X, \|\cdot\|)$ be a normed space, and let $f: X \to \mathbb{R}$ be the function such that $f(u) = \|u\|$ for every $u \in X$. Prove that the Gâteaux derivative of f at $u = 0$ does not exist.

14.2 Let $\sigma: \mathbb{R}^n \setminus \{0\} \to \mathbb{R}^n$ be the function such that $\sigma(x) = x/\|x\|$ for every $x \in \mathbb{R}^n \setminus \{0\}$. Prove that

$$D\sigma(x): h \in \mathbb{R}^n \mapsto -\frac{1}{\|x\|^3}\langle x \mid h\rangle x + \frac{h}{\|x\|},$$

where $\langle \cdot \mid \cdot \rangle$ denotes the usual inner product of $\mathbb{R}^n$ and $\|\cdot\|$ the associated Euclidean norm.

14.3 Let X and Y be normed spaces, U an open subset of X, and $F: U \to Y$ a Fréchet-differentiable function. Prove that F is a Lipschitz function if and only if there exists a number $M > 0$ such that $\|DF(u)\|_{L(X,Y)} \le M$ for every $u \in U$.

14.4 A subset C of a vector space X is conic if and only if $tC = \{tx \mid x \in C\} \subset C$ for every $t > 0$. A function $f: C \to \mathbb{R}$ defined on a conic subset C is homogeneous of degree $p \in \mathbb{Z}$ if and only if $f(tx) = t^p f(x)$ for every $x \in C$ and for every $t > 0$. Prove the following result, due to Eulero: suppose that C is open and conic in $\mathbb{R}^n$. A differentiable function $f: C \to \mathbb{R}$ is homogeneous of degree p if and only if

$$pf(x) = \sum_{k=1}^{n} x_k \partial_k f(x) \quad \text{for every } x \in C.$$

14.5 A function $F: X \to Y$ between two real Banach spaces is Hadamard-differentiable at a point $u \in X$ if and only if there exists $A \in L(X, Y)$ such that

$$\lim_{n \to +\infty} \frac{F(u + t_n v_n) - F(u)}{t_n} = Av$$

for all $v \in X$, for all sequences $\{t_n\}_n$ in $\mathbb{R}\backslash\{0\}$ such that $t_n \to 0$ and for all sequences $\{v_n\}_n$ in X such that $v_n \to v$. Similarly, F is weakly Hadamard-differentiable at $u \in X$ if and only if there exists $A \in L(X, Y)$ such that

$$\lim_{n \to +\infty} \Lambda \left(\frac{F(u + t_n v_n) - F(u)}{t_n} \right) = \Lambda (Av)$$

for all $v \in X$ and all $\Lambda \in Y^*$, for all sequences $\{t_n\}_n$ in $\mathbb{R} \setminus \{0\}$ such that $t_n \to 0$ and for all sequences $\{v_n\}_n$ in X such that $v_n \rightharpoonup v$. Prove the following statements:

1. F is Fréchet-differentiable at $u \in X$ if and only if there exists $A \in L(X, Y)$ such that

$$\lim_{t \to 0} \frac{F(u + tv) - F(u)}{t} = Av$$

   uniformly with respect to v in bounded subsets of X.
2. F is Hadamard-differentiable at $u \in X$ if and only if there exists $A \in L(X, Y)$ such that

$$\lim_{t \to 0} \frac{F(u + tv) - F(u)}{t} = Av$$

   uniformly with respect to v in compact subsets of X.

3. F is weakly Hadamard-differentiable at $u \in X$ if and only if there exists $A \in L(X, Y)$ such that for all $\Lambda \in Y^*$,

$$\lim_{t \to 0} \Lambda \left(\frac{F(u + tv) - F(u)}{t} \right) = \Lambda (Av)$$

uniformly with respect to v in weaklycompact subsets of X.

4. Let $X = \mathbb{R}$ and $Y = L^2(\mathbb{R})$. Pick $z \in Y \setminus \{0\}$ and define $F : \mathbb{R} \to L^2(\mathbb{R})$ by

$$F(t) = \begin{cases} tz(\cdot + t^{-1}) & \text{for } t \neq 0 \\ 0 & \text{for } t = 0. \end{cases}$$

Prove that F is weakly Hadamard-differentiable at $t = 0$ with $F'(0) = 0$. Prove that

$$\left\| \frac{F(t)}{t} \right\|_Y^2 = \|z\|_Y^2 \quad \text{for all } t \neq 0,$$

and deduce that F is not Gâteaux-differentiable at $t = 0$.

14.18   Comments

Linear Functional Analysis is precisely concerned with topological vector spaces. i.e. vector spaces endowed with a topology compatible with the two algebraic operations. This means that the sum and the product with scalar numbers must be continuous functions. Normed vector spaces are important examples of topological vector spaces, and many mathematical analysts do use them in everyday life. We refer to [5, 7] for a thorough study of topological vector spaces.

Our definition of tangent functions and of the Fréchet derivative follows [4]. It seems that the French school of Bourbaki popularized the theory of differential calculus in normed spaces. We have followed the elegant survey of [1]. The interested reader should refer to [2] for a treatise on differential calculus in infinite-dimensional spaces.

The remarkable global inversion theorem appears in [3], and generalizes similar results of [1].

References

1. A. Ambrosetti, G. Prodi, *A Primer of Nonlinear Analysis* (Cambridge University Press, 1995)
2. H. Cartan, *Cours de calcul différentiel*. Collection Méthodes (Hermann, Paris, 1997)
3. G. De Marco, G. Gorni, G. Zampieri, Global inversion of functions: an introduction. NoDEA **1**, 229–248 (1994)
4. J. Dieudonné, *Foundations of Modern Analysis*. Pure and Applied Mathematics, vol. 10 (Academic Press, New York, 1960)
5. J.L. Kelley, I. Namioka, *Linear Topological Spaces*. Graduate Texts in Mathematics 36 (Springer, New York, 1976)
6. J. Mawhin, M. Willem, *Critical Point Theory and Hamiltonian Systems*. Applied Mathematical Sciences, vol. 74 (Springer, 1989)
7. W. Rudin, *Functional Analysis*. International Series in Pure and Applied Mathematics (McGraw-Hill, New York, 1991)

Chapter 15
A Functional Approach to Lebesgue Integration Theory

Abstract The standard approach to the Lebesgue integral is via measure theory: we must define a set function—called a *measure*—on a set of suitable sets—called *measurable* sets, then we can define measurable functions, and finally integrable functions. The main advantage of this approach is that at the end we have the highest generality. On the other hand, such a construction requires a good amount of mathematical education before it can be understood.

15.1   The Riemann Integral in Higher Dimension

Let us present the basic construction of the Riemann integral in $\mathbb{R}^n$. We will see that much the same ideas that we developed in $\mathbb{R}$ can be adapted. We start with some notation.

Definition 15.1 (Cell) An n-cell is a cartesian product

$$B = [a_1, b_1] \times \cdots \times [a_n, b_n]$$

of closed and bounded intervals of $\mathbb{R}$. The volume of B is defined as in elementary geometry:

$$\text{Vol}(B) = \prod_{k=1}^{n} (b_k - a_k).$$

We often fix a *basic cell*, which we denote by **B**.

Let $f: \mathbf{B} \to \mathbb{R}$ be a bounded function. A partition P of **B** is a splitting of **B** into sub-cells (i.e. each interval $[a_k, b_k]$ is partitioned exactly as we did in our construction of the Riemann integral in one dimension) $B_1, \ldots B_p$. To each partition

S. Secchi, *A Circle-Line Study of Mathematical Analysis*,
La Matematica per il 3+2 141, https://doi.org/10.1007/978-3-031-19738-3_15

P we associated the numbers

$$M_k = \sup\{f(x) \mid x \in B_k\}$$
$$m_k = \inf\{f(x) \mid x \in B_k\},$$

for $k = 1, \ldots, p$. Again we define

$$U(P, f) = \sum_{k=1}^{p} M_k \, \text{Vol}(B_k)$$

$$L(P, f) = \sum_{k=1}^{p} m_k \, \text{Vol}(B_k),$$

and again it is easy to check that if P' is finer than P, then

$$L(P, f) \le L(P', f), \quad U(P', f) \le U(P, f).$$

Furthermore, if P_1 and P_2 are partitions of $\mathbf{B}$, then

$$L(P_1, f) \le U(P_2, f).$$

The number

$$\underline{\int}_{\mathbf{B}} f = \sup\{L(P, f) \mid P \text{ is a partition of } \mathbf{B}\}$$

is the *lower Riemann integral* of f over $\mathbf{B}$, and

$$\overline{\int}_{\mathbf{B}} f = \inf\{L(P, f) \mid P \text{ is a partition of } \mathbf{B}\}$$

is the *upper Riemann integral* of f over $\mathbf{B}$.

Definition 15.2 The function f is R-integrable on $\mathbf{B}$ if

$$\underline{\int}_{\mathbf{B}} f = \overline{\int}_{\mathbf{B}} f.$$

If this is the case, the common value is denoted by

$$\int_{\mathbf{B}} f,$$

and is called the *Riemann integral* of f on $\mathbf{B}$.

The theory of the Riemann integral can now be developed as it was developed in $\mathbb{R}$, but this is not our aim. We are going to construct a *different* integral, called the *Lebesgue integral*.

15.2 Elementary Integrals

Definition 15.3 Suppose that H is a set of bounded real-valued functions defined on a set X, and suppose that

(a) H is a vector space with the usual addition and multiplication by real numbers;
(b) if $h \in H$, then $|h| \in H$.

The elements of H will be called *elementary functions*.

Exercise 15.1 If $h \in H$, prove that $h^+ = \max\{h, 0\}$ and $h^- = \max\{0, -h\}$ belong to H. *Hint:* recall that $2x^+ = |x| + x$ and $2x^- = |h| - x$ for each $x \in \mathbb{R}$.

Definition 15.4 An *elementary integral* on H is a function $I: H \to \mathbb{R}$ such that

(1) (linearity) if $h, k \in H, \alpha, \beta \in \mathbb{R}$, then

$$I(\alpha h + \beta k) = \alpha I h + \beta I k.$$

(2) (positivity) If $h \in H$ and $h \geq 0$, then $Ih \geq 0$.
(3) (continuity) If $\{h_p\}_p$ is a non-increasing sequence of elementary functions which converges pointwise to zero on X, then $\lim_{p \to +\infty} I h_n = 0$.

Exercise 15.2 Let h, k be elementary functions. Prove that $h \leq k$ implies $Ih \leq Ik$. Hence positivity of I is actually monotonicity of I.

15.3 Null and Full Sets

Definition 15.5 (Measure Zero) A subset Z of X is a set of measure zero, or a null set, if and only if for every $\varepsilon > 0$ there exists a non-decreasing sequence $\{h_p\}_p$ of elementary functions such that

$$I h_p < \varepsilon,$$

$$\sup_p h_p(x) \geq 1 \quad \text{for each } x \in Z.$$

On the opposite side, a subset E of X is a set of full measure, if $X \setminus E$ is a set of measure zero.

Definition 15.6 Let $P(x)$ be a logical statement depending on the free variable x. If

$$\{x \in X \mid P(x)\}$$

is a set of full measure, we say that the property P holds for almost every $x \in X$, or that P holds almost everywhere in X.

Example 15.1 Let $\{h_p\}_p$ be a sequence of elementary functions. We say that $h_p \to 0$ almost everywhere on X if $h_p(x) \to 0$ (as $p \to +\infty$) for every x is a subset of full measure of X.

Here is a first application of our definitions. The reader is invited to compare the next statement with the third property of the elementary integral.

Proposition 15.1 *Suppose that a sequence $\{h_p\}_p$ of non-negative elementary functions is non-increasing and converges to zero almost everywhere. Then*

$$\lim_{p \to +\infty} Ih_p = 0.$$

Proof Recalling that elementary functions are bounded, we set

$$M_1 = \sup\{h_1(x) \mid x \in X\}.$$

Let Z be the subset of X on which the sequence $\{h_p\}_p$ does not converge to zero: by assumption, Z is a null set. If $\varepsilon > 0$, there exists a non-decreasing sequence $\{k_p\}_p$ of elementary functions such that

$$Ik_p < \frac{\varepsilon}{M_1}$$

and $\sup_p k_p(x) \geq 1$ for every $x \in Z$. The two limits

$$\lim_{p \to +\infty} Ih_p \geq 0, \qquad \lim_{p \to +\infty} Ik_p \leq \frac{\varepsilon}{M_1}$$

exist by monotonicity and boundedness. Furthermore the sequence $\{h_p - M_1 k_p\}_p$ is non-increasing and has a non-positive limit everywhere. By the continuity property of I,

$$I\left(h_p - M_1 k_p\right) \leq I\left(h_p - M_1 k_p\right)^+ \to 0,$$

and therefore

$$\lim_{p \to +\infty} Ih_p - M_1 \lim_{p \to +\infty} Ik_p \leq 0.$$

But then

$$0 \le \lim_{p \to +\infty} Ih_p \le M_1 \lim_{p \to +\infty} Ik_p \le M_1 \cdot \frac{\varepsilon}{M_1} = \varepsilon.$$

Since $\varepsilon > 0$ was arbitrary, we deduce that $\lim_{p \to +\infty} Ih_p = 0$.   □

Example 15.2 A countable union of subsets of measure zero is a subset of measure zero. Indeed, if $Z_1, \ldots, Z_n, \ldots$ is a countable family of null subsets, for every $\varepsilon > 0$ and every n there exists a non-decreasing sequence $\{h_{n,p}\}_p$ of elementary functions such that $Ih_{n,p} < 2^{-n}\varepsilon$ and $\sup_p h_{n,p} \ge 1$ on Z_n. Then the sequence defined by

$$h_p = \max\{h_{1,p}, \ldots, h_{n,p}\}$$

is non-decreasing, $\sup_p h_p \ge 1$ on $\bigcup_{n=1}^{\infty} Z_n$, and

$$Ih_p \le \sum_{n=1}^{p} Ih_{n,p} \le \varepsilon.$$

Exercise 15.3 Prove that a countable intersection of subsets of full measure is a subset of full measure.

> We will write $h_p \nearrow f$ to mean that $h_1 \le h_2 \le h_3 \le \ldots$ and $h_p(x)$ converges to $f(x)$ for almost every $x \in X$. The symbol $h_p \searrow f$ has an analogous definition.

Remark 15.1 Quite often the arrows $\nearrow$ and $\searrow$ mean monotone convergence *at every point*. Be careful, since we will always use them with almost everywhere convergence.

15.4   The Class L^+

Definition 15.7 A function $f : X \to (-\infty, +\infty]$ belongs to the class L^+ (or $L^+(X)$ when confusion may arise) if there exists a sequence $\{h_p\}_p$ in H such that $h_p \nearrow f$ and

$$C = \sup_p Ih_p \in \mathbb{R}.$$

Proposition 15.2 *If* $f \in L^+$, *then* $f(x) < +\infty$ *for almost every* $x \in X$.

Proof Let Z be the set of all $x \in X$ such that $f(x) = +\infty$. Replacing h_p with $h_p - h_1$, we may assume that each h_p is non-negative. Discarding a set of measure zero, we may then assume that $\{h_p\}_p$ is non-decreasing and converges to $+\infty$ on the whole set Z.

Pick $\varepsilon > 0$ and $x \in Z$; then the inequality

$$h_p(x) \geq \frac{C}{\varepsilon}$$

must holds from some value of p onwards. Hence Z is covered by the countable family of sets

$$\left\{ x \;\middle|\; h_p(x) \geq \frac{C}{\varepsilon} \right\}, \quad p \in \mathbb{N}.$$

Hence

$$\sup_p \frac{\varepsilon h_p(x)}{C} \geq 1$$

and

$$I\left(\frac{\varepsilon h_p}{C}\right) = \frac{\varepsilon}{C} I h_p \leq \varepsilon.$$

Therefore Z is a set of measure zero.   □

Let $f \in L^+$; we know that f is almost everywhere the increasing limit of a sequence $\{h_p\}_p$ of elementary functions. Suppose that f is also almost everywhere the increasing limit of another sequence $\{k_p\}_p$ of elementary functions. We claim that

$$\lim_{p \to +\infty} I h_p = \lim_{p \to +\infty} I k_p.$$

Since the two sequences can be swapped, it suffices to prove that

$$\lim_{p \to +\infty} I h_p \leq \lim_{p \to +\infty} I k_p. \tag{15.1}$$

Fix an index m, and consider the non-increasing sequence

$$n \mapsto h_m - k_n.$$

As $n \to +\infty$, this sequence converges to $h_m - f \le f - f \le 0$. Therefore $(h_m - k_n)^+ \searrow 0$, and by the continuity property of the integral

$$I(h_m - k_n)^+ \searrow 0.$$

But $I(h_m - k_n) \le I(h_n - k_m)^+$, it follows that

$$n \mapsto I(h_m - k_n) = Ih_m - Ik_n$$

is non-increasing and has a non-positive limit, which in turn implies that $Ih_m \le \lim_{n \to +\infty} Ik_n$. Since m was arbitrary, we deduce that (15.1) holds. It is now legitimate to propose the following definition.

Definition 15.8 The integral of a function $f \in L^+$ is defined as

$$If = \lim_{p \to +\infty} Ih_p,$$

where $\{h_p\}_p$ is any sequence as in Definition 15.7.

Exercise 15.4 More generally, suppose that $\{h_p\}_p$ and $\{k_p\}_p$ are two sequences of elementary functions such that $h_p \nearrow f$, $k_p \nearrow g$, $f \le g$ almost everywhere, and

$$\sup_p Ih_p \in \mathbb{R}$$

$$\sup_p Ik_p \in \mathbb{R}.$$

Prove that $If \le Ig$.

The elements of L^+ have some useful features that we now describe. The standard proofs are left to the reader.

1. If $f \in L^+$ and $g \in L^+$, then $f + g \in L^+$ and there results

$$I(f + g) = If + Ig.$$

2. If $f \in L^+$ and $\alpha \in [0, +\infty)$, then $\alpha f \in L^+$ and there results

$$I(\alpha f) = \alpha If.$$

3. If $f \in L^+$ and $g \in L^+$, then $\max\{f, g\} \in L^+$ and $\min\{f, g\} \in L^+$.

Important: Instability Under Multiplication

On the contrary, the class L^+ is not stable under multiplication by *negative* numbers, since such a multiplication turns increasing sequences into decreasing sequences.

As a deeper property, we prove that the class L^+ is closed under limits of increasing sequences with bounded integrals.

Theorem 15.1 *Suppose* $f_n \in L^+$ *for* $n = 1, 2, \ldots$, $f_n \nearrow f$ *and* $\sup_n I f_n \in \mathbb{R}$. *Then* $f \in L^+$ *and*

$$If = \lim_{n \to +\infty} I f_n.$$

Proof Given $n \in \mathbb{N}$, we construct sequences of elementary functions such that

$$h_{11} \leq \ldots \leq h_{1n} \leq \ldots, \quad h_{1n} \nearrow f_1$$
$$h_{21} \leq \ldots \leq h_{2n} \leq \ldots, \quad h_{2n} \nearrow f_2$$
$$\ldots$$
$$h_{k1} \leq \ldots \leq h_{kn} \leq \ldots, \quad h_{kn} \nearrow f_k$$
$$\ldots$$

according to the definition of L^+. We then set

$$h_n = \max \{h_{1n}, \ldots, h_{nn}\}.$$

It is clear that h_n is an elementary function, and that $h_n \leq h_{n+1}$ for every n. Since $h_n \leq \max \{f_1, \ldots, f_n\} = f_n$, we see that

$$\sup_n I h_n \leq \sup_n I f_n \in \mathbb{R}.$$

Setting $f^* = \lim_{n \to +\infty} h_n$, by definition of L^+ there results $f^* \in L^+$, and

$$If^* = \lim_{n \to +\infty} I h_n.$$

We claim that $f^* = f$ almost everywhere in X. Indeed, for any $k \in \mathbb{N}$ and any $n \geq k$, we have $h_{kn} \leq h_n \leq f_n$, so that $f_k \leq f^* \leq f$. But $f_n \nearrow f$ by assumption, and the claim is proved. Finally, $I h_{kn} \leq I h_n \leq I f_n \leq If$ and $I h_n \nearrow If^* = If$ imply $If = \lim_{n \to +\infty} I f_n$. □

A typical and nice application of the previous convergence result is to series of functions.

Theorem 15.2 *Suppose that $g_k \in L^+$ and $g_k \geq 0$ in X for every $k \in \mathbb{N}$. If there exists a constant $C \in \mathbb{R}$ such that*

$$I\left(\sum_{k=1}^{n} g_k\right) \leq C \quad n \in \mathbb{N},$$

then $f = \sum_{k=1}^{\infty} g_k \in L^+$ and there results

$$If = \lim_{n \to +\infty} \sum_{k=1}^{n} Ig_k = \sum_{k=1}^{\infty} Ig_k.$$

Proof Since each $g_k \geq 0$, the sequence

$$n \mapsto \sum_{k=1}^{n} g_k$$

is non-decreasing, and Theorem 15.1 applies.   □

15.5   The Class L of Integrable Functions

We can now complete the last step of our construction. We have seen that the class L^+ is not closed under multiplication by negative numbers, so that we cannot *subtract* elements of L^+. This is a gap we need to fill by enlarging the class L^+.

Definition 15.9 The class $L = L(X)$ of *integrable* functions on X is the set of all functions φ on X which can be represented almost everywhere as $\varphi = f - g$, for some $f \in L^+$ and $g \in L^+$.

It is evident from the previous discussion that L enjoys the following properties:

1. if $\varphi_1 = f_1 - g_1$ and $\varphi_2 = f_2 - g_2$ are elements of L, then $\varphi_1 + \varphi_2 = (f_1 + f_2) - (g_1 + g_2)$, and therefore $\varphi_1 + \varphi_2 \in L$.
2. If $\varphi = f - g$ and $\alpha \in \mathbb{R}$, then $\alpha\varphi \in L$, Indeed, if $\alpha \geq 0$, then $\alpha\varphi = \alpha f - \alpha g$ and αf, αg belong to L^+. If $\alpha < 0$, then $-\alpha > 0$ and $\alpha\varphi = (-\alpha)g - (-\alpha f)$, and it follows again that $\alpha\varphi \in L$.
3. If $\varphi \in L$, then φ^+, φ^- and $|\varphi|$ belong to L. Indeed, from $\varphi = f - g$ it follows that

$$|\varphi| = \max\{f, g\} - \min\{f, g\}$$

belongs to L. Then

$$\varphi^+ = \frac{|\varphi| + \varphi}{2}$$

$$\varphi^- = \frac{|\varphi| - \varphi}{2}$$

also belong to L by linearity.

We propose a formal definition, which will immediately be justified.

Definition 15.10 The integral $I\varphi$ of a function $\varphi \in L$ is defined as

$$I\varphi = If - Ig,$$

where $\varphi = f - g$, $f \in L^+$ and $g \in L^+$.

Proposition 15.3 *The integral of $\varphi \in L$ is independent on the representation $\varphi = f - g$.*

Proof Let $\varphi = f - g = f_1 - g_1$, for suitable functions f, f_1, g and g_1 in L^+. Since $f + g_1 = g + f_1$, we have $I(f + g_1) = I(g + f_1)$, or $If + Ig_1 = Ig + If_1$. This is equivalent to

$$I(f - g) = I(f_1 - g_1),$$

which completes the proof.   □

Remark 15.2 It should be noted that mathematical analysts do not usually appreciate definitions based on arbitrary choices of something. These definitions are typical of abstract algebra, whilst analysis tends to be more constructive. The previous definition of the (abstract!) integral is an exception to the rule.

Exercise 15.5 Prove that I is a linear operator: $I(f + g) = If + Ig$ and $I(\alpha f) = \alpha If$ for every f and g in L, and every $\alpha \in \mathbb{R}$.

Proposition 15.4 *If $\varphi \in L$ and $\varphi \geq 0$ in X, then $I\varphi \geq 0$.*

Proof If $\varphi = f - g$, we must have $f \geq g$, and therefore $If \geq Ig$ by monotonicity.   □

Corollary 15.1 *For every $\varphi \in L$, there results $|I\varphi| \leq I(|\varphi|)$.*

Proof Since $\varphi \leq |\varphi|$ and $-\varphi \leq |\varphi|$, the conclusion follows from the previous Corollary.   □

Important: Notation

We have carefully avoided the use of the integral symbol $\int$ for our abstract integral. The use of $\int_X \varphi$ is clearly possible—and often useful—, but our choice of the letter I highlights the *functional* nature of our integral. In other words, in our approach the integral is a linear operator which acts on the vector space $L(X)$, and we prefer to encourage this viewpoint.

15.6   Taking Limits Under the Integral Sign

When we were dealing with sequences and series of functions, we realized quite easily that a pointwise convergence of the integrands does not imply the convergence of the integrals. Uniform convergence was a successful replacement, but the weakness of the Riemann integral with respect to limits remains a matter of facts. We want to convince the reader that the Lebesgue (abstract) integral I is much more flexible. We begin with a technical lemma.

Lemma 15.1 *Any $\varphi \in L$ admits a representation with the following property: for every $\varepsilon > 0$, there exist $f \in L^+$ and $g \in L^+$ such that $\varphi = f - g$, $g \geq 0$ and $Ig < \varepsilon$.*

Proof By definition, $\varphi = f - g$ for some f and g in L^+. Let $h_n \nearrow g$ be a sequence of elementary functions such that $Ig = \lim_{n\to+\infty} Ih_n$. We can write

$$\varphi = f - g = (f - h_n) - (g - h_n) = f_n - g_n.$$

Now $f_n \in L^+$ since $f_n = f - h_n$ is the sum of two elements of L^+, and similarly $g_n = g - h_n \in L^+$. Clearly $g_n \geq 0$ for n sufficiently large, and $Ig_n < \varepsilon$ because $\lim_{n\to+\infty} Ig_n = 0$.   □

Theorem 15.3 (Beppo Levi) *Assume that $\varphi_k \in L$, $\varphi_k \geq 0$ for every $k \in \mathbb{N}$, and*

$$I\left(\sum_{k=1}^{n} \varphi_k\right) \leq C, \quad n \in \mathbb{N}$$

for some suitable constant C. Then $\varphi = \sum_{k=1}^{\infty} \varphi_k$ belongs to L, and

$$I\varphi = \sum_{k=1}^{\infty} I\varphi_k.$$

Proof We use Lemma 15.1 to decompose $\varphi_k = f_k - g_k$, where $f_k, g_k \in L^+$, $g_k \geq 0$ and $I g_k < 2^{-k}$, for every $k = 1, 2, \ldots$. Since $\varphi_k \geq 0$, it turns out that $f_k \geq 0$. It is easy to check that Theorem 15.2 applies, hence $g = \sum_{k=1}^{\infty} g_k$ belongs to L^+ and

$$
I g = \sum_{k=1}^{\infty} I g_k .
$$

The same conclusion holds also for f_k, and then $f = \sum_{k=1}^{\infty} f_k \in L^+$, $If = \sum_{k=1}^{\infty} I f_k$. Putting everything together we see that

$$
\varphi = \sum_{k=1}^{\infty} \varphi_k = \sum_{k=1}^{\infty} f_k - \sum_{k=1}^{\infty} g_k = f - g \in L
$$

and

$$
I\varphi = If - Ig = \sum_{k=1}^{\infty} I\left(f_k - g_k\right) = \sum_{k=1}^{\infty} I \varphi_k .
$$

□

Recalling that sequences and series are the same mathematical object, we deduce the following statement.

Theorem 15.4 *If $\psi_n \in L$ for $n = 1, 2, \ldots$, $\psi_n \nearrow \psi$ and $I\psi_n \leq C$ for every n, then $\psi \in L$ and $I\psi = \lim_{n \to +\infty} I\psi_n$.*

Proof We set $\varphi_1 = \psi_1$, $\varphi_n = \psi_n - \psi_{n-1}$, and the conclusion follows from Theorem 15.3. □

A very useful consequence of Beppo Levi's Convergence Theorem is a sort of characterization of nonnegative functions whose integral vanishes.

Exercise 15.6 Prove that the integral of a function $\varphi \in L$ such that $\varphi = 0$ almost everywhere is zero.

The converse implication is contained in the next result.

Theorem 15.5 *Let $\varphi_0 \in L$ be a non-negative function such that $I\varphi_0 = 0$. Then $\varphi_0 = 0$ almost everywhere in X.*

Proof Define the sequence $n \mapsto \varphi_n = n\varphi_0$. Clearly $\varphi_n \geq 0$ and $I\varphi_n = nI\varphi_0 = 0$ for each n, and $\varphi_n \nearrow \varphi$, where $\varphi = 0$ on the set where $\varphi_0 = 0$, and $\varphi = +\infty$ on the set where $\varphi_0 > 0$. By Beppo Levi's Theorem, $\varphi \in L$, so that $\varphi = +\infty$ only on a set of measure zero, and in particular $\varphi_0 > 0$ only on a set of measure zero. □

Exercise 15.7 Show that the previous result is generally false if we drop the non-negativity of φ_0.

Corollary 15.2 *If $Z \subset X$, suppose that for every $\varepsilon > 0$ there exists a sequence of integrable functions*

$$0 \le \varphi_{\varepsilon,1} \le \varphi_{\varepsilon,2} \le \ldots \le \varphi_{\varepsilon,n} \le \ldots$$

such that $I\varphi_{\varepsilon,n} < \varepsilon$ for each n and $\sup_n \varphi_{\varepsilon,n} \ge 1$ on Z. The Z is a set of measure zero.

Proof Of course there is nothing to prove if each φ_n is an elementary function. In the general case, let $\varphi_\varepsilon = \lim_{n \to +\infty} \varphi_{\varepsilon,n}$. By Beppo Levi's Theorem, φ is integrable and

$$I\varphi_\varepsilon = \lim_{n \to +\infty} I\varphi_{\varepsilon,n} \le \varepsilon.$$

We discretize $\varepsilon = 1/m$, for $m \in \mathbb{N}$, so that we have the functions

$$\psi_1 = \varphi_1$$

$$\psi_2 = \min\{\varphi_1, \varphi_2\}$$

$$\ldots$$

$$\psi_n = \min\{\varphi_1, \varphi_2, \ldots, \varphi_n\}$$

$$\ldots$$

Each function ψ_m is non-negative and $\psi_m \ge 1$ on the set Z. Furthermore $\psi_1 \ge \psi_2 \ge \psi_3 \ge \ldots$ and

$$I\psi_m \le I\varphi_{1/m} \le \frac{1}{m}.$$

Setting $\psi = \lim_{m \to +\infty} \psi_m$, we see that $\psi \in L$ and $I\psi = \lim_{m \to +\infty} I\psi_m = 0$. The limit ψ is non-negative, and $\psi \ge 1$ on Z. Hence the set $Z' = \{x \in X \mid \psi(x) > 0\}$ has measure zero. But then Z has measure zero.   □

The most powerful Convergence Theorem for our integral is due to Lebesgue.

Theorem 15.6 (Dominated Convergence Theorem) *Suppose that $\varphi_n \in L$ for every n, $\varphi_n \to \varphi$ almost everywhere, and there exists a function $\varphi_0 \in L$ such that*

$$|\varphi_n(x)| \le \varphi_0(x), \quad x \in X, \ n = 1, 2, \ldots \tag{15.2}$$

Under these assumptions, $\varphi \in L$ and $I\varphi = \lim_{n \to +\infty} I\varphi_n$.

Proof The assumptions imply that $-\varphi_0 \leq \varphi_n \leq \varphi_0$, and in turn $-I\varphi_0 \leq I\varphi_n \leq I\varphi_0$. On a set of full measure we have $-\varphi_0 \leq \varphi \leq \varphi_0$, and hence $\varphi \in L$. Indeed, we can set

$$\psi_n(x) = \sup\{\varphi_n(x), \varphi_{n+1}(x), \ldots\}$$
$$\chi_n(x) = \inf\{\varphi_n(x), \varphi_{n+1}(x), \ldots\}.$$

These functions are integrable and satisfy (almost everywhere) $-\varphi_0 \leq \psi_n \leq \varphi_0$, $-\varphi_0 \leq \chi_n \leq \varphi_0$. Restricting x to the subset of X on which $\varphi_n(x)$ converges to φ, we have

$$\psi_n(x) \geq \lim_{p \to +\infty} \varphi_{n+p}(x) = \varphi(x)$$
$$\chi_n(x) \leq \lim_{p \to +\infty} \varphi_{n+p}(x) = \varphi(x).$$

Since $\psi_{n+1} \leq \psi_n$, $\chi_{n+1} \geq \chi_n$, the assumption $\varphi_n \nearrow \varphi$ implies $\psi_n \searrow \varphi$, $\chi_n \nearrow \varphi$, and in conclusion a set of full measure we have $-\varphi_0 \leq \varphi \leq \varphi_0$.

Observing that $I\chi_n \nearrow I\varphi$, $I\psi_n \searrow I\varphi$, $I\chi_n \leq I\varphi_n \leq I\psi_n$, we conclude that $I\varphi_n \to I\varphi$. The proof is complete.   □

A natural question is what happens if we weaken (15.2). Since (15.2) implies that $I(|\varphi_n|) \leq I\varphi_0 = C$ for every n, a good replacement is

$$I(|\varphi_n|) \leq C.$$

Theorem 15.7 (Fatou's Lemma) *Suppose that $\varphi_n \in L$, $\varphi_n \geq 0$, $\varphi_n \to \varphi$ almost everywhere, and for some constant $C \geq 0$ we have $I(|\varphi_n|) \leq C$ for every n. Then $\varphi \in L$ and*

$$0 \leq I\varphi \leq C.$$

Proof We define $\chi_n = \inf\{\varphi_n, \varphi_{n+1}, \ldots\}$, observing that $\chi_n \leq \chi_{n+1}$ and $\chi_n \to \varphi$ almost everywhere. Furthermore $\chi_n \leq \varphi_n$, $I\chi_n \leq I\varphi_n \leq C$, and Beppo Levi's Theorem for sequences implies that $\varphi \in L$. Since $I\chi_n \nearrow I\varphi$, it follows in particular that $0 \leq I\varphi = \lim_{n \to +\infty} I\varphi_n \leq C$.   □

15.7   Measurable Functions and Measurable Sets

The *leitmotiv* of Measure Theory is the adjective *measurable*: measurable sets, measurable functions. Abstract Measure Theory resembles General Topology, with the difference that open sets are replaced by the class of measurable sets. Once these are defined, measurable functions come out naturally.

On the contrary, we have developed the Theory of the Integral by embedding "good" functions into the class $L(X)$. The concept of measurability becomes less evident, and we need to define it from scratch.

Definition 15.11 A function $\varphi\colon X \to (-\infty, +\infty]$ is a measurable function if $\varphi < +\infty$ almost everywhere, and φ is almost everywhere the limit of a sequence of elementary functions.

Theorem 15.8 *Suppose that φ is a measurable function such that $|\varphi(x)| \le \varphi_0(x)$ for every $x \in X$, where $\varphi_0 \in L$, then $\varphi \in L$.*

Proof Since φ is the almost everywhere limit of a sequence $\{h_n\}_n$ of elementary functions, we just observe that elementary functions are integrable, and the conclusion follows from the Dominated Convergence Theorem.   □

The class of measurable function is an open door on a fascinating new world, which we try to introduce in the next pages. First of all we list a few immediate properties of this class.

1. if f and g are measurable, then $\alpha f + \beta g$ is measurable for every choice of $\alpha \in \mathbb{R}$, $\beta \in \mathbb{R}$.
2. If f is measurable, so are $|f|$, f^+ and f^-.
3. Every $f \in L^+$ is measurable, since it is the limit of a sequence of elementary functions. In particular every integrable function is measurable.

Coming to convergence, we have the following result

Theorem 15.9 *If $\{f_n\}_n$ is a non-decreasing sequence of integrable functions converging almost everywhere to a finite limit f, then f is measurable.*

Proof We consider first the case $f_n \in L^+$, and pick $h_{nk} \in H$ such that $h_{nk} \nearrow f_n$. If

$$h_n = \max\{h_{1n}, h_{2n}, \dots, h_{nn}\},$$

then $\{h_n\}_n$ is a non-decreasing sequence of elementary functions which converges to a limit f^*. Since $n > k$ implies $h_{nk} \le h_n \le f_n$, letting $n \to +\infty$ yields $f_k \le f^* \le f$. This shows that f^* is almost everywhere finite. Hence f^* is a measurable function, but then $f^* = f$ almost everywhere, since $f_n \nearrow f$, and f is measurable.   □

Much the same conclusion holds for sequences of measurable functions.

Theorem 15.10 *If a sequence $\{f_n\}_n$ of measurable functions converges almost everywhere to a function f, then f is a measurable function.*

Proof Splitting $f_n = f_n^+ - f_n^-$ and $f = f^+ - f^-$, we may assume without loss of generality that $f_n \ge 0$ and $f \ge 0$. Each f_n is the limit of a sequence $\{h_{np}\}_p$ of elementary functions which may be assumed to be non-negative with positive integrals. Consider then the function

$$\varphi_o = \sum_{n=1}^{\infty} \sum_{p=1}^{\infty} c_{np} \frac{h_{np}}{I h_{np}},$$

where the real coefficients c_{np} are chosen so that the series $\sum_{n=1}^{\infty} \sum_{p=1}^{\infty} c_{np}$ converges. By Beppo Levi's Theorem, φ_0 is integrable, and $\varphi_0(x) > 0$ whenever $f_n(x) > 0$. It follows that f is also the limit of the sequence

$$g_n(x) = \min \{f(x), n\varphi_0(x)\}.$$

According to Theorem 15.9, we need to show that the measurability of f_n implies the measurability of g_n. But

$$g_n(x) = \min \{f(x), n\varphi_0(x)\}$$

$$= \lim_{m \to +\infty} \min \{f_m(x), n\varphi_0(x)\}$$

and each function $\min \{f_m(x), n\varphi_0(x)\}$ is measurable and bounded by the integrable function $n\varphi_0$. The Dominated Convergence Theorem then yields the integrability of g_n, and the proof is complete.   □

Corollary 15.3 *If $\{f_n\}_n$ is a sequence of measurable functions, then $\inf_n f_n$ and $\sup_n f_n$ are measurable functions. If each f_n is finite almost everywhere, then $\liminf_{n \to +\infty} f_n$ and $\limsup_{n \to +\infty} f_n$ are measurable.*

Once measurable functions have been defined, measurable sets do not come as a surprise.

Definition 15.12 A subset $E \subset X$ is measurable if and only if its characteristic function χ_E is a measurable function.

We recall that

$$\chi_E(x) = \begin{cases} 1 & \text{if } x \in E \\ 0 & \text{if } x \in X \setminus E. \end{cases}$$

Definition 15.13 A subset E of X has finite measure if and only if $\chi_E \in L(X)$. In this case, the measure of E is the number

$$\mu(E) = I \chi_E.$$

If χ_E is not integrable, we set $\mu(E) = +\infty$. In a conventionally way, we set $\mu(\emptyset) = 0$.

Exercise 15.8

1. Prove that a measurable subset of a set of finite measure, is a set of finite measure.
2. Prove that any subset of a set of measure zero is a set of measure zero.

Important: For Experts of Measure Theory

If Measure Theory is introduced *before* the integral, as is customary in Geometric Measure Theory, the fact that a subset of a set of measure zero has measure zero is not assumed. Then the issue of *completeness* of a measurable space comes into play. In this sense, the Measure Theory we are constructing is slightly less general, and always produces *complete* measurable spaces. We will come back to this topic in the next chapter.

Theorem 15.11 *The union, intersection and difference of two measurable sets are measurable sets.*

Proof Let E, F be measurable sets. The conclusion follows from the identities

$$\chi_{E \cup F} = \max\{\chi_E, \chi_F\}$$

$$\chi_{E \cap F} = \min\{\chi_E, \chi_F\}$$

$$\chi_{E \setminus F} = \chi_E - \chi_F.$$

$\square$

The previous result can be generalized to countable unions as follows.

Theorem 15.12 *If each set E_n, $n \in \mathbb{N}$, is measurable, then $E = \bigcup_{n=1}^{\infty} E_n$ is measurable. Moreover, if $E_i \cap E_j = \emptyset$ whenever $i \neq j$, then*

$$\mu(E) = \sum_{n=1}^{\infty} \mu(E_n). \tag{15.3}$$

The case $+\infty = +\infty$ is not excluded in (15.3)

Proof It follows from Corollary 15.3 that

$$\chi_E = \sup\{\chi_{E_1}, \chi_{E_2}, \ldots\} = \lim_{n \to +\infty} \sup\{\chi_{E_1}, \ldots, \chi_{E_n}\}$$

is a measurable function, hence E is a measurable set. To prove (15.3), we remark that if $\mu(E_n) = +\infty$ for some n, then $\mu(E) = +\infty$ because $E \supset E_n$. On the other hand, if $\mu(E_n) \in \mathbb{R}$ for each n, then $\chi_E = \sum_{n=1}^{\infty} \chi_{E_n}$. Beppo Levi's Theorem implies that χ_E is integrable, and

$$I\chi_E = \sum_{n=1}^{\infty} I\chi_{E_n},$$

provided that the series $\sum_{n=1}^{\infty} I\chi_{E_n} = \sum_{n=1}^{\infty} \mu(E_n)$ converges. Conversely, if χ_E is integrable, then $\sum_{k=1}^{n} I\chi_{E_k} \leq I\chi_E$ for each n, and hence the series $\sum_{n=1}^{\infty} I\chi_{E_n} = \sum_{n=1}^{\infty} \mu(E_n)$ converges. We have proved that (15.3) holds in any case. □

Corollary 15.4 *Suppose that E_n, $n \in \mathbb{N}$, is a measurable set.*

(a) If $E_1 \subset E_2 \subset \ldots$, then $E = \bigcup_{n=1}^{\infty} E_n$ is measurable, and $\mu(E) = \lim_{n\to+\infty} \mu(E_n)$.
(b) The set $F = \bigcap_{n=1}^{\infty} E_n$ is measurable. If $E_1 \supset E_2 \supset \ldots$ and $\mu(E_1) \in \mathbb{R}$, then $\mu(F) = \lim_{n\to+\infty} \mu(E_n)$.

Proof If some E_n has infinite measure, then $\mu(E) = +\infty$, and there is nothing to prove. Otherwise, we write

$$E = E_1 \cup (E_2 \setminus E_1) \cup (E_3 \setminus E_2) \cup \ldots$$

so that E is a countable union of disjoint subsets. By Theorem 15.12 we see that

$$\mu(E) = \lim_{n\to+\infty} \sum_{k=1}^{n} \mu(E_k \setminus E_{k-1}) = \lim_{n\to+\infty} \mu(E_n),$$

having set $E_0 = \emptyset$ for convenience. This proves (a).

To prove (b) we take complements and recall that the complement of an intersection is the union of the complements. The measurability of F follows from (a) in this way. As before, we write

$$E_1 = F \cup (E_1 \setminus E_2) \cup (E_2 \setminus E_3) \cup \ldots$$

and apply (a) again. The term $\mu(E_1)$ cancels since it is finite, and the proof is complete. □

15.8   Integration Over Measurable Sets

Up to now, all our integrals have been computed "on the whole space X." In many applications it would be convenient to integrate functions over subsets of X. Although we already have the best possible candidate, i.e. measurable subsets, a problem arises: consider indeed $\mu(X)$, the measure of the whole space. There is no need for X to be a measurable set; but even if this were the case, $\mu(X)$ should be defined as $I\chi_X$, or equivalently as $I(1)$, and in this case $\mu(X)$ would be a *finite* value.

As we will soon see, this fact would prevent us from constructing the (concrete) Lebesgue measure on $\mathbb{R}^n$, since $\mathbb{R}^n$ is a set of *infinite* measure to any reasonable mind. To overcome this obstacle, we need to introduce a new axiom.

Stone's Axiom The collection H of elementary functions satisfies

(c) If $h \in H$, then $\min\{h, 1\} \in H$.
(d) There exists a sequence $\{h_n\}_n$ of non-negative elementary functions such that $I h_n > 0$ and $\sup_n h_n(x) > 0$ for every $x \in X$.

Example 15.3 Axiom (c) extends to measurable functions: if $\varphi = \lim_{n\to+\infty} h_n$ is a measurable function, then $\min\{\varphi, 1\} = \lim_{n\to+\infty} \min\{h_n, 1\}$ is measurable.

Example 15.4 Axiom (d) yields the existence of an integrable function φ_0 such that $\varphi_0(x) > 0$ at every $x \in X$. Indeed,

$$\varphi_0(x) = \sum_{n=1}^{\infty} \frac{1}{n^2} \frac{h_n(x)}{I h_n}$$

does the job, where $\{h_n\}_n$ is the sequence considered in Axiom (d).

Theorem 15.13 *The function χ_X is measurable, so that X is a measurable set. In particular, the set $X \setminus E$ is measurable for every measurable set $E \subset X$.*

Proof Actually, $1 = \lim_{n\to+\infty} \min\{1, n\varphi_0\}$, and Axiom (c) ensures that $\{1, n\varphi_0\}$ is measurable. $\square$

Corollary 15.5 *If φ is a measurable function and a, b, c are real numbers, then*

$$\min\{\varphi, c\}, \quad \max\{\varphi, c\}, \quad \max\{\min\{\varphi, b\}, a\}$$

are measurable functions.

The next result contains the characterization of measurable functions which is introduced in Geometric Measure Theory.

Theorem 15.14 *Let φ be a function, almost everywhere finite. The set*

$$[\varphi > c] = \{x \in X \mid \varphi(x) > c\}$$

is measurable for every $c \in \mathbb{R}$ if and only if the function φ is measurable.

Proof If φ is measurable, then so is the function

$$\varphi_{n,c} = \frac{\min\{\varphi, c + 1/n\} - \min\{\varphi, c\}}{1/n}$$

for every c and n. Keeping c fixed, we see that

$$\lim_{n \to +\infty} \varphi_{n,c}(x) = \begin{cases} 0 & \text{if } \varphi(x) \le c \\ 1 & \text{if } \varphi(x) > c. \end{cases}$$

Hence $\chi_{[\varphi > c]}$ is the limit of measurable functions, and this implies the measurability of $[\varphi > c]$.

Conversely, suppose that $[\varphi > c]$ is measurable for every c. Then

$$\{x \in X \mid c < \varphi(x) \le d\} = [\varphi > c] \setminus [\varphi > d]$$

is also measurable for every $c < d$. Given $n \in \mathbb{N}$, we consider the function φ_n equal to k/n on the measurable sets

$$E_{k,n} = \left\{ x \in X \ \middle|\ \frac{k}{n} < \varphi(x) \le \frac{k+1}{n} \right\},$$

$k \in \mathbb{Z}$. The function φ_n is defined almost everywhere, and differs from φ by at most $1/n$. Moreover

$$\varphi_n = \sum_{k=-\infty}^{\infty} \frac{k}{n} \chi_{E_{k,n}},$$

and hence φ_n is measurable. Since $\varphi_n \to \varphi$ uniformly on X, φ is a measurable function, and the proof is complete. □

Proposition 15.5 *The product of two measurable functions is a measurable function.*

Proof We suppose that f and g are measurable functions. Without loss of generality, we may assume that $f \ge 0$ and $g \ge$ (the general case follows by writing $f = f^+ - f^-$ and $g = g^+ - g^-$). Given $c \in \mathbb{R}$, we observe that

$$\{x \in X \mid f(x)g(x) > c\} = \bigcup \left\{ \left\{ x \in X \ \middle|\ f(x) > r, \ g(x) > \frac{c}{r} \right\} \ \middle|\ r \in \mathbb{Q}, \ r > 0 \right\},$$

and the conclusion follows from the fact that $\mathbb{Q}$ is a countable set. □

Corollary 15.6 *If f is a measurable function and E is a measurable set, then $f \chi_E$ is a measurable function. In particular, if f is integrable and E is measurable, then $f \chi_E$ is integrable.*

The integral of a function on a subset is now defined naturally.

Definition 15.14 A function φ defined on X is integrable (or measurable) on a measurable $E \subset X$ if $\varphi \chi_E$ is integrable (or measurable). In the first case, the integral of φ on E is defined to be

$$\int_E \varphi \, d\mu = I \, (\varphi \chi_E).$$

At this point, we will freely write $\int_X \varphi \, d\mu$ for $I\varphi$, as a particular case. Of course this is just a customary piece of notation that adds no content to the formal $I \, (\varphi \chi_E)$.

We collect several easy properties of the integral over measurable sets.

Proposition 15.6 *If φ is integrable on E and $|\varphi| \leq M$ on E, then*

$$\int_F |\varphi| \, d\mu \leq M\mu(E).$$

Proof Since $\chi_E |\varphi| \leq M \chi_E$ on X, we see that

$$\int_E |\varphi| \, d\mu = I \, (\chi_E |\varphi|) \leq M I \chi_E = M\mu(E).$$

$\square$

Proposition 15.7 *If $\{E_n \mid n \in \mathbb{N}\}$ is a countable family of mutually disjoint measurable sets, and if φ is integrable (resp. measurable) on $E = \bigcup_{n \in \mathbb{N}} E_n$, then φ is integrable (resp. measurable) on each E_n. If φ is integrable on E, then*

$$\int_E \varphi \, d\mu = \sum_{n=1}^{\infty} \int_{E_n} \varphi \, d\mu.$$

Proof If $\chi_E \varphi$ is measurable (resp. integrable) on X, then so is the product $\chi_{E_n} \chi_E \varphi$. Moreover $\chi_{E_1} + \chi_{E_2} + \cdots = \chi_E$, so that $\chi_{E_1} \varphi + \chi_{E_2} \varphi + \cdots = \chi_E \varphi$. If φ is integrable, the partial sums of the series in the left-hand side are bounded by the integrable function $\chi_E \varphi$, and we can integrate term by term. This completes the proof. $\square$

Proposition 15.8 *If $\{E_n \mid n \in \mathbb{N}\}$ is a countable family of measurable sets, and if φ is measurable on each E_n, then φ is measurable on $E = \bigcup_{n=1}^{\infty} E_n$.*

Proof Indeed,

$$\chi_E \varphi = \varphi \chi_{E_1} + \varphi \chi_{E_2 \setminus E_1 \cap E_2} + \varphi \chi_{E_3 \setminus E_1 \cap E_3 \setminus E_2 \cap E_3} + \cdots$$

and each function on the right-hand side is measurable. By Theorem 15.10 the function $\chi_E\varphi$ is measurable. □

Theorem 15.15 (Absolute Continuity of the Integral) *Suppose that φ is an integrable function on X. For every $\varepsilon > 0$ there exists $\delta > 0$ such that if E is a measurable set and $\mu(E) < \delta$, then*

$$\left|\int_E \varphi\,d\mu\right| < \varepsilon.$$

Proof Pick an elementary function h such that

$$I(||\varphi| - h|) < \frac{\varepsilon}{2}.$$

Since the function h is bounded, say $0 \leq h \leq M$, we set $\delta = M\varepsilon/2$ and compute

$$\left|\int_E \varphi\,d\mu\right| \leq \int_E |\varphi|\,d\mu$$

$$\leq \int_E ||\varphi| - h|\,d\mu + \int_E h\,d\mu$$

$$\leq \frac{\varepsilon}{2} + M\delta < \varepsilon$$

whenever E is a measurable set with $\mu(E) < \delta$. □

15.9   The Concrete Lebesgue Integral

So far we have proposed a completely abstract construction of the integral. However, there is no reason why we should call I an *integral*, since it appears as a map with some selected properties. In this section we want to show that a true integration theory on $\mathbb{R}^n$ can be constructed by applying our abstract scheme to the Riemann integral in n dimensions.

To this aim we need a good class H of elementary functions, and of course a good map I on H which satisfies the necessary axioms. The sketch we proposed in Sect. 15.1 is not a complete solution yet, so how do we select the vector space H?

At least two possible constructions are possible: the first one is based on *step functions*, the second one on *continuous functions*. The resulting integrals are equivalent, but this must be proved. We begin with step functions.

Definition 15.15 Let $\mathbf{B} = B_1 \cup B_2 \cup \cdots \cup B_n$ be a partition of the basic n-cell $\mathbf{B}$, and we suppose that the different sub-cells B_j do not have interior points in common.[1] Any function h such that

$$h(x) = \begin{cases} h_1 & \text{if } x \in B_1 \\ h_2 & \text{if } x \in B_2 \\ \vdots \\ h_n & \text{if } x \in B_n \end{cases}$$

for suitable real numbers $h_1, \ldots, h_n$ is a step function. The collection of all step functions on $\mathbf{B}$ is denoted by $H(\mathbf{B})$.

Remark 15.3 The previous definition is somehow troublesome, since h might be defined in different ways on the layer $B_i \cap B_j$, for $i \neq j$. This is irrelevant to us, and we might even leave step functions undefined on the interface layers of the partition. As we will see in a moment, their values on such layers plays no role at all in our construction.

Definition 15.16 Let $h \in H(\mathbf{B})$ be a step function, and consider the sets

$$B_j = \{x \in \mathbf{B} \mid h(x) = h_j\}, \qquad j = 1, \ldots, n.$$

The integral of h is

$$Ih = \sum_{j=1}^{n} h_j \, \text{Vol}(B_j).$$

It is now easy to check that $I: H(\mathbf{B}) \to \mathbb{R}$ is an elementary integral in the sense of Definition 15.4. Therefore our abstract extension produces a class $L(\mathbf{B})$ of integrable functions and an integral associated to each integrable function on $\mathbf{B}$.

What happens if we start with the collection $\tilde{H}$ of continuous functions defined on $\mathbf{B}$? Clearly enough, this is a vector spaces that satisfies Definition 15.3, and each continuous function has a Riemann integral as described in Sect. 15.1. The only non-trivial fact to check is the continuity axiom, which follows from the next result.

Theorem 15.16 (Dini's Lemma) *A non-increasing sequence $\{f_n\}_n$ of continuous functions converging pointwise to zero on a compact set K, converges uniformly to zero on K.*

Proof In Problem 11.5 a more general statement was proposed. For the reader's convenience we write here the proof. Fix any $\varepsilon > 0$; to any point $x_0 \in K$ there

---

[1] Geometrically, we are assuming here that the cells B_j can touch each other only at their boundaries.

corresponds an integer $m = m(x_0)$ such that $f_m(x_0) < \varepsilon$. By continuity, there exists a neighborhood $U(x_0)$ of x_0 such that $f_m(x) < \varepsilon$ for every $x \in U(x_0)$. By monotonicity, if $p > m$ then $f_p(x) \leq f_m(x) < \varepsilon$ for every $x \in U(x_0)$. As x_0 ranges over K, the neighborhoods $U(x_0)$ form an open cover of K: let $\{U(x_1), U(x_2), \ldots, U(x_\nu)\}$ be a finite subcover of the compact set K. If q denotes the smallest index of the functions which participate in this subcover, we see that $f_r(x) < \varepsilon$ for every $x \in K$ and every $r > q$, and the proof is complete.   □

As before, we can now turn on the engine of our abstract extension machine, and obtain a space $\tilde{L}(\mathbf{B})$ of integrable functions and a corresponding integral $\tilde{I}$. We now show that $\tilde{L} = L$ and $\tilde{I} = I$, so that the concrete Lebesgue integral can be equivalently defined in two ways.

Theorem 15.17 *There results* $\tilde{L}(\mathbf{B}) = L(\mathbf{B})$ *and* $\tilde{I} = I$.

Proof Since the proof is long, we split it into several steps.

Step 1.   Every continuous function f belongs to L. Indeed, let $\varepsilon > 0$ be given, and we select a partition $P = \{B_1, \ldots, B_m\}$ of $\mathbf{B}$ such that

$$\left| \int_{\mathbf{B}} f - \sum_{j=1}^{m} f(\xi_j) \operatorname{Vol}(B_j) \right| < \varepsilon,$$

whenever $\xi_j \in B_j$. This is equivalent to

$$\left| \int_{\mathbf{B}} f - I h_P \right| < \varepsilon,$$

where h_P is the step function whose value on B_j is $f(\xi_j)$. Since h_P converges uniformly to f as the partition P is indefinitely refined, it follows from the Dominated Convergence Theorem that $f \in L(\mathbf{B})$ and

$$If = \tilde{I}f.$$

Step 2.   Every step function h belongs to $\tilde{L}$. This is actually a density statement about the approximation of step functions by continuous function. We consider a function h which is equal to 1 on a cell B and to 0 outside B. If the dimension n of the space is 1, a "trapezoidal" graph shows that there exists a continuous (actually *piecewise affine*) function that approximates h with any prescribed precision. If $n > 1$, we just pick such an approximating function $f_i = f_i(x_i)$ in each variable x_i, and define $(x_1, \ldots, x_n) \mapsto f_1(x_1) f_2(x_2) \cdots f_n(x_n)$. As a consequence, each step function h can be expressed as the limit of a sequence $\{f_m\}_m$ of continuous functions such that, as we have shown above, $\tilde{I} f_m = I f_m$ for each m. Using again the Dominated Convergence Theorem, $h \in \tilde{L}$ and $\tilde{I}h = \lim_{m \to +\infty} \tilde{I} f_m = \lim_{m \to +\infty} I f_m = Ih$.

Step 3.   Both constructions lead to the same sets of measure zero.[2] Indeed, let $\tilde{Z}$ be a set of measure zero with respect to the integral $\tilde{I}$. Given $\varepsilon > 0$, there exists a non-decreasing sequence of non-negative continuous functions f_m such that $\tilde{I} f_m < \varepsilon$ and $\sup_m f_m(x) \geq 1$ for every $x \in Z$. By Step 1, $\tilde{I} f_m = I f_m$ for each m. By Beppo Levi's Theorem, $\tilde{Z}$ is a set of measure zero with respect to the integral I. Conversely, let Z be a set of measure zero with respect to the integral I. For this reason, there exists a non-decreasing sequence of step functions h_m such that $I h_m < \varepsilon$ and $\sup_m h_m(x) \geq 1$ for every $x \in Z$. By Step 2, $h_m \in \tilde{L}$ and $\tilde{I} h_m = I h_m$, so that we conclude again that Z is a set of measure zero with respect to $\tilde{I}$ as well.

Step 4.   Monotone passages to the limit and formation of differences. Pick any $f \in L^+$, so that f is the limit (almost everywhere) of a non-decreasing sequence of step functions h_m with bounded integrals $I h_m$. Then the integrals $\tilde{I} h_m = I h_m$ remain bounded, too, and hence by Beppo Levi's Theorem $f \in \tilde{L}$, $\tilde{I} f = I f$. On the other hand, if $f \in \tilde{L}^+$, then f is the limit (almost everywhere) of a non-decreasing sequence of continuous functions f_m with bounded integrals $\tilde{I} f_m$. As before the integrals $I f_m = \tilde{I} f_m$ remain bounded, $f \in L$ and $I f = \tilde{I} f$. By taking differences, $\tilde{L}$ contains every function of L, and vice-versa, with equal integrals. The proof is now complete.

$\square$

For the concrete Lebesgue integral, a geometric characterization of sets of measure zero is possible. We remark that the next result is usually taken as the *definition* of measure zero in Euclidean spaces.

Theorem 15.18 *Let* $Z \subset \mathbf{B}$ *be a subset of the basic block* $\mathbf{B}$. *The following statements are equivalent:*

(a) *for every* $\varepsilon > 0$ *there exists a countable (i.e. finite or countably infinite) collection of n-cells* $B_1, B_2, \ldots,$ *such that* $Z \subset \bigcup_{k=1}^{\infty} B_k$ *and* $\sum_{k=1}^{\infty} \mathrm{Vol}(B_k) < \varepsilon$;

(b) *for every* $\varepsilon > 0$ *there exists a non-decreasing sequence of non-negative step functions*

$$h_1^{(\varepsilon)} \leq \cdots \leq h_m^{(\varepsilon)} \leq \cdots$$

such that $I h_m^{(\varepsilon)} < \varepsilon$ *for every* m *and*

$$\sup_m h_m^{(\varepsilon)}(x) \geq 1 \quad \text{for every } x \in Z.$$

Proof Assume that (a) holds, and call $h_m^{(\varepsilon)}$ the step function which is equal to 1 on the cells $B_1, \ldots, B_m$ and equal to 0 outside these cells. Now, every point $x_0 \in$

---

[2] Hence the sentence "almost everywhere" can be interpreted with respect to both integrals.

Z belongs to some cell B_m, hence $h_m^{(\varepsilon)}(x_0) = 1$. The remaining properties of the sequence $\{h_m^{(\varepsilon)}\}_m$ are trivial, and therefore (b) holds.

Conversely, suppose that (b) holds. Let $B_1, \dots, B_{r_1}$ be the collection of cells on which the function $h_1^{(\varepsilon)} \geq 1/2$. Then $h_2^{(\varepsilon)}$ is larger than $1/2$ on the same cells, and also on some other cells $B_{r_1+1}, \dots, B_{r_2}$. Iterating this construction, we obtain an infinite collection of cells $B_1, \dots, B_{r_1}, \dots, B_{r_2}, \dots$ with no interior points in common. Since

$$\sup_m h_m^{(\varepsilon)}(x) \geq 1 \quad \text{for every } x \in Z,$$

the set Z is covered by all these cells. To complete the proof, we need to compute the sum of the volumes of such cells. We consider only the cells $B_1, \dots, B_{r_m}$ on which $h_m^{(\varepsilon)} \geq 1/2$, it follows from $I h_m^{(\varepsilon)} < \varepsilon$ that

$$\sum_{k=1}^{r_m} \text{Vol}(B_k) \leq 2\varepsilon.$$

Letting $m \to +\infty$ we derive $\sum_{k=1}^{\infty} \text{Vol}(B_k) \leq 2\varepsilon$. We must now remark that the cells B_k just considered may not cover Z, since the points of Z need not lie in the interior of such cells. But this is not an obstruction, since we may replace B_k by a concentric cell B_k' such that $\text{Vol } B_k' = 2\,\text{Vol}(B_k)$, $Z \subset \bigcup_{k=1}^{\infty} B_k'$ and $\sum_{k=1}^{\infty} \text{Vol}(B_k') \leq 4\varepsilon$. Hence (a) holds, and the proof is complete.   □

15.10   Integration on Product Spaces

Every student knows that $\mathbb{R}^n = \mathbb{R} \times \mathbb{R} \times \cdots \times \mathbb{R}$ (n times). So a natural question is whether the integral on $\mathbb{R}^n$ can be reduced to the integral on $\mathbb{R}$ by some "product rule." The answer is essentially affirmative, and we present it in our style: first an abstract statement, then a concrete construction.

Theorem 15.19 (Abstract Fubini Theorem) *Let X and Y be two sets, and let $W = X \times Y$ be their cartesian product. Assume that $L(X)$, $L(Y)$, $L(W)$ are spaces of integrable functions on each set, and that I_X, I_Y, $I_W = I$ are the corresponding abstract integrals. Assume moreover that the family $H(W)$ of elementary functions which generate $L(W)$ has the following properties:*

(a) for every function $h \in H(W)$, the function $x \mapsto h(x, y)$ belongs to $L(X)$ for almost every $y \in Y$;

(b) for every $x \in X$, the function $y \mapsto I_X h(x, y)$ belongs to $L(Y)$;

(c) $Ih = I_Y(I_X h)$.

The family $L(W)$ has the same properties: every function $\varphi \in L(W)$ is integrable in the first variable for almost every value of the second variable, the integral $I_X\varphi(x, \cdot)$ is integrable on Y, and $I\varphi = I_Y(I_X\varphi)$.

Proof The proof is long, so we split it into several steps. We define the class Φ of all functions $\varphi \in L(W)$ for which the conclusion is true. We will eventually show that $\Phi = L(W)$. Since Φ contains all elementary functions by assumption, we only need to prove the inclusion $L(W) \subset \Phi$.

Step 1. Φ is closed under the formation linear combinations. This is trivial, and we omit the details.

Step 2. Φ is closed under monotonic limits. Let $\{\varphi_n\}_n$ be a sequence in Φ which is monotonic, and suppose that the sequence of the integrals $I\varphi_n$ remains bounded. Then the pointwise limit $\varphi = \lim_{n\to+\infty} \varphi_n$ belongs to Φ. Indeed, suppose for the sake of definiteness that $\{\varphi_n\}_n$ is non-decreasing, and put $g_n(y) = I_N\varphi_n(\cdot, y)$. The the sequence $\{g_n\}_n$ is also non-decreasing and the sequence of integrals $I_Y g_n$ remains bounded. Furthermore

$$I_Y g_n = I_Y(I_X\varphi_n) = I\varphi_n \nearrow I\varphi.$$

An argument based on Beppo Levi's Theorem shows that g_n converges to an integrable function g which is almost everywhere finite, and moreover $I_Y g = \lim_{n\to+\infty} I_Y g_n = I\varphi$. Let E be the subset of Y of full measure on which the function g is finite, and fix $y \in E$. The sequence $\{\varphi(\cdot, y)\}_n$ is non-decreasing and the sequence of integrals $I_X\varphi_n(\cdot, y)$ remains bounded: $I_X\varphi_n(\cdot, y) = g_n \nearrow g$. Again the limit $\varphi(\cdot, y)$ is integrable on X and

$$\lim_{n\to+\infty} I_X\varphi_n(\cdot, y) = g(y) = I_X\varphi(\cdot, y).$$

But then $I\varphi = I_Y g = I_Y(I_X\varphi)$, and thus $\varphi \in \Phi$.

Step 3. Φ contains every function z which is almost everywhere equal to zero. Let $Z \subset W$ be the set of measure zero on which z is not zero. As a first step, we assume that z takes on values between 0 and 1. For every $m \in \mathbb{N}$ we select a non-decreasing sequence $n \mapsto h_{m,n}$ of non-negative elementary functions such that

$$I h_{m,n} < \frac{1}{m}, \quad \lim_{n\to+\infty} h_{m,n} \geq 1 \text{ on } Z.$$

Replacing $h_{m+1,n}$ by $\min\{h_{m+1,n}, h_{m,n}\}$ we may also assume that $h_{m+1,n} \leq h_{m,n}$. The limit $h_m = \lim_{n\to+\infty} h_{m,n}$ is a monotonic limit of elementary functions of Φ, and then belongs to Φ by Step 2. For the very same reason,

$h = \lim_{m \to +\infty} h_m$ belongs to Φ, and

$$Ih_m = \lim_{n \to +\infty} Ih_{m,n} \le \frac{1}{m}$$

$$Ih = \lim_{m \to +\infty} Ih_m = 0.$$

Now $h_m \ge z$ on Z implies $h \ge z$ on Z, and setting $y \mapsto g(y) = I_X h(\cdot, y)$ we see that

$$I_Y g = I_Y(I_X h) = Ih = 0$$

by Step 2. Therefore $g = 0$ almost everywhere. For almost every $y \in Y$ we deduce that $h(\cdot, y)$ is almost everywhere equal to zero, and the same must hold for z. It follows that $I_X z(\cdot, y)$, and hence

$$Iz = 0 = I_Y(I_X z).$$

We need to remove the condition on the range of z. Let $z \ge 0$ be an arbitrary function that vanishes outside Z, and let

$$\ell = \begin{cases} 1 & \text{on } Z \\ 0 & \text{on } W \setminus Z. \end{cases}$$

There results

$$z = \lim_{n \to +\infty} n \min \left\{ \ell, \frac{1}{n} z \right\},$$

so that $z \in \Phi$ by the previous case. Finally, writing $z = z^+ - z^-$ recovers the case of variable sign.

Step 4.   Every element of $L^+(W)$ belongs to Φ. Indeed, by definition every $f \in L^+(W)$ is the almost everywhere non-decreasing limit of a sequence $\{h_n\}_n$ of elementary functions:

$$h_n \nearrow f, \quad Ih_n \nearrow If.$$

We call $\hat{f}$ the limit of the sequence defined by

$$\hat{h}_1 = h$$

$$\hat{h}_2 = \max\{h_1, h_2\}$$

$$\cdots$$

$$\hat{h}_n = \max\{h_1, h_2, \ldots, h_n\}$$

$$\cdots$$

This sequence is everywhere non-decreasing, and the functions $\hat{h}_n$, h_n coincide almost everywhere, so that $I\hat{h}_n = Ih$. As a result, the function f and the function $\hat{f}$ coincide almost everywhere, and we can write $\hat{f} = f + z$ for some function z which is almost everywhere equal to zero. By Steps 2 and 3, both $\hat{f}$ and z belong to Φ, so that $f \in \Phi$ by Step 1.

Step 5. Φ contains every function $\varphi \in L(W)$. Indeed, φ can be written as the difference of two elements of $L^+(W)$.

$\square$

Theorem 15.20 (Tonelli)   *Suppose that φ is a measurable function on W such that $\varphi \geq 0$. If the iterated integral $I_Y(I_X\varphi)$ exists, then $\varphi \in L(W)$ and*

$$I\varphi = I_Y(I_X\varphi).$$

Proof Call $A = I_Y(I_X\varphi)$, where $\varphi = \lim_{n\to+\infty} h_n$ for some elementary functions h_n. Then the functions

$$\varphi_n = \min\{\varphi, \max\{h_1, \ldots, h_n\}\}$$

are measurable; furthermore the function φ_n is dominated by the integrable function $\max\{h_1, \ldots, h_n\}$. Hence φ_n is integrable on W, and

$$I\varphi_n = I_Y(I_X\varphi_n) \leq A$$

by Fubini's Theorem. Since $\varphi_n \nearrow \varphi$ and $I\varphi_n \leq A$, it follows from Beppo Levi's Theorem that $\varphi \in L(W)$. A second application of Fubini's Theorem completes the proof.

$\square$

Exercise 15.9   Consider the integrals

$$\int_{(0,+\infty)\times(0,+\infty)} e^{-xy} \sin x \sin y \, dx \, dy \tag{15.4}$$

$$\int_{(0,1)\times(0,1)} \frac{x^2 - y^2}{(x^2 + y^2)^2} \, dx \, dy, \tag{15.5}$$

with integrals computed in the concrete Lebesgue sense. Prove that the corresponding iterated integrals exist for either order of integration, and that they coincide for (15.4) but differ for (15.5). Finally, prove that both integrands are not integrable.

Taking account of the last exercise, we may say that Theorem 15.20 is a partial converse to Theorem 15.19, and that the non-negativity assumption in Theorem 15.20 is optimal.

Let us stop for a moment. In the two main theorems of this Section, the existence of a class of integrable functions was *assumed*. The main issue is now to *construct* such a class and the corresponding integral.

Definition 15.17 Let X, Y be sets equipped with abstract integrals I_X and I_Y. If $W = X \times Y$, the class $H(W)$ of elementary functions on the product W consists of all functions of the form

$$h: (x, y) \in W \mapsto \sum_{j=1}^{m} \alpha_j \chi_{E_j}(x) \chi_{F_j}(y),$$

where $m \in \mathbb{N}$, $\alpha_j \in \mathbb{R}$, and the sets $E_j \subset X$, $F_j \subset Y$ are integrable sets (i.e. $\chi_{E_j} \in L(X)$, $\chi_{F_j} \in L(Y)$) for every $j = 1, \ldots, m$.

Remark 15.4 It is clear that we can always assume $E_i \cap E_j = \emptyset$, $F_i \cap F_j = \emptyset$ whenever $i \neq j$.

Exercise 15.10 Prove that $H(W)$ satisfies the axioms of Definition 15.3. *Hint:* if $h(x, y) = \sum_{j=1}^{m} \alpha_j \chi_{E_j}(x) \chi_{F_j}(y)$ is an element of $H(W)$ such that $E_i \cap E_j = \emptyset$, $F_i \cap F_j = \emptyset$ whenever $i \neq j$, observe that $|h(x, y)| = \sum_{j=1}^{m} |\alpha_j| \chi_{E_j}(x) \chi_{F_j}(y)$.

Definition 15.18 The elementary integral I on $L(W)$ is defined by

$$Ih = \sum_{j=1}^{m} \alpha_j \mu_X(E_j) \mu_Y(F_j)$$

for every $h \in H(W)$ of the form $h(x, y) = \sum_{j=1}^{m} \alpha_j \chi_{E_j}(x) \chi_{F_j}(y)$. Of course $\mu_X(E_j) = I_X \chi_{E_j}$ and $\mu_Y(F_j) = I_Y \chi_{F_j}$.

Concerning Definition 15.4, we prove that $Ih_n \to 0$ if $h_n \searrow 0$ in $H(W)$, since the linearity and the positivity of I are trivial. We claim that it suffices to prove that $I_X h_n(\cdot, y) \searrow 0$ for every y. Indeed if this is the case, then $I_Y(I_X h_n) \to 0$ by Beppo Levi's Theorem. But the assumption $h_n \searrow 0$ (everywhere) implies $I_X h_n(\cdot, y) \searrow$ for every $y \in Y$ by Levi's Theorem, and the claim is proved.

Once the elementary integral on $W = X \times Y$ has been defined, our abstract machinery yields a class $L(W)$ of integrable functions and an associated integral I which satisfies all the assumptions of Fubini's Theorem.

15.11 Spaces of Integrable Functions

Let us start with an easy remark: if a function f is measurable, then $|f|^p$ is measurable for every fixed $p > 0$. Indeed, if c denotes any (positive) real number, then

$$\left\{ x \in X \mid |f(x)|^p > c \right\} = \left\{ x \in X \mid |f(x)| > c^{1/p} \right\}.$$

Definition 15.19 For every $p > 0$, the set $L^p = L^p(X)$ is defined to be the set of all measurable functions f for which

$$I\left(|f|^p\right) = \int_X |f|^p \, d\mu \in \mathbb{R}.$$

Theorem 15.21 *For every $p > 0$, $L^p(X)$ is a vector space.*

Proof Suppose that $f \in L^p$, $g \in L^p$. We already know that $f + g$ is measurable; since

$$|f + g| \leq \left(|f|^p + |g|^p\right) \leq (2 \sup\{|f|, |g|\})^p$$
$$= 2^p \sup\left\{|f|^p, |g|^p\right\}$$
$$\leq 2^p \left(|f|^p + |g|^p\right),$$

we see that $f + g \in L^p$. The fact that any real multiple of f belongs to L^p is trivial, and the proof is complete.   □

Definition 15.20 The standard norm of $L^p(X)$ is defined by

$$\|f\|_p = \left(I \, |f|^p\right)^{1/p} = \left(\int_X |f|^p \, d\mu\right)^{1/p}.$$

The fact that $\|\cdot\|_p$ is actually a norm follows from two fundamental inequalities.

Theorem 15.22

(a) *(Hölder's inequality) If $f \in L^p$ and $g \in L^q$ for some numbers $p > 1$, $q > 1$ such that*

$$\frac{1}{p} + \frac{1}{q} = 1,$$

then

$$\int_X |fg| \, d\mu \leq \|f\|_p \|g\|_q.$$

(b) *(Minkowski's inequality) If $f \in L^p$, $g \in L^p$, then*

$$\|f + g\|_p \leq \|f\|_p + \|g\|_p.$$

Proof Let us write

$$A = \|f\|_p, \quad B = \|g\|_q.$$

If $A = 0$ then $f = 0$ almost everywhere, and $fg = 0$ almost everywhere. Therefore we need to consider only $A > 0$ and $B > 0$. We introduce

$$F = \frac{|f|}{A}, \quad G = \frac{|g|}{B},$$

so that $\int_X F^p \, d\mu = 1 = \int_X G^q \, d\mu$. If $x \in X$ is such that $0 < F(x) < +\infty$ and $0 < G(x) < +\infty$, then there exist real numbers s and t such that

$$F(x) = e^{s/p}, \quad G(x) = e^{t/q}.$$

The convexity of the exponential function yields

$$e^{\frac{s}{p}+\frac{t}{q}} \leq \frac{e^s}{p} + \frac{e^t}{q}.$$

It follows that

$$F(x)G(x) \leq \frac{F(x)^s}{p} + \frac{G(x)^q}{q}.$$

We integrate this inequality and get (a). The proof of (b) follows easily from (a). Indeed, let us write

$$|f + g|^p = |f + g|^{p-1} \cdot |f + g| \leq |f| \, |f + g|^{p-1} + |g| \, |f + g|^{p-1}.$$

By Hölder's inequality,

$$\int_X |f| \, |f + g|^{p-1} \, d\mu \leq \left(\int_X |f|^p \, d\mu \right)^{1/p} \left(\int_X |f + g|^{(p-1)q} \, d\mu \right)^{1/q}.$$

Since $(p - 1)q = p$, we conclude that

$$\left(\int_X |f + g|^p \, d\mu \right)^{1/p} \leq \left(\int_X |f|^p \, d\mu \right)^{1/p} + \left(\int_X |g|^p \, d\mu \right)^{1/p}.$$

□

Theorem 15.23 (Riesz-Fischer) *The space $L^p(X)$ is complete.*

Proof Let $\{\varphi_n\}_n$ be a Cauchy sequence in L^p. It is sufficient to prove that some subsequence converges in L^p, since the whole sequence will then converge to the same limit. We first find indices $n_1 < n_2 < n_3 < \ldots$ of positive integers such that

$$\left\| \varphi_{n_{k+1}} - \varphi_{n_k} \right\| < \frac{1}{2^k}, \quad k = 1, 2, \ldots.$$

Hence the series $\sum_{k=1}^{\infty} |\varphi_{n_{k+1}} - \varphi_{n_k}|$ converges almost everywhere. Indeed,

$$\left\| \sum_{k=1}^{N} |\varphi_{n_{k+1}} - \varphi_{n_k}| \right\|_p < \sum_{k=1}^{N} \|\varphi_{n_{k+1}} - \varphi_{n_k}\|_p < \sum_{k=1}^{N} \frac{1}{2^k} < 1$$

for every $N > 1$, and the claim follows from Beppo Levi's Theorem.

As a consequence, the series $\sum_{k=1}^{\infty} (\varphi_{n_{k+1}} - \varphi_{n_k})$ converges almost everywhere with partial sums

$$\sum_{k=1}^{N} (\varphi_{n_{k+1}} - \varphi_{n_k}) = \varphi_{n_{N+1}} - \varphi_{n_1}.$$

This means that the sequence $\{\varphi_{n_k}\}_k$ has a limit φ almost everywhere. For any fixed k, the function $\varphi_{n_j} - \varphi_{n_k}$ approaches $\varphi - \varphi_{n_k}$ almost everywhere as $j \to +\infty$. Since

$$\|\varphi_{n_j} - \varphi_{n_k}\|_p < \frac{1}{2^k}, \quad k = 1, 2, \ldots$$

it follows that $\varphi - \varphi_{n_k} \in L^p$, and then $\varphi \in L^p$. Taking the limit as $j \to +\infty$ we also derive

$$\|\varphi - \varphi_{n_k}\| < \frac{1}{2^k}.$$

Letting $k \to +\infty$ we see that $\varphi_{n_k} \to \varphi$ in L^p, and the proof is complete.   □

The previous proof hides a statement of fundamental importance in several aspects of Functional Analysis. We record it as follows.

Theorem 15.24 (Partial Converse of the Dominated Convergence Theorem)
Let $p \geq 1$ and let $\varphi_n \to \varphi$ in $L^p(X)$. There exists a subsequence $\psi_k = \varphi_{n_k}$ and a function $g \in L^p(X)$ such that $|\psi_k| \leq g$ and $\psi_k \to \varphi$ almost everywhere.

Proof Since $\{\varphi_n\}_n$ converges in L^p, it is a Cauchy sequence. The subsequence $\{\psi_k\}_k = \{\varphi_{n_k}\}_k$ of the previous proof converges almost everywhere to φ, and for every index k,

$$|\psi_k| \leq |\psi_1| + \sum_{k=1}^{\infty} |\psi_{k+1} - \psi_k| \in L^p(X).$$

□

Theorem 15.25 (Density of Elementary Functions) *The collection of elementary functions H is dense in $L^p(X)$.*

Proof Let $h \in H$. We first prove that $h \in L^p$. Since elementary functions are bounded, we assume that $|h(x)| \leq M$ for every $x \in X$, and we consider the set $E = \{x \in X \mid |h(x)| > 1\}$. Then E is measurable with a finite measure, and

$$|h(x)|^p \leq \begin{cases} M^p & \text{if } x \in E \\ |h(x)| & \text{if } x \notin E. \end{cases}$$

In other words,

$$|h(x)|^p \leq M^p \chi_E(x) + |h(x)| \chi_{X \setminus E}(x) \leq M^p \chi_E(x) + |h(x)|.$$

It follows at once that $|h|^p$ is integrable.

Let now $f \in L^p(X)$ be given. Since both f^+ and f^- belong to L^p, without loss of generality we suppose that $f \geq 0$. Define the increasing sequence of measurable sets

$$E_n = \left\{ x \in X \mid \frac{1}{n} < f(x) < n \right\}, \quad n = 1, 2, \ldots$$

and set

$$f_n(x) = \begin{cases} f(x) & \text{if } x \in E_n \\ 0 & \text{otherwise.} \end{cases}$$

Obviously $f_n \nearrow f$ and $(f - f_n)^p \searrow 0$, and therefore Beppo Levi's Theorem yields $f_n \to f$ in L^p. Given $\varepsilon > 0$ we can choose a positive integer n such that

$$\|f - f_n\|_p < \frac{\varepsilon}{2}.$$

Since $\chi_{E_n} \leq \chi_{E_n}^p \leq n^p f^p$, the function f_n is integrable by Hölder's inequality:

$$\int_X f_n \, d\mu = \int_X \chi_{E_n} f \, d\mu \leq \left\| \chi_{E_n}^q \right\|_q \left\| f^p \right\|_p.$$

Assume that H is dense in L^1. Then a sequence of elementary functions h_k exists such that $h_k \to f_n$ in L^1 as $k \to +\infty$. Replacing h_k by h_k^+ we may also assume that $h_k \geq 0$. Replacing h_k by

$$\min \{h_n, n\} = n \min \left\{ \frac{1}{n} h_k, 1 \right\},$$

we may also assume that $|h_k| \leq n$. Here we are using Stone's axiom. We claim that $h_k \to f_n$ in L^p. Indeed,

$$\|f_n - h_k\|_p^p = \int_X |f_n - h_k|^p \, d\mu = \int_X |f_n - h_k|^{p-1} |f_n - h_k| \, d\mu$$

$$< n^{p-1} \int_X |f_n - h_k| \to 0$$

as $k \to +\infty$, and we can choose k so large that

$$\|f_n - h_k\|_p < \frac{\varepsilon}{2}.$$

We conclude that

$$\|f - h_k\|_p < \|f_n - h_k\|_p + \|f - f_n\|_p < \varepsilon.$$

To complete the proof, it remains to show that H is dense in $L^1(X) = L(X)$. But every function in L is the difference of two functions in L^+, and we need to prove the claim for functions $f \subset L^+$ which are limits (in the norm of L^1) of a sequence of functions $h_n \in H$. The natural choice for this sequence is the sequence which defines f. Then $h_n \nearrow f$, $Ih_n \nearrow If$, and

$$\|f - h_n\|_1 = I(f - h_n) = If - Ih_n \to 0$$

as $n \to +\infty$.   □

Theorem 15.26 (Generalized Hölder's Inequality) *Suppose* $1 < p_j < +\infty$, $u_j \in L^{p_j}(X)$ *for* $1 \leq j \leq k$. *If*

$$\frac{1}{p_1} + \frac{1}{p_2} + \cdots + \frac{1}{p_k} = 1,$$

then $u_1 u_2 \cdots u_k \in L^1$ *and*

$$\int_X u_1 u_2 \cdots u_k \, d\mu \leq \|u_1\|_{p_1} \cdots \|u_k\|_{p_k}.$$

Proof Exercise.   □

Theorem 15.27 (Interpolation Inequality) *Suppose that* $1 \leq p < q < r < +\infty$,

$$\frac{1}{q} = \frac{1 - \lambda}{p} + \frac{\lambda}{r}$$

and $u \in L^p(X) \cap L^r(X)$. *Then* $u \in L^q(X)$ *and*

$$\|u\|_p \le \|u\|_p^{1-\lambda} \|u\|_r^{\lambda}.$$

Proof Exercise. □

Theorem 15.28 *Suppose that* $p \ge 1$ *and* $\{u_n\}_n$ *is a sequence in* $L^p(X)$ *such that*

(a) $\|u_n\|_p \to \|u\|_p$ *as* $n \to +\infty$,
(b) $u_n \to u$ *almost everywhere as* $n \to +\infty$.

Then $\|u_n - u\|_p \to 0$ *as* $n \to +\infty$.

Proof We have almost everywhere

$$0 \le 2^p \left(|u_n|^p + |u|^p\right) - |u_n - u|^p.$$

Fatou's Lemma implies

$$2^{p+1} \int_X |u|^p \, d\mu \le \liminf_{n \to +\infty} \int_X \left[2^p \left(|u_n|^p + |u|^p\right) - |u_n - u|^p \right] d\mu$$

$$= 2^{p+1} \int_X |u|^p \, d\mu - \limsup_{n \to +\infty} \int_X |u_n - u|^p \, d\mu.$$

We conclude that $\limsup_{n \to +\infty} \int_X |u_n - u|^p \, d\mu \le 0$, and the proof is complete. □

Theorem 15.29 (Brexis-Lieb) *Suppose that* $p \ge 1$ *and* $\{u_n\}_n$ *is a sequence in* $L^p(X)$ *such that*

(a) $c = \sup_n \|u_n\|_p \in \mathbb{R}$,
(b) $u_n \to u$ *almost everywhere as* $n \to +\infty$.

Then $u \in L^p(X)$ *and*

$$\lim_{n \to +\infty} \left(\|u_n\|_p^p - \|u_n - u\|_p^p \right) = \|u\|_p^p.$$

Proof The assumptions imply $|u\|_p \le c$. Fix any $\varepsilon > 0$. By homogeneity, there exists a number $C(\varepsilon) > 0$ such that for every a, b in $\mathbb{R}$,

$$\left| |a + b|^p - |a|^p - |b|^p \right| \le \varepsilon |a|^p + C(\varepsilon)|b|^p.$$

Again by Fatou's Lemma,

$$\int_X C(\varepsilon)|u|^p \, d\mu \le \liminf_{n \to +\infty} \int_X \varepsilon |u_n - u|^p + C(\varepsilon)|u|^p$$

$$- \left| |u_n|^p - |u_n - u|^p - |u|^p \right| d\mu$$

$$\leq (2c)^p \varepsilon + \int_X C(\varepsilon)|u|^p \, d\mu$$

$$- \limsup_{n \to +\infty} \int_X \Big| |u_n|^p - |u_n - u|^p - |u|^p \Big| \, d\mu,$$

which means

$$\limsup_{n \to +\infty} \int_X \Big| |u_n|^p - |u_n - u|^p - |u|^p \Big| \, d\mu \leq (2c)^p \varepsilon.$$

Since $\varepsilon > 0$ is arbitrary, the proof is complete.   □

15.12   The Space L^∞

The formal case $p = \infty$ of L^p gives rise to a very different space of functions.

Definition 15.21 Suppose that $f : X \to [0, +\infty]$ be a measurable function. We consider the set

$$S = \Big\{ \alpha \in \mathbb{R} \ \Big| \ \mu(g^{-1}((\alpha, +\infty])) = 0 \Big\}.$$

If $S = \emptyset$, we put $\beta = +\infty$. If $S \neq \emptyset$, we put $\beta = \sup S$. The number β is called the essential supremum of f.

Exercise 15.11 Prove that $\beta \in S$. *Hint:* $g^{-1}((\beta, +\infty]) = \bigcup_{n=1}^{\infty} g^{-1}\left(\left(\beta + \frac{1}{n}, +\infty\right]\right)$.

Definition 15.22 A measurable function $f : X \to \mathbb{R}$ belongs to $L^\infty(X)$ if and only if the essential supremum of $|f|$ is a finite real number. In this case we write $\|f\|_\infty$ to denote this essential supremum. In symbols,

$$L^\infty(X) = \{ f \mid \|f\|_\infty < +\infty \}$$

$$= \left\{ f \ \middle| \ \begin{array}{l} f \text{ is measurable and there exists} \\ C \in \mathbb{R} \text{ such that } |f| \leq C \text{ almost} \\ \text{everywhere} \end{array} \right\}.$$

The inequalities of Hölder and Minkowski can be extended to the L^∞-case without much pain. For example,

Theorem 15.30 *If $f \in L^1(X)$ and $g \in L^\infty(X)$, then $fg \in L^1(X)$ and $\|fg\|_1 \leq \|f\|_1 \|g\|_\infty$.*

Proof For almost every $x \in X$ we have $|f(x)g(x)| \leq \|f(x)\| \|g\|_\infty$. The conclusion follows by integration of this inequality.   □

The completeness of L^∞ is a deeper result.

Theorem 15.31 *The space $L^\infty(X)$ is a Banach space.*

Proof Suppose that $\{f_n\}_n$ is a Cauchy sequence in $L^\infty(X)$, and set

$$A_k = \{x \in X \mid |f_k(x)| > \|f_k\|_\infty\}$$
$$B_{m,n} = \{x \in X \mid |f_m(x) - f_n(x)| > \|f_m - f_n\|_\infty\}$$
$$E = \bigcup \{A_k \cup B_{n,m} \mid k, m, n \in \mathbb{N}\}.$$

As a countable union of sets of measure zero, we have $\mu(E) = 0$. On $X \setminus E$ we have that $f_n \to f$ uniformly to a bounded limit f. If we set $f(x) = 0$ for every $x \in E$, the function f belongs to $L^\infty(X)$, and $\|f_n - f\|_\infty \to 0$ as $n \to +\infty$. The proof is complete. □

15.13   Changing Variables in Multiple Integrals

Following [2], we quickly discuss an important technique of Advanced Calculus.

Definition 15.23 Let Ω and ω be open subsets of $\mathbb{R}^n$. A diffeomorphism $f: \Omega \to \omega$ is a continuously differentiable bijective function such that for every $x \in \Omega$,

$$J_f(x) = \det Df(x) \neq 0.$$

Here $Df(x)$ denotes the *Jacobian matrix* of f at $x \in \Omega$.

Theorem 15.32 (Change of Variables) *Let $f: \Omega \to \omega$ be a diffeomorphism and $u \in C_c(\omega)$. The $u \circ f \in C_c(\Omega)$, and*

$$\int_\Omega u(f(x))|J_f(x)|\, dx = \int_\omega u(y)\, dy.$$

Notice that we are denoting by a symbol like $\int \ldots dx$ the Lebesgue integral in $\mathbb{R}^n$.

Proof We proceed by induction on the dimension $n \geq 1$.

1. $n = 1$. In this case we may assume that $\Omega = (a, b)$. By the Fundamental Theorem of Calculus,

$$\int_a^b u(f(x))f'(x)\, dx = \int_{f(a)}^{f(b)} u(y)\, dy.$$

If $f' > 0$, then $\omega = (f(a), f(b))$. If $f' < 0$, then $\omega(f(b), f(a))$. In both cases the claim is proved.

2. The induction step. We assume the result has been proved in dimension $n-1$. Let $a \in \Omega$ be a fixed point. Since f is a diffeomorphism, $(\partial_1 f_1(a), \ldots, \partial_n f_n(a)) \neq 0$. Without loss of generality we may assume that $\partial_n f_n(a) \neq 0$. By the Implicit Function Theorem, there exist $r > 0$, an open set $U \subset \Omega$ such that $a \in U$, an open set $V \subset \mathbb{R}^{n-1}$, and a function $\beta \in C^1(V \times (f_n(a) - r, f_n(a) + r))$ such that for $|t - f_n(a)| < r$ there results

$$\{f_n = t\} \cap U = \left\{(x', \beta(x', t)) \mid x' \in V\right\}. \tag{15.6}$$

We now split[3]

$$f = (f', f_n)$$
$$h(x', x_n) = (x', f_n(x', x_n))$$
$$\Phi_t(x') = f'(x', \beta(x', t))$$
$$g(x', t) = (\Phi_t(x'), t).$$

We add the assumption that $\operatorname{supp} u \subset U$. Fubini's Theorem and the induction hypothesis ensure that[4]

$$\int u(g(x))|J_g(x)|\, dx = \int dt \int_V u(\Phi_t(x'), t)|J_{\Phi_t}(x')|\, dx'$$
$$= \int dt \int u(y', t)\, dy'$$
$$= \int u(y)\, dy.$$

Now let $v = (u \circ g)|J_g|$. Fubini's Theorem and the proof in dimension $n = 1$ imply that

$$\int v(h(x))|J_h(x)|\, dx = \int dx' \int v(x', f_n(x', x_n))|\partial_n f_n(x', x_n)|\, dx_n$$
$$= \int dx' \int v(x', t)\, dt$$
$$= \int u(g(x))|J_g(x)|\, dx.$$

---

[3] The prime $'$ does not mean differentiation. It is used to group the first $n-1$ components of a vector in $\mathbb{R}^n$.

[4] All the integrals in the next equations may be computed on $\mathbb{R}^n$, since the integrand functions have compact support.

Recalling that $f = g \circ h$ on U, we get $Df = Dg(h)Dh$ and $J_f = J_g(h)J_h$. Therefore

$$\int u(f(x))|J_f(x)|\,\mathrm{d}x = \int u(g(h(x)))|J_g(x)|\,|J_h(x)|\,\mathrm{d}x$$

$$= \int v(h(x))|J_h(x)|\,\mathrm{d}x$$

$$= \int u(y)\,\mathrm{d}y.$$

To conclude the proof, we need to remove the condition that the support of u be contained in U. By assumption the support of u is a compact subset of $\mathbb{R}^n$, and we can cover it by a finite collection of open sets U_j which satisfy (15.6). If $\{\psi_j\}_j$ is a finite partition of unity subordinated to the covering $\{U_j\}_j$, we have $u = \sum_j \psi_j u$, and the general case follows from the linearity of the integral.

$\square$

15.14   Comments

The construction of the abstract Lebesgue integral dates back to P.J. Daniell in 1918. It has the typical elegance of formal axiomatizations, which isolate the essential properties of the object we want to define. A short but complete reference is [1]. The clean approach of [2] is another interesting source.

To be fairly honest, many analysts are satisfied with the basic Lebesgue integral. If this is the main purpose, the approach we present in the next chapter might be preferable. It is nonetheless interesting that the theory of integration can be completely developed in terms of a functional-analytic *completion* process: we construct the Lebesgue integral from the Riemann integral in the same way as we construct the real numbers from the rational ones.

References

1. G.E. Shilov, B.L. Gurevich, *Integral, Measure and Derivative: A Unified Approach* (Dover Publications, New York, 1977)
2. M. Willem, *Functional Analysis: Fundamentals and Applications* (Birkhäuser/Springer, New York, 2013)

Chapter 16
Measures Before Integrals

Abstract This chapter presents a straightforward approach to abstract measure theory. We will see that an abstract Lebesgue integral can be defined after introducing the idea of measuring sets. In the end we will connect these two approaches to integration theory.

The popular approach to the Lebesgue integral is via abstract measure theory: we first define a set function—called a *measure*—on a set of suitable sets—called *measurable* sets, then we define measurable functions, and finally integrable functions. The main pro of this approach is that at the end we have the highest generality. A troublesome con is that such a construction requires a good amount of mathematical education before it can be understood.

Our approach follows closely [3]. Historically, the Bourbaki group rejected abstract measure theory for a long time, since they decided to focus only on Radon measures defined as continuous linear functionals. There is some reason for doing this, but the stiffness of Radon measures is an obstacle to several investigations in Calculus of Variations, Geometric Measure Theory, Differential Geometry, and so on.

16.1 General Measure Theory

Definition 16.1 Let X be a set. A family Σ of subsets of X is a σ-algebra if and only if

1. $X \in \Sigma$,
2. if $A \in \Sigma$, then $X \setminus A \in \Sigma$,
3. if $A = \bigcup_{n=1}^{\infty} A_n$ and $A_n \in \Sigma$ for every n, then $A \in \Sigma$.

The elements of a σ-algebra Σ are called measurable sets, and X is a measure space.

© The Author(s), under exclusive license to Springer Nature Switzerland AG 2022
S. Secchi, *A Circle-Line Study of Mathematical Analysis*,
La Matematica per il 3+2 141, https://doi.org/10.1007/978-3-031-19738-3_16

Definition 16.2 Let X be a measure space and Y be a topological space. A function $f : X \to Y$ is measurable if and only if $f^{-1}(V)$ is measurable whenever V is open in Y.

Exercise 16.1 Let Σ be a σ-algebra on a set X. Prove the following statements:

1. $\emptyset \in \Sigma$.
2. Σ is closed under the formation of finite and countable intersections.
3. If $A \in \Sigma$ and $B \in \Sigma$, then $A \setminus B \in \Sigma$.

Theorem 16.1 *Let Y and Z be topological spaces, and let $g : Y \to Z$ be a continuous function. If X is a measure space, if $f : X \to Y$ is measurable, and if $h = g \circ f$, then $h : X \to Z$ is measurable.*

Proof If V is open in Z, then $g^{-1}(V)$ is open in Y, and $h^{-1}(V) = f^{-1}(g^{-1}(V))$. Since f is measurable, it follows that $h^{-1}(V)$ is measurable, and the proof is complete.   □

Theorem 16.2 *Let u and v be measurable functions on a measure space X, and let Φ be a continuous function from $\mathbb{R}^2$ to a topological space Y. Define*

$$h(x) = \Phi(u(x), v(x)), \qquad x \in X.$$

Then $h : X \to Y$ is measurable.

Proof Let us write $f(x) = (u(x), v(x))$ for $x \in X$. Since $h = \Phi \circ f$, Theorem 16.1 shows that we only need to prove that f is measurable. To this aim, we suppose that $R = (a, b) \times (c, d)$. It follows that

$$f^{-1}(R) = u^{-1}((a, b)) \cap v^{-1}((c, d)),$$

which is a measurable set by assumption. Since any open set $V \subset \mathbb{R}^2$ is a countable union of rectangles of the form $(a, b) \times (c, d)$, we deduce that $f^{-1}(V)$ is measurable, and the proof is complete.   □

Corollary 16.1 *Let X be a measure space. If f, g are real-valued measurable functions, then $f + g$, fg, $|f|$ are measurable.*

Proof We just consider $\Phi(s, t) = s + t$, $\Phi(s, t) = st$ in Theorem 16.2, or $g(x) = |x|$ in Theorem 16.1.   □

Theorem 16.3 *Let X be a measure space. A function $f : X \to \mathbb{R}$ is measurable if and only if any of the following statements holds true:*

1. *the set $\{x \in X \mid f(x) > a\}$ is measurable for every $a \in \mathbb{R}$;*
2. *the set $\{x \in X \mid f(x) \geq a\}$ is measurable for every $a \in \mathbb{R}$;*
3. *the set $\{x \in X \mid f(x) < a\}$ is measurable for every $a \in \mathbb{R}$;*
4. *the set $\{x \in X \mid f(x) \leq a\}$ is measurable for every $a \in \mathbb{R}$.*

Proof This follows immediately from the remark that the collection of all open half-lines of the form $(a, +\infty)$ or $(-\infty, b)$ is a sub-basis of the standard topology of $\mathbb{R}$.   □

Theorem 16.4 *If $\{f_n\}_n$ is a sequence of real-valued measurable functions on X, then $\inf_n f_n$, $\sup_n f_n$, $\liminf_{n\to+\infty} f_n$, $\limsup_{n\to+\infty} f_n$ are measurable.*

Proof For example,

$$\left\{ x \in X \ \Big| \ \sup_n f_n(x) > a \right\} = \bigcup_{n=1}^{\infty} \{x \in X \mid f_n(x) > a\},$$

so that $\sup_n f_n$ is measurable. Since $-\sup_n(-f_n) = \inf_n f_n$, $\limsup_{n\to+\infty} f_n = \inf_m \sup_{n\geq m} f_n$ and $\liminf_{n\to+\infty} f_n = \sup_m \inf_{n\geq m} f_n$, the conclusion follows.   □

Exercise 16.2 If f and g are real-valued measurable functions, prove that $\max\{f, g\}$ and $\min\{f, g\}$ are measurable.

Theorem 16.5 (Existence of σ-Algebras) *If $\mathcal{F}$ is any collection of subsets of X, then there exists a smallest σ-algebra Σ^* in X such that $\mathcal{F} \subset \Sigma^*$. This is called the σ-algebra generated by $\mathcal{F}$.*

Proof Since 2^X is trivially a σ-algebra, the set Ω of all σ-algebras containing $\mathcal{F}$ is not empty. Define

$$\Sigma^* = \bigcap \{\Sigma \mid \Sigma \in \Omega\}.$$

It is obvious that $\mathcal{F} \subset \Sigma^*$ and that Σ^* is contained in every σ-algebra which contains $\mathcal{F}$. The proof will be completed once we show that Σ^* is a σ-algebra.
    Suppose $A_n \in \Sigma^*$ for $n \in \mathbb{N}$. For every $\Sigma \in \Omega$, $A-n \in \Sigma$, so that $\bigcup_{n=1}^{\infty} A_n \in \Sigma$. But Σ is an arbitrary σ-algebra containing $\mathcal{F}$, and we conclude that $\bigcup_{n=1}^{\infty} A_n \in \Sigma^*$. A similar argument shows that $X \in \Sigma^*$ and that $A \in \Sigma^*$ implies $X \setminus A \in \Sigma^*$. Hence Σ^* is a σ-algebra, and the proof is complete.   □

If the set X is endowed with a topology, the previous result provides us with a σ-algebra compatible with open subsets.

Definition 16.3 Suppose that X is a topological space. The Borel σ-algebra $\mathcal{B}$ in X is the σ-algebra generated by the topology of X, i.e. by the collection of open subsets of X. Every member of $\mathcal{B}$ is a Borel set.

Definition 16.4 Suppose that X is a topological space. If a subset of X can be expressed as a countable union of closed sets, it is called a F_σ set. If a subset of X can be expressed as a countable intersection of open sets, it is called a G_δ set.

Exercise 16.3 Prove that any F_σ and any G_δ set are Borel sets.

Definition 16.5 Let (X, Σ) be a measure space. A function $\mu : \Sigma \to [0, +\infty]$ is a measure if and only if

1. $\mu(\emptyset) = 0$,
2. if $\{A_n \mid n \in \mathbb{N}\}$ is a sequence of measurable sets such that $A_i \cap A_j = \emptyset$ for $i \neq j$, then

$$\mu \left(\bigcup_{n=1}^\infty A_n \right) = \sum_{n=1}^\infty \mu(A_n).$$

The space X is of finite measure if and only if $\mu(X) < +\infty$. The space X is σ-finite if and only if it is the countable union of measurable subsets of finite measure: $X = \bigcup_{n=1}^\infty X_n$ with $\mu(X_n) < +\infty$ for every n.

Example 16.1 (Counting Measure) For every subset A of a set X, we define $\mu(A) = +\infty$ if A contains infinitely many elements, and $\mu(A) = \#A$ if A contains finitely many elements.

Example 16.2 (Dirac Measure) Let X be a set, and let $x_0 \in X$ be a fixed point. If $A \subset X$ contains x_0, we set $\mu(A) = 1$; otherwise we set $\mu(A) = 0$. This is the Dirac measure concentrated at x_0.

Definition 16.6 A subset E of a measure space (X, Σ, μ) has measure zero if and only if $E \in \Sigma$ and $\mu(E) = 0$. A property holds almost everywhere (a.e. for short) if and only if it holds in X except for a subset of measure zero.

Theorem 16.6 *Let (X, Σ, μ) be a measure space.*

(a) *If $\{A_n\}_n$ is a sequence of measurable subsets such that $A_n \subset A_{n+1}$ for every n, then*

$$\mu \left(\bigcup_{n=1}^\infty A_n \right) = \lim_{n \to +\infty} \mu(A_n).$$

(b) *If $\{A_n\}_n$ is a sequence of measurable subsets such that $A_n \supset A_{n+1}$ for every n, and if $\mu(A_1) < +\infty$, then*

$$\mu \left(\bigcap_{n=1}^\infty A_n \right) = \lim_{n \to +\infty} \mu(A_n).$$

Proof We may assume that $\mu(A_n) < +\infty$ for every n, otherwise the conclusion is trivial. From the identity

$$\bigcup_{n=1}^{\infty} A_n = A_1 \cup (A_2 \setminus A_1) \cup (A_3 \setminus A_2) \cup \cdots$$

we derive

$$\mu\left(\bigcup_{n=1}^{\infty} A_n\right) = \mu(A_1) + \mu(A_2 \setminus A_1) + \mu(A_3 \setminus A_2) + \cdots$$

$$= \mu(A_1) + \mu(A_2) - \mu(A_1) + \mu(A_3) - \mu(A_2) + \cdots$$

$$= \lim_{n \to +\infty} \mu(A_n).$$

This proves (a). To prove (b) we set $B_n = A_1 \setminus A_n$, so that $B_n \subset B_{n+1}$ for every n. Furthermore $\mu(B_n) = \mu(A_1) - \mu(A_n)$, and

$$\bigcup_{n=1}^{\infty} B_n = A_1 \setminus \bigcap_{n=1}^{\infty} A_n.$$

Hence (a) implies

$$\mu(A_1) - \mu\left(\bigcap_{n=1}^{\infty} A_n\right) = \mu\left(A_1 \setminus \bigcap_{n=1}^{\infty} A_n\right)$$

$$= \mu\left(\bigcup_{n=1}^{\infty} B_n\right) = \lim_{n \to +\infty} \mu(B_n)$$

$$= \mu(A_1) - \lim_{n \to +\infty} \mu(A_n).$$

Since $\mu(A_1) \in \mathbb{R}$, the conclusion follows.   □

Definition 16.7 A function $s \colon X \to \mathbb{R}$ defined on a set X is a simple function if and only if the image $s(X)$ is a finite set.

Exercise 16.4 Prove that any simple function s is a linear combination of characteristic functions. *Hint:* assume that $s(X) = \{c_1, c_2, \ldots, c_n\}$. For $1 \le i \le n$, set $E_i = \{x \in X \mid s(x) = c_i\}$. Conclude that $s = \sum_{i=1}^{n} c_i \chi_{E_i}$.

Theorem 16.7 *Let X be a measure space, and let s be a simple function whose range is $\{c_1, \ldots, c_n\}$. The function s is measurable if and only if the sets $E_i = \{x \in X \mid s(x) = c_i\}$ are measurable for $1 \le i \le n$.*

Proof This follows immediately from Theorem 16.3.   □

Simple function are the bricks with which we build integration theory. Let us start with an approximation result.

Theorem 16.8 *If $f : X \to \mathbb{R}$ is a function, then there exists a sequence $\{s_n\}_n$ of simple functions such that $s_n(x) \to f(x)$ for every $x \in X$. If f is measurable, then each s_n can be taken measurable. If $f \geq 0$, then we may suppose that $s_n \leq s_{n+1}$ for every n.*

Proof We begin with the case $f \geq 0$. For any $n \in \mathbb{N}$ and $i \in \{1, 2, \dots, n \cdot 2^n\}$ we define

$$E_{ni} = \left\{ x \in X \;\middle|\; \frac{i-1}{2^n} \leq f(x) \leq \frac{i}{2^n} \right\}$$

$$F_n = \{x \in X \mid f(x) \geq n\}$$

$$s_n(x) = \sum_{i=1}^{n \cdot 2^n} \frac{i-1}{2^n} \chi_{E_{ni}}(x) + n \chi_{F_n}(x).$$

It is easy to check that $s_n \to f$ pointwise in X, and that $s_n \leq s_{n+1}$ for every n. In the general case, we split $f = f^+ - f^-$ and repeat the same construction for f^+ and f^-. $\square$

Exercise 16.5 If f is a bounded function, prove that the sequence $\{s_n\}_n$ constructed above converges *uniformly* to f.

We can now introduce the (abstract) integral in a measure space (X, Σ, μ).

Definition 16.8 (Integral of Simple Functions) Let $E \subset X$ be a measurable set. The integral on E of a simple measurable function s such that

$$s = \sum_{i=1}^{n} c_i \chi_{E_i},$$

$c_i \geq 0$ for every i, and $c_i \neq c_j$ for $i \neq j$, is the number

$$I_E(s) = \sum_{i=1}^{n} c_i \mu(E \cap E_i).$$

We agree that $0 \cdot \infty = 0$. At this stage this is actually harmless: we have in mind the situation in which a set $E \cap E_i$ has infinite measure, while $c_i = 0$.

Definition 16.9 (Integral of Measurable Functions) Let $E \subset X$ be a measurable set. The integral on E of a measurable function $f : X \to [0, +\infty)$ is defined as

$$\int_E f \, d\mu = \sup \{ I_E(s) \mid s \text{ is a simple function s.t. } 0 \le s \le f \text{ in } E \}.$$

Definition 16.10 Let $f : X \to \mathbb{R}$ be a measurable function. We say that f is integrable on E if and only if

$$\int_E f^+ \, d\mu < +\infty, \qquad \int_E f^- \, d\mu < +\infty.$$

In this case we write $f \in L^1(E) = L^1(E, \mu)$, and

$$\int_E f \, d\mu = \int_E f^+ \, d\mu - \int_E f^- \, d\mu.$$

We collect several properties of the abstract integral. The easy proofs can be provided by the reader as an exercise.

Proposition 16.1 *Let X be a measurable space, and let E be a measurable subset of X.*

1. *If f is measurable and bounded on E, and if $\mu(E) < +\infty$, then $f \in L^1(X)$.*
2. *If $\mu(E) < +\infty$ and $a \le f \le b$ on E, then $a\mu(E) \le \int_E f \, d\mu \le b\mu(E)$.*
3. *If $f \in L^1(E)$ and $c \in \mathbb{R}$, then $cf \in L^1(E)$, and $\int_E (cf) \, d\mu = c \int_E f \, d\mu$.*
4. *If $f \in L^1(E)$, $g \in L^1(E)$ and $f \le g$ on E, then $\int_E f \, d\mu \le \int_E g \, d\mu$.*
5. *If $f \in L^1(E)$, A is measurable and $A \subset E$, then $f \in L^1(A)$.*
6. *If $\mu(E) = 0$ and f is measurable, then $\int_E f \, d\mu = 0$.*

Any measurable function induces a new measure by means of the integral. Here is the precise statement.

Theorem 16.9 *Suppose that f is a measurable function on a measurable space (X, Σ, μ) such that $f \ge 0$ on X. Then the function*

$$\nu : A \in \Sigma \mapsto \int_A f \, d\mu$$

is a measure on X.

Proof Consider any sequence $\{A_n\}_n$ of measurable sets such that $A_i \cap A_j = \emptyset$ for $i \ne j$, and $A = \bigcup_{n=1}^{\infty} A_n$. We need to prove that

$$\nu(A) = \sum_{n=1}^{\infty} \nu(A_n).$$

We suppose that $f = \chi_E$, where E is a measurable set. In this case the conclusion follows from the identity

$$\int_B \chi_E \, d\mu = \mu(B \cap E)$$

for every measurable B and from countable additivity of μ. It then follows that the same conclusion holds for every simple function f.

In the general case, let s be a simple measurable function such that $0 \le s \le f$ on X. We have

$$\int_A s \, d\mu = \sum_{n=1}^{\infty} \int_{A_n} s \, d\mu \le \sum_{n=1}^{\infty} v(A_n).$$

Hence

$$\int_A f \, d\mu = v(A) \le \sum_{n=1}^{\infty} v(A_n).$$

To prove the reversed inequality, we may clearly restrict to the case $v(A_n) \in \mathbb{R}$ for every n. Pick $\varepsilon > 0$ and two measurable simple functions s_1, s_2 such that

$$\int_{A_1} s_1 \, d\mu > \int_{A_1} f \, d\mu - \frac{\varepsilon}{2}$$

$$\int_{A_2} s_2 \, d\mu > \int_{A_2} f \, d\mu - \frac{\varepsilon}{2}.$$

Then

$$v(A_1 \cup A_2) \ge \int_{A_1 \cup A_2} \left(s_1 \chi_{A_1} + s_2 \chi_{A_2} \right) d\mu$$

$$= \int_{A_1} s_1 \, d\mu + \int_{A_2} s_2 \, d\mu \ge v(A_1) + v(A_2) - \varepsilon.$$

Since this holds for any $\varepsilon > 0$, we see that $v(A_1 \cup A_2) \ge v(A_1) + v(A_2)$. By induction

$$v(A_1 \cup \cdots \cup A_n) \ge \sum_{k=1}^{n} v(A_k) \quad n \ge 1,$$

and finally $v(A) \ge \sum_{n=1}^{\infty} v(A_n)$, since $A \supset \bigcup_{k=1}^{n} A_k$. Since the other properties of a measure are trivially satisfied, the proof is complete.   □

Exercise 16.6 Suppose that A and B are measurable sets such that $\mu(A \setminus B) = 0$. If f is integrable, prove that $\int_A f\, d\mu = \int_B f\, d\mu$.

Theorem 16.10 *Let f be a measurable function on X.*

(a) *If $f \in L^1(X)$, then $|f| \in L^1(X)$, and $\left|\int_X f\, d\mu\right| \le \int_X |f|\, d\mu$.*
(b) *If $g \in L^1(X)$ and $|f| \le g$ on X, then $f \in L^1(X)$.*

Proof Let $A = \{x \in X \mid f(x) \ge 0\}$ and $B = \{x \in X \mid f(x) < 0\}$. Clearly X is the disjoint union of A and B, and therefore

$$\int_X |f|\, d\mu = \int_A |f|\, d\mu + \int_B |f|\, d\mu$$

$$= \int_A f^+\, d\mu + \int_B f^-\, d\mu,$$

and the last two integrals are finite. Hence $|f| \in L^1(X)$. Recalling that $f \le |f|$ and $-f \le |f|$, we see that

$$\int_X f\, d\mu \le \int_X |f|\, d\mu, \quad -\int_X f\, d\mu \le \int_X |f|\, d\mu.$$

The proof of (a) is thus complete. To prove (b) we remark that $f^+ \le g$, $f^- \le g$, so that $f = f^+ - f^-$ is the difference of two integrable functions.   □

16.2   Convergence Theorems

We consider a fixed measurable space (X, Σ, μ).

Theorem 16.11 (Beppo Levi) *Suppose $\{f_n\}_n$ is a sequence of measurable functions such that $f_n \ge 0$ and $f_n \le f_{n+1}$ for every n. Let $E \in \Sigma$ and $f(x) = \lim_{n \to +\infty} f_n(x)$ for every $x \in E$. Then*

$$\int_E f\, d\mu = \lim_{n \to +\infty} \int_E f_n\, d\mu.$$

Proof The sequence $n \mapsto \int_E f_n\, d\mu$ is increasing, so there exists $\alpha \in [0, +\infty]$ such that $\alpha = \lim_{n \to +\infty} \int_E f_n\, d\mu$. Furthermore

$$\alpha \le \int_E f\, d\mu.$$

Let $0 < c < 1$ and s be a measurable simple function such that $0 \le s \le f$ on E. We define

$$E_n = \{x \in E \mid f_n(x) \ge cs(x)\}, \quad n \ge 1.$$

Of course each E_n is a measurable set, and $E_1 \subset E_2 \subset E_3 \subset \cdots$ Since $f_n \to f$ on E, $E = \bigcup_{n=1}^{\infty} E_n$. On the other hand, for each $n \ge 1$ we have

$$\int_E f_m \, d\mu \ge \int_{E_n} f_n \, d\mu \ge c \int_{E_n} s \, d\mu.$$

By Theorem 16.9 we may let $n \to +\infty$ and conclude that $\alpha \ge c \int_E s \, d\mu$. Since this holds for every $0 < c < 1$, we also have $\alpha \ge \int_E s \, d\mu$. Now the conclusion follows from the arbitrariness of the simple function s. $\qquad\square$

Remark 16.1 The integrals appearing in Beppo Levi's Theorem may well be infinite. It should be remembered as a statement about limit of measurable functions, without any reference to integrability properties.

Theorem 16.12 (Fatou's Lemma) *Suppose $f_n \ge 0$ is a measurable function on a set $E \in \Sigma$ for every n, and let $f(x) = \liminf_{n \to +\infty} f_n(x)$ for every $x \in E$. Then*

$$\int_E f \, d\mu \le \liminf_{n \to +\infty} \int_E f_n \, d\mu.$$

Proof For every $x \in E$ we set $g_n(x) = \inf_{m \ge n} \sup_{n \in \mathbb{N}} f_n(x)$. Clearly

$$0 \le g_1 \le g_2 \le g_3 \le \cdots$$

$$g_n \le f_n$$

$$\lim_{n \to +\infty} g_n(x) = f(x)$$

for every $x \in E$. Beppo Levi's Theorem yields

$$\int_E f \, d\mu = \lim_{n \to +\infty} \int_E g_n \, d\mu \le \liminf_{n \to +\infty} \int_E f \, d\mu.$$

$\qquad\square$

Beppo Levi's Theorem is a fundamental result for proving basic statements about integrable functions. We just provide an example.

Proposition 16.2 *Suppose that f_1 and f_2 are integrable functions on a measurable set E. If $f = f_1 + f_2$, then $f \in L^1(E)$ and*

$$\int_E f \, d\mu = \int_E f_1 \, d\mu + \int_E f_2 \, d\mu.$$

Proof Suppose initially that $f_1 \geq 0$ and $f_2 \geq 0$. Let $\{s_n^1\}_n$ and $\{s_n^2\}_n$ be sequences of measurable simple functions such that $s_n^1 \to f_1$, $s_n^2 \to f_2$ on E. If we define $s_n = s_n^1 + s_n^2$, then

$$\int_E s_n \, d\mu = \int_E s_n^1 \, d\mu + \int_E s_n^2 \, d\mu.$$

Beppo Levi's Theorem yields now that $\int_E f \, d\mu = \int_E f_1 \, d\mu + \int_E f_2 \, d\mu$. Suppose now that $f_1 \geq$ and $f_2 \leq 0$. We construct the sets

$$A = \{x \in E \mid f(x) \geq 0\}, \quad B = \{x \in E \mid f(x) < 0\}.$$

Since f, f_1 and $-f_2$ are non-negative on A, we have

$$\int_A f_1 \, d\mu = \int_A f \, d\mu + \int_A (-f_2) \, d\mu = \int_A f \, d\mu - \int_A f_2 \, d\mu.$$

Since $-f$, f_1 and $-f_2$ are non-negative on B, we see that

$$\int_B (-f_2) \, d\mu = \int_B f_1 \, d\mu + \int_B (-f) \, d\mu.$$

But this means that

$$\int_B f_1 \, d\mu = \int_B f \, d\mu - \int_B f_2 \, d\mu,$$

and the conclusion follows in this case. The other cases are treated similarly. $\quad\square$

Theorem 16.13 (Dominated Convergence Theorem) *Let E be a measurable set, and suppose that $\{f_n\}_n$ is a sequence of measurable functions such that*

$$f(x) = \lim_{n \to +\infty} f_n(x) \quad \text{for every } x \in E.$$

If there exists a function $g \in L^1(E)$ such that

$$|f_n(x)| \leq g(x) \quad \text{for every } x \in E \text{ and every } n \in \mathbb{N},$$

then $f \in L^1(E)$ and

$$\int_E f \, d\mu = \lim_{n \to +\infty} \int_E f_n \, d\mu.$$

Proof By Theorem 16.10 we have $f_n \in L^1(E)$ and $f \in L^1(E)$. Furthermore

$$f_n + g \geq 0 \quad \text{on } E,$$

so that Fatou's Lemma yields

$$\int_E (f + g)\, d\mu \leq \liminf_{n \to +\infty} \int_E (f_n + g_n)\, d\mu,$$

or

$$\int_E f\, d\mu \leq \liminf_{n \to +\infty} \int_E f_n\, d\mu.$$

Similarly, $g - f_n \geq 0$, and therefore

$$\int_E f\, d\mu \leq \liminf_{n \to +\infty} \left(-\int_E f_n\, d\mu \right).$$

This is equivalent to

$$\int_E f\, d\mu \geq \limsup_{n \to +\infty} \int_E f_n\, d\mu,$$

and the proof is complete.   □

16.3   Complete Measures

We have already remarked that sets of measure zero should be completely invisible in integration theory. Formally, we may agree that two measurable functions f and g are equivalent if and only if the set

$$\{x \in X \mid f(x) \neq g(x)\}$$

has measure zero in X. Then integrable equivalent functions have the same integral. Furthermore, it would be quite natural to relax the assumptions of Beppo Levi's Theorem, or of the Dominated Convergence Theorem, to allow pointwise limits almost everywhere. Unfortunately this may be troublesome, since a subset of a set of measure zero need not be measurable. We thus introduce a reasonable definition.

Definition 16.11 A measurable space is complete if and only if the following condition is satisfied: if E is a measurable set of measure zero and if $F \subset E$, then F is a measurable set (and of course the measure of F is then zero).

Theorem 16.14 *Suppose that (X, Σ, μ) is a measurable space. We define Σ^* as the set of all $E \subset X$ such that there exist measurable sets A and B such that $A \subset E \subset B$ and $\mu(B \setminus A) = 0$. For every $E \in \Sigma^*$ we define $\mu(E) = \mu(A)$. Then Σ^* is a σ-algebra and μ is a measure on Σ^*.*

Proof First of all we check that μ is well-defined on Σ^*. Assume that $A \subset E \subset B$ and $A_1 \subset E \subset B_1$, and $\mu(B_1 \setminus A_1) = \mu(B \setminus A) = 0$. From the inclusions $A \setminus A_1 \subset E \setminus A_1 \subset B_1 \setminus A_1$ we see that $\mu(A \setminus A_1) = 0$, hence $\mu(A) = \mu(A \cap A_1)$. For the very same reason, $\mu(A_1) = \mu(A \cap A_1)$, and we conclude that $\mu(A) = \mu(A_1)$.

Now we have to prove that Σ^* is a σ-algebra. Clearly enough, $X \in \Sigma^*$ since $X \in \Sigma$ and $\Sigma \subset \Sigma^*$. Furthermore, if $A \subset E \subset B$, then $X \setminus B \subset X \setminus E \subset X \setminus A$. Hence $E \in \Sigma^*$ implies $X \setminus E \in \Sigma^*$, because $(X \setminus A) \setminus (X \setminus B) = (X \setminus A) \cap B = B \setminus A$.

Finally, if the sets $A_n \subset E_n \subset B_n$ for every $n \in \mathbb{N}$, and if $E = \bigcup_{n=1}^{\infty} E_n$, $A = \bigcup_{n=1}^{\infty} A_n$, $B = \bigcup_{n=1}^{\infty}$, then $A \subset E \subset B$ and

$$B \setminus A = \bigcup_{n=1}^{\infty} (B_n \setminus A) \subset \bigcup_{n=1}^{\infty} (B_n \setminus A_n).$$

Since countable unions of sets of measure zero are sets of measure zero, it follows that $E \in \Sigma^*$ is each $E_n \in \Sigma^*$.

If, in particular, the sets E_n are disjoint, then the sets A_n are disjoint, so that

$$\mu(E) = \mu(A) = \sum_{n=1}^{\infty} \mu(A_n) = \sum_{n=1}^{\infty} \mu(E_n),$$

and μ is a measure on Σ^*. This completes the proof. □

Remark 16.2 The previous theorem shows a remarkable property of measurable spaces. If a *given* measure may fail to be complete, it is however true that this measure can be suitably *completed*.

With Theorem 16.14 in mind, we can propose a relaxed definition of measurable functions which takes into account sets of measure zero.

Definition 16.12 A function f defined on a set $E \in \Sigma$ is measurable if and only if $\mu(X \setminus E) = 0$ and $f^{-1}(V) \cap E$ is measurable for every open set V.

We may then *define* $f = 0$ on $X \setminus E$, which is a measurable function according to our previous definition. And if the measure is complete, we may even define f on $X \setminus E$ in an arbitrary manner, and still we get a measurable function.

In this way our statements can take into accounts function defined almost everywhere, without altering the conclusions. Let us prove a useful example.

Theorem 16.15 *Suppose that $\{f_n\}_n$ is a sequence of measurable functions defined almost everywhere on X. If*

$$\sum_{n=1}^{\infty} \int_X |f_n|\, d\mu < +\infty,$$

then the series $f(x) = \sum_{n=1}^{\infty} f_n(x)$ converges for almost every $x \in X$, $f \in L^1(X)$, and

$$\int_X f\, d\mu = \sum_{n=1}^{\infty} \int_X f_n\, d\mu$$

Proof Each function f_n is defined on a subset S_n of X such that $\mu(X \setminus S_n) = 0$. For every $x \in S = \bigcap_{n=1}^{\infty} S_n$ we set $\varphi(x) = \sum_{n=1}^{\infty} |f_n(x)|$. Then[1] $\mu(X \setminus S) = 0$. By Beppo Levi's Theorem, $\varphi \in L^1(S)$. If $E = \{x \in S \mid \varphi(x) < +\infty\}$, then $\mu(X \setminus E) = 0$.

The series $\sum_{n=1}^{\infty} f_n(x)$ converges absolutely for every $x \in E$, and $|f(x)| \le \varphi(x)$ on E. Hence $f \in L^1(E)$. Introducing the partial sum $g_n = f_1 + \cdots + f_n$, we see that $|g_n| \le \varphi$ on E, that $g_n \to f$ on E, and the Dominated Convergence Theorem yields

$$\int_E f\, d\mu = \sum_{n=1}^{\infty} \int_E f_n\, d\mu.$$

But $\mu(X \setminus E) = 0$, and the conclusion follows. $\square$

Theorem 16.16 *If $f: X \to [0, +\infty]$ is measurable, $E \in \Sigma$ and $\int_E f\, d\mu = 0$, then $f = 0$ almost everywhere on E.*

Proof For every positive integer n, We define $A_n = \{x \in E \mid f(x) > 1/n\}$. The set A_n is measurable, and

$$\frac{1}{n}\mu(A_n) \le \int_{A_n} f\, d\mu \le \int_E f\, d\mu = 0,$$

which implies $\mu(A_n) = 0$ for every n. Since $\{x \in E \mid f(x) > 0\} = \bigcup_{n=1}^{\infty} A_n$, it follows that $f = 0$ almost everywhere. $\square$

Theorem 16.17 *If $f \in L^1(X)$ and $\int_E f\, d\mu = 0$ for every measurable $E \subset X$, then $f = 0$ almost everywhere on X.*

---

[1] With our original definition of measurable functions, this implication would require the completes of the measurable space X.

Proof We apply the assumption to $E = \{x \in X \mid f(x) \geq 0\}$, so that $\int_E f^+ \, d\mu = 0$. Hence $f^+ = 0$ almost everywhere on X. Similarly $f^- = 0$ almost everywhere on X, and the proof is complete. $\qquad\square$

16.4 Different Types of Convergence

For a sequence of functions, pointwise convergence is often too less, while uniform convergence is usually too much in applications. Measure Theory provides some intermediate types of convergence which are of great convenience in mathematical analysis. In this section we want to review some of them. As a rule, all functions will be defined on a measurable space (X, Σ, μ).

Definition 16.13 A sequence of measurable functions $\{f_n\}_n$ converges almost uniformly to a measurable limit f if and only if for every $\varepsilon > 0$ there exists a measurable set E such that $\mu(X \setminus E) < \varepsilon$ and $f_n \to f$ uniformly on E.

Exercise 16.7 Prove that almost uniform convergence implies almost everywhere convergence.

For a converse of the previous exercise, an additional assumption is needed.

Theorem 16.18 (Severini-Egorov) *Suppose that $\mu(X) < +\infty$. If a sequence of measurable functions $\{f_n\}_n$ converges almost everywhere to f in X, then $f_n \to f$ almost uniformly.*

Proof Removing a set of measure zero, we may assume that $f_n \to f$ pointwise in the whole X. For every $k \in \mathbb{N}$ and $m \in \mathbb{N}$ we define the set

$$E(m, k) = \bigcap_{n > m} \left\{ x \in X \,\middle|\, |f_n(x) - f(x)| < \frac{1}{k} \right\}.$$

We have $E(1, k) \subset E(2, k) \subset E(3, k) \subset \cdots$, and $\bigcup_{m \in \mathbb{N}} E(m, k) = X$. As a consequence, to each $k \in \mathbb{N}$ there corresponds an integer $m_k \in \mathbb{N}$ such that

$$\mu\left(\bigcap_{n > m_k} \left\{ x \in X \,\middle|\, |f_n(x) - f(x)| < \frac{1}{k} \right\} \right) < \frac{\varepsilon}{2^k}.$$

If we set

$$E = \bigcap_{k \in \mathbb{N}} \bigcap_{n > m_k} \left\{ x \in X \,\middle|\, |f_n(x) - f(x)| < \frac{1}{k} \right\},$$

it follows that $\mu(X \setminus E) \leq \sum_{k=1}^{\infty} \varepsilon \cdot 2^{-k} = \varepsilon$ and $f_n \to f$ uniformly on E. The proof is complete. $\qquad\square$

Exercise 16.8 When did we use the assumption $\mu(X) < +\infty$ in the previous proof?

Definition 16.14 A sequence $\{f_n\}_n$ of measurable functions converges in measure to the limit f if and only if for every $\varepsilon > 0$ there results

$$\lim_{n \to +\infty} \mu(\{x \in X \mid |f_n(x) - f(x)| > \varepsilon\}) = 0.$$

Example 16.3 Let $X = [0, +\infty)$ with the concrete Lebesgue measure, and define f_n on X such that

$$f_n(x) = \begin{cases} 0 & \text{if } 0 \le x < n \\ 1 & \text{if } n \le x < +\infty. \end{cases}$$

Then $f_n \to 0$ pointwise in X, but it does not converge to zero in measure. Indeed, for $0 < \varepsilon < 1$, $f_n(x) > \varepsilon$ if and only if $x > n$.

Theorem 16.19 *If $f_n \to f$ in measure, then there exists a subsequence $\{f_{n_k}\}_k$ which converges to f almost everywhere. If, in addition, $\mu(X) < +\infty$, then convergence almost everywhere implies convergence in measure.*

Proof Pick a sequence $n_1 < n_2 < n_3 < \dots$ of positive integers such that the measure of the set

$$E_k = \left\{x \in X \,\middle|\, |f_{n_k}(x) - f(x)| > \frac{1}{k}\right\}$$

is smaller than $1/k^2$. Thus

$$\mu\left(\bigcup_{k=m}^{\infty} E_k\right) \le \sum_{k-n}^{\infty} \frac{1}{k^2}$$

and

$$E = \bigcap_{m=1}^{\infty} \bigcup_{k=m}^{\infty} E_k$$

has measure zero. Since

$$X \setminus E = \bigcup_{m=1}^{\infty} \bigcap_{k=m}^{\infty} \left\{x \in X \,\middle|\, |f_{n_k}(x) - f(x)| \le \frac{1}{k}\right\},$$

we see that $f(x) = \lim_{k \to +\infty} f_{n_k}(x)$ for every $x \in X \setminus E$, and the first assertion is proved.

To complete the proof, we assume that X has finite measure and we use Theorem 16.18. For each $\eta > 0$ there exists a measurable set E such that $\mu(X \setminus E) < \eta$ and $f_n \to f$ uniformly on E. Then

$$\{x \in X \mid |f_n(x) - f(x)| > \varepsilon\}$$
$$= \{x \in E \mid |f_n(x) - f(x)| > \varepsilon\} \cup \{x \in X \setminus E \mid |f_n(x) - f(x)| > \varepsilon\}$$

is a measurable set whose measure if smaller than η provided that n is sufficiently large. Since $\eta > 0$ was arbitrary, the proof is complete.   □

In many applications a direct proof of L^1-convergence remains out-of-reach. We now introduce a fundamental result due to Vitali which may be useful in such situations.

Definition 16.15 A sequence $\{f_n\}_n$ in $L^1(X)$ is called equi-integrable if and only if the following condition is satisfied: for every $\varepsilon > 0$ there exist a number $\delta > 0$ and a measurable set A such that $\mu(A) < \delta$ and

1. for every $n \in \mathbb{N}$ there results $\int_{X \setminus A} |f_n| \, d\mu < \varepsilon$;
2. for every measurable set E such that $\mu(E) < \delta$ there results $\int_E |f_n| \, d\mu < \varepsilon$ for every n.

The first condition is clearly empty when the whole space has finite measure.
     We first establish a general result about integrable functions.

Proposition 16.3 (Absolute Continuity of the Integral) *Let* $f \in L^1(X)$. *For every* $\varepsilon > 0$ *there exists* $\delta > 0$ *such that for any measurable set* E, $\mu(E) < \delta$ *implies* $\int_E |f| \, d\mu < \varepsilon$. *Furthermore, for every* $\varepsilon > 0$ *there exists a measurable subset* X_0 *of* X *such that* X_0 *has finite measure and* $\int_{X \setminus X_0} |f| \, d\mu < \varepsilon$.

Proof We may assume $f \geq 0$ without loss of generality, so that $|f| = f$. In the general case we split $f = f^+ - f^-$. Fix any $\varepsilon > 0$. By definition of $\int_X f \, d\mu$, there exists a measurable simple function s such that $0 \leq s \leq f$ and $0 \leq \int_X f \, d\mu - \int_X s \, d\mu < \varepsilon/2$. Since s is a bounded function, there exists $M > 0$ such that $0 \leq s \leq M$ on X. Therefore, for every measurable $E \subset X$ of finite measure, we have

$$\int_E f \, d\mu = \int_E s \, d\mu + \int_E (f - s) \, d\mu \leq \int_E s \, d\mu + \frac{\varepsilon}{2} \leq M\mu(E) + \frac{\varepsilon}{2}.$$

The conclusion follows with $\delta = \varepsilon/(2M)$.
     To prove the second statement, we remark that s is integrable, so that $X_0 = \{x \in X \mid s(x) > 0\}$ has finite measure. Moreover

$$\int_{X \setminus X_0} f \, d\mu = \int_{X \setminus X_0} (f - s) \, d\mu \leq \int_X (f - s) \, d\mu < \varepsilon.$$

The proof is complete.   □

Theorem 16.20 (Vitali) *Let $\{f_n\}_n$ be a sequence of equi-integrable functions. If $f_n \to f$ almost everywhere and if $f \in L^1(X)$, then $f_n \to f$ in $L^1(X)$.*

Proof Trivially, $|f_n - f| \le |f_n| + |f|$. If X_0 and X_1 are measurable sets with $X_1 \subset X_0$, then

$$\left| \int_X (f_n - f) \, d\mu \right| \le \int_{X_1} |f_n - f| \, d\mu + \int_{X_0 \setminus X_1} (|f_n| + |f|) \, d\mu + \int_{X \setminus X_0} (|f_n| + |f|) \, d\mu.$$

Fix $\varepsilon > 0$. By Proposition 16.3 and the first condition of equi-integrability, there exists a measurable set $X_0 \subset X$ of finite measure such that

$$\int_{X \setminus X_1} (|f_n| + |f|) \, d\mu = \int_{X \setminus X_0} |f_n| \, d\mu + \int_{X \setminus X_0} |f| \, d\mu < \frac{\varepsilon}{3}.$$

By Proposition 16.3 again, we may find a number $\delta > 0$ such that for every measurable $E \subset X$, $\mu(E) < \delta$ implies

$$\int_E (|f_n| + |f|) \, d\mu = \int_E |f_n| \, d\mu + \int_E |f| \, d\mu < \frac{\varepsilon}{3}.$$

By assumption $f \in L^1(X)$. In particular $f < +\infty$ almost everywhere, and $\mu(X_0) < +\infty$. It follows from Theorem 16.18 that there exists a measurable subset X_1 of X_0 such that $\mu(X_0 \setminus X_1) < \delta$ and $f_n \to f$ uniformly on X_1. Hence

$$\int_{X_0 \setminus X_1} (|f_n| + |f|) \, d\mu < \frac{\varepsilon}{3}$$

and there exists $N \in \mathbb{N}$ such that $n \ge N$ implies

$$\int_{X_1} |f_n - f| \, d\mu \le \sup \{|f_n(x) - f(x)| \mid x \in X_1\} \mu(X_1) < \frac{\varepsilon}{3}.$$

Collecting these estimates we conclude that $n \ge N$ implies

$$\left| \int_X (f_n - f) \, d\mu \right| < \frac{\varepsilon}{3} + \frac{\varepsilon}{3} + \frac{\varepsilon}{3} = \varepsilon.$$

The proof is complete.   □

Exercise 16.9 Suppose that for every $n \in \mathbb{N}$, $h_n \in L^1(X)$, $h_n \ge 0$ and $h_n \to 0$ almost everywhere on X. Prove that $\int_X h_n \, d\mu \to 0$ if and only if $\{h_n\}_n$ is equi-integrable. What can you deduce in the particular case $h_n = |f_n - f|$?

16.5   Measure Theory on Product Spaces

Let Σ be a σ-algebra on a set X and T be a σ-algebra on a set Y.[a]   A *measurable rectangle* is a set $A \times B$ such that $A \in \Sigma$ and $B \in T$. The class $\mathcal{E}$ of *elementary sets* consists of all sets $Q = R_1 \cup \cdots \cup R_n$ where each R_j is a measurable rectangle and $R_i \cap R_j = \emptyset$ when $i \neq j$.

[a]Here T should be seen as the capital τ of the Greek alphabet.

Definition 16.16 The product σ-algebra $\Sigma \times T$ is the smallest σ-algebra which contains all measurable rectangles.

A *monotone class* $\mathcal{M}$ is a collection of sets with the following properties: if $A_i \in \mathcal{M}$, $B_i \in \mathcal{M}$, $A_i \subset A_{i+1}$, $B_i \supset B_{i+1}$ for every $i = 1, 2, 3, \ldots$ and if

$$A = \bigcup_{i=1}^{\infty} A_i, \quad B = \bigcap_{i=1}^{\infty} B_i,$$

then $A \in \mathcal{M}$ and $B \in \mathcal{M}$. To summarize, monotone classes are closed with respect to countable increasing unions and countable decreasing intersections.

Theorem 16.21 $\Sigma \times T$ *is the smallest monotone class which contains all elementary sets.*

Proof Let $\mathcal{M}$ be the intersection of all monotone classes which contain $\mathcal{E}$. It is clear that $\mathcal{M} \neq \emptyset$, since $X \times Y$ is a monotone class which contains $\mathcal{E}$. But $\Sigma \times T$ is a monotone classes, so that $\mathcal{M} \subset \Sigma \times T$. For $A_1 \in \Sigma$, $A_2 \in \Sigma$, $B_1 \in T$, $B_2 \in T$ we have

$$(A_1 \times B_1) \cap (A_2 \times B_2) = (A_1 \cap A_2) \times (B_1 \cap B_2)$$

$$(A_1 \times B_1) \setminus (A_2 \times B_2) = ((A_1 \setminus A_2) \times B_1) \cup ((A_1 \cap A_2) \times (B_1 \setminus B_2)) .$$

It follows that $P \in \mathcal{E}$, $Q \in \mathcal{E}$ imply $P \cap Q \in \mathcal{E}$ and $P \setminus Q \in \mathcal{E}$. Furthermore $P \cup Q = (P \setminus Q) \cup Q$ and $(P \setminus Q) \cap Q = \emptyset$, hence $P \cup Q \in \mathcal{E}$.

For every $P \subset X \times Y$, call $\Omega(P)$ the family of all $Q \subset X \times Y$ such that $P \setminus Q \in \mathcal{M}$, $Q \setminus P \in \mathcal{M}$, and $P \cup Q \in \mathcal{M}$. It is evident that $Q \in \Omega(P)$ if and only if $P \in \Omega(Q)$, and that each $\Omega(P)$ is a monotone class.

Now select $P \in \mathcal{E}$. As we have just seen, $Q \in \mathcal{E}$ implies $Q \in \Omega(P)$, so that $\mathcal{E} \subset \Omega(P)$ and finally $\mathcal{M} \subset \Omega(P)$.

Select $Q \in M$. Since $P \in \mathcal{E}$ implies $Q \in \Omega(P)$, we also have $P \in \Omega(Q)$. Hence $\mathcal{E} \subset \Omega(Q)$ and we conclude again that $M \subset \Omega(Q)$.

We have proved that $P \setminus Q \subset M$ and $P \cup Q \in M$ provided that $P \in M$ and $Q \in M$. It is easy to deduce that M is a σ-algebra on $X \times Y$. Indeed $X \times Y \in M$ because $X \times Y \in \mathcal{E}$.

If $Q \in M$, then $(X \times Y) \setminus Q \in M$ since the difference of any two elements of M is an element of M.

Finally, if $P_i \in M$ for $i = 1, 2, 3, \ldots$ and $P = \bigcup_{i=1}^{\infty} P_i$, we set $Q_n = P_1 \cup \cdots \cup P_n$ and remark that M is closed under the formation of finite unions. Hence $Q_n \in M$. But $Q_n \subset Q_{n+1}$ and $P = \bigcup_{n=1}^{\infty} Q_n$, and the monotonicity of M shows that $P \in M$.

We have proved that M is a σ-algebra such that $\mathcal{E} \subset M \subset \Sigma \times T$. Hence $M = \Sigma \times T$, and the proof is complete. $\qquad\square$

The introduction of a measure on $X \times Y$ is associated to the *sections* of a measurable subset.

Definition 16.17 If $E \subset X \times Y$, $x \in X$ and $y \in Y$, we set

$$E_x = \{y \in Y \mid (x, y) \in E\}$$
$$E^y = \{x \in X \mid (x, y) \in E\}.$$

Proposition 16.4 *If $E \in \Sigma \times T$, then $E_x \in T$ and $E^y \in \Sigma$ for every $x \in X$, $y \in Y$.*

Proof We temporarily say that $E \in \Omega$ if and only if $E_x \in T$ for every $x \in X$. If $E = A \times B$, then

$$E_x = \begin{cases} B & \text{if } x \in A \\ \emptyset & \text{if } x \notin A. \end{cases}$$

Hence every measurable rectangle belongs to Ω. Observe now that $X \times Y \in \Omega$, that $((X \times Y) \setminus E)_x = (X \times Y) \setminus E_x$, and that

$$\left(\bigcup_{i=1}^{\infty} E_i \right)_x = \bigcup_{i=1}^{\infty} (E_i)_x.$$

Hence Ω is a σ-algebra which must coincide with $\Sigma \times T$. A completely analogous proof can be repeated for E^y. $\qquad\square$

We now deal with functions of two variables in a similar manner.

Definition 16.18 Let f be a function defined on $X \times Y$. For every $x \in X$ and every $y \in Y$ we denote

$$f_x = f(x, \cdot), \quad f^y = f(\cdot, y).$$

Proposition 16.5 *Let f be measurable in $\Sigma \times T$. Then f_x is measurable in T and f^y is measurable in Σ for every $x \in X$, $y \in Y$.*

Proof Consider an open set V (contained in the unnamed codomain of f), and set $Q = \{(x, y) \mid f(x, y) \in V\} = f^{-1}(V)$. By assumption $Q \in \Sigma \times T$, and $Q_x = \{y \in Y \mid f_x(y) \in V\}$. We have already proved that $Q_x \in T$, and this shows that f_x is measurable in T. The same reasoning can be applied to f^y, and the proof is complete.   □

We can define the product measure on $X \times Y$ in terms of integrals.

Definition 16.19 Let (X, Σ, μ) and (Y, T, λ) be σ-finite measurable spaces. For every $Q \in \Sigma \times T$ we define

$$(\mu \times \lambda)(Q) = \int_X \lambda(Q_x)\, d\mu(x) = \int_Y (Q^y)\, d\lambda(y).$$

This is the product measure of μ and λ.

Of course the previous definition must be confirmed by proving that indeed $\int_X \lambda(Q_x)\, d\mu(x) = \int_Y (Q^y)\, d\lambda(y)$. This is the content of the next result.

Theorem 16.22 *Let (X, Σ, μ) and (Y, T, λ) be σ-finite measurable spaces. Suppose that $Q \in \Sigma \times T$. If*

$$\varphi(x) = \lambda(Q_x), \quad \psi(y) = \mu(Q^y) \tag{16.1}$$

for every $x \in X$, $y \in Y$, then φ is measurable in T, ψ is measurable in Σ, and

$$\int_X \lambda(Q_x)\, d\mu(x) = \int_Y (Q^y)\, d\lambda(y).$$

Proof We call Ω the collection of all subsets Q of $X \times Y$ for which the conclusion of the theorem holds.

(a) Every measurable rectangle belongs to Q. Indeed, if $Q = A \times B$ is a measurable rectangle, then $\lambda(Q_x) = \lambda(B)\chi_A(x)$, $\mu(Q^y) = \mu(A)\chi_B(y)$, and both integrals are equal to $\mu(A)\lambda(B)$.

(b) If $Q_1 \subset Q_2 \subset Q_3 \subset \dots$, if $Q_i \in \Omega$ for every i and if $Q = \bigcup_{i=1}^{\infty} Q_i$, then $Q \in \Omega$. Indeed, let φ_i and ψ_i be defined as in (16.1) with Q_i instead of Q. It follows from the countable additivity of each measure that $\varphi_i \nearrow \varphi$, $\psi_i \nearrow \psi$ pointwise as $i \to +\infty$. Beppo Levi's Theorem yields the conclusion.

(c) If $\{Q_i\}_i$ is a disjoint countable collection of elements of Ω, and if $Q = \bigcup_{i=1}^{\infty} Q_i$, then $Q \in \Omega$. Indeed, this is trivial for a finite collection, since the characteristic function of a union of disjoint sets is just the sum of their characteristic functions. In the general case we may use (b).

(d) If $\mu(A) < +\infty$, $\lambda(B) < +\infty$, if $A \times B \supset Q_1 \supset Q_2 \supset Q_3 \supset \ldots$, if $Q = \bigcap_{i=1}^{\infty} Q_i$ and $Q_i \in \Omega$ for each i, then $Q \in \Omega$. Indeed, we can repeat the proof of (b) after replacing Beppo Levi's Theorem with the Dominated Convergence Theorem.

Now, recall that $X = \bigcup_{n=1}^{\infty} X_n$ and $Y = \bigcup_{m=1}^{\infty} Y_m$ for suitable disjoint sets X_n, Y_m of finite measure. For every $m \in \mathbb{N}$, $n \in \mathbb{N}$, we define $Q_{mn} = Q \cap (X_n \times Y_m)$. Let us agree that $Q \in \mathcal{M}$ if and only if $Q_{mn} \in \Omega$ for every m and n. We have proved in (b) and (d) that $\mathcal{M}$ is a monotone class. Furthermore $\mathcal{E}$ is contained in $\mathcal{M}$ by (a) and (c). Since $\mathcal{M} \subset \Sigma \times T$, it follows from Theorem 16.21 that $\mathcal{M} = \Sigma \times T$.

We have proved that for every $Q \in \Sigma \times T$ we have $Q_{mn} \in \Omega$ for every choice of m and n. But $Q = \bigcup_{m,n=1}^{\infty} Q_{mn}$, and the sets Q_{mn} are disjoint. Hence (c) implies that $Q \in \Omega$. The proof is complete.   □

Exercise 16.10 Prove that $\mu \times \lambda$ is indeed a measure on $(X \times Y, \Sigma \times T)$, and that it is a σ-finite measure.

Remark 16.3 We notice that we have defined the product measure $\mu \times \lambda$ by means of two (equal) *iterated* integrals. This might look strange, since the integral should come after the measure. It would be possible to define $\mu \times \lambda$ without any reference to integration, and we would all agree that $(\mu \times \lambda)(A \times B) = \mu(A)\lambda(B)$ for any measurable rectangle; but a valid formula for any measurable subset of $\Sigma \times T$ would be much harder to guess.

The equality of the two integrals in Theorem 16.22 suggests a further step. This is the content of a celebrated result that we discussed from an abstract viewpoint in Daniell's approach to integration theory.

Theorem 16.23 (Fubini-Tonelli) *Let (X, Σ, μ) and (Y, T, λ) be σ-finite measurable spaces, and let f be a measurable function in $\Sigma \times T$.*

(a) If $0 \le f \le +\infty$, and if

$$\varphi(x) = \int_Y f_x \, d\lambda, \quad \psi(y) = \int_X f^y \, d\mu$$

for every $x \in X$, $y \in Y$, then φ is measurable in Σ, ψ is measurable in Y, and

$$\int_X \varphi \, d\mu = \int_{X \times Y} f \, d(\mu \times \lambda) = \int_Y \psi \, d\lambda.$$

(b) If f is real-valued, if

$$\varphi^*(x) = \int_Y |f|_x \, d\lambda$$

and if $\int_X \varphi^* \, d\mu < +\infty$, then $f \in L^1(X \times Y)$.

(c) *If $f \in L^1(X \times Y)$, then $f_x \in L^1(Y)$ for almost every $x \in X$, $f^y \in L^1(X)$ for almost every $y \in Y$; the functions φ and ψ are defined almost everywhere and belong to $L^1(X)$ and $L^1(Y)$ respectively. Furthermore,*

$$\varphi(x) = \int_Y f_x \, d\lambda, \quad \psi(y) = \int_X f^y \, d\mu.$$

Proof We consider (a). The conclusion holds if f is a simple function. In the general case, there exists a sequence $\{s_n\}_n$ of non-negative simple functions such that $s_n \nearrow f$ pointwise on $X \times Y$. We associate to each s_n a function φ_n in the same way as we associated φ to f. Then

$$\int_X \varphi_n \, d\mu = \int_{X \times Y} s_n \, d(\mu \times \lambda)$$

for $n = 1, 2, \ldots$ We can apply Beppo Levi's Theorem in Y to conclude that $\varphi_n(x) \nearrow \varphi(x)$ for every $x \in X$. A second application of Lev's Theorem yields $\int_X \varphi \, d\mu = \int_{X \times Y} f \, d(\mu \times \lambda)$. If we swap the roles of x and y, we also get $\int_{X \times Y} f \, d(\mu \times \lambda) = \int_Y \psi \, d\lambda$.

The proof of (b) follows from (a) applied to $|f|$. To prove (c), we first observe that (a) applies to f^+ and f^-. Let us denote by φ_1 and φ_2 the functions associated to f^+ and f^- in the same way as φ was associated to f. The assumption $f \in L^1(X \times Y)$ and the trivial inequality $f^+ \leq |f|$ imply that $\varphi_1 \in L^1(X)$. Similarly, $\varphi_2 \in L^1(Y)$.

Notice now that $f_x = (f^+)_x - (f^-)_x$, so that $f_x \in L^1(Y)$ for every $x \in X$ such that $\varphi_1(x) < +\infty$ and $\varphi_2(x) < +\infty$. This must hold for almost every $x \in X$, since φ_1, φ_2 are integrable. For any such value of x, we have $\varphi(x) = \varphi_1(x) - \varphi_2(x)$. In particular $\varphi \in L^1(X)$. The conclusion of (c) thus holds both with φ_1 and f^+, and with φ_2 and f^- in place of φ and f. Subtracting these two equalities we conclude the proof for f_x. By the same token, the conclusion holds for f^y. $\square$

16.6 Measure, Topology, and the Concrete Lebesgue Measure

It is a matter of fact that no description of Measure Theory can be complete without a construction of the concrete Lebesgue measure in Euclidean spaces. However, this is a rather complicated construction, since the topology of $\mathbb{R}^n$ plays a fundamental role. The first step is a version of Urysohn's Theorem 13.60 for locally compact Hausdorff spaces. The proof we present here is more direct than the proof of Theorem 13.65.

Definition 16.20 We write $K \prec f$ to mean that K is a compact subset of X, that $f \in C_c(X)$, $0 \leq f \leq 1$ on X, and that $f = 1$ on K.

Similarly, we write $f \prec V$ to mean that V is an open subset of X, that $f \in C_c(X)$, $0 \le f \le 1$ on X, and that supp $f \subset V$.

Theorem 16.24 (Urysohn's Lemma for Locally Compact Hausdorff Spaces)
Suppose that X is a locally compact Hausdorff space, V is open in X, $K \subset V$, and K is compact. Then there exists $f \in C_c(X)$ such that $K \prec f \prec V$.

Proof Let $r_1 = 0$, $r_2 = 1$, and $\{r_3, r_4, r_5, \ldots\}$ be an enumeration of $\mathbb{Q} \cap (0, 1)$. By Proposition 13.1, there exist open sets V_0 and V_1 such that $\overline{V_0}$ is compact and

$$K \subset V_1 \subset \overline{V_1} \subset V_0 \subset \overline{V_0} \subset V.$$

Suppose $n \ge 2$ and that $V_{r_1}, V_{r_2}, \ldots, V_{r_n}$ have been chosen so that $r_i < r_j$ implies $\overline{V_{r_i}} \subset V_{r_j}$. Among the rational numbers $r_1, \ldots, r_n$, choose the largest one, say r_i, such that $r_i < r_{n+1}$, and the smallest one, say r_j, such that $r_j > r_{n+1}$. Arguing again as before, we select an open set $V_{r_{n+1}}$ such that

$$\overline{V_{r_j}} \subset V_{r_{n+1}} \subset \overline{V_{r_{n+1}}} \subset V_{r_i}.$$

By induction we construct a countable family $\{V_r \mid r \in \mathbb{Q} \cap [0, 1]\}$ of open sets, such that $K \subset V_1$, $\overline{V_0} \subset V$, each V_r has compact closure, and $s > r$ implies $\overline{V_s} \subset V_r$.
Define

$$f_r(x) = \begin{cases} r & \text{if } x \in V_r \\ 0 & \text{otherwise,} \end{cases}$$

$$g_s(x) = \begin{cases} 1 & \text{if } x \in \overline{V_s} \\ s & \text{otherwise,} \end{cases}$$

and $f = \sup\{f_r \mid r \in \mathbb{Q} \cap [0, 1]\}$, $g = \inf\{g_s \mid s \in \mathbb{Q} \cap [0, 1]\}$. It is easy to check that f is lower semicontinuous and that g is upper semicontinuous. Clearly $0 \le f \le 1$, $f(x) = 1$ if $x \in K$, and the support of f is contained in V. To conclude, we prove that $f = g$, getting the continuity of f.

The inequality $f_r(x) > g_s(x)$ is possible only if $r > s$, $x \in V_r$, and $x \notin \overline{V_s}$. But $r > s$ implies $V_r \subset V_s$. Hence $f_r \le f_s$ for all r and s, which yields $f \le g$. Suppose that $f(x) < g(x)$ for some x. There exist rational numbers r and s such that $f(x) < r < s < g(x)$. The inequality $f(x) < r$ implies $x \notin V_r$. The inequality $g(x) > s$ implies $x \in \overline{V_s}$. This is clearly a contradiction to the property that $s > r$ implies $\overline{V_s} \subset V_r$.

We conclude that $f = g$, and the proof is complete.   □

Theorem 16.25 (Finite Partition of Unity) *Suppose $V_1, \ldots, V_n$ are open subsets of a locally compact Hausdorff space X. If K is compact and $K \subset V_1 \cup \cdots \cup V_n$, then there exist functions $h_i \prec V_i$, $i = 1, \ldots, n$, such that $\sum_{i=1}^{n} h_i(x) = 1$ for every $x \in K$.*

Proof For every $x \in K$ there exists an open neighborhood W_x with compact closure $\overline{W_x} \subset V_i$ for some index i which depends on x. By compactness, we select points $x_1, \ldots, x_n$ such that $K \subset W_{x_1} \cup \cdots \cup W_{x_n}$. For $1 \leq i \leq n$, let

$$H_i = \bigcup \left\{ \overline{W_{x_j}} \mid \overline{W_{x_j}} \subset V_i \right\}.$$

By Theorem 16.24, there exist functions g_i such that $H_i \prec g_i \prec V_i$. We now define

$$h_1 = g_1$$
$$h_2 = (1 - g_1)g_2$$
$$\vdots$$
$$h_n = (1 - g_1)(1 - g_2) \cdots (1 - g_{n-1})g_n.$$

It is now easy to check that $h_i \prec V_i$ and

$$\sum_{i=1}^{n} h_i = 1 - (1 - g_1) \cdots (1 - g_n).$$

Since $K \subset H_1 \cup \cdots H_n$, at least on $g_i(x) = 1$ at each point $x \in K$. The proof is complete. $\qquad \square$

We are ready to state and proof the cornerstone of Topological Measure Theory.

Theorem 16.26 (Riesz Representation Theorem for Positive Measures) *Let X be a locally compact Hausdorff space, and let $\Lambda \colon C_c(X) \to \mathbb{R}$ be positive and linear.[2] Then there exists a σ-algebra Σ in X which contains all Borel sets, and there exists a unique positive measure μ on Σ such that*

(a) $\Lambda f = \int_X f \, d\mu$ for all $f \in C_c(X)$.

Furthermore

(b) $\mu(K)$ is finite for every compact K.
(c) For every $E \in \Sigma$,

$$\mu(E) = \inf \{ \mu(V) \mid E \subset V, \ V \text{ open} \}.$$

(d) The equality

$$\mu(E) = \sup \{ \mu(K) \mid K \subset E, \ K \text{ compact} \}$$

---

[2] No continuity is assumed here. The term *positive* means that $f \in C_c(X)$ and $f(X) \subset [0, +\infty)$ imply $\Lambda f \geq 0$.

holds for every open set E and for every measurable set E of finite measure.
(e)   The measure μ *is complete.*

Proof   The proof is rather long, so we state some general principle: K will always denote a compact set of X, and V an open set of X.

The uniqueness of the measure μ is rather easy. Indeed, properties (c) and (d) imply that μ is determined by its values on compact sets. Let μ_1 and μ_2 two measures for which the theorem holds, and assume that $\mu_1(K) = \mu_2(K)$ for every K. Fix any such K, and let $\varepsilon > 0$. By (b) and (c) there is V such that $\mu_2(V) < \mu_2(K) + \varepsilon$. Theorem 16.24 provides a function f such that $K \preccurlyeq f \preccurlyeq V$, hence

$$\mu_1(K) = \int_X \chi_K \, d\mu_1 \leq \int_X f \, d\mu_1 = \Lambda f = \int_X f \, d\mu_2$$

$$\leq \int_X \chi_V \, d\mu_2 = \mu_2(V) < \mu_2(K) + \varepsilon.$$

Letting $\varepsilon \to 0$ we find $\mu_1(K) \leq \mu_2(K)$. Exchanging μ_1 and μ_2 yields $\mu_1(K) = \mu_2(K)$. Hence uniqueness of μ is proved.

We now construct Σ and μ. For every open V, define

$$\mu(V) = \sup \{\Lambda f \mid f \preccurlyeq V\}.$$

Clearly $V_1 \subset V_2$ implies $\mu(V_1) \leq \mu(V_2)$; thus we define

$$\mu(E) = \inf \{\mu(V) \mid E \subset V, \ V \text{ open}\}$$

for every $E \subset X$. This definition agrees with the previous one if E is open. But beware! Our set function μ is not countably additive on the whole 2^X, and for this reason we need to introduce a good σ-algebra. Let Σ_F be the collection of all $E \subset X$ such that $\mu(E)$ is finite and

$$\mu(E) = \sup \{\mu(K) \mid K \subset E, \ K \text{ compact}\}. \tag{16.2}$$

Our σ-algebra is

$$\Sigma = \{E \subset X \mid E \cap K \in \Sigma_F \text{ for every compact } K\}.$$

The rest of the proof consists in showing that Σ and μ have the desired properties.

Evidently μ is monotone, i.e. $\mu(A) \leq \mu(B)$ if $A \subset B$, and $\mu(E) = 0$ implies $E \in \Sigma_F$ and then $E \in \Sigma$. Hence μ is a complete measure, and (c) holds true by definition.

Claim 1.   If E_i are subsets of X, $i \in \mathbb{N}$, then

$$\mu \left(\bigcup_{i=1}^{\infty} E_i \right) \leq \sum_{i=1}^{\infty} \mu(E_i).$$

We first prove that $\mu(V_1 \cup V_2) \leq \mu(V_1) + \mu(V_2)$ if V_1 and V_2 are open sets. Pick any $g \prec V_1 \cup V_2$. There exists a partition of unity consisting of two functions h_1, h_2 such that $h_i \prec V_i$ and $h_1(x) + h_2(x) = 1$ for all $x \in \text{supp } g$. Hence $h_i g \prec V_i$, $g = g h_1 + g h_2$, and therefore

$$\Lambda g = \Lambda(g h_1) + \Lambda(g h_2) \leq \mu(V_1) + \mu(V_2).$$

Since g was arbitrary, it follows that $\mu(V_1 \cup V_2) \leq \mu(V_1) + \mu(V_2)$. Let us now consider the general case, and we assume without loss of generality that $\mu(E_i)$ is finite for every index i (otherwise the inequality is trivial). Let $\varepsilon > 0$, and consider open sets $V_i \supset E_i$ such that

$$\mu(V_i) \leq \mu(E_i) + \frac{\varepsilon}{2^i}.$$

We set $V = \bigcup_{i=1}^{\infty} V_i$ and we choose any $f \prec V$. This says that f has compact support, so that $f \prec V_1 \cup \cdots \cup V_n$ for some n. Iterating the previous case we get

$$\Lambda f \leq \mu(V_1 \cup \cdots \cup V_n) \leq \mu(V_1) + \cdots + \mu(V_n)$$

$$\leq \sum_{i=1}^{\infty} \mu(E_i) + \varepsilon.$$

Since f was arbitrary and $\bigcup_{i=1}^{\infty} E_i \subset V$, we may conclude that

$$\mu \left(\bigcup_{i=1}^{\infty} E_i \right) \leq \mu(V) \leq \sum_{i=1}^{\infty} \mu(E_i) + \varepsilon.$$

We prove the Claim by letting $\varepsilon \to 0$.

Claim 2.   If K is compact then $K \in \Sigma_F$ and

$$\mu(K) = \inf \{ \Lambda f \mid K \prec f \}.$$

Of course this claim proves part (b) of the theorem. So, let $K \prec f$ and $0 < \alpha < 1$. Define $V_\alpha = \{ x \in X \mid f(x) > \alpha \}$. It is clear that $K \subset V_\alpha$ and that $\alpha g \leq f$ as soon as $g \prec V_\alpha$. As a consequence

$$\mu(K) \leq \mu(V_\alpha) - \sup \{ \Lambda g \mid g \prec V_\alpha \} \leq \frac{1}{\alpha} \Lambda f.$$

Letting $\alpha \to 1^-$ we conclude that $\mu(K) \leq \Lambda f$. In particular the measure of K is finite. Now let $\varepsilon > 0$. There exists an open set V such that $K \subset V$ and $\mu(V) < \mu(K) + \varepsilon$. By Urysohn's Lemma, for a suitable f we have $K \prec f \prec V$. Therefore $\Lambda f \leq \mu(V) < \mu(K) + \varepsilon$, and the proof of the claim is complete.

Claim 3. Every open set V satisfies (16.2). Hence Σ_F containers every open set of finite measure. Indeed, we fix a real number α such that $\alpha < \mu(V)$, and a function $f \prec V$ such that $\alpha < \Lambda f$. If W is any open set such that $K = \mathrm{supp}\, f \subset W$, then $f \prec W$ and $\Lambda f \leq \mu(W)$. Hence $\Lambda f \leq \mu(K)$, and this argument provides a compact $K \subset V$ such that $\alpha < \mu(K)$, so that (16.2) holds for V.

Claim 4. Suppose that $E_i \in \Sigma_F$ for $i \in \mathbb{N}$, and that $E_i \cap E_j = \emptyset$ when $i \neq j$. If $E = \bigcup_{i=1}^{\infty} E_i$, then $\mu(E) = \sum_{i=1}^{\infty} \mu(E_i)$. If, in addition, $\mu(E)$ is finite, then also $E \in \Sigma_F$.

We first prove that $\mu(K_1 \cup K_2) = \mu(K_1) + \mu(K_2)$ if K_1 and K_2 are disjoint compact sets. Let $\varepsilon > 0$. Urysohn's Lemma provides $f \in C_c(X)$ such that $f \equiv 1$ on K_1, $f \equiv 0$ on K_2, and $0 \leq f \leq 1$. By Claim 2 there exists a function g such that $K_1 \cup K_2 \prec g$ and $\Lambda g < \mu(K_1 \cup K_2) + \varepsilon$. We remark that $K_1 \prec fg$ and $K_2 \prec (1-f)g$. The linearity of Λ implies

$$\mu(K_1) + \mu(K_2) \leq \Lambda(fg) + \lambda(g - fg) = \Lambda g < \mu(K_1 \cup K_2) + \varepsilon.$$

Letting $\varepsilon \to 0$ we deduce $\mu(K_1 \cup K_2) = \mu(K_1) + \mu(K_2)$. Now, if $\mu(E) = +\infty$, the claim follows from Claim 1. We assume that $\mu(E)$ is finite, and we choose $\varepsilon > 0$. For every index i, $E_i \in \Sigma_F$ implies the existence of a compact set $H_i \subset E_i$ such that

$$\mu(H_i) > \mu(E_i) - \frac{\varepsilon}{2^i}.$$

The set $K_n = H_1 \cup \cdots \cup H_n$ is compact, and we deduce by induction that

$$\mu(E) \geq \mu(K_n) = \sum_{i=1}^{n} \mu(H_i) > \sum_{i=1}^{n} \mu(E_i) - \varepsilon.$$

Letting first $n \to +\infty$ and then $\varepsilon \to 0$, we see that $\mu(E) \geq \sum_{i=1}^{\infty} \mu(E_i)$, and the conclusion follows from Claim 1. To prove that $E \in \Sigma_F$ we recall the definition of convergent sequence, and deduce that

$$\mu(E) \leq \sum_{i=1}^{N} \mu(E_i) + \varepsilon$$

for some positive integer N. Hence $\mu(E) \leq \mu(K_N) + 2\varepsilon$, so that E satisfies (16.2). The claim is now proved.

Claim 5. If $E \in \Sigma_F$ and $\varepsilon > 0$, there exist a compact set K and an open set V such that $K \subset E \subset V$ and $\mu(V \setminus K) < \varepsilon$.

Indeed, our definitions show that there are $K \subset E$ and $V \supset E$ such that

$$\mu(V) - \frac{\varepsilon}{2} < \mu(E) < \mu(K) + \frac{\varepsilon}{2}.$$

Now, $V \setminus K$ is open and $V \setminus K \in \Sigma_F$ by Claim 3. Hence Claim 4 yields

$$\mu(K) + \mu(V \setminus K) = \mu(V) < \mu(K) + \varepsilon.$$

Claim 6.   If $A \in \Sigma_F$ and $B \in \Sigma_F$, then $A \setminus B \in \Sigma_F$, $A \cup B \in \Sigma_F$, and $A \cap B \in \Sigma_F$.
Indeed, given $\varepsilon > 0$, Claim 5 shows that there exist sets K_i and V_i such that
$K_1 \subset A \subset V_1$, $K_2 \subset B \subset V_2$, and $\mu(V_i \setminus K_i) < \varepsilon$ for $i = 1, 2$. Since

$$A \setminus B \subset V_1 \setminus K_1 \subset (V_1 \setminus K_1) \cup (K_1 \setminus V_2) \cup (V_2 \setminus K_2),$$

it follows from Claim 1 that $\mu(A \setminus B) \le \varepsilon + \mu(K_1 \setminus V_2) + \varepsilon$. Observing that
$K_1 \setminus V_2$ is a compact subset of $A \setminus B$, we see that $A \setminus B \in \Sigma_F$. Furthermore,
$A \cup B = (A \setminus B) \cup B$ and $A \cap B = A \setminus (A \setminus B)$, and an application of Claim 4
leads to the conclusion.

Claim 7. Σ is a σ-algebra which contains all Borel sets.
Indeed, let K be any compact set in X. For every $A \in \Sigma$, $(X \setminus A) \cap K = K \setminus (A \cap K)$, hence $(X \setminus A) \cap K$ is the difference of two elements of Σ_F. Hence
$A \in \Sigma$ implies $X \setminus A \in \Sigma$. Next we suppose that $A = \bigcup_{i=1}^{\infty} A_i$, where $A_i \in \Sigma$.
Let $B_1 = A_1 \cap K$ and

$$B_n = (A_n \cap K) \setminus (B_1 \cup \cdots \cup B_{n-1})$$

for $n \ge 2$. The collection $\{B_i\}_i$ is a disjoint collection of elements of Σ_F by
Claim 6, and $A \cap K = \bigcup_{i=1}^{\infty} B_i$. Claim 4 implies $A \cap K \in \Sigma_F$, hence $A \in \Sigma$.
Finally, let C be a closed subset of X. Then $C \cap K$ is compact, hence $C \cap K \in \Sigma_F$,
so $C \in \Sigma$. As a particular case, $X \in \Sigma$. Since Σ is a σ-algebra which contains
all closed sets, it must contain all the Borel sets.

Claim 8. $E \in \Sigma_F$ if and only if $E \in \Sigma$ and $\mu(E)$ is finite.
Indeed, let $E \in \Sigma_F$. Claims 2 and 6 imply $E \cap K \in \Sigma_F$ for every compact K,
hence $E \in \Sigma$. Conversely, let $E \in \Sigma$, $\varepsilon > 0$. If $\mu(E)$ is finite, then there exists
and open set V such that $E \subset V$ and $\mu(V) < +\infty$. By Claims 3 and 5, there
exists a compact $K \subset V$ such that $\mu(V \setminus K) < \varepsilon$. Since $E \cap K \in \Sigma_F$, there
exists a compact $H \subset E \cap K$ such that $\mu(E \cap K) < \mu(H) + \varepsilon$. Now

$$E \subset (E \cap K) \cup (V \setminus K)$$

implies

$$\mu(E) \le \mu(E \cap K) + \mu(V \setminus K) < \mu(H) + 2\varepsilon.$$

which implies $E \in \Sigma_F$. The claim is proved, and part (d) of the theorem as well.

Claim 9. μ is a measure on Σ. Indeed, this follows at once from Claim 4 and 8.
Claim 10. μ represents Λ in the sense that $\Lambda f = \int_X f \, d\mu$ for every $f \in C_c(X)$.
This is part (a) of the theorem.
Indeed, we observe that we may prove the inequality $\Lambda f \leq \int_X f \, d\mu$, since by
linearity

$$- \Lambda f = \Lambda(-f) \leq \int_X (-f) \, d\mu = - \int_X f \, d\mu,$$

and the equality follows. So, we call K the support of $f \in C_c(X)$, and let $[a, b]$
be an intervals which contains the range of f. For every $\varepsilon > 0$ we choose points
$y_0, y_1, \ldots, y_n$ such that $y_i - y_{i-1} < \varepsilon$ and $y_0 < a < y_1 < \ldots < y_n = b$. We
define the sets

$$E_i = \{x \in X \mid y_{i-1} < f(x) \leq y_i\} \cap K.$$

As a continuous function, f is Borel measurable, so that every E_i is a Borel set.
Furthermore $E_i \cap E_j = \emptyset$ if $i \neq j$, and $K = \bigcup_{i=1}^n E_i$. Thus there exist open sets
$V_i \supset E_i$ such that

$$\mu(V_i) < \mu(E_i) + \frac{\varepsilon}{n}$$

and such that $f(x) < y_i + \varepsilon$ for every $x \in V_i$. We introduce a partition of unity
$\{h_i\}_i$ such that $h_i \prec V_i$ and $\sum_{i=1}^n h_i = 1$ on K. Hence $f = \sum_{i=1}^n f h_i$ and
Claim 2 yields

$$\mu(K) \leq \Lambda \left(\sum_{i=1}^n h_i \right) = \sum_{i=1}^n \Lambda h_i.$$

Observing that $h_i f < (y_i + \varepsilon) h_i$ and that $y_i - \varepsilon < f(x)$ for $x \in E_i$, we have

$$\Lambda f = \sum_{i=1}^n \Lambda(h_i f) \leq \sum_{i=1}^n (y_i + \varepsilon) \Lambda h_i$$

$$= \sum_{i=1}^n (|a| + y_i + \varepsilon) \, \Lambda h_i - |a| \sum_{i=1}^n \Lambda h_i$$

$$\leq \sum_{i=1}^n (|a| + y_i + \varepsilon) \left(\mu(E_i) + \frac{\varepsilon}{n} \right) - |a| \mu(K)$$

$$= \sum_{i=1}^n (y_i - \varepsilon) \mu(E_i) + 2\varepsilon \mu(K) + \frac{\varepsilon}{n} \sum_{i=1}^n (|a| + y_i + \varepsilon)$$

$$\leq \int_X f \, d\mu + \varepsilon \left(2\mu(K) + |a| + b + \varepsilon \right).$$

The arbitrariness of $\varepsilon > 0$ proves that $\Lambda f \leq \int_X f \, d\mu$. The theorem is completely proved.

□

The Riesz Representation Theorem will be our access point to the concrete Lebesgue measure in $\mathbb{R}^n$. Although our approach might be considered rather abstract, it has some advantages over the usual approach via an outer measure and a Carathéodory completion.

Definition 16.21 A measure μ defined on the σ-algebra of Borel sets in a locally compact Hausdorff space X is called a Borel measure on X.

Definition 16.22 A Borel set E is outer regular if

$$\mu(E) = \inf \{\mu(V) \mid E \subset V, \ V \text{ open}\}.$$

The Borel set E is inner regular if

$$\mu(E) = \sup \{\mu(K) \mid K \subset E, \ K \text{ compact}\}.$$

Finally, the measure μ is regular if every Borel set is both inner and outer regular.

Remark 16.4 An inspection of Theorem 16.26 shows that the measure induced by the positive linear functional Λ is not regular, in general. Indeed, the inner regularity holds for open sets and for Borel sets of finite measure.

Definition 16.23 A set E in a topological space is σ-compact if E is a countable union of compact sets.

We can now show that σ-compactness fills the gap of inner regularity.

Theorem 16.27 *Let X be a locally compact, σ-compact Hausdorff space, and let Σ, μ be defined according to Theorem 16.26.*

(i) *If $E \in \Sigma$ and $\varepsilon > 0$, there exist a closed set F and an open set V such that $F \subset E \subset V$ and $\mu(V \setminus F) < \varepsilon$.*
(ii) *μ is a regular measure on X.*
(iii) *If $E \in \Sigma$, there exist sets A and B such that A is F_σ, B is G_δ, $A \subset E \subset B$, and $\mu(B \setminus A) = 0$.*

Proof By assumption, $X = \bigcup_{i=1}^{\infty} K_i$, where each K_i is compact. Pick $E \in \Sigma$ and $\varepsilon > 0$. There results $\mu(K_n \cap E) < +\infty$ and there exist open sets $V_n \supset K_n \cap E$ such that

$$\mu\left(V_n \setminus (K_n \cap E)\right) < \frac{\varepsilon}{2^{n+1}}$$

for $n \in \mathbb{N}$. We define $V = \bigcup_{i=1}^{\infty} V_i$ so that $V \setminus E \subset \bigcup_{i=1}^{\infty} (V_n \setminus (K_n \cap E))$ and $\mu(V \setminus E) < \varepsilon/2$. The very same construction applies to $X \setminus E$ in place of E, yielding an open set $W \supset X \setminus E$ such that $\mu(W \setminus (X \setminus E)) < \varepsilon/2$. With $F = X \setminus W$ we get $F \subset E$ and $E \setminus F = W \setminus (X \setminus E)$. Conclusion (i) follows at once.

Every closed set F is σ-compact, since $F = \bigcup_{i=1}^{\infty} (F \cap K_i)$. Hence (i) implies that every set $E \in \Sigma$ is inner regular, and (ii) follows.

The proof of (iii) is now easy, choosing $\varepsilon = 1/j$, $j = 1, 2, \ldots$ This provides us with closed sets F_j and open sets V_j such that $F_j \subset E \subset V_j$ and $\mu(V_j \setminus F_j) < 1/j$. We call $A = \bigcup_{j=1}^{\infty} F_j$, $B = \bigcap_{j=1}^{\infty} V_j$, to get $A \setminus E \setminus B$ and $\mu(B \setminus A) = 0$. By definition A is F_σ and B is G_δ. The proof is complete.   □

The following regularity result will be used in the construction of the Lebesgue measure.

Theorem 16.28 *Let X be a locally compact Hausdorff space in which every open set is σ-compact. If λ is a Borel measure on X such that $\lambda(K)$ is finite for every compact set K, then λ is regular.*

Proof We introduce the positive linear functional Λ defined on $C_c(X)$ by $\Lambda f = \int_X f \, d\lambda$. The assumption on λ implies that Λ is well-defined, and Theorem 16.26 provides us with a regular measure μ such that

$$\int_X f \, d\lambda = \int_X f \, d\mu$$

for every $f \in C_c(X)$. We will prove that $\lambda = \mu$, so that λ is a regular measure as a corollary.

Let V be an open set, so that $V = \bigcup_{i=1}^{\infty} K_i$, where every K_i is compact. By Urysohn's Lemma we can choose $f_i \in C_c(X)$ such that $K_i \preccurlyeq f_i \preccurlyeq V$. Let $g_n = \max\{f_1, \ldots, f_n\}$. Clearly $g_n \nearrow \chi_V$ pointwise, and Beppo Levi's Theorem ensures that

$$\lambda(V) = \lim_{n \to +\infty} \int_X g_n \, d\lambda = \lim_{n \to +\infty} \int_X g_n \, d\mu = \mu(V).$$

For a generic Borel set E, we fix any $\varepsilon > 0$. By Theorem 16.27, there exist a closed set F and an open set V such that $F \subset E \subset V$ such that $\mu(V \setminus F) < \varepsilon$. In particular $\mu(V) \leq \mu(F) + \varepsilon \leq \mu(E) + \varepsilon$. We apply the previous considerations to the open set $V \setminus F$, and we get $\lambda(V \setminus V) < \varepsilon$ and thus $\lambda(V) < \lambda(E) + \varepsilon$. We conclude that

$$\lambda(E) \leq \lambda(V) = \mu(V) \leq \mu(E) + \varepsilon$$

$$\mu(E) \leq \mu(V) = \lambda(V) \leq \lambda(E) + \varepsilon.$$

Hence $|\lambda(E) - \mu(E)| < \varepsilon$ for every $\varepsilon > 0$, and therefore $\mu(E) = \lambda(E)$. The proof is complete.   □

Since continuous functions with compact support appear as the basic ingredient of the Riesz Representation Theorem, we investigate their role in Measure Theory.

The following is a basic result in the approximation of measurable functions by means of more regular functions.

Theorem 16.29 (Lusin) *Let X be a locally compact Hausdorff space, and let μ be a measure on X which has the properties described in Theorem 16.26. Suppose f is a real-valued measurable function on X, that A is a measurable set such that $\mu(A)$ is finite and $f(x) = 0$ whenever $x \notin A$. For every $\varepsilon > 0$ there exists $g \in C_c(X)$ such that*

$$\mu(\{x \in X \mid f(x) \neq g(x)\}) < \varepsilon.$$

Furthermore, we may select g so that

$$\sup\{|g(x)| \mid x \in X\} \leq \sup\{|f(x)| \mid x \in X\}.$$

Proof Suppose initially that $0 \leq f < 1$ and that A is compact. We consider a sequence $\{s_n\}_n$ of simple functions as in Theorem 16.8. We define recursively $t_1 = s_1$ and $t_n = s_n - s_{n-1}$. The function $2^n t_n$ is the characteristic function of a set $T_n \subset A$, and

$$f(x) = \sum_{n=1}^{\infty} t_n(x), \quad x \in X.$$

Fix an open set V such that $A \subset V$ and $\overline{V}$ is compact. There exist compact sets K_n and open sets V_n such that $K_n \subset T_n \subset V_n$ and $\mu(V_n \setminus K_n) < \varepsilon/2^n$. By Urysohn's Lemma there are functions h_n such that $K_n \prec h_n \prec V_n$. We define

$$g(x) = \sum_{n=1}^{\infty} \frac{h_n(x)}{2^n}$$

for every $x \in X$. This series converges uniformly on X, so that g is a continuous function. The support of g is contained in $\overline{V}$. Since $2^{-n} h_n = t_n$ except in $V_n \setminus K_n$, we must have $g = f$ except in $\bigcup_{n=1}^{\infty}(V_n \setminus K_n)$. The measure of this latter set is smaller than ε. The first statement is proved if $0 \leq f < 1$ and if A is compact.

A simple scaling argument shows that the conclusion holds for every bounded measurable f if A is compact. To remove the compactness condition, we remark that the set A has finite measure, hence it must contain a compact set K such that $\mu(A \setminus K)$ is as small as we wish.

To remove the boundedness condition on f, we set $B_n = \{x \in X \mid |f(x)| > n\}$, so that $\bigcap_{n=1}^{\infty} B_n = \emptyset$. Hence $\mu(B_n) \to 0$ as $n \to +\infty$. But f coincides with the bounded function $(1 - \chi_{B_n})f$ except on B_n, and the proof of the first statement of the theorem is complete in the general case.

Finally, let $R = \sup\{|f(x)| \mid x \in X\}$, and define

$$
\varphi(t) = \begin{cases} t & \text{if } -R \leq t \leq R \\ \frac{Rt}{|t|} & \text{if } |t| > R. \end{cases}
$$

The function φ is continuous, and if g satisfies the first statement of the theorem, then $g_1 = \varphi \circ g$ satisfies the second statement as well. The proof is complete. □

16.6.1 The Concrete Lebesgue Measure

From now on, we want to construct the concrete Lebesgue measure using Theorem 16.26 as a starting point. This allows us to deduce some additional (regularity) properties of the Lebesgue measure almost for free.

Due to some possible conflict with the index of several sequences, we will denote by $k \geq 1$ the dimension of our basic Euclidean space $\mathbb{R}^k$. Let us recall that a k-cell is any set of the form

$$
W = \left\{ x = (x_1, \ldots, x_k) \in \mathbb{R}^k \ \middle|\ \alpha_i < x_i < \beta_i, \ 1 \leq i \leq k \right\}
$$
$$
= (\alpha_1, \beta_1) \times \cdots \times (\alpha_k, \beta_k), \tag{16.3}
$$

where α_i and β_i are given real numbers. Sometimes we can replace some or all inequality signs $<$ by $\leq$: we already suspect that the difference will be a set of measure zero. Finally, recall that the volume of the k-cell W is defined to be

$$
\mathrm{Vol}(W) = \prod_{i=1}^{k} (\beta_i - \alpha_i).
$$

For our purposes, it will be better to replace the basic spherical neighborhood $B(a, \delta)$ by a box. More precisely, for $a = (a_1, \ldots, a_k) \in \mathbb{R}^k$ and $\delta > 0$, we define the δ-box with corner at a as

$$
Q(a, \delta) = \{x \mid a_i \leq x_i < a_i + \delta, \ 1 \leq i \leq k\}.
$$

Definition 16.24 For $n \in \mathbb{N}$ we define P_n as the set of all points $x \in \mathbb{R}^k$ whose coordinates are integral multiples of 2^{-n}, and we define Ω_n as the collection of all 2^{-n}-boxes with corners at points of P_n.

Theorem 16.30 *Every non-empty open set of $\mathbb{R}^k$ is a countable union of disjoint boxes belonging to $\bigcup_{i=1}^{\infty} \Omega_i$.*

Proof Let V be an open set. Every point $x \in V$ lies in an open ball which lies, in turn, in V. Hence $x \in Q \subset V$ for some suitable Q which belongs to some Ω_n. Equivalently, V is the union of all boxes which lie in V and which belong to some Ω_n.

From this collection of boxes we first select those boxes belonging to Ω_1, and we remove those in $\Omega_2, \Omega_3, \dots$ which lie in any of the selected boxes. From the remaining collection, we select those boxes of Ω_2 which lie in V, and we remove those in $\Omega_3, \Omega_4, \dots$ which lie in any of the selected boxes. This procedure provides a countable collection of disjoint boxes in $\Omega_1 \cup \Omega_2 \cup \Omega_3 \cup \cdots$ whose union is V. The proof is complete.   □

The next theorem defines the concrete Lebesgue measure in $\mathbb{R}^k$.

Theorem 16.31 *There exists a positive complete measure m defined on a σ-algebra $\mathcal{M}$ of $\mathbb{R}^k$, with the following properties:*

(a) $m(W) = \mathrm{Vol}(W)$ *for every k-cell W.*

(b) *$\mathcal{M}$ contains all Borel sets; more precisely, $E \in \mathcal{M}$ if and only if there exist sets A and B such that $A \subset E \subset B$, A is F_σ, B is G_δ, and $m(B \setminus A) = 0$. Moreover, m is a regular measure.*

(c) *m is invariant under translations.*

(d) *If μ is any Borel measure on $\mathbb{R}^k$, invariant under translations and such that $\mu(K)$ is finite for every compact K, then there exists a constant c such that $\mu(E) = cm(E)$ for every $E \in \mathcal{M}$.*

Proof Our proof, as we said, constructs $\mathcal{M}$ and m via the Riesz Representation Theorem. Clearly enough, we need a positive linear functional to begin with. This functional is precisely the Riemann (or Cauchy) integral in $\mathbb{R}^k$, which was already studied in Sect. 15.1. For the reader's sake we recall here the basic ideas.

Let $f \in C_c(\mathbb{R}^k)$, and define for $n \in \mathbb{N}$ the functional

$$\Lambda_n f = \frac{1}{2^{nk}} \sum \{ f(x) \mid x \in P_n \},$$

where P_n was introduced in Definition 16.24. Let W be an open k-cell containing the support of f. By uniform continuity, there exist an integer N and functions g, h with support in W, such that (i) g and h are constant on each box of Ω_N, (ii) $g \leq f \leq h$, (iii) $h - g < \varepsilon$.

It is easy to see that

$$\Lambda_N g = \Lambda_n g \leq \Lambda_n f \leq \Lambda_n h = \Lambda_N h$$

for every $n > N$. As a consequence,

$$\limsup_{n \to +\infty} \Lambda_n f - \liminf_{n \to +\infty} \Lambda_n f < \varepsilon \, \mathrm{Vol}(W),$$

which implies that the limit $\Lambda f = \lim_{n\to+\infty} \Lambda_n f$ exists (as a finite real number). As a simple exercise, the reader can prove that Λ is a positive linear operator on $C_c(\mathbb{R}^k)$. The σ-algebra $\mathcal{M}$ and the measure m are now defined according to Theorem 16.26. Since $\mathbb{R}^k$ is σ-compact, (b) is automatically true.

To prove (a) we use the k-cell W of (16.3). For any $r \in \mathbb{N}$, we call E_r the union of those boxes of Ω_r whose closure belongs is contained in W. We select f_r such that $\overline{E_r} \prec f_r \prec W$ and we put $g_r = \max\{f_1, \ldots, f_r\}$. It follows from the definition of Λ that

$$\mathrm{Vol}(E_r) \leq \Lambda f_r \leq \Lambda g_r \leq \mathrm{Vol}(W).$$

When we let $r \to +\infty$, $\mathrm{Vol}(E_r) \to \mathrm{Vol}(W)$ and

$$\Lambda g_r = \int_{\mathbb{R}^k} g_r \, dm \to m(W)$$

by Beppo Levi's Theorem, observing that $g_r \nearrow \chi_W$. Thus $m(W) = \mathrm{Vol}(W)$ for every open k-cell W. Since any k-cell is the intersection of a decreasing sequence of open k-cells, the proof of (a) is complete.

Let us now remark what follows: if λ is a Borel measure on $\mathbb{R}^k$ and $\lambda(E) = m(E)$ for every box E, then the same equality holds for every open set E. This is indeed a consequence of Theorem 16.30. Once this is established, it follows from Theorem 16.28 that the equality holds for every Borel set, since λ and m are regular.

Consider statement (c). Let $x \in \mathbb{R}^k$ and define $\lambda(E) = m(E + x)$. Clearly λ is a measure, and (a) implies that λ coincides with m on all boxes, and thus on all Borel sets: this means that $m(E) = m(E + x)$. The same equality holds for every $E \in \mathcal{M}$ because of (b). Hence (c) is proved.

Finally, suppose that μ satisfies the hypotheses of (d). Let Q_0 be a 1-box, and set $c = \mu(Q_0)$. Since Q_0 is the union of 2^{nk} disjoint 2^{-n}-boxes which are translates of each other, we have

$$2^{nk}\mu(Q) = \mu(Q_0) = cm(Q_0) = c \cdot 2^{nk} m(Q)$$

for every 2^{-n}-box Q. Theorem 16.30 implies that $\mu(E) = cm(E)$ for all open sets E of $\mathbb{R}^k$, and the proof of (d) is complete.   □

Remark 16.5 The symbolism for the concrete Lebesgue measure m is very rich. Firstly, it may be useful to write m_k in order to denote the dimension of the Euclidean space. But sometimes λ_k is preferred[3] to m_k. Some books use $\mathcal{L}^k$, and even $|E|$ to denote the Lebesgue measure of E. When integrals come into play, the

---

[3] λ is reminiscent of the "L" of Lebesgue.

pedantic notation

$$\int_{\mathbb{R}^k} f \, dm_k$$

is often replaced by

$$\int_{\mathbb{R}^k} f(x) \, dx,$$

where x is a dummy variable. However we need to point out that the bad habit of calling dx the Lebesgue measure, which occurs in several Calculus texts.

It is interesting to show that non-measurable sets exist for the concrete Lebesgue measure.

Theorem 16.32 *Every set of positive Lebesgue measure contains a non-measurable subset.*

This is indeed a corollary of a more general result.

Theorem 16.33 *If $A \subset \mathbb{R}$ and if every subset of A is Lebesgue measurable, then $m(A) = 0$.*

Proof The basic tool in the proof is the structure of $\mathbb{R}$ as a group relative to addition. Consider $\mathbb{Q}$ as a subgroup of $\mathbb{R}$, and introduce an equivalence relation as follows: $x \sim y$ if and only if $x - y \in \mathbb{Q}$. For $x \in \mathbb{R}$, we write $[x]_\sim = \{y \in \mathbb{R} \mid y \sim x\}$, the *equivalence class* of x. Using the Axiom of Choice, we construct a set E which contains exactly one point from each equivalence class of $\mathbb{Q}$ in $\mathbb{R}$.

Claim 1.   If $r \in \mathbb{Q}, s \in \mathbb{Q}$ and $r \neq s$, then $(E + r) \cap (E + s) = \emptyset$. Indeed, suppose $x \in (E + r) \cap (E + s)$. Then $x = y + r = z + s$ for some $y \in E, z \in E, y \neq z$. But $y - z = s - r \in \mathbb{Q}$, so that y and z lie in the same equivalence class of $\mathbb{Q}$, a contradiction.

Claim 2.   Every point $x \in \mathbb{R}$ belongs to $E + r$, for some $r \in \mathbb{Q}$. Indeed, there exists one and only one point y of E such that $y \sim x$. We merely define $r = x - y$, and the claim is proved.

Let now $t \in \mathbb{Q}$ and put $A_t = A \cap (E + t)$. By assumption, A_t is Lebesgue measurable. Let K be a compact subset of A_t, and let H be the union of all translates $K + r$, as r ranges over $\mathbb{Q} \cap [0, 1]$. The set H is bounded, hence $m(H)$ is finite. Since $K \subset E + t$, Claim 1 shows that the sets $K + r$ are pairwise disjoint. Thus

$$m(H) = \sum_r m(K + r).$$

But $M(K + r) = m(K)$, hence $m(K) = 0$. Since this holds for every compact $K \subset A_t$, we deduce that $m(A_t) = 0$. Claim 2 shows that $A = \bigcup \{A_t \mid t \in \mathbb{Q}\}$. Since $\mathbb{Q}$ is a countable set, we conclude that $m(A) = 0$, and the proof is complete.   □

Lusin's theorem is a powerful result that allows us to approximate different classes of functions by means of continuous functions.

Theorem 16.34 *Let X be a locally compact Hausdorff space, endowed with a measure as in Theorem 16.26. If $1 \leq p < +\infty$, then $C_c(X)$ is dense in $L^p(X)$.*

Proof Let S be the set of all the measurable simple functions s on X such that $\mu(\{x \in X \mid s(x) \neq 0\}) < +\infty$. We claim that S is dense in $L^p(X)$. Firstly, $S \subset L^p$. Suppose that $f \geq 0$ is a function in L^p, and let $\{s_n\}_n$ be a sequence of measurable simple functions which monotonically converges to f. Since $0 \leq s_n \leq f$, we have $s_n \in L^p$ for every n, hence $s_n \in S$. But $|f - s_n|^p \leq f^p$, hence the Dominated Convergence Theorem yields $\|f - s_n\|_p \to 0$ as $n \to +\infty$, and f belongs to the closure of S in L^p. The case of a function f with variable sign follows from the decomposition $f = f^+ - f^-$.

Now let $\varepsilon > 0$ and $s \in S$. By Lusin's Theorem, there exists $g \in C_c(X)$ such that $g = s$ except on a set of measure $< \varepsilon$, and $|g| \leq \sup_X |s|$. Hence

$$\|g - s\|_p \leq 2\varepsilon^{1/p}\|s\|_p.$$

Since S is dense in L^p, the proof is complete.   □

We conclude this section with the answer to a basic question: what functions are Riemann-integrable?

Theorem 16.35 (Lebesgue) *A bounded function f on an interval $[a, b]$ is Riemann-integrable if and only if the set of points at which f is discontinuous has measure zero.*

Proof We introduce the lower and the upper envelope of f as follows:

$$g(y) = \sup\{\inf\{f(x) \mid |x - y| < \delta\} \mid \delta > 0\}$$
$$h(y) = \inf\{\sup\{f(x) \mid |x - y| < \delta\} \mid \delta > 0\}.$$

The following statements are easy exercises:

(a) for every $x \in [a, b]$, $g(x) \leq f(x) \leq h(x)$. More precisely, $f(x) = g(x)$ if and only if f is lower semicontinuous at x, and $f(x) = h(x)$ if and only if f is upper semicontinuous at x. In particular, $g(x) = h(x)$ if and only if f is continuous at x.

(b) The function g is lower semicontinuous, and the function h is upper semicontinuous.

(c) If φ is any lower semicontinuous function such that $\varphi \leq f$ on $[a, b]$, then $\varphi \leq g$ on $[a, b]$. Similarly, if ψ is any upper semicontinuous function such that $f \leq \psi$ on $[a, b]$, then $\psi \leq h$ on $[a, b]$.

Consider now any step function $\varphi \geq f$; in particular $\varphi \geq h$ except at a finite number of points. Hence

$$\int_a^b h \leq \overline{\int_a^b} f \, dx.$$

On the other hand, by definition of the upper envelope of f, there exists a sequence $\{\psi_n\}_n$ of step functions such that $\varphi_n \searrow f$. Since f is bounded on $[a, b]$, The Dominated Convergence Theorem implies

$$\int_a^b h \, dx = \lim_{n \to +\infty} \int_a^b \varphi_n \, dx \geq \overline{\int_a^b} f \, dx.$$

We have thus proved that

$$\overline{\int_a^b} f \, dx = \int_a^b h \, dx.$$

By the same token,

$$\underline{\int_a^b} f \, dx = \int_a^b g \, dx.$$

It follows that f is R-integrable on $[a, b]$ if and only if $\int_a^b (h - g) \, dx = 0$. Since $h \geq g$, we deduce that this happens if and only if $h = g$ a.e. on $[a, b]$. Recalling (a) above, this is equivalent to the fact that the set of points at which f is discontinuous has measure zero. The proof is complete.   □

16.7   Mollifiers and Regularization

In the whole section, we will be working on the measurable space $X = \mathbb{R}^N$ with the standard Lebesgue measure.[4]

Our aim is to introduce a general technique for *regularizing* integrable functions. As a by-product, we will provide a compactness result in L^p. Let us start with some notation.

Definition 16.25   Let Ω be an open subset of $\mathbb{R}^N$, $N \geq 1$. We write

$$\mathcal{D}(\Omega) = \left\{ u \in C^\infty(\Omega) \mid \text{the support of } u \text{ is a compact subset of } \Omega \right\}.$$

---

[4] The choice of N as the fixed dimension of the Euclidean space allows us to use freely the index n.

A multi-index is an element $\alpha = (\alpha_1, \ldots, \alpha_N)$ of $\mathbb{N}^N$. Its length is

$$|\alpha| = \sum_{j=1}^{N} \alpha_j,$$

and we will write

$$\partial_j = \frac{\partial}{\partial x_j}, \qquad D^\alpha = \partial_1^{\alpha_1} \partial_2^{\alpha_2} \cdots \partial_N^{\alpha_N}.$$

Exercise 16.11 Prove that the function $f: \mathbb{R} \to \mathbb{R}$ such that

$$f(x) = \begin{cases} e^{1/x} & \text{if } x < 0 \\ 0 & \text{if } x \geq 0 \end{cases}$$

belongs to $C^\infty(\mathbb{R})$. *Hint:* show by induction that for every $n \in \mathbb{R}$ and every $x < 0$ there results $D^n f(0) = 0$ and $D^n f(x) = P_n(1/x)e^{1/x}$, where P_n is a polynomial.

Definition 16.26 (Standard Mollifiers) Let $\varrho: \mathbb{R}^N \to \mathbb{R}$ such that

$$\varrho(x) = \begin{cases} \dfrac{e^{\frac{1}{|x|^2 - 1}}}{\int_{B(0,1)} e^{\frac{1}{|x|^2 - 1}} \, dx} & \text{if } |x| < 1 \\ 0 & \text{if } |x| \geq 1. \end{cases}$$

The standard mollifiers are the functions $\varrho_n: \mathbb{R}^N \to \mathbb{R}$ such that $\varrho_n(x) = n^N \varrho(nx)$ for every $x \in \mathbb{R}^N$ and every $n \in \mathbb{N}$.

Exercise 16.12 Prove that for every $n \in \mathbb{N}$, $\varrho_n \in \mathcal{D}(\mathbb{R}^N)$, $\mathrm{supp}\, \varrho_n \subset \overline{B(0,1)}$, $\varrho_n \geq 0$ and $\int_{\mathbb{R}^N} \varrho_n(x) \, dx = 1$.

Since we will often need to restrict functions to relatively compact subsets of Ω, we introduce a new space.

Definition 16.27 (Local Lebesgue Spaces) Let Ω be an open subset of $\mathbb{R}^N$, and let ω be such that ω is open and $\overline{\omega}$ is a compact subset of Ω. For brevity, we will write $\omega \subset\subset \Omega$.

For every $1 \leq p < \infty$ we define

$$L_{\mathrm{loc}}^p(\Omega) = \left\{ u \in \mathbb{R}^\Omega \mid u_{|\omega} \in L^p(\omega) \text{ for every } \omega \subset\subset \Omega. \right\}$$

Although local Lebesgue spaces are not normed spaces,[5] yet we introduce a definition of convergent sequences.

Definition 16.28 (Convergence in L^p_{loc}) A sequence $\{u_n\}_n$ in $L^p_{\text{loc}}(\Omega)$ converges to u if and only if for every $\omega \subset\subset \Omega$

$$\lim_{n \to +\infty} \int_\omega |u_n - n|^p \, dx = 0.$$

Here comes the most useful tool of Harmonic Analysis: the convolution. We consider a particular case, which however is sufficient for our purposes.

Definition 16.29 (Convolution in Local Lebesgue Spaces) Suppose $u \in L^1_{\text{loc}}(\Omega)$ and $v \in C_c(\mathbb{R}^N)$ are such that

$$\operatorname{supp} v \subset \overline{B\left(0, \frac{1}{n}\right)}.$$

For every $n \in \mathbb{N}$, the convolution $v * u$ is defined on the set

$$\Omega_n = \left\{ x \in \Omega \;\middle|\; d(x, \partial\Omega) > \frac{1}{n} \right\}$$

by

$$v * u(x) = \int_\Omega v(x - y)u(y) \, dy = \int_{B(0,1/n)} v(y)u(x - y) \, dy.$$

Definition 16.30 (Translation Operator) If $|y| < 1/n$, the translation of $u \in L^1_{\text{loc}}(\Omega)$ by y is defined on Ω_n by

$$\tau_y u(x) = u(x - y).$$

The use of convolution products in regularization is explained by the next result.

Theorem 16.36 *Suppose $u \in L^1_{\text{loc}}(\Omega)$ and $v \in \mathcal{D}(\mathbb{R}^N)$ are such that*

$$\operatorname{supp} v \subset \overline{B\left(0, \frac{1}{n}\right)}.$$

---

[5] The intuitive reason is that there is no norm that can take into account all possible sets $\omega \subset\subset \Omega$ at the same time.

Then $v * u \in C^\infty(\Omega_n)$, and for every multi-index α there results

$$D^\alpha(v * u) = (D^\alpha v) * u.$$

Proof We prove the result under the additional assumption $|\alpha| = 1$. The general case follows by induction on $|\alpha|$. Fix $x \in \Omega_n$: there exists $r > 0$ such that[6] $\overline{B(0, r)} \subset \Omega_n$. Hence

$$\omega = B\left(x, r + \frac{1}{n}\right) \subset\subset \Omega,$$

and if $0 < |\varepsilon| < r$ we have

$$\frac{v * u(x + \varepsilon\alpha)}{\varepsilon} = \int_\omega \frac{v(x + \varepsilon\alpha - y) - v(x - y)}{\varepsilon} u(y)\, dy.$$

Since $v(x + \varepsilon\alpha - y) - v(x - y) = \varepsilon D^\alpha v(x - y) + o(\varepsilon)$ and

$$\sup\left\{\left|\frac{v(x + \varepsilon\alpha - y) - v(x - y)}{\varepsilon}\right| \,\middle|\, y \in \omega,\ 0 < |\varepsilon| < r\right\},$$

the Dominated Convergence Theorem yields

$$D^\alpha(v * u)(x) = \int_\omega D^\alpha v(x - y) u(y)\, dy = (D^\alpha v) * u(x).$$

$\square$

Theorem 16.37 (Continuity of Translations) *Let $\omega \subset\subset \Omega$.*

(a) If $u \in C(\Omega)$, then

$$\lim_{y \to 0} \sup_{x \in \omega} |\tau_y u(x) - u(x)| = 0.$$

(b) If $u \in L^p_{loc}(\Omega)$ for some $1 \le p < \infty$, then

$$\lim_{y \to 0} \|\tau_y u - u\|_p = 0.$$

Proof

(a) Choose an open set U such that $\omega \subset\subset U \subset\subset \Omega$. Since u is uniformly continuous on U, the conclusion follows immediately from (16.4).

---

[6] $\mathbb{R}^N$ is locally compact!

(b) Pick $\varepsilon > 0$, and choose now an open set U such that $\omega \subset\subset U \subset\subset \Omega$. By Theorem 16.34 a function $v \in C_c(U)$ exists such that $\|u - v\|_{L^p(U)} \le \varepsilon$. By (a) there exists $0 < \delta < d(\omega, \partial U)$ such that, if $|y| < 1/n$, then

$$\sup_{x \in \omega} |\tau_y u(x) - u(x)| \le \varepsilon.$$

Hence, if $|y| < 1/n$,

$$\|\tau_y u - u\|_{L^p(\omega)} \le \|\tau_y u - \tau_y v\|_{L^p(\omega)} + \|\tau_y v - v\|_{L^p(\omega)} + \|v - u\|_{L^p(\omega)}$$

$$\le 2 \|u - v\|_{L^p(U)} + m_N(\omega)^{1/p} \sup_{x \in \omega} |\tau_y u(x) - v(x)|$$

$$\le \left(2 + m_N(\omega)^{1/p}\right) \varepsilon,$$

where m_N denotes the Lebesgue measure in $\mathbb{R}^N$, as usual. Since $\varepsilon > 0$ is arbitrary, the proof is complete.

$\square$

Theorem 16.38 (Regularization Theorem)

(a) *If* $u \in C(\Omega)$, *then* $\{\varrho_n * u\}_n$ *converges uniformly to* u *on every compact subset of* Ω.
(b) *If* $u \in L^p_{\text{loc}}(\Omega)$, $1 \le p < \infty$, *then* $\{\varrho_n * u\}_n$ *converges to* u *in* $L^p(\Omega)$.

Proof

(a) We claim that, for $n \in \mathbb{N}$ sufficiently large,

$$\sup_{x \in \omega} |\varrho_n * u(x) - u(x)| \le \sup_{|y| < \frac{1}{n}} \sup_{x \in \omega} |\tau_y u(x) - u(x)|. \qquad (16.4)$$

Indeed, for n sufficiently large, $\omega \subset\subset \Omega_n$. From the properties of the mollifiers, for every $x \in \omega$ we have

$$|\varrho_n * u(x) - u(x)| = \left| \int_{B(0,1/n)} \varrho_n(y) (u(x - y) - u(x)) \, dy \right|$$

$$\le \sup_{|y| < 1/n} \sup_{x \in \omega} |u(x - y) - u(x)|,$$

and (16.4) is proved. The conclusion follows from the continuity of translations.
(b) We claim that, for every $n \in \mathbb{N}$ sufficiently large,

$$\|\varrho_n * u - u\|_{L^p(\omega)} \le \sup_{|y| < 1/n} \|\tau_y u - u\|_{L^p(\omega)}. \qquad (16.5)$$

Indeed, by Hölder's inequality, for every $x \in \omega$ we have

$$|\varrho_n * u(x) - u(x)| = \left| \int_{B(0,1/n)} \varrho_n(y) \, (u(x-y) - u(x)) \, dy \right|$$

$$\leq \left(\int_{B(0,1/n)} \varrho_n(y) \, |u(x-u) - u(x)|^p \, dy \right)^{1/p}.$$

Interchanging the order of integration,

$$\int_\omega |\varrho_n(y) * u(x) - u(x)|^p \, dx \leq \int_\omega dx \int_{B(0,1/n)} \varrho_n(y) \, |u(x-y) - u(x)|^p \, dy$$

$$= \int_{B(0,1/n)} dy \int_\omega \varrho_n(y) \, |u(x-y) - u(x)|^p \, dx$$

$$\leq \sup_{|y|<1/n} \int_\omega |u(x-y) - u(x)|^p \, dx,$$

and (16.5) follows. The conclusion follows from the continuity of translations.
□

As we promised, the convolution with mollifiers allows us to approximate integrable functions with smooth functions.

Theorem 16.39 (Smooth Functions Are Dense in Lebesgue Spaces) *If $1 \leq p < \infty$, then $\mathcal{D}(\Omega)$ is dense in $L^p(\Omega)$.*

Proof Theorem 16.34 ensures the density of $C_c(\Omega)$ in $L^p(\Omega)$. Fix $u \in C_c(\Omega)$ and an open set ω such that supp $u \subset\subset \omega \subset\subset \Omega$. Taking n sufficiently large, the support of $u_n = \varrho_n * u$ is contained in ω, and $u_n \in C^\infty(\mathbb{R}^N)$. It follows that $u_n \in \mathcal{D}(\Omega)$, and the conclusion follows from Theorem 16.38.
□

It is not too difficult to convince ourselves that we cannot approximate *globally* a continuous function with a smooth function, or equivalently that the uniform convergence of $\varrho_n * u$ to u in part (a) of Theorem 16.38 is optimal.

On the other hand, part (b) extends to the case $\Omega = \mathbb{R}^N$.

Theorem 16.40 *Let $1 \leq p < \infty$. If $u \in L^p(\mathbb{R}^N)$, then $\|\varrho_n * u\|_p \leq \|u\|_p$, and $\varrho_n * u \to u$ in $L^p(\mathbb{R}^N)$.*

Proof By Hölder's inequality,

$$|\varrho_n * u(x)| = \left| \int_{\mathbb{R}^N} u(y)\varrho_n(x-y) \, dy \right| \leq \left| \int_{\mathbb{R}^N} |u(y)|^p \, \varrho_n(x-y) \, dy \right|^{1/p}.$$

Using Fubini's Theorem we see that

$$\int_{\mathbb{R}^N} |\varrho_n * u(x)|^p \, dx \leq \int_{\mathbb{R}^N} dx \int_{\mathbb{R}^N} |u(y)|^p \, \varrho_n(x-y) \, dy$$

$$= \int_{\mathbb{R}^N} dy \int_{\mathbb{R}^N} |u(y)|^p \, \varrho_n(x-y) \, dx$$

$$= \int_{\mathbb{R}^N} |u(y)|^p \, dy.$$

This proves the first part of the theorem. Let now $u \in L^p(\mathbb{R}^N)$ and $\varepsilon > 0$. By Theorem 16.34 there exists $v \in C_c(\mathbb{R}^N)$ such that $\|v - u\|_p \leq \varepsilon$. By Theorem 16.38, $\varrho_n * v \to v$ in $L^p(\mathbb{R}^N)$. Fix $m \in \mathbb{N}$ such that, for every $n \geq m$, there results $\|\varrho_n * v - v\|_p \leq \varepsilon$. For these values of n,

$$\|\varrho_n * u - u\|_p \leq \|\varrho_n * (u - v)\|_p + \|\varrho_n * v - v\|_p + \|v - u\|_p \leq 3\varepsilon,$$

and the proof is complete. $\square$

16.8 Compactness in Lebesgue Spaces

Theorem 16.41 (M. Riesz) *Let Ω be an open subset of $\mathbb{R}^N$, $1 \leq p < \infty$ and suppose that $S \subset L^p(\Omega)$ satisfies*

(a) $c = \sup\{\|u\|_{L^p(\Omega)} \mid u \in S\} \in \mathbb{R}$;
(b) for every $\varepsilon > 0$ there exists $\omega \subset\subset \Omega$ such that $\sup\{\|u\|_{L^p(\Omega \setminus \omega)} \mid u \in S\} \leq \varepsilon$;
(c) for every $\omega \subset\subset \Omega$, $\lim_{y \to 0} \sup\{\|\tau_y u - u\|_{L^p(\omega)} \mid u \in S\} = 0$.

Then S is relatively compact in $L^p(\Omega)$.

Proof Fix $\varepsilon > 0$, and let ω be as in condition (b). Using (c) we see that there exists $\delta \in (0, d(\omega, \partial\Omega))$ such that if $|y| < \delta$, then

$$\sup\left\{\|\tau_y u - u\|_{L^p(\omega)} \mid u \in S\right\} \leq \varepsilon.$$

Choose now an integer $n > 1/\delta$. It follows from (16.5) that

$$\sup\left\{\|\tau_y u - u\|_{L^p(\omega)} \mid u \in S\right\} \leq \sup\left\{\sup_{|y|<1/n} \|\tau_y u - u\|_{L^p(\omega)} \mid u \in S\right\} \leq \varepsilon.$$

$$(16.6)$$

Define

$$U = \left\{ x \in \mathbb{R}^N \,\middle|\, d(x, \omega) < \frac{1}{n} \right\} \subset\subset \Omega.$$

We claim that the collection $\mathcal{F} = \{\varrho_n * u|_\omega \mid u \in S\}$ satisfies the assumptions of Corollary 13.3.

Indeed, by condition (a), for every $u \in S$ and every $x \in \omega$, we have

$$|\varrho_n * u(x)| \le \int_U \varrho_n(x - z)\,|u(z)|\,dz \le \|\varrho_n\|_\infty \|u\|_{L^1(U)} \le c_1.$$

Similarly, for every $x \in \omega$ and $y \in \omega$ we have

$$|\varrho_n * u(x) - \varrho_n * u(y)| \le \int_U |\varrho_n(x - z) - \varrho_n(y - z)|\,|u(z)|\,dz$$

$$\le \sup\left\{|\varrho_n(x - z) - \varrho_n(y - z)| z \in \mathbb{R}^N\right\}$$

$$\|u\|_{L^1(U)} \le c_2|x - y|.$$

Hence $\mathcal{F}$ is relatively compact in the space of bounded continuous functions on ω. But

$$\|v\|_{L^p(\omega)} \le m_N(\omega)^{1/p}\|v\|_{L^\infty(\omega)},$$

$\mathcal{F}$ is also relatively compact in $L^p(\omega)$.[7] Now (16.6) implies the existence of a finite cover of $\mathcal{F}|_\omega$ in $L^p(\omega)$ by balls of radius 2ε. We finally use assumption (b) to ensure the existence of a finite cover of $\mathcal{F}$ in $L^p(\Omega)$ by balls of radius 3ε. We have thus proved that $\mathcal{F}$ is totally bounded in the complete metric space $L^p(\Omega)$, hence its closure is relatively compact by Theorem 13.78.   □

16.9 The Radon-Nykodim Theorem

Let us consider a σ-finite measurable space (X, Σ) together with a measure μ, and let $f \in L^1(X)$ be a non-negative function. It is easy to check that

$$\nu : A \in \Sigma \mapsto \int_A f\,d\mu$$

---

[7] We have proved that the space of bounded continuous functions on ω is continuously embedded into $L^p(\omega)$.

is a measure on X which satisfies the property

$$\mu(A) = 0 \implies \nu(A) = 0.$$

This motivates the following definition.

Definition 16.31 A measure ψ on (X, Σ) is absolutely continuous with respect to μ if and only if for every $A \in \Sigma$ such that $\mu(A) = 0$ there results $\psi(A) = 0$.

The main result of this section is the following description of absolutely continuous measures.

Theorem 16.42 *Let (X, Σ) be a σ-finite measurable space with measure μ. If ν is a finite measure on (X, Σ), absolutely continuous with respect to μ, then there exists a measurable function f on X such that $f \geq 0$, $\int_X f \, d\mu < \infty$ and*

$$\nu(E) = \int_E f \, d\mu \quad \text{for every } E \in \Sigma.$$

The proof of this theorem requires some technical results. Let $\varphi : \Sigma \to \mathbb{R}$ a countably additive function, i.e. $\varphi(\emptyset) = 0$ and

$$\varphi\left(\bigcup_{n=1}^{\infty} E_n\right) = \sum_{n=1}^{\infty} \varphi(E_n)$$

for every sequence $\{E_n\}_n$ of pairwise disjoint sets in Σ. We define

$$\varphi^+(X) = \sup \{\varphi(A) \mid A \in \Sigma\}.$$

Lemma 16.1 *There exists a set $E \in \Sigma$ such that $\varphi^+(X) = \varphi(E)$. Moreover*

$$\varphi(A) \geq 0 \ \text{ if } A \subset E$$

$$\varphi(A) \leq 0 \ \text{ if } A \subset X \setminus E.$$

Proof We consider a minimizing sequence $\{E_n\}_n$ for $\varphi^+(X)$ such that

$$\varphi^+(X) \geq \varphi(E_n) \geq \varphi^+(X) - \frac{1}{2^n} \quad \text{for } n \geq 1.$$

If $A \subset E_n$, writing $\varphi(E_n) = \varphi(A) + \varphi(E_n \setminus A)$, we see that

$$\varphi^+(X) - \frac{1}{2^n} \leq \varphi(E_n) \leq \varphi(A) + \varphi^+(X).$$

Hence $\varphi(A) \geq -2^{-n}$. Now, for every $n \geq 1$,

$$\varphi\left(\bigcup_{m=n}^{\infty} E_m\right) = \varphi\left(E_n \cup (E_{n+1} \setminus E_n) \cup (E_{n+2} \setminus E_{n+1}) \cup \cdots\right)$$

$$\geq \varphi(E_n) - \frac{1}{2^{n+1}} - \frac{1}{2^{n+2}} - \cdots$$

$$\geq \varphi^+(X) - \frac{2}{2^n}.$$

It follows that

$$\varphi\left(\bigcap_{n=1}^{\infty} \bigcup_{m=n}^{\infty} E_m\right) = \lim_{n \to +\infty} \varphi\left(\bigcup_{m=n}^{\infty} E_m\right) \geq \varphi^+(X).$$

Hence the set

$$E = \bigcap_{n=1}^{\infty} \bigcup_{m=n}^{\infty} E_m$$

satisfies $\varphi(E) = \varphi^+(X)$. To prove the second part, we notice that if $A \subset E$, then

$$\varphi^+(X) = \varphi(E) = \varphi(A) + \varphi(E \setminus A) \leq \varphi(A) + \varphi^+(X),$$

or $\varphi(A) \geq 0$. If $A \subset X \setminus E$, then

$$\varphi^+(X) \geq \varphi(A \cup E) = \varphi(A) + \varphi(E) = \varphi(A) + \varphi^+(X),$$

or $\varphi(A) \leq 0$. The proof is complete.   □

Proof of Theorem 16.42 We first consider the case $\mu(X) < \infty$. We propose a *variational* proof, based on a minimization problem. Consider indeed the set

$$\mathcal{I} = \left\{ g \geq 0 \,\middle|\, \int_E g \, d\mu \leq \nu(E) \text{ for every } E \in \Sigma \right\}.$$

We claim that there exists $f \in \mathcal{I}$ such that

$$\int_X f \, d\mu = \sup\left\{ \int_X g \, d\mu \,\middle|\, g \in \mathcal{I} \right\} = S.$$

For every integer $n \geq 1$, let $g_n \in I$ be such that

$$\int_X g_n \, d\mu \geq S - \frac{1}{n}.$$

We define

$$f_n = \max \{g_1, \ldots, g_n\}, \qquad f = \lim_{n \to +\infty} f_n.$$

It is not difficult to check that $f_n \in I$ for every $n \geq 1$. For instance, in the case $n = 2$, for every $E \in \Sigma$ we have

$$\int_E f_2 \, d\mu = \int_{E \cap \{g_1 \geq g_2\}} g_1 \, d\mu + \int_{E \cap \{g_2 > g_1\}} g_2 \, d\mu$$
$$\leq v(E \cap \{g_1 \geq g_2\}) + v(E \cap \{g_2 > g_1\})$$
$$= v(E).$$

The general case is similar. Since $f_n \nearrow f$, Beppo Levi's Theorem yields $f \in I$ and $\int_X f \, d\mu = S$. The claim is thus proved.

Let us now show that for every $E \in \Sigma$ there results $v(E) = \int_E f \, d\mu$. By definition of I, we only need to show that $v(E) \leq \int_E f \, d\mu$. Suppose not, so that there exist $\varepsilon > 0$ and $E_0 \in \Sigma$ such that

$$\int_{E_0} (f + \varepsilon) \, d\mu < v(E_0).$$

Since v is absolutely continuous with respect to μ, we see that $\mu(E_0) > 0$. If we define

$$\varphi(E) = v(E) - \int_E (f + \varepsilon) \, d\mu,$$

Lemma 16.1 yields two subset F, G of E_0 such that $E_0 = F \cup G$,

$$v(A) - \int_A (f + \varepsilon) \, d\mu \geq 0 \ \text{ if } A \subset F$$

$$v(A) - \int_A (f + \varepsilon) \, d\mu \leq 0 \ \text{ if } A \subset G.$$

The function g defined as

$$g(x) = \begin{cases} f(x) & \text{if } x \notin F \\ f(x) + \varepsilon & \text{if } x \in F \end{cases}$$

belongs to I and

$$\int_X g \, d\mu > \int_X f \, d\mu = S.$$

This is impossible, and the proof is complete in the particular case $\mu(X) < \infty$.

To deal with the general case, let $\{X_i\}_i$ be a sequence of pairwise disjoint measurable sets such that $\mu(X_i)$ is finite for every i and

$$X = \bigcup_{i=1}^{\infty} X_i.$$

Applying the previous step in each X_i, we see that there exist functions $f_i \in L^1(X_i, \mu)$ such that $f_i \geq 0$ and

$$\nu(E) = \int_E f_i \, d\mu \quad \text{for every } E \subset X_i.$$

If we set $f = f_i$ on X_i, we define a function f on X. If E is contained in a finite union of the sets X_i, we have

$$\nu(E) = \int_E f \, d\mu.$$

Furthermore, for every $n \geq 1$,

$$\int_{\bigcup_{i=1}^{\infty} X_i} f \, d\mu \leq \nu(X),$$

hence $f \in L^1(X, \mu)$. If now E is any measurable set, then

$$\int_E f \, d\mu = \lim_{n \to +\infty} \int_{E \cap \bigcup_{i=1}^{n} X_i} f \, d\mu$$

$$= \lim_{n \to +\infty} \nu \left(E \cap \bigcup_{i=1}^{n} X_i \right) = \nu(E),$$

and the proof is complete.   □

Exercise 16.13 By inspection of the previous proof, deduce that if ν is just σ-finite instead of finite, the conclusion of the Theorem holds, but f need not be integrable with respect to μ.

16.10   A Strong Form of the Fundamental Theorem of Calculus

The basic result of Riemann integration theory is for sure the formula

$$\int_a^b f'(x)\, dx = f(b) - f(a),$$

which expresses a function as the integral of its derivative. We refer back to Theorem 10.12 for a precise statement.

Important: Question

Does there exist a Fundamental Theorem of Calculus for Lebesgue integrals?

Such a natural question requires several new ideas to be answered. In the rest of the section, $\lambda = m_1$ will always denote the Lebesgue measure in $\mathbb{R}$, and "almost everywhere" will always refer to this measure.

Definition 16.32   Let $a \in \mathbb{R}$ and $\delta > 0$. If $\varphi \colon (a, a + \delta) \to \mathbb{R}$, we define

$$\liminf_{h \to a} \varphi(h) = \sup\{\inf\{\varphi(h) \mid a < h < t\} \mid a < t \le a + \delta\}$$

$$\limsup_{h \to a} \varphi(h) = \inf\{\sup\{\varphi(h) \mid a < h < t\} \mid a < t \le a + \delta\}.$$

These quantities are called respectively the lower right limit and the upper right limit of the function φ at the point a. Similarly, if $\varphi \colon (a - \delta, a) \to \mathbb{R}$, we define the lower left limit and the upper left limit of φ at a as

$$\liminf_{h \to a} \varphi(h) = \sup\{\inf\{\varphi(h) \mid t < h < a\} \mid a - \delta \le t < a\}$$

$$\limsup_{h \to a} \varphi(h) = \inf\{\sup\{\varphi(h) \mid t < h < a\} \mid a - \delta \le t < a\}.$$

Definition 16.33 (Dini's Derivatives)   Let $a \in \mathbb{R}$ and $\delta > 0$. If $f \colon [a, a + \delta) \to \mathbb{R}$, we define

$$D_+ f(a) = \liminf_{h \to 0} \frac{f(a + h) - f(a)}{h}$$

$$D^+ f(a) = \limsup_{h \to 0} \frac{f(a + h) - f(a)}{h}.$$

If $f : (a - \delta, a] \to \mathbb{R}$, we define

$$D_- f(a) = \liminf_{h \to 0} \frac{f(a + h) - f(a)}{h}$$

$$D^- f(a) = \limsup_{h \to 0} \frac{f(a + h) - f(a)}{h}.$$

Exercise 16.14 Prove that $D^+ f(a)$ is the largest limit of a sequence

$$\left\{ \frac{f(a + h_n) - f(a)}{h_n} \right\}_n$$

where $h_n > 0$ and $\lim_{n \to +\infty} h_n = 0$. Conjecture and prove similar statements for the remaining Dini's derivatives.

The four Dini's derivatives describe the lack of differentiability of f at a, since it is clear that f is differentiable at a if and only if the four Dini's derivatives are finite and coincide.

Theorem 16.43 *Let (a, b) be an open interval, and let f be a real-valued function defined on (a, b). There exist at most countably many points $x \in (a, b)$ such that*

$$D_+ f(x) = D^+ f(x)$$

and

$$D_- f(x) = D^- f(x)$$

both exist in $[-\infty, +\infty]$ but are different.

Proof In case $D_+ f(x) = D^+ f(x)$, we call $f'_+(x)$ the common value. and similarly for $f'_-(x)$. These quantities may well be infinite. Let

$$A = \left\{ x \in (a, b) \mid f'_+(x) < f'_-(x) \right\}$$
$$B = \left\{ x \in (a, b) \mid f'_+(x) > f'_-(x) \right\}.$$

For each point $x \in A$ we select a rational number $r(x)$ such that $f'_+(x) < r(x) < f'_-(x)$. Next we select rational numbers $s(x)$ and $t(x)$ such that $a < s(x) < x < t(x) < b$,

$$\frac{f(y) - f(x)}{y - x} > r(x) \quad \text{if } s(x) < y < x$$

and

$$\frac{f(y) - f(x)}{y - x} < r(x) \quad \text{if } x < y < t(x).$$

Then $y \neq x$ and $s(x) < y < t(x)$ imply

$$f(y) - f(x) < r(x)(y - x).$$

Hence we have defined a function $\varphi \colon A \rightarrow \mathbb{Q} \times \mathbb{Q} \times \mathbb{Q}$ such that $\varphi(x) = (r(x), s(x), t(x))$. We claim that φ is injective.

For the sake of contradiction, assume that there exist points $y \neq x$ in A such that $\varphi(x) = \varphi(y)$. This implies $(s(y), t(y)) = (s(x), t(x))$, and x, y both lie in this interval. It follows that

$$f(y) - f(x) < r(x)(y - x)$$
$$f(x) - f(y) < r(y)(x - y).$$

But $r(x) = r(y)$, hence $0 < 0$. This contradiction proves that φ is injective, and thus A is a countable set. The proof that B is also countable is similar. □

Definition 16.34 Let E be a subset of $\mathbb{R}$. A collection $\mathcal{V}$ of closed intervals, each having positive measure, is a Vitali cover of E if and only if for every $x \in E$ and for every $\varepsilon > 0$ there exists an interval $I \in \mathcal{V}$ such that $x \in I$ and $\lambda(I) < \varepsilon$.

Roughly speaking, Vitali covers consist of closed interval of arbitrarily small lengths.

Theorem 16.44 (Vitali's Covering Theorem) *Let $\mathcal{V}$ be a non-empty Vitali cover of a set $E \subset \mathbb{R}$. Then there exists a pairwise disjoint countable collection $\{I_n\}_n \subset \mathcal{V}$ such that*

$$\lambda \left(E \cap \left(\mathbb{R} \setminus \bigcup_{n=1}^{\infty} I_n \right) \right) = 0.$$

If $\lambda(E) \in \mathbb{R}$ and if $\varepsilon > 0$, there exists a finite pairwise disjoint collection $\{I_1, \ldots, I_p\} \subset \mathcal{V}$ such that

$$\lambda \left(E \cap \left(\mathbb{R} \setminus \bigcup_{n=1}^{p} I_n \right) \right) = 0.$$

Proof

First case: $\lambda(E) \in \mathbb{R}$. We fix an open set V which contains E and such that $\lambda(V) \in \mathbb{R}$. Let

$$\mathcal{V}_0 = \{I \in \mathcal{V} \mid I \subset V\}.$$

The fact that $\mathcal{V}_0$ is a Vitali cover of E is clear. If $I_1 \in \mathcal{V}_0$ and $E \subset I_1$, the proof is complete. Otherwise we proceed by induction. Assume that $I_1, I_2, \ldots, I_n$ have been chosen and are pairwise disjoint. If $E \subset \bigcup_{k=1}^{n} I_k$, the proof is complete. Otherwise we write

$$A_n = \bigcup_{k=1}^{n} I_k$$

$$U_n = V \cap (\mathbb{R} \setminus A_n).$$

The set A_n is closed as a finite union of closed sets, U_n is open, and $U_n \cap E \neq \emptyset$. Define

$$\delta_n = \sup\{\lambda(I) \mid I \in \mathcal{V}_0,\ I \subset U_n\}.$$

Next we select $I_{n+1} \in \mathcal{V}_0$ such that $I_{n+1} \subset U_n$ and $\lambda(I_{n+1}) > \delta_n/2$.

If this procedure continuous indefinitely, we get an infinite sequence $\{I_n\}_n$ of pairwise disjoint elements of $\mathcal{V}_0$. Let $A = \bigcup_{n=1}^{\infty} I_n$, and we claim that $\lambda(E \cap (\mathbb{R} \setminus A)) = 0$. Indeed, for every positive integer n there exists a unique closed interval J_n having the same mid-point as I_n and such that $\lambda(J_n) = 5\lambda(I_n)$. Since

$$\lambda\left(\bigcup_{n-1}^{\infty} J_n\right) \leq \sum_{n=1}^{\infty} \lambda(J_n) = 5\sum_{n=1}^{\infty} \lambda(I_n) = 5\lambda(A) \leq 5\lambda(V),$$

we see that

$$\lim_{p \to +\infty} \lambda\left(\bigcup_{n=p}^{\infty} J_n\right) = 0.$$

As a consequence, it suffices to show that $E \cap (\mathbb{R} \setminus A) \subset \bigcup_{n=p}^{\infty} J_n$ for every $p \in \mathbb{N}$.

Fix $p \in \mathbb{N}$ and $x \in E \cap (\mathbb{R} \setminus A)$. Then $x \in E \cap (\mathbb{R} \setminus A_p) \subset U_p$, hence there exists $I \in \mathcal{V}_0$ such that $x \in I \subset U_p$. Clearly $\delta_n < 2\lambda(I_{n+1})$, and $\lambda(I_n) \to 0$ as $n \to +\infty$. Hence there exists $n \in \mathbb{N}$ such that I is not contained in U_n. Call q the smallest such integer. It is obvious that $p < q$, so that

$$I \cap A_q \neq \emptyset, \qquad I \cap A_{q-1} = \emptyset.$$

This yields $I \cap I_q \neq \emptyset$ and, since I is a subset of U_{q-1}, we find $\lambda(I) \leq \delta_{q-1} < 2\lambda(I_q)$. Recalling that $\lambda(J_q) = 5\lambda(I_q)$, we see that

$$I \subset J_q \subset \bigcup_{n=p}^{\infty} J_n,$$

hence $x \in \bigcup_{n=p}^{\infty} J_n$. In conclusion $E \cap (\mathbb{R} \setminus A) \subset \bigcup_{n=p}^{\infty} J_n$, and this shows that $\lambda(E \cap (\mathbb{R} \setminus A)) = 0$. To conclude the first part of the proof, let $\varepsilon > 0$ be given and select a positive integer p such that

$$\sum_{n=p+1}^{\infty} \lambda(I_n) < \varepsilon.$$

But

$$E \cap (\mathbb{R} \setminus A_p) \subset (E \cap (\mathbb{R} \setminus A)) \cup \bigcup_{n=p+1}^{\infty} I_n,$$

hence

$$\lambda\left(E \cap (\mathbb{R} \setminus A_p)\right) \leq 0 + \lambda\left(\bigcup_{n=p+1}^{\infty} I_n\right) < \varepsilon.$$

Second case: $\lambda(E) = +\infty$. For every $n \in \mathbb{N}$ we introduce $E_n = E \cap (n, n+1)$ and

$$\mathcal{V}_n = \{I \in \mathcal{V} \mid I \subset (n, n+1)\}.$$

It is a simple exercise to check that $\mathcal{V}_n$ is a Vitali cover of E_n. Since $\lambda(E_n)$ is finite, the first part of the proof applies and yields a finite pairwise disjoint collection $\mathcal{I}_n \subset \mathcal{V}_n$ such that

$$\lambda\left(E_n \cap \left(\mathbb{R} \setminus \bigcup \mathcal{I}_n\right)\right) = 0 \quad \text{for every } n \in \mathbb{Z}.$$

To conclude, let $\mathcal{I} = \bigcup_{n \in \mathbb{Z}} \mathcal{I}_n$. Then $\mathcal{I}$ is a countable pairwise disjoint subcollection of $\mathcal{V}$ and

$$E \cap \left(\mathbb{R} \setminus \bigcup \mathcal{I}\right) \subset \mathbb{Z} \cup \left(\bigcup_{n \in \mathbb{Z}} E_n \cap (\mathbb{R} \setminus \mathcal{I}_n)\right).$$

Since

$$\lambda \left(E \cap \left(\mathbb{R} \setminus \bigcup \mathcal{I} \right) \right) \leq \lambda(\mathbb{Z}) + \sum_{n=-\infty}^{\infty} 0 = 0,$$

the proof is complete.

□

A fundamental result of Lebesgue measure theory is the differentiability almost everywhere of monotone functions.

Theorem 16.45 (Lebesgue) *If* $f : [a, b] \to \mathbb{R}$ *is a monotone function on a closed interval* $[a, b]$, *then* f *has a finite derivative almost everywhere in* $[a, b]$.

Proof Replacing f with $-f$, we may assume that f is monotone increasing on $[a, b]$. Let us define

$$E = \left\{ x \in [a, b) \mid D_+ f(x) < D^+ f(x) \right\}.$$

We claim that $\lambda(E) = 0$. Indeed, we can write $E = \bigcup \{E(u, v) \mid u \in \mathbb{Q}, v \in \mathbb{Q}, 0 < u < v\}$, where

$$E(u, v) = \left\{ x \in [a, b) \mid D_+ f(x) < u < v < D^+ f(x) \right\}.$$

If we show that $\lambda(E(u, v)) = 0$ for every such u and v, the claim follows.

Arguing by contradiction, we assume that for some $0 < u < v$ in $\mathbb{Q}$, $\lambda(E(u, v)) = \alpha > 0$. Fix $\varepsilon > 0$ so small that

$$0 < \varepsilon < \frac{\alpha(v - u)}{u + 2v}.$$

By the regularity properties of the Lebesgue measure, there exists an open set U such that $E(u, v) \subset U$ and $\lambda(U) < \alpha + \varepsilon$. By definition, to each $x \in E(u, v)$ there correspond arbitrarily small numbers h such that $[x, x + h] \subset U \cap [a, b]$ and

$$f(x + h) - f(x) < uh.$$

The collection of all such intervals $[x, x + h]$ is a Vitali cover $\mathcal{V}$ of $E(u, v)$, and therefore there exists a finite pairwise disjoint sub-collection $\{[x_i, x_i + h_i] \mid i = 1, \ldots, m\}$ of $\mathcal{V}$ such that

$$\lambda \left(E(u, v) \cap \left(\mathbb{R} \setminus \bigcup_{i=1}^{m} [x_i, x_i + h_i] \right) \right) < \varepsilon.$$

If $V = \bigcup_{i=1}^{m}(x_i, x_i + h_i)$, then $\lambda(E \cap (\mathbb{R} \setminus V)) < \varepsilon$. Since $V \subset U$, we have

$$\sum_{i=1}^{m} h_i = \lambda(V) \leq \lambda(U) < \alpha + \varepsilon,$$

and thus

$$\sum_{i=1}^{m} (f(x_i + h_i) - f(x_i)) < u \sum_{i=1}^{m} h_i < u(\alpha + \varepsilon).$$

We now apply again a similar reasoning: to each $y \in E(u, v) \cap V$ there correspond arbitrarily small numbers k such that $[y, y + k] \subset V$ and

$$f(y + k) - f(y) > vk.$$

As before, there exists a finite pairwise disjoint collection $\{[y_j, y_j + k_j] \mid j = 1, \ldots, n\}$ such that

$$\lambda\left(E(u, v) \cap V \cap \left(\mathbb{R} \setminus \bigcup_{j=1}^{n}[y_j, y_j + k_j]\right)\right) < \varepsilon.$$

Summing up, we see that

$$\alpha = \lambda(E(u, v)) \leq \lambda(E(u, v) \cap (\mathbb{R} \setminus V)) + \lambda(E(u, v) \cap V) < \varepsilon + \varepsilon + \sum_{j=1}^{n} k_j.$$

Next, the previous inequalities yield

$$v(\alpha - 2\varepsilon) < v \sum_{j=1}^{n} k_j < \sum_{j=1}^{n} f(y_j + k_j) - f(y_j).$$

Recalling that $\bigcup_{j=1}^{n}[y_j, y_j+k_j] \subset \bigcup_{i=1}^{m}[x_i, x_i+h_i]$, the monotonicity of f implies

$$\sum_{j=1}^{m} f(y_j + k_j) - f(y_j) \leq \sum_{i=1}^{m} f(x_i + h_i) - f(x_i).$$

In conclusion $v(\alpha - 2\varepsilon) < u(\alpha + \varepsilon)$, which is a contradiction. We have thus proved that $\lambda(E) = 0$, so that the right derivative of f exists as a real number at almost every point of $[a, b]$. By the same token, the left derivative of f exists and is finite at almost every point of $[a, b]$. Theorem 16.43 implies that the derivative $f'(x)$

exists for almost every $x \in [a, b]$. To complete the proof, we must show that the set F of points $x \in (a, b)$ where $f'(x) = +\infty$ is a set of measure zero.[8]

Let $\beta > 0$ be given. For every $x \in F$ there exist arbitrarily small numbers h such that $[x, x + h] \subset (a, b)$ and $f(x + h) - f(x) > \beta h$. We can therefore construct a countable pairwise disjoint collection $\{[x_n, x_n + h_n] \mid n = 1, 2, \ldots\}$ such that

$$\lambda \left(F \cap \left(\mathbb{R} \setminus \bigcup_{n=1}^{\infty} [x_n, x_n + h_n] \right) \right) = 0.$$

We derive that

$$\beta \lambda(F) \leq \beta \sum_{n=1}^{\infty} h_n < \sum_{n=1}^{\infty} f(x_n + h_n) - f(x_n) \leq f(b) - f(a).$$

Since $\beta > 0$ can be arbitrarily large, we see that $\lambda(F) = 0$. The proof of the theorem is complete.   □

Definition 16.35 Let $f : [a, b] \to \mathbb{R}$ be a function. We define the total variation of f over $[a, b]$ as

$$V_a^b f = \sup \left\{ \sum_{k=1}^{n} |f(x_k) - f(x_{k-1})| \ \Big| \ a = x_0 < x_1 < \ldots < x_n = b \right\}.$$

The function f is a function of bounded variation over $[a, b]$ if and only if $V_a^b f \in \mathbb{R}$.

Remark 16.6 The previous definition can be extended to functions of several variables, but the language of Measure Theory becomes necessary, and the development is much more involved. We will not enter into the details in this book.

Exercise 16.15 Prove the identity $V_a^b f + V_b^c f = V_a^c f$ for every $a < b < c$.

Exercise 16.16 Prove that the function $x \mapsto V_a^x f$ is non-decreasing on $[a, b]$.

These exercises inspire the next result.

Theorem 16.46 (Jordan Decomposition Theorem) *A function of bounded variation is the difference of two non-decreasing functions.*

Proof If f is of bounded variation over $[a, b]$, we write

$$f(x) = V_a^x f - \left(V_a^x f - f(x) \right).$$

---

[8] Recall that f is a monotone increasing function, hence the derivative $f'(x)$ cannot equal $-\infty$.

Here we set $V_a^a f = 0$. The function $x \mapsto V_a^x f$ is non-decreasing. If $x_1 < x_2$, then

$$V_a^{x_2} f - f(x_2) - \left(V_a^{x_1} f - f(x_1)\right) = V_{x_1}^{x_2} f - (f(x_2) - f(x_1)) \geq 0$$

by definition of $V_{x_1}^{x_2} f$. Hence $x \mapsto V_a^x f - f(x)$ is also non-decreasing, and the proof is complete. $\qquad\qquad\qquad\qquad\qquad\qquad\qquad\qquad\qquad\qquad\qquad\qquad\square$

We are ready to state the fundamental result of differentiability for functions of bounded variation. The proof is an immediate consequence of the previous theorems.

Theorem 16.47 (Lebesgue) *A function of bounded variation has finite derivative almost everywhere.*

Let us take a break. We have so far introduced a class of real-valued functions whose derivative exists at almost every point. But our goal was much more ambitious: we wanted to characterize functions which satisfy the Fundamental Theorem of Calculus.

It turns out the road is still long, and we present an interesting tool first.

Theorem 16.48 (Fubini) *Suppose $\{f_n\}_n$ is a sequence of monotone functions on a common interval $[a, b]$. If $\sum_{n=1}^{\infty} f_n(x) = s(x)$ exists and is finite for every $x \in [a, b]$, then*

$$s'(x) = \sum_{n=1}^{\infty} f_n'(x) \quad \text{for a.e. } x \in [a, b].$$

Proof We will suppose without loss of generality that all functions f_n are non-decreasing. Replacing f_n with $f_n - f_n(a)$, we also assume that $f_n \geq 0$. Thus $s = \sum_{n=1}^{\infty} f_n$ is non-negative and non-decreasing. Therefore s' exists as a finite number almost everywhere in (a, b). Let

$$s_n = f_1 + \cdots + f_n, \qquad r_n = s - s_n.$$

Each function f_j has finite derivative almost everywhere: there exists a set $A \subset (a, b)$ such that $\lambda((\mathbb{R} \setminus A) \cap (a, b)) = 0$,

$$s_n'(x) = f_1'(x) + \cdots + f_n'(x) < +\infty$$

for every $x \in A$ and every $n \in \mathbb{N}$, and $s'(x)$ exists in $\mathbb{R}$ for every $x \in A$.
    If $x \in (a, b)$ and $h > 0$ is such that $x + h \in (a, b)$, then

$$\frac{s(x+h) - s(x)}{h} = \frac{s_n(x+h) - s_n(x)}{h} + \frac{r_n(x+h) - r_n(x)}{h},$$

hence

$$\frac{s_n(x+h) - s_n(x)}{h} \le \frac{s(x+h) - s(x)}{h}.$$

As $h \to 0$, this yields $s_n'(x) \le s'(x)$ for every $x \in A$. Since $s_n'(x) \le s_{n+1}'(x)$ is trivial, we deduce that

$$s_n'(x) \le s_{n+1}'(x) \le s'(x) \quad \text{for every } x \in A \text{ and } n \in \mathbb{N}.$$

As a consequence $\lim_{n \to +\infty} s_n'(x) = \sum_{j=1}^\infty f_j'(x)$ exists almost everywhere. It remains to show that $\lim_{n \to +\infty} s_n'(x) = s'(x)$ almost everywhere.

The sequence $\{s_n'(x)\}_n$ is non-decreasing for every $x \in A$, so that it suffices to prove that $\{s_n'\}_n$ has a subsequence converging to s' almost everywhere. We select $n_1 < n_2 < n_3 < \cdots$ in $\mathbb{N}$ such that

$$\sum_{k=1}^\infty \left[s(b) - s_{n_k}(b) \right] < +\infty.$$

We remark that for every k and for every $x \in (a, b)$ we have $0 \le s(x) - s_{n_k}(x) \le s(b) - s_{n_k}(b)$. Therefore the series $\sum_{k=1}^\infty \left[s(x) - s_{n_k}(x) \right]$ converges. Since the terms of this series are monoton functions with finite derivative almost everywhere, the same reasoning as above shows that $\sum_{k=1}^\infty \left[s'(x) - s_{n_k}'(x) \right]$ converges almost everywhere. It is now clear that $\lim_{k \to +\infty} s_{n_k}'(x) = s'(x)$ almost everywhere in (a, b). The proof is complete.   □

Let us now face the main problem of identifying those functions F of the form

$$F(x) = \int_a^x f(t) \, dt$$

for some $f \in L^1(a, b)$.

Theorem 16.49 *If $f \in L^1(a, b)$, we define its integral function F on $[a, b]$ as*

$$F: x \mapsto \int_a^x f(t) \, dt.$$

The function F is uniformly continuous and of bounded variation $V_a^b F = \int_a^b |f(t)| \, dt$. The same conclusion holds if $f \in L^1(\mathbb{R})$ and $F(x) = \int_{-\infty}^x f(t) \, dt$.

Proof If $x_1 < x_2$ are points of $[a, b]$, we compute

$$|F(x_2) - F(x_1)| = \left| \int_{x_1}^{x_2} f(t) \, dt \right|.$$

It follows from Proposition 16.3 that F is uniformly continuous. If

$$a = x_0 < x_1 < \ldots < x_n = b,$$

then

$$\sum_{k=1}^{n} |F(x_k) - F(x_{k-1})| = \sum_{k=1}^{n} \left| \int_{x_{k-1}}^{x_k} f(t) \, dt \right| \leq \sum_{k=1}^{n} \int_{x_{k-1}}^{x_k} |f(t)| \, dt = \int_{a}^{b} |f(t)| \, dt.$$

This shows that $V_a^b F \leq \|f\|_1$, and in particular F is a function of bounded variation. To prove the reversed inequality, we use the density of simple functions

$$s = \sum_{k=1}^{n} \alpha_k \chi_{[x_{k-1}, x_k)}$$

in $L^1(a, b)$, where $a = x_0 < x_1 < \ldots < x_n = b$. We consider the function sign f defined by

$$\operatorname{sign} f(x) = \begin{cases} 1 & \text{if } f(x) > 0 \\ 0 & \text{if } f(x) = 0 \\ -1 & \text{if } f(x) < 0. \end{cases}$$

Let $\{s_m\}_m$ be a sequence of simple functions such that

$$\|s_m - \operatorname{sign} f\|_1 \leq \frac{1}{m}.$$

This inequality is preserved if we replace every α_k such that $|\alpha_k| > 1$ with $\alpha_k |\alpha_k|^{-1}$. In other words, we may always assume that $|s_m| \leq 1$ for every m. Since s_m converges to sign f in measure, there exists a subsequence $\{s_{m_j}\}_j$ such that $s_{m_j} \to \operatorname{sign} f$ pointwise almost everywhere in $[a, b]$. The Dominated Convergence Theorem implies now that

$$\int_{a}^{b} |f(t)| \, dt = \int_{a}^{b} f(t) \operatorname{sign} f(t) \, dt = \lim_{j \to +\infty} \int_{a}^{b} f(t) s_{m_j}(t) \, dt.$$

But

$$\left| \int_{a}^{b} f(t) s_{m_j}(t) \, dt \right| = \left| \sum_{k=1}^{n} \alpha_k \int_{x_{k-1}}^{x_k} f(t) \, dt \right|$$

$$= \left| \sum_{k=1}^{n} \alpha_k (F(x_k) - F(x_{k-1})) \right|$$

$$\leq \sum_{k=1}^{n} |\alpha_k| |F(x_k) - F(x_{k-1})|$$

$$\leq \sum_{k=1}^{n} |F(x_k) - F(x_{k-1})|$$

$$\leq V_a^b F.$$

This shows that $\int_a^b |f(t)| \, dt \leq V_a^b F$, and the proof is complete.   □

Theorem 16.50 *If A is a subset of $\mathbb{R}$, then*

$$\lim_{k \to 0+} \frac{\lambda(A \cap (x, x+k))}{k} = \lim_{h \to 0+} \frac{\lambda(A \cap (x-h, x))}{h}$$

$$= \lim_{\substack{(h,k) \to (0,0) \\ h>0 \\ k>0}} \frac{\lambda(A \cap (x-h, x+k))}{h+k} = 1$$

for almost every $x \in A$. If A is a Lebesgue-measurable set, then all the previous limits are equal to zero for almost every $x \in \mathbb{R} \setminus A$.

Proof Since we always intersect A with a bounded interval, we may assume without loss of generality that A is itself a bounded subset. Consider a sequence of bounded open sets U_n such that

$$U_1 \supset U_2 \supset \cdots \supset U_n \supset \cdots \supset A$$

and $\lambda(U_n) - 2^{-n} < \lambda(A)$. We call $a = \inf U_1$ and consider the functions

$$\varphi_n(x) = \lambda(U_n \cap (a, x)), \qquad \varphi(x) = \lambda(A \cap (a, x)).$$

For every $x \in U_n$ and $h > 0$ sufficiently small, it is easy to check that

$$\frac{\varphi_n(x+h) - \varphi_n(x)}{h} = \frac{\varphi_n(x) - \varphi_n(x-h)}{h} = 1.$$

Hence $\varphi_n'(x)$ exists at all $x \in U_n$ and $\varphi_n'(x) = 1$. We consider the series

$$(\varphi_1 - \varphi) + (\varphi_2 - \varphi) + \cdots + (\varphi_n - \varphi) + \cdots$$

We claim that $\varphi_n - \varphi$ is monotone for every n. Indeed, if $h > 0$, then

$$\varphi_n(x+h) - \varphi(x+h) - (\varphi_n(x) - \varphi(x))$$
$$= \lambda(U_n \cap [x, x+h)) - \lambda(A \cap (a, x+h)) + \lambda(A \cap (a, x))$$
$$\geq \lambda(U_n \cap [x, x+h)) - \lambda(A \cap [x, x+h)) \geq 0,$$

as a consequence of the inequality

$$\lambda(A \cap (a, x+h)) \leq \lambda(A \cap (a, x)) + \lambda(A \cap [x, x+h))$$

and of the inclusion $A \cap [x, x+h) \subset U_n \cap [x, x+h)$. The claim is proved.
    Let $b = \sup U_1$, so that

$$\varphi_n(b) - \varphi(b) = \lambda(U_n) - \lambda(A) < \frac{1}{2^n}.$$

For every $x \in [a, b]$ we then have

$$\sum_{n=1}^{\infty} (\varphi_n(x) - \varphi(x)) \leq \sum_{n=1}^{\infty} (\varphi_n(b) - \varphi(b)) \leq \sum_{n=1}^{\infty} \frac{1}{2^n} < +\infty.$$

We may set $s(x) = \sum_{n=1}^{\infty} (\varphi_n(x) - \varphi(x))$. By Theorem 16.48 we have that

$$s'(x) = \sum_{n=1}^{\infty} \left(\varphi_n'(x) - \varphi'(x)\right) \in \mathbb{R}$$

for almost every $x \in (a, b)$, and so also $\lim_{n \to +\infty} \varphi_n'(x) = \varphi'(x)$ for almost every $x \in (a, b)$. To summarize, we see that $\varphi' = 1$ on $\bigcap_{n=1}^{\infty} U_n$ except on a set of measure zero, and the first statement of the theorem is proved.
    If A is also Lebesgue-measurable, then

$$1 = \frac{\lambda(A \cap (x-h, x+k))}{h+k} + \frac{\lambda((\mathbb{R} \setminus A) \cap (x-h, x+k))}{h+k}$$
$$= \psi_A(x) + \psi_{\mathbb{R} \setminus A}(x).$$

As $h \to 0$ and $k \to 0$ along positive values, the first part of the theorem applied to $\mathbb{R} \setminus A$ yields that $\psi_{\mathbb{R} \setminus A}(x) \to 1$ for a.e. $x \in \mathbb{R} \setminus A$. Hence $\psi_A(x) \to 0$ for a.e. $x \in \mathbb{R} \setminus A$, and the proof is complete. $\square$

Theorem 16.51 *If $f \in L^1(a, b)$ and $F(x) = \int_a^x f(t)\, dt$, then $F'(x) = f(x)$ for almost every $x \in (a, b)$.*

Proof If $f = \chi_A$ for some measurable subset A of (a, b), then conclusion follows from Theorem 16.50. Let $s = \sum_{k=1}^{n} \alpha_k \chi_{A_k}$ be a non-negative simple measurable function, so that

$$S(x) = \int_a^x s(t)\,dt = \sum_{k=1}^{n} \alpha_k \int_a^x \chi_{A_k}(t)\,dt.$$

Again Theorem 16.50 shows that $S'(x) = s(x)$ for a.e. $x \in (a, b)$. If $f \in L^1(a, b)$, there exists a sequence $\{s_n\}_n$ of simple measurable functions such that $s_n \le s_{n+1}$ and $s_n(x) \to f(x)$ for all $x \in [a, b]$. If $S_n(x) = \int_a^x s_n(t)\,dt$, Beppo Levi's Theorem yields

$$F(x) = \int_a^x f(t)\,dt = \lim_{n \to +\infty} \int_a^x s_n(t)\,dt = \lim_{n \to +\infty} S_n(x)$$

$$= S_1(x) + \sum_{n=1}^{\infty} (S_{n+1}(x) - S_n(x))$$

for every $x \in [a, b]$. We apply Theorem 16.48 and obtain

$$F'(x) = S_1'(x) + \sum_{n=1}^{\infty} \left(S_{n+1}'(x) - S_n'(x)\right)$$

$$= \lim_{n \to +\infty} S_{n+1}'(x)$$

for a.e. $x \in (a, b)$. Therefore $\lim_{n \to +\infty} S_n'(x) = \lim_{n \to +\infty} s_n(x)$ for a.e. $x \in (a, b)$. Since $s_n \to f$, we conclude that $F'(x) = f(x)$ for a.e. $x \in (a, b)$. □

The hard question is whether the last theorem can be reversed: if φ is a continuous function whose derivative exists almost everywhere, is it true that

$$\varphi(x) - \varphi(a) = \int_a^x \varphi'(t)\,dt\ ?$$

Very technical examples show that this is typically false. The validity of the Fundamental Theorem of Calculus is however true for a restricted class of functions.

Definition 16.36 Let J be an interval (of any kind, even unbounded), and let $f : J \to \mathbb{R}$ be a function. We say that f is absolutely continuous on J if and only if for every $\varepsilon > 0$ there exists $\delta > 0$ such that

$$\sum_{k=1}^{n} |f(d_k) - f(c_k)| < \varepsilon$$

for every finite pairwise disjoint family $\{(c_k, d_k)\}_{k=1}^n$ of open intervals contained in J such that $\sum_{k=1}^n |d_k - c_k| < \delta$.

Proposition 16.6 *Any absolutely continuous function f defined on $[a, b]$ has finite variation on $[a, b]$.*

Proof This is almost trivial: consider $\varepsilon = 1$ and $\delta > 0$ according to the definition of absolute continuity. Fix a positive integer $n > (b-a)/\delta$ and split $[a, b]$ by points $a = x_0 < \ldots < x_n = b$ such that $x_k - x_{k-1} = (b-a)/n < \delta$ for every k. It follows that $V_{x_{k-1}}^{x_k} f \leq 1$ for every k, hence

$$V_a^b f \leq \sum_{k=1}^n V_{x_{k-1}}^{x_k} f \leq n.$$

$\square$

Proposition 16.7 *Every absolutely continuous function f on $[a, b]$ is continuous, and can be written as the difference $f_1 - f_2$ of two non-decreasing absolutely continuous functions f_1 and f_2.*

Proof It is obvious that absolutely continuous functions are continuous. We set $f_1(x) = V_a^x f$ and $f_2 = f_1 - f$. We need to prove that f_1 is absolutely continuous.
Pick $\varepsilon > 0$ and $\delta > 0$ so small that

$$\sum_{k=1}^n |f(d_k) - f(c_k)| < \frac{\varepsilon}{2}$$

whenever the pairwise disjoint intervals (c_k, d_k) satisfy $\sum_{k=1}^n (d_k - c_k) < \delta$. Since f is of bounded variation, each (c_k, d_k) admits a subdivision

$$c_k = a_0^{(k)} < a_1^{(k)} < \ldots < a_{\ell_k}^{(k)} = d_k$$

such that

$$V_{c_k}^{d_k} f < \sum_{j=0}^{\ell_k - 1} \left| f(a_{j+1}^{(k)}) - f(a_j^{(k)}) \right| + \frac{\varepsilon}{2n}.$$

Hence

$$\sum_{k=1}^n |f_1(d_k) - f_1(c_k)| = \sum_{k=1}^n V_{c_k}^{d_k} f < \sum_{k=1}^n \sum_{j=0}^{\ell_k - 1} \left| f(a_{j+1}^{(k)}) - f(a_j^{(k)}) \right| + \frac{\varepsilon}{2}$$

$$< \frac{\varepsilon}{2} + \frac{\varepsilon}{2} = \varepsilon,$$

and the proof is complete.

$\square$

Theorem 16.52 *if f is a non-decreasing function defined on [a, b], then f' is Lebesgue-measurable and*

$$\int_a^b f'(x)\, dx \le f(b) - f(a).$$

If g is a real-valued function of bounded variation on [a, b], then $g' \in L^1([a, b])$.

Proof We extend f by setting $f(x) = f(b)$ when $x > b$. For $n \in \mathbb{N}$ and $a \le x \le b$, we define

$$f_n(x) = \frac{f\left(x + \frac{1}{n}\right) - f(x)}{\frac{1}{n}}.$$

Then $\{f_n\}_n$ is a sequence of non-negative measurable functions such that $f_n(x) \to f'(x)$ for a.e. $x \in (a, b)$. In particular f' is measurable. We apply Fatou's Lemma:

$$\int_a^b f'(x)\, dx = \int_a^b \lim_{n\to+\infty} f_n(x)\, dx \le \liminf_{n\to+\infty} \int_a^b f_n(x)\, dx$$

$$= \liminf_{n\to+\infty} n \int_a^b \left(f\left(x + \frac{1}{n}\right) - f(x)\right) dx$$

$$= \liminf_{n\to+\infty} \left(n \int_b^{b+\frac{1}{n}} f(x)\, dx - n \int_a^{a+\frac{1}{n}} f(x)\, dx\right)$$

$$\le \liminf_{n\to+\infty} \left(n \int_b^{b+\frac{1}{n}} f(b)\, dx - n \int_a^{a+\frac{1}{n}} f(a)\, dx\right)$$

$$= f(b) - f(a).$$

To prove the second statement, we decompose g as a linear combination of two non-decreasing functions, and we apply to each one the first statement. The proof is complete. □

We can now prove a generalization of the basic principle for differentiable functions.

Theorem 16.53 *Suppose that $f : [a, b] \to \mathbb{R}$ is an absolutely continuous function. If $f' = 0$ a.e. on [a, b], then f is a constant function.*

Proof Let $c \in (a, b]$ be an arbitrary point; we claim that $f(c) = f(a)$. To prove this claim, we pick any $\varepsilon > 0$. By absolute continuity, there exists $\delta > 0$ as in Definition 16.36. If $E = \{x \in (a, c) \mid f'(x) = 0\}$, by assumption $\lambda(E) = c - a$. Hence to every $x \in E$ there correspond arbitrarily small values $h > 0$ such that $[x, x + h] \subset (a, c)$ and

$$|f(x + h) - f(x)| < \frac{h}{c - a}\varepsilon.$$

Since the collection of all such intervals $[x, x+h]$ is a Vitali cover of E, there exists a finite pairwise disjoint collection $\{[x_k, x_k + h_k] \mid k = 1, \ldots, n\}$ such that

$$\lambda \left(E \cap \left(\mathbb{R} \setminus \bigcup_{k=1}^{n} [x_k, x_k + h_k] \right) \right) < \delta.$$

Hence $\lambda((a, c)) = \lambda(E) < \delta + \sum_{k=1}^{n} h_k$. Assuming that $x_1 < x_2 < \cdots < x_n$, it follows that the sum of the lengths of the intervals

$$(a, x_1), (x_1 + h_1, x_2), \ldots, (x_n + h_n, c)$$

is smaller than δ; our choice of δ yields now

$$|f(a) - f(x_1)| + \sum_{k=1}^{n-1} |f(x_k + h_k) - f(x_k)| + |f(x_n + h_n) - f(c)| < \varepsilon.$$

We now conclude by the triangle inequality:

$$|f(a) - f(c)| \leq |f(a) - f(x_1)| + \sum_{k=1}^{n-1} |f(x_k + h_k) - f(x_k)|$$

$$+ |f(x_n + h_n) - f(c)| + \sum_{k=1}^{n} |f(x_k + h_k) - f(x_k)|$$

$$< \varepsilon + \sum_{k=1}^{n} \frac{h_k}{c - a} \varepsilon \leq 2\varepsilon.$$

Since $\varepsilon > 0$ is arbitrary, we conclude that $f(a) = f(c)$, and f is constant on $[a, b]$. □

Theorem 16.54 (Fundamental Theorem of Calculus) *If $f : [a, b] \to \mathbb{R}$ is an absolutely continuous function, then $f \in L^1([a, b])$ and*

$$f(x) = f(a) + \int_a^x f(t) \, dt$$

for every $x \in [a, b]$.

Proof We know that f is a function of bounded variation, hence f' exists a.e. in $[a, b]$ and $\int_a^b f'(x) \, dx \leq f(b) - f(a)$. Hence $f' \in L^1([a, b])$. Setting $g(x) = \int_a^x f'(t) \, dt$, we see that g is absolutely continuous and $g' = f$ almost every by

Theorem 16.51. It follows that $h = f - g$ is absolutely continuous and $h' = 0$ almost everywhere, hence h is a constant function. Thus

$$f(x) = h(x) + g(x) = h(a) + \int_a^x f'(t)\,dt = f(a) + \int_a^x f'(t)\,dt$$

for every $x \in [a, b]$. The proof is complete.   □

Since it is an easy exercise to prove that the indefinite integral of a function in $L^1([a, b])$ is absolutely continuous, we can summarize our results in the following statement.

Theorem 16.55 *A function $f : [a, b] \to \mathbb{R}$ has the form*

$$f(x) = f(a) + \int_a^x \varphi(t)\,dt$$

for some $\varphi \in L^1([a, b])$ if and only if f is absolutely continuous on $[a, b]$. In this case there results $\varphi = f'$ almost everywhere in $[a, b]$.

Exercise 16.17 Prove that a function $f : \mathbb{R} \to \mathbb{R}$ has the form $f(x) = \int_{-\infty}^x \varphi(t)\,dt$ for some $\varphi \in L^1(\mathbb{R})$ if and only if f is absolutely continuous on $[-A, A]$ for every $A > 0$, $V_{-\infty}^{+\infty} f \in \mathbb{R}$ and $\lim_{x \to -\infty} f(x) = 0$. To this aim, proceed as follows.

(1) If f has the form $f(x) = \int_{-\infty}^x \varphi(t)\,dt$, prove that f is absolutely continuous on every interval $[-A, A]$ and that $V_{-\infty}^{+\infty} f = \int_{\mathbb{R}} \varphi(t)\,dt$. Use the Dominated Convergence Theorem to show that $f(x) \to 0$ as $x \to -\infty$.
(2) Conversely, apply the previous theorem to deduce that $f(x) = f(-A) + \int_{-A}^x f'(t)\,dt$ for every $A > 0$ and every $x > -A$.
(3) Let $A \to +\infty$ and deduce that $f(x) = \lim_{A \to +\infty} \int_{-A}^x f'(t)\,dt$.
(4) Prove that $\int_{-\infty}^{+\infty} |f'(t)|\,dt = \lim_{n \to +\infty} \int_{-n}^n |f'(t)|\,dt = \lim_{n \to +\infty} V_{-n}^n f \le V_{-\infty}^{+\infty} f \in \mathbb{R}$.
(5) Deduce that $f' \in L^1(\mathbb{R})$, so that (3) gives $f(x) = \int_{-\infty}^x f'(t)\,dt$.

16.11 Problems

16.1 Let $\{E_n\}_n$ be a sequence of measurable sets in a measurable space X such that $\sum_{n=1}^\infty \mu(E_n)$ converges. Prove that almost every $x \in X$ lie in at most finitely many of the sets E_n. Hint: show that the set A of all x which lie in infinitely many E_n coincides with $\bigcap_{n=1}^\infty \bigcup_{k=n}^\infty E_k$.

16.2 Suppose $f_n : X \to [0, +\infty]$ is measurable for every n, that $f_1 \ge f_2 \ge f_3 \ge \dots$ and that $f_n \to f$ for every $x \in X$. If $f_1 \in L^1(X)$, prove that $\int_X f_n d\mu \to \int_X f\, d\mu$. Is the assumption that $f_1 \in L^1(X)$ really necessary?

16.3 In a metric setting, the proof of Urysohn's Lemma is much easier. Let (X, d) be a metric space. For every non-empty subset E of X we define for all $x \in X$

$$d_E(x) = \inf \{d(x, y) \mid y \in E\}.$$

Prove that d_E is uniformly continuous on X. Now let A and B be disjoint non-empty closed subsets, and derive Urysohn's Lemma from the function

$$f(x) = \frac{d_A(x)}{d_A(x) + d_B(x)}.$$

16.4 Prove that Theorem 16.18 extends to the situation in which the sequence $\{f_n\}_n$ is replaced by a family $\{f_t \mid t \in \mathbb{R}\}$ such that (i) $f_t(x) \to f(x)$ as $t \to +\infty$, for every $x \in X$; (b) $t \mapsto f_t(x)$ is continuous, for every $x \in X$.

16.5 Let $u \in L^1_{\text{loc}}(\Omega)$ be such that $\int_\Omega u(x)v(x)\,dx = 0$ for every $v \in \mathcal{D}(\Omega)$. Prove that $u = 0$ almost everywhere on Ω. *Hint:* observe that $\varrho_n * u = 0$ for every n.

16.12 Comments

As we said, Measure Theory is usually introduced as the study of certain functions defined on suitable collections of sets, called σ-algebras. In this book we did not consider signed measures or complex measures, for which we refer to [3] or [2]. Algebras and σ-algebras of sets are the natural families of measurable sets, but sometimes it is natural to introduce "measures" which are defined on every subset of a gives set. For instance one might define the "measure" of an arbitrary subset E of $\mathbb{R}$ by covering E with sequences of pair-wise disjoint intervals I_k, summing the lengths of all the I_k, and taking the infimum over all such coverings of E. This leads to the concept of *outer measures*, and measurable sets a those sets which satisfy a particular condition introduced by C. Carathéodory. The interested reader is referred to [1].

References

1. L.C. Evans, R.F. Gariepy, *Measure Theory and Fine Properties of Functions*. Textbooks in Mathematics Series Profile, 2nd edn. (CRC Press, Boca Raton, 2015)
2. G.B. Folland, *Real Analysis: Modern Techniques and Their Applications*. Pure and Applied Mathematics. A Wiley-Interscience Series of Texts, Monographs, and Tracts, 2nd edn. (Wiley, New York, 1999)
3. W. Rudin, *Real and Complex Analysis*. McGraw-Hill Series in Higher Mathematics, xi, 412 p. (McGraw-Hill Book Company, New York, 1966)

Printed in the United States
by Baker & Taylor Publisher Services